MATHEMATICS AND ASTRONOMY

FROM ANCIENT GREECE TO NEWTON

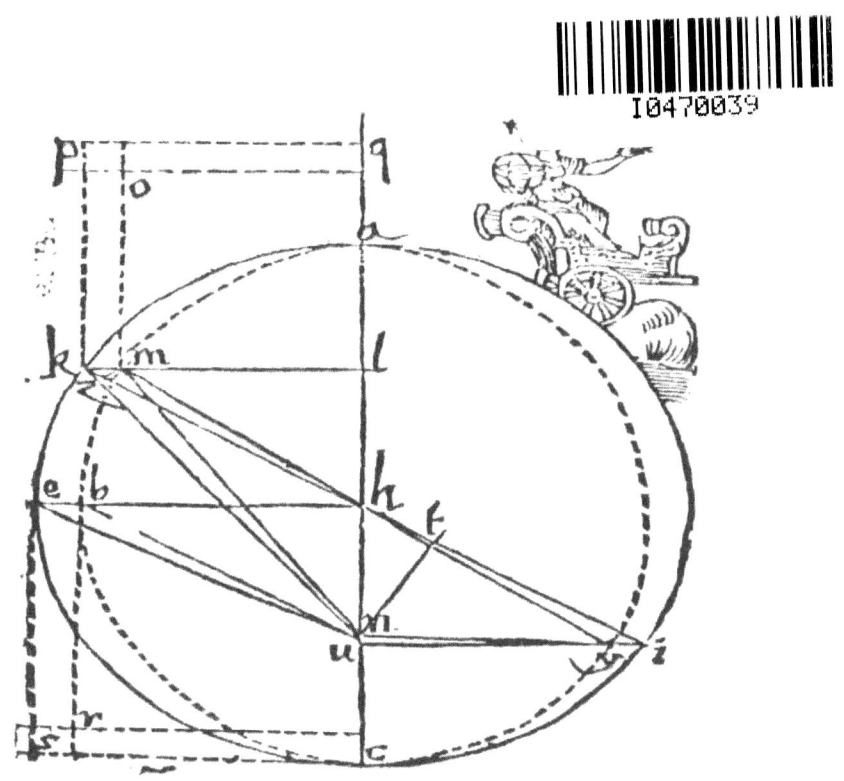

Charles Holmburg. MD

VOLUME 2

CHAPTERS 28-47

ISBN 978-1492774358

Mathematics and Astronomy from Ancient Greece to Newton

Volume 2: Chapters 28-47

Chapter 28

From Ptolemy to Copernicus

History teaches us that Ptolemy used his ability to predict planetary positions for more than the advancement of mathematics and science. He like most astronomers until the 17th century was also involved in astrology.   Ptolemy also wrote a book on astrology which was considered quite important in his era. The celestial longitudes of the planets at the moment of birth could be used to predict an individual's destiny. Ptolemy's book, "Tetrabiblos", gave rules on how to predict an individual's fate based on the planetary positions at the time of birth. Astronomers often turned to astrology and earned more from this endeavor than their mathematical work.

Later in the Roman Empire the emperors tried to eliminate astrology as a superstition. This had the effect of discouraging astronomy and mathematics.   Thus, after Ptolemy no new work of note occurred in mathematics and astronomy in the Roman period and the Middle Ages following the collapse of the Roman Empire.

By 640 A.D. the Moslems conquered Alexandria, and scholars migrated to Constantinople.   The ancient Greek texts on mathematics and astronomy were lost to Europe for almost 800 years. It was the Moslem or Arabic world which continued the study of mathematics and astronomy. The Arab conquests spread to both

India in the east and Spain in the west after the death of Mohammed in 632 A.D. There were two parts to this vast Arab empire; one headquartered in Bagdad for the east, and the other centered in Cordova in Spain for the west. The Greek texts were translated into Arabic and studied in these centers of learning in the period after 800 A.D.

We recall Ptolemy's "Almagest" received its name from the Arabs who considered the book almost a divine work. The Arabs translated Euclid, Aristotle, Archimedes, and Apollonius as well. We will consider certain of the Arab contributions to mathematics in the period from 800 A.D. to 1300 A.D.

In 830 A.D. an Arab astronomer, Al-Khowarizini, wrote a book called "Al-jabr w'al muqubala". Our word 'algebra' comes from "Al-jabr" which means 'restoring'. "Al muqubala" means simplification. The book shows how to solve and simplify algebraic expressions. A pair of equations in two unknowns can be solved by methods in the book. For example, in modern symbolism,

1 . $x + y = 10$

2. $xy = 24$.   We solve the second equation for x,

3. $x = 24/y$.   We use this value in equation 1. Note y cannot be 0.

4. $24/y + y = 10$

5. $24 + y^2 = 10y$,

after multiplying both sides by y.

6. $y^2 - 10y + 24 = 0$

7. $(y - 6)(y - 4) = 0$, so $y = 6$, and

y = 4 satisfies this equation and are the solutions of the two   equations, 1 and 2.

If y = 6 then

x = 4, and if y = 4 then x = 6.

Prior to this book a systematic approach to equations had not been developed.

An Arab astronomer named Tabit ben Kora, who had translated "The Almagest" into Arabic, found the rate of precession to be greater than the value of 1° per 100 years given by Ptolemy. He argued that Ptolemy could not have been incorrect in his value, and that therefore the rate of precession was not constant but variable. This became known as the theory of the trepidation of the equinoxes. We know today that the rate of precession is constant and that Ptolemy's value is incorrect, but the theory of trepidation was accepted by Copernicus which made his theory more complicated than necessary.

Another Arab mathematician and astronomer of great repute was Al-Battani whose name in Latin became Albategnius. He determined with the use of his armillary spheres, which were more accurate than Ptolemy' s, that the obliquity of the ecliptic was 23° 35', a more accurate value than Ptolemy's (23° 51' 20"). He also found the angle of the line of apsides had been increasing from Ptolemy's value of 65° 30'. He had a more accurate value than Hipparcus and Ptolemy for the length of the tropical year.   He improved the table of chords and arcs found in "The Almagest" using the half chord. See Figure 28-1.

Figure 28-1

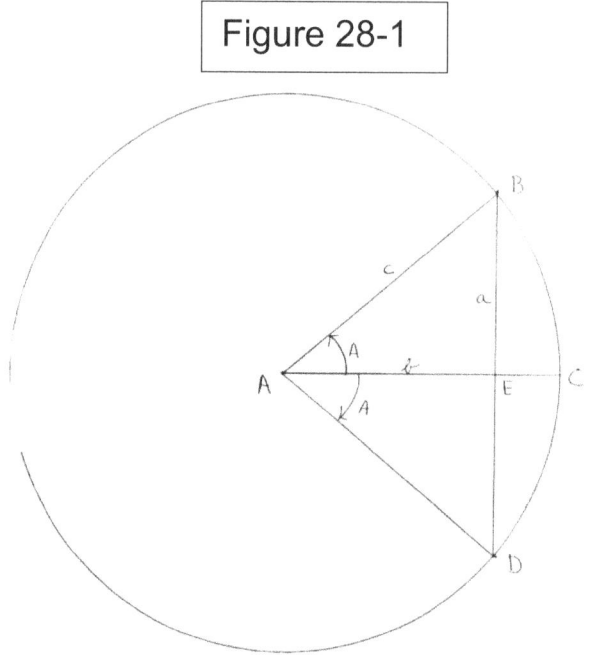

Instead of calculating chord BD for angle 2A = arc BD, he calculated the length of the half chord, ½ BD, for arc BC = angle A. He assigned a value to radius AB, side c of right triangle AEB, and then found side a. From modern trigonometry we know,

sin A = a/c, and if c equals 1 unit, ½chord BD = a = sin A,

Thus Al-Battani was the first to really develop a table of sines.   Furthermore, he calculated results for every 10' of angle from 0° to 90°. He defined for the first time the tangent as the ratio of side a to b, and calculated a table of tangents.

The Arab astronomer and mathematician Nasir-Eddin worked on the solution of spherical triangles and improved on Ptolemy's method of calculation by developing the essentials of our modern laws of sines and cosines for spherical triangles. However, Europeans were not aware of this work until around 1450 A.D.

Nassir-Eddin built a great observatory at Maragha, near the northwest border of modern Iran. He was able to find the rate of precession to a very accurate level. He suspected the theory of trepidation was incorrect and that Ptolemy did not have accurate measurements and was in error.

In 1420 A.D. Ulugh Begh, the grandson of the Tartar conqueror Tamerlane, built an observatory at Samarcand in Russia near the modern Afghanistan border. He observed star positions and made a new star catalog.

Despite all this work Ptolemy's basic theory prevailed and no one came up with a major new theory. It is true some of the Arab astronomers felt the concept of the equant could not be correct, and that the uniform motion of the epicycle center should take place around the center of the deferent. They suggested an epicycle on the epicycle so uniform motion of the main epicycle center would occur around the center of the deferent. This idea never gained wide acceptance. However, the idea of uniform motion around the center of the deferent was very important to Copernicus who studied the work of the Arab astronomers working in Maragha.

Northern Europeans knew little or nothing of Greek mathematics and astronomy until the 12th century when Spain began to eliminate Moslem rule.   As the Latin scholars came to Cordova and Toledo they began to translate with the help of some remaining Arabs the Arabic translations of the Greek classics into Latin.

The King of Leon and Castille in Spain from 1223 to 1224 was Alfonsine X. He had conquered Toledo from the Arabs and under his supervision the tables of Ptolemy were revised and made into the Alfonsine Tables in 1252.   These tables of stellar and planetary positions greatly revived interest in astronomy and mathematics in Europe. Gherardo of Germona (1114-1187) had translated "The Almagest" from Arabic into Latin, and by the 15th century German astronomers translated "The Almagest" directly from the original Greek into Latin.

Two mathematicians, George Purbach and John Muller, worked on Ptolemy's text and Muller brought out after Purbach's death the "Epitome of Astronomy" based on "The Almagest", which was the major text studied by Copernicus as he worked on his new theory in the 16th century. Muller, who called himself Regiomontanus from the district in which he lived, found inaccuracies in the Alfonsine Tables. For example, the position predicted for the position of Mars was in error by 2° by the 16th century. Regiomontanus still believed in the theory of Ptolemy and gave means to correct the positions predicted by the Ptolemaic theory. He also produced a textbook of trigonometry and a table of sines for every minute of arc from 0° to 90°.

One of the debates in the 15th and 16th centuries among astronomers was the validity of the celestial spheres. Aristotle long before Ptolemy had stated that each planet was attached to a crystalline sphere which rotated around the earth. Combinations of spheres were necessary to account for the apparent irregularities of motion as seen from earth. All the stars were attached to the celestial sphere known as the firmament, which rotated around the earth once a day. Outside of the celestial sphere was the *primum mobile*, a sphere which supplied the energy of motion to all the other spheres. Aristotle felt that these motions were heavenly and not natural. Natural motion was motion directed to the center of the universe which was the center of the earth. The stars, planets, sun and moon being attached to spheres and under heavenly motion would not fall towards the earth. The earth itself was judged motionless, since if it were moving objects on the surface would be whirled into space. Ptolemy, who wrote centuries after Aristotle, basically ignored the concept of spheres in producing his orbits for the planets sun and moon. However, he retained the celestial sphere as we have seen.

Astronomers of the 15th and 16th centuries mixed the Ptolemaic and the Aristotelian concepts so that the centers of the epicycles were considered as attached to crystalline spheres. Copernicus also used the idea of spheres in developing his theory.

15

We can now turn to the advent of Copernicus who grew up and was educated at a time when Greek astronomy and mathematics was rediscovered along with the Arabic ideas. Copernicus was born as Niklas Koppernigk on Feb 19, 1473 in Thorn on the Vistula River in northern Poland. When he was 18 he studied at the University of Cracow where he began his work in mathematics and astronomy.

His father had died and his uncle Lucas Watzelrode, Bishop of Ermland, provided for his education and support. Copernicus was educated to be the Canon, or supervisor, of the cathedral of Frauenberg. After studying at Cracow, Copernicus was sent to Italy for more training in theology, philosophy, medicine, and astronomy. It was not unusual for a medieval scholar to study all these subjects in preparation for a career.

In 1497 Copernicus enrolled at the University of Bologna where he stayed 3 years. Of interest was his study of Plato. Copernicus believed Plato was correct in his view that the sphere was the perfect shape and all heavenly bodies were spheres. He also believed the perfect motion was circular, and uniform circular motion about the center of the circle was the perfect motion in which the celestial spheres participated. Later when he became acquainted with Ptolemy's theory, he could not accept Ptolemy's creation of the equant which to Copernicus violated the fundamental ideals of Plato and Aristotle.

In 1501 Copernicus became Canon of the Cathedral at Frauneberg. He again returned to Italy this time to Padua where he studied law and medicine. In 1506 he returned to Fraunenberg to begin his work as Canon. He had enough time in the years from 1506 to 1543 to work on his mathematical and astronomical ideas so that he could publish on his theory of planetary motion.

The idea for reforming the Ptolemaic theory was probably begun in the period of 1510 to 1514. By 1514 historians have shown that a treatise reached Cracow outlining a theory wherein the earth moves and the sun is at rest.   This treatise was not really published as general knowledge in Copernicus's lifetime. Today we know the treatise as the "*Commentariolus*", or the hypothesis of heavenly motions. The work was circulated privately. When Copernicus's great work was published in 1543 no mention is made of the "*Commentariolus*".   Fortunately, a copy of the "*Commentariolus*" did reach Tycho Brahe and copies made from it have survived so we have some idea about Copernicus's early thinking on the heliocentric theory.

Copernicus does not offer a physical proof of the earth's motion, nor does he give new observational data from which the theory came. He has to argue from a set of postulates, which he feels must be true much like Euclid's axioms.   He then can develop his system and show how it resolves some of the peculiarities of Ptolemy's system such as why retrograde arcs for the superior planets are only seen when these planets are in opposition to the sun, while for the inferior planets the retrograde arcs occur when the planets are in conjunction with the sun.

In the "*Commentariolus*" Copernicus lists seven axioms on which his system is based. He admits that Ptolemy's system allows adequate prediction of planetary motion with the corrections as given by Regiomontanus.   However, Ptolemy's system requires the use of the equant and non uniform motion of the epicycle center around the center of the deferent. Copernicus wanted to eliminate non uniform motion about the center of motion as it had violated the ideas of Aristotle and Plato.

We will summarize the seven axioms of Copernicus.

1. There is no single center for the motions of the planets and the celestial sphere.

2. The earth's center is not the center of the universe.

3. All the spheres revolve about the sun as the center (later this axiom was altered).

4. The distance from the earth to the sun is so small in comparison to the distance from the sun to the firmament (the celestial sphere) that an observer on earth cannot detect the earth's movement in its orbit by observations of the celestial sphere. This means there is no detectable stellar parallax (We will discuss this later).

5. The motion of the celestial sphere as seen from earth is only apparent, and is explained by the earth rotating on its own axis. The celestial sphere and all the stars are fixed and not in motion.

6. The motion of the sun around the ecliptic from west to east as seen from earth is only apparent and is caused by the earth orbiting the sun.

7. The retrograde arcs of the planets arise from the motion of the earth around the sun and the motion of the planets around the sun.

The remainder of the "*Commentariolus*" goes into greater explanation of the seven axioms. There is not a detailed geometric and mathematical presentation of Copernicus' system. Copernicus spent the years from 1514 to 1543 working out the mathematical details of his heliocentric system, and published a large book called, "*De revolutionibus orbium coelestium, Libra VI*", in 1543 which gives a detailed mathematical presentation of his system. The title in English is "The Revolutions of the Celestial Spheres in Six Books".

Copernicus did set up an observatory in Fraunenberg, which he used from 1517 to 1529 to acquire observational data to use in his book. However, Copernicus is not regarded as a great observer of planetary motion, and he used very few of his own observations to derive the parameters of his theory. He had great regard for the ancient Greek data and believed Ptolemy's observations and data were correct. He did not fully realize that new observational data was needed to truly reform astronomy. It is said, for example, he did not observe and determine the position of Mercury in all the years he worked blaming atmospheric haze of his region for making determinations impossible (we recall Mercury is so close to the sun it can only be seen shortly before or after sunset near the horizon).

By the early 1530's mathematicians and astronomers in Europe heard that Copernicus was working on a new planetary theory. By 1535 Copernicus had completed a set of tables on solar, lunar, and planetary positions indicating that he had worked out most of his parameters. A noted mathematician named Georg Joachim who called himself Rheticus was very interested in Copernicus's system and came to Fraunenberg to learn the details in 1539.

Rheticus took his Latinized name from the district where he was born in 1514. He lived until 1576. He was the first mathematician to define the trigonometric functions from a right triangle as is the modern practice. He assigned the length of the hypotenuse 10,000,000 units and calculated the sines of angles from 0° to 90° in 10' intervals to seven figure accuracy. He also calculated values for the cosine, tangent, cotangent, cosecant, and secant to this accuracy. It is said it took hired mathematicians 12 years to do these calculations.

After he had completed these trigonometric tables he undertook to do another book of trigonometric tables with the length of the hypotenuse of 10,000,000,000 units (ten billion). The sines were calculated for every 10 seconds of arc from 0° to 90°. This massive book was published after his death in 1596 as the "*Opus Palatinum*", and ran 732 pages, each page 9 by 15 inches. All the multiplications for these tables had to be done by hand.

Rheticus came to Frauenberg and was so impressed with Copernicus' work that he urged Copernicus to publish as soon as possible. Copernicus was very reluctant to proceed because he felt he would be criticized by philosophers, who in that era were Aristotelian, and by theologians who had come to accept the Ptolemaic system as most compatible with Holy Scripture.

Rheticus in his great enthusiasm for the heliocentric system, wrote his own short treatise called the "*Narritio Prima*", which explained features of the system giving credit to Copernicus for the work. By 1540 and 1541 Copernicus agreed to publish after the continual urging of Rheticus and Copernicus' friend Tiedemann Giese, the Bishop of Kulm who was familiar with Copernicus' theory.

In the final stages of publication Copernicus became ill and entrusted the project to a Lutheran preacher named Osiander. Osiander took the position that because Copernicus could not produce proof of the earth's motion, the book was only a hypothesis of planetary motion useful for calculations but not to be considered as reality. It is clear, however, that Copernicus did believe in the reality of the earth's motion and the immobility of the sun around which the planets and the earth moved.   Osiander was concerned that if the book were published as representing reality it would be denounced by the church authorities.   Osiander inserted a preface to the work which was unsigned so readers might believe it was written by Copernicus. The preface presents the work as a hypothesis for the mathematicians to do their calculations.   Copernicus was presented a copy of the published work on his deathbed and probably did not realize what Osiander had done.   Rheticus and Giese, however, became aware of what was done and were enraged, but nothing was done to correct the insertion by Osiander. It wasn't until Kepler found out the truth years later that it became known the preface was written by Osiander.

Copernicus had a stroke in Dec 1542 and was too ill to really make sure the final stages of the publication were carried out according to his wishes. He died shortly after the publication of "*De revolutionibus*". He had urged that mathematicians carefully consider his work to determine its importance and validity.

In our work we will consider selected aspects of Copernicus's system. A detailed mathematical treatment of "De revolutionibus" is now available in the book by N.M. Swerdlow and O. Neugebauer, "Mathematical Astronomy in Copernicus's De Revolutionibus".

To understand how Copernicus came to adopt the heliocentric theory we turn first to a book written by

Ptolemy after "The Almagest" called the "Planetary Hypotheses". In this book Ptolemy gave his ideas on the order of the planets from the earth, and estimated their nearest and farthest distances. He felt no space from the earth to the celestial sphere should be wasted. This concept is shown in Figure 28-2.

Figure 28-2

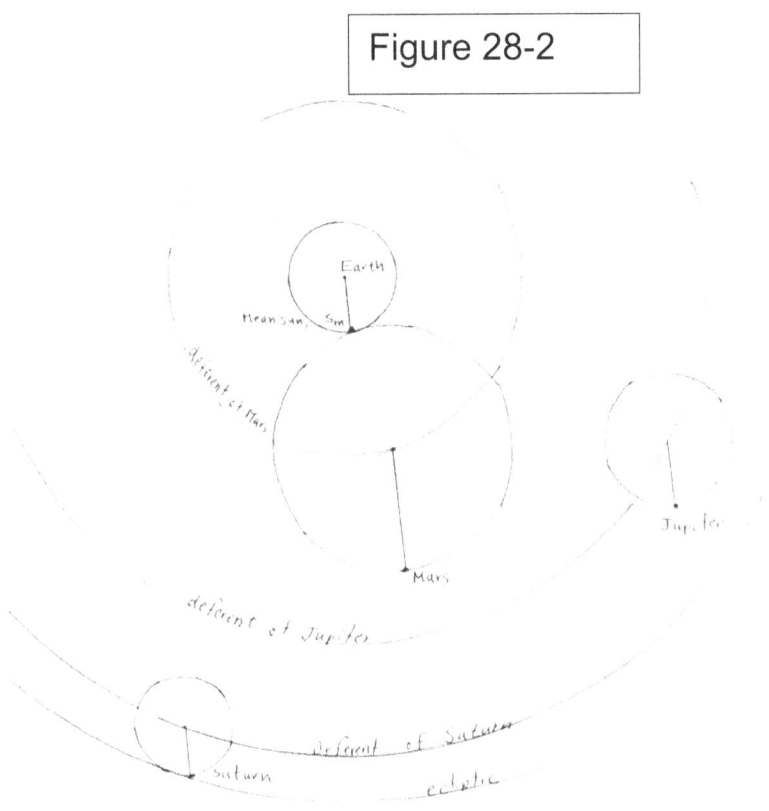

In this figure we see the earth, the mean sun and three superior planets Mars, Jupiter and Saturn as viewed from the north ecliptic pole. The circular orbit of the mean sun is shown closest to the earth and reaches to the innermost point of the epicycle of Mars. The outermost point on the epicycle of Mars reaches to the innermost point of the epicycle of Jupiter. The outermost point of the epicycle of Jupiter reaches to the innermost point of the epicycle of Saturn, and the outermost point of the epicycle of Saturn reaches to the celestial sphere. We note the epicycle radius of Mars is larger than the distance from the earth to the sun and larger than the epicycle radius of Jupiter, which in turn is larger than the epicycle radius of Saturn. This became important to Copernicus as we shall soon show.

The situation for the inferior planets is shown in Figure 28-3.

Figure 28-3

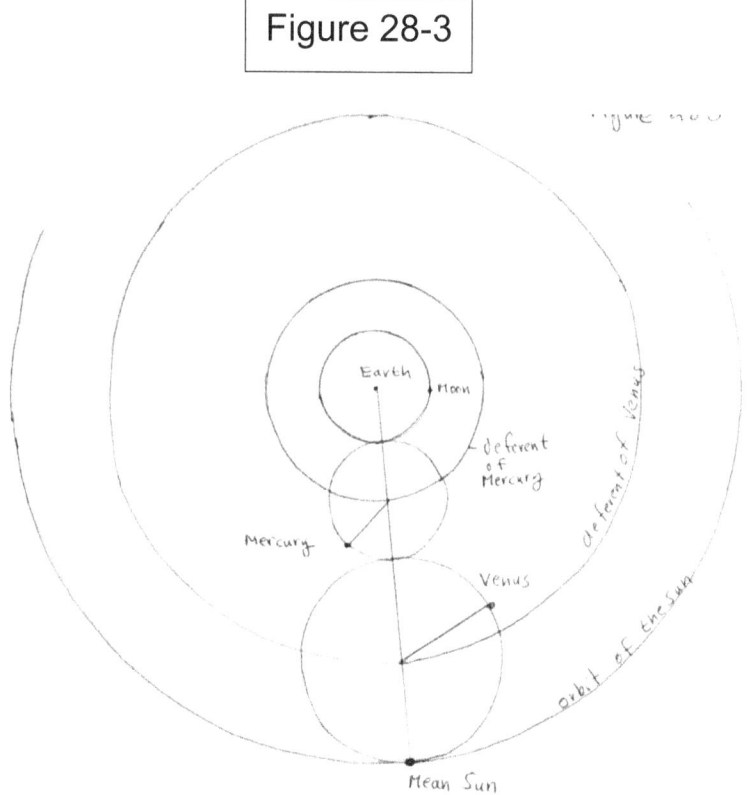

The view is again from the north ecliptic pole and is simplified from the theory in "The Almagest". The moon's orbit is pictured as circular and nearest to the earth. Next comes the epicycle of Mercury pictured on a circular deferent such that the innermost point reached by Mercury reaches to the orbit of the moon, while the outermost point reaches to the innermost point on the epicycle of Venus. The outermost point of the epicycle of Venus reaches to the orbit of the mean sun.

These ideas of Ptolemy were not based on observation but were given by Ptolemy as the most reasonable hypothesis for the space between earth and the celestial sphere.

Throughout the period of the middle ages, as we have mentioned, it was believed the planets' epicycle centers were attached to crystalline spheres which were in the arrangement given by Ptolemy's "Planetary Hypotheses". When Copernicus studied this system he read of the work of the Arabic astronomers of the Maragha school who objected to the idea of the equant and non-uniform motion around the deferent's center. Copernicus also read Regiomontanus' work on Ptolemy and became convinced that the equant must be wrong, and that a reconsideration of planetary theory was in order. He felt that the planets were attached to crystalline spheres which were free to rotate around their axes without interacting with each other. Furthermore, he felt the rotation occurred with uniform motion as seen from the center of the spheres as originally postulated by Plato and Aristotle. Kepler once said that Copernicus' work was to interpret Ptolemy and not nature, since so many of Copernicus' ideas are derived from removing the equant from Ptolemy's theory.

Noel Swerdlow, in the book to which we referred, calls attention to Copernicus's copy of the Alfonsine tables. He argues that Copernicus first worked on his theory in the period of 1510 to 1514, and left notes on his copy of the Alfonsine Tables indicating some preliminary models of planetary motion before settling on the heliocentric model. Regiomontanus had pointed out that Apollonius' theory of the epicycle and deferent is equivalent to a moving eccentric circle. We already know that Ptolemy chose an eccentric circle theory for the sun and the epicycle theory for the planets. Regiomontanus suggested Ptolemy could have used a moving eccentric circle for the superior planets. In Figure 28-4 we show how this theory for the superior planets might have been tried by Copernicus as a preliminary model.

Figure 28-4

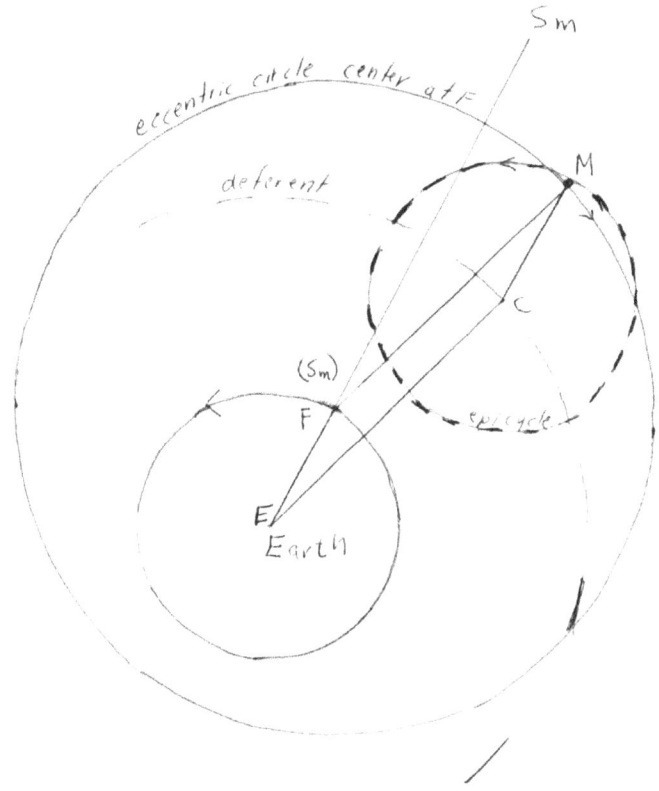

The earth is at the center of the celestial sphere and viewed from the north celestial pole. The planet is shown at M moving on an eccentric circle with a center at F. The circle moves as its center rotates around the earth with a radius, EF, equal to the radius of the epicyle in Ptolemy's theory. The epicycle and deferent of Ptolemy's theory are shown in broken lines. In this figure there is no equant since Copernicus was determined to find a system without an equant. We can complete the parallelogram as shown so radius EC equals eccentric radius FM. We recall Ptolemy used 60 units for this

27

radius and then found radius CM. We recall also that the line of sight to the mean sun, S_m , is always parallel to radius CM for the superior planets. Copernicus could next try moving the mean sun to position F, since the observer on earth cannot tell where along line EF the mean sun actually lies. If the mean sun were at position F then the superior planet would be orbiting the sun while the mean sun orbits the earth.

This type of system where the earth is stationary and the planets orbit the sun while the sun orbits the earth is known to us as the Tychonian system, because after Copernicus, Tycho Brahe used this system in his planetary theory. We see here that Copernicus most likely tried this system, but rejected it for reasons we will give shortly. Regiomontanus had also discussed a Tychonic system for the inferior planets in his "Epitome of the Almagest", which Copernicus had studied. See Figure 28-5.

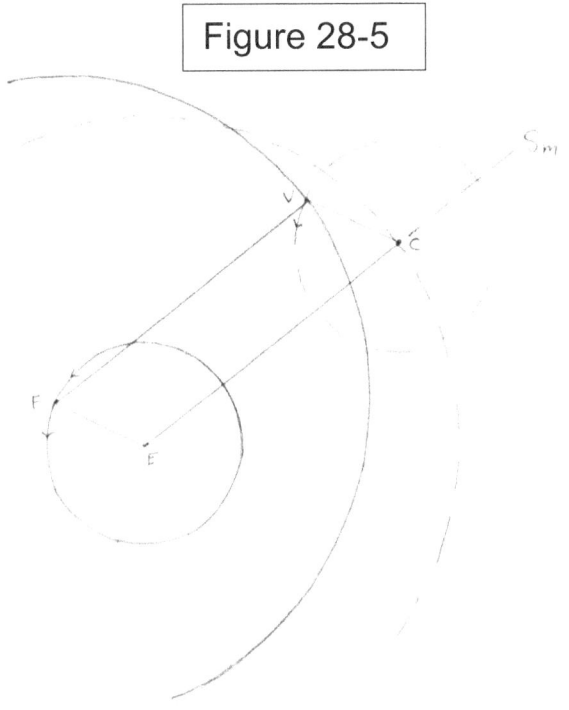

Figure 28-5

Here we see the earth at E from the north celestial pole with Venus at position V moving on an eccentric circle with center at F. The center of the eccentric circle rotates around the earth with a radius equal to the epicycle radius in Ptolemy's theory of epicycle and deferent shown in broken lines. Again there is no equant and radius EC equals radius FV from the parallelogram shown. For the inferior planets the line of sight to the mean sun, S_m, always lies along radius EC. Copernicus tried placing the mean sun at position C, the center of Ptolemy's epicycle, since the observer at earth cannot tell where along line EC the mean sun lies.

If the mean sun is at position C for the inferior planets then the effect will be that of Venus orbiting the sun while the sun orbits the earth, i.e. the Tychonic

29

system. However, Figure 28-5 differs from Figure 28-4 in that in Figure 28-5 the mean sun is not at position F, which is really an empty center of motion for the eccentric circle. Copernicus then tried transforming Figure 28-5 into Figure 28-6

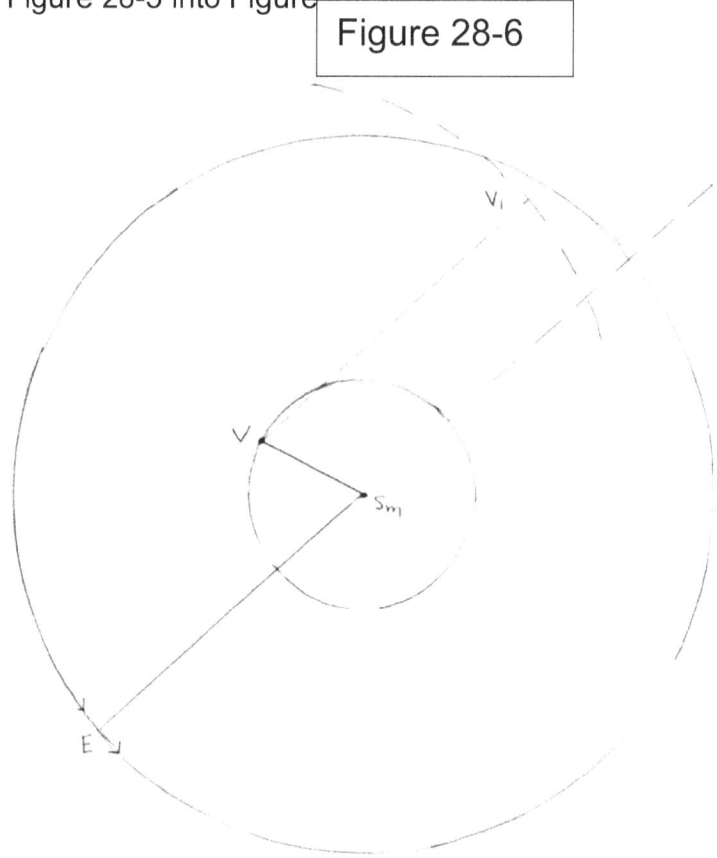

Figure 28-6

Here the mean sun, S_m , has been moved to position E in Figure 28-5 and placed at the center of the celestial sphere, while the earth moves in orbit around the sun with radius equal to EC in Figure 28-5. Venus is moved the corresponding distance V_1 to V as shown by the broken lines. Thus Venus orbits the sun inside of the earth's orbit. Thus Copernicus has made a heliocentric system for the inferior planets starting with the eccentric theory of Apollonius which was the equivalent of the epicycle deferent theory used by Ptolemy. Recall Ptolemy modified the epicycle deferent theory with the equant which Copernicus rejected.

At this point Copernicus had two models. For the superior planets he had the Tychonian system where the earth is stationary while the superior planets orbit the sun as the sun orbits the earth. For the inferior planets he had a heliocentric system. What convinced him to reject the Tychonian system for the superior planets? The answer historians feel is related to his idea that the spheres on which the planets are attached should not cross or interact with each other. When he tested the Tychonic system for Mars using the data provided by Ptolemy he found the sphere with Mars would intersect the sphere which holds the sun.

We consider the Tychonic system for Mars in Figure 28-7.

Figure 28-7

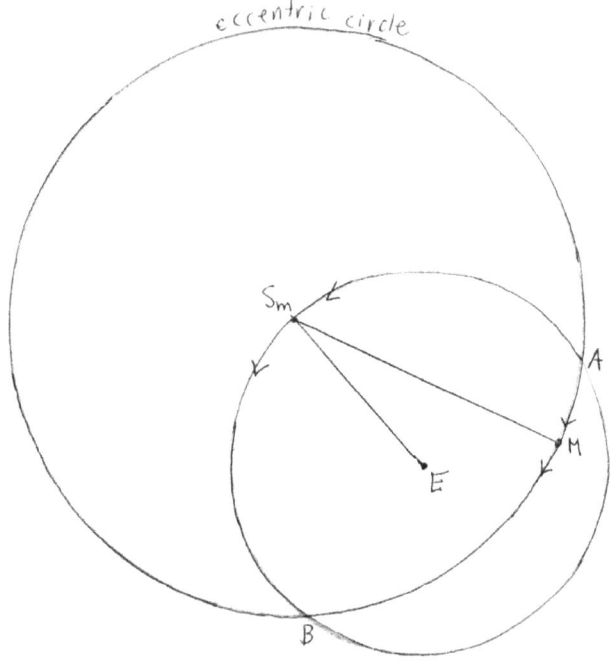

eccentric circle

The radius of the moving eccentric circle from center S_m to M is about 1.5 times the radius from the stationary earth to the mean sun (ES_m). Copernicus can derive this from the parameters of Mars given in "The Almagest". Since this value is less than 2 there will be periods where Mars will be within the orbit of the sun as shown in Figure 28-7. Remember the eccentric circle moves as the mean sun orbits the earth, and we show only one point in time where the eccentric circle, which Copernicus considered as being on the sphere containing Mars, cuts the sphere containing the sun at A and B in the plane we are viewing. To Copernicus such a phenomenon was unacceptable because this would mean the spheres would be interacting and interfering with each other's motion, which violated his fundamental ideas of celestial motion.

Now this interaction of spheres would not take place with Jupiter or Saturn since the ratio of the radius from the sun to the planet taken to the radius of earth to the sun is greater than 2 using the parameters in "The Almagest".

Let us see how Copernicus arrived at the value of 1.5 for the ratio of the radii in the case of Mars as this would lead to adopting the heliocentric system for the superior planets. We recall from our work with Aristarchus in chapter 13 that we had calculated the distance from the earth to the sun was about 760 earth diameters or 1,520 earth radii.   In "The Almagest" Ptolemy had recalculated the value by a method similar to Aristarchus and found the mean distance to the sun was 1,210 earth radii. In chapter 23 we learned in the epicycle model for Mars the epicycle radius is equal to .658(R), where R is taken as approximately the distance EC in Figure 28-4. Actually in chapter 23, R is the radius of the deferent and varies from (.1(R) + R) to (-.1(R) + R) from C.   See Figure 23-2.

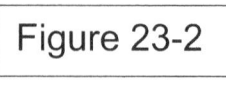

Figure 23-2

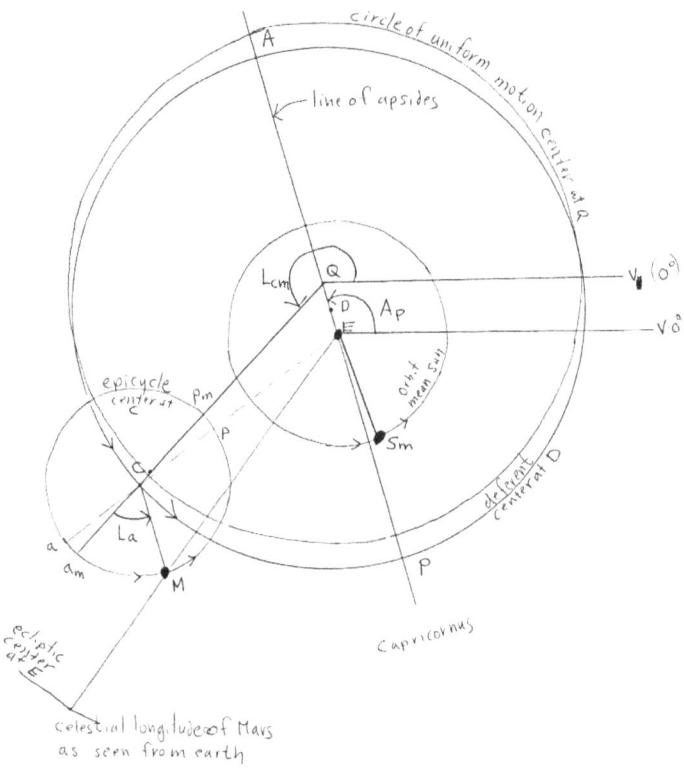

In Figure 28-4, CM = ES$_m$, ES$_m$ = 1,210,   CM = .658(R). Thus, .658(R) = 1,210, or R = 1,839 earth radii.

In Figure 28-4, R is the distance of the radius of Mars orbit with the mean sun as center, S$_m$M. Thus,

$$\frac{S_m}{ES_m} = \frac{1,839}{1,210} = 1.52 \cong 1.5.$$

If the ratio is greater than 2 in Figure 28-7 then S$_m$M is greater than the diameter of the sun's orbit in the Tychonic system so the eccentric circle with radius S$_m$M will always be outside of the circle of the sun's orbit. This is the case for Jupiter and Saturn using the data in "The Almagest". In chapter 23 we found for Jupiter, CM = 11.5 units, so,

$$\frac{CM}{R} = \frac{11.5}{60} = .192, or\ CM = .192R.$$

In Figure 28-4, CM = ES_m =1,210 earth radii. Thus, .192(R) = 1,210, or R = 6,302, where R is S_mM in Figure 28-7. The ratio of S_mM to ES_m, then is 6,302 to 1,210, or 5.2 to 1. For Saturn the epicycle radius, CM,   is even smaller at 6.5 units so the ratio S_mM to ES_m calculates out to 9.2 to 1.

Because Mars behavior in the Tychonic system meant a crossing of the spheres Copernicus rejected the Tychonic system for the superior planets.

He then went to the heliocentric model for the outer planets and chose this model when he found it kept all the spheres from interacting with each other. We note particularly that Copernicus developed these ideas from his study of Ptolemy and the exposition of Ptolemy given by Regiomontanus. In addition the work of the Arab astronomers, who had sought to eliminate the equant, played a role in his thinking.   It was not the collection of new observations or new developments in mathematics that led to the Copernican system. Copernicus worked with the data provided by Ptolemy with the goals of uniform circular motion about the centers of rotation, and the goal of finding a system wherein the spheres to which the planets were attached would not interact with each other. The heliocentric model then comes from consideration of the mathematical details of planetary motion as given in "The Almagest" combined with an analysis of the eccentric circle theory of Apollonius.

The years from 1510 to 1543 were used by Copernicus to work out the parameters of his system so that his system could predict the future positions of the planets. He used data from the ancients and added some observations made with his own instruments. His instruments were the armillary sphere similar to the one used by Ptolemy and the quadrant.

In the next chapter we will take up some of the details of the Copernican system.

Bibliography for Chapter 28

1. Dreyer, J.L.E., "A History of Astronomy from Thales to Kepler", Dover Publications, Inc., New York, 1953.
2. Swerdlow, N.M. and Neugebauer, 0., "Mathematical Astronomy in Copernicus's De Revolutionibus", Springer-Verlag, New York, 1984.

Chapter 29

Copernicus: The Earth Moves

...I too began to consider the mobility of the earth. And even though the idea seemed absurd, nevertheless I knew that others before me had been granted the freedom to imagine any circles whatever for the purpose of, explaining the heavenly phenomena.   Hence I thought that I too would be readily permitted to ascertain whether explanations sounder than those of my predecessors could be found for the revolution of the celestial spheres on the assumption of some motion of the earth.

Having thus assumed the motions which I ascribe to the earth later on in the volume, by long and intense study, I finally found that if the motions of the other planets are correlated with the orbiting of the earth, and are computed for the revolution of each planet, not only do their phenomena follow therefrom but also the order and size of all the planets and spheres, and heaven itself, is so linked together that in no portion of it can any be shifted without disrupting the remaining parts and the universe as a whole.

Copernicus, "On the Revolutions"

In Copernicus' system the celestial sphere to

which the fixed stars are attached is stationary and at rest. Further, the sun is also at rest and the planets orbit around the sun, but as we shall see the sun is not the center of motion for the movement of the earth and planets. Copernicus put the earth in motion around the sun. He explained what we see on earth is the consequence of the several motions which the earth has.   However, Copernicus argued that the celestial sphere with the fixed stars is so far away from the earth, sun, and planets that the solar system can be regarded as a point when compared to the distance from any object in the solar system to the celestial sphere.

The consequence of this idea is that astronomers on earth would not be able to detect stellar parallax due to the earth's motion around the sun. In theory if the earth moves relative to a fixed celestial sphere parallax must exist, and if powerful telescopes could be developed parallax would be detected. To see what we mean by stellar parallax consider Figure 29-1.

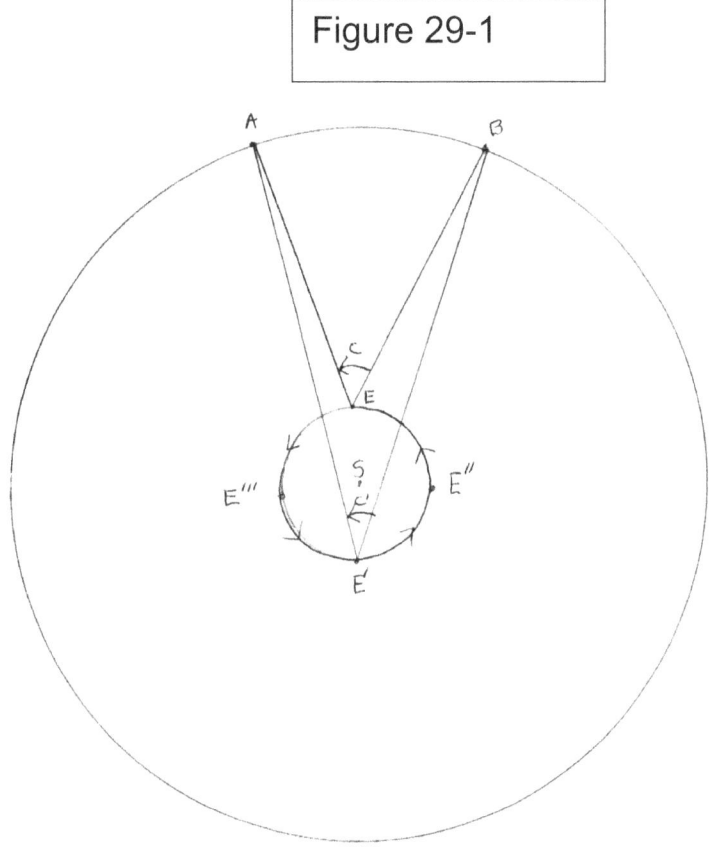

Figure 29-1

In the figure we see the view from the north celestial pole with the earth, sun, and celestial sphere shown such that the earth and sun are not treated as a point. The sun at S and the circle representing a cross section of the celestial sphere are stationary while the earth moves in a circular orbit around the sun such that the sun is the center of motion for the earth (this is an approximation of Copernicus' actual theory, but does not affect the arguments on stellar parallax).

We show the earth first at point E. A

hypothetical observer on earth notes the angle between two stars, A and B near the ecliptic on the celestial sphere, which we designate as angle C. We next note the position of the earth at point E', after the earth has moved 180° from point E. The earth is seen to be at its greatest distance from the stars A and B. The angle between the stars is now angle C'. It is true that at position E' an observer on earth could not observe these stars because the sun will be close to the two stars and obscure their view. Nonetheless if the earth moves as Copernicus claimed angle C' must be less than angle C from geometrical considerations.

Now as the earth moves from E' toward E eventually the stars will be seen at night so the angular distance between them can be ascertained at a position such as E". As the earth moves from E" to E the angular distance between the stars should become larger until the earth reaches point E. After that the earth moves towards position E"'. As the angle is measured during this period it should become smaller due to the motion of the earth relative to the celestial sphere in Copernicus' theory.   Stellar parallax is the variation in angles between stars at different points of the earth's orbit.

In Copernicus' time no astronomer with naked eye instruments had measured stellar parallax. This was one of the major arguments in favor of Ptolemy's theory of a stationary earth. After

Copernicus the Danish astronomer Tycho Brahe, 1546-1601, greatly improved observational instruments so that the angles between celestial objects could be ascertained in many cases to an accuracy of 2 minutes of arc, the highest degree of accuracy obtainable without a telescope. In contrast the best Ptolemy could do was an accuracy of 10 minutes of arc, and as we have mentioned, many of Ptolemy's observations are off by 1 to 2 degrees of arc. Tycho was so sure that if stellar parallax truly existed he would have found it. As a consequence of his failure to detect stellar parallax, he rejected the motion of the earth (he also gave other reasons to reject the earth's motion including Biblical citations), and created the Tychonian theory which we presented in the previous chapter.

After 1609 when the telescope was invented Galileo, 1564-1642, also tried to find stellar parallax with his telescope, but was unsuccessful. Nonetheless, he supported Copernicus' theory and predicted that stellar parallax would eventually be discovered as the resolving power of telescopes improved. By Galileo's time the concept that all the stars were attached to a celestial sphere the same distance from the sun was abandoned, and it was believed that some stars are considerably closer to our solar system and some considerably farther away. Galileo postulated a situation as shown in Figure 29-2 where two stars, A and B, at different distances from the solar system are observed from earth at positions E and E' with a stationary sun at S.

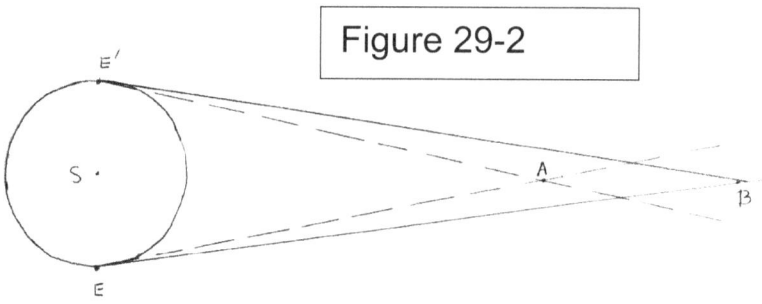

Figure 29-2

We see stars A and B would have essentially the same celestial longitude. Star A is much closer to the solar system than B. At position E the observer on earth sees star A as having a greater celestial longitude than B (A to the east of B), when at position E', the observer sees star B as having the greater celestial longitude due to stellar parallax. The angles are markedly exaggerated in the figure for clarity of exposition, since the angles involved would be much less than a minute of arc. Galileo predicted such phenomena would be observed to prove Copernicus' contention that the earth moves around the sun. This method of detecting stellar parallax became known as the differential method of parallax.

The first actual detection of stellar parallax came almost two centuries after Copernicus with the work of astronomer Fredrich Wilhelm Bessel, 1784-1846. In 1838 Bessel working with greatly improved telescopes compared to Galileo's found parallax for a star in the constellation Cygnus called 61 Cygni. This star was observed over a period of years and found to have a yearly variation of angular distances between two other stars in the constellation thus confirming the prediction of Galileo and the theory of earth's motion indicated by Copernicus. The situation described by Bessel may be simplified by the constructions in Figure 29-3.

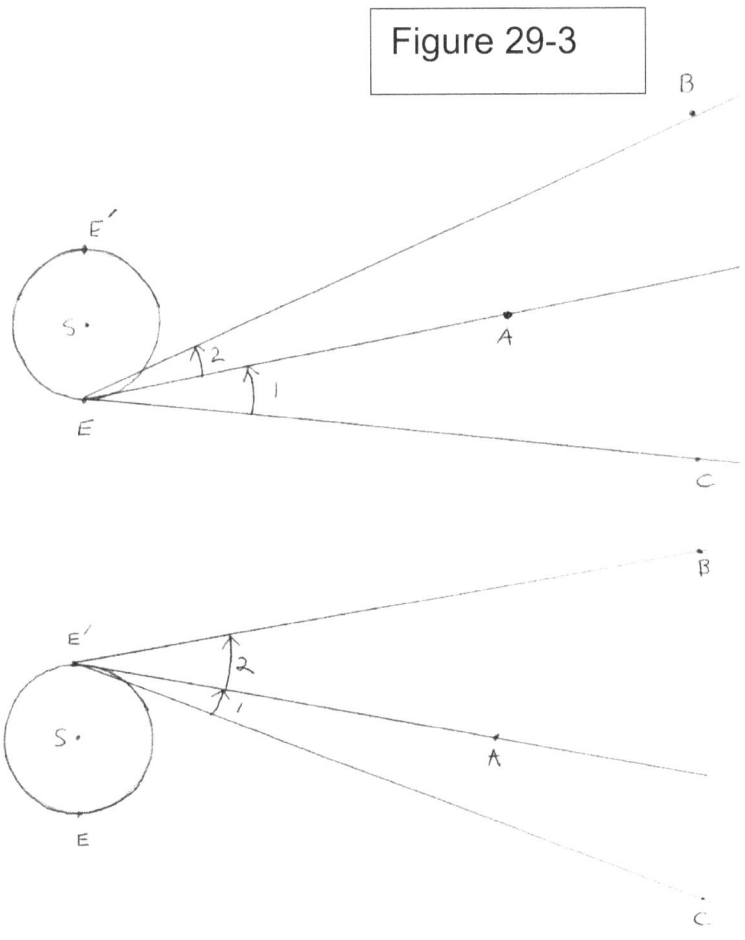

Figure 29-3

The situation is similar to the previous figures.

The angular distances between 3 stars A, B, and C, are measured as the earth moves in its orbit from E to E'. Star A (61 Cygni) is considerably closer to the solar system than stars B and C. At position E angle 1 is the angle between A and C, while angle 2 is the angle between A and B.   After 6 months the earth has moved 180° in its orbit. We see in the bottom view in the figure that angle 2 has increased while angle 1 has decreased due to the earth's relative motion with these stars. In the following 6 months the situation reverses as angle 2 decreases and angle 1 increases. The net effect after years of observation is that Bessel found star A (61 Cygni) had a yearly variation of position of .5 seconds of arc (.008 minute of arc, or .000138 degrees of arc). This is a very small angle and required very powerful telescopes to detect.

Soon after Bessel's great discovery astronomer Thomas Henderson found a stellar parallax of almost 1 second of arc for the star alpha Centauri in 1839.

The modern method to diagram stellar parallax is shown in Figure 29-4

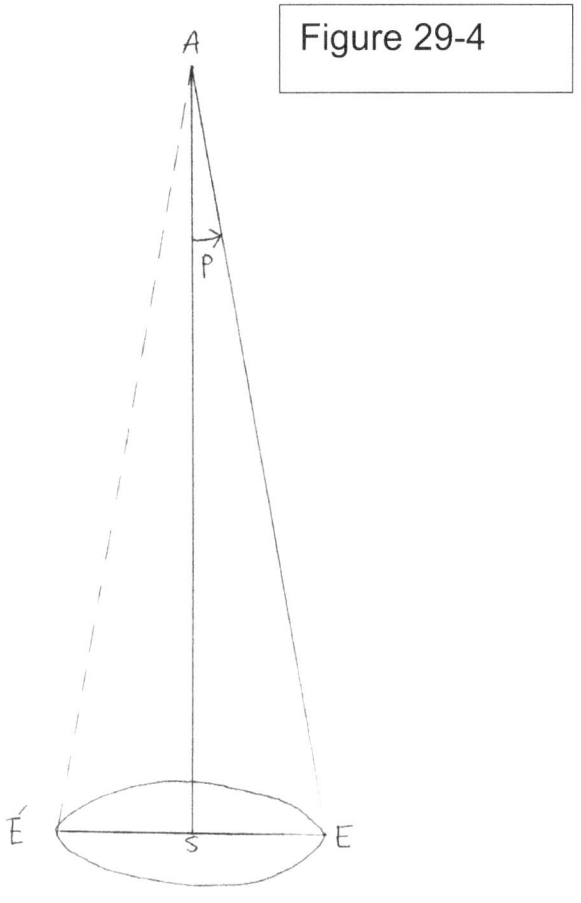

Figure 29-4

The angle of parallax is angle SAE = angle SAE', if we regard the earth's orbit as circular with the sun as the center. The angle of parallax is designated as angle p. In Figure 29-5 we show how to determine parallax by observing the change in the relative position of star A with star B as seen from earth at the two positions E and E' of the earth's orbit 180° apart.

Figure 29-5

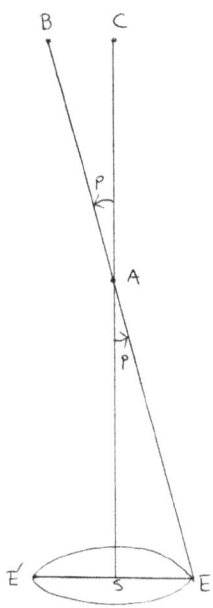 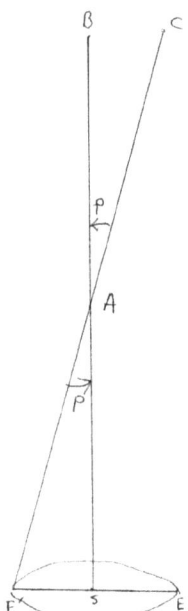

From Figure 29-4 we see tan p = ES/AS.

From this we can determine how far away star A is

compared with the distance of the earth to the sun.

If p = .5" then,

tan .5" = tan .000139° = .00000242.

ES/AS = .00000242, or AS = 412,530 (ES).

Thus the star is over 412 thousand times as far

away from the sun as the sun is from the earth.

These vast distances were not conceived of in

Copernicus' era.

In addition to the earth's motion around the

sun Copernicus stated the earth rotates once a day

around its polar axis which is tilted to the plane which includes the sun and earth. See Figure 29-6.

Figure 29-6

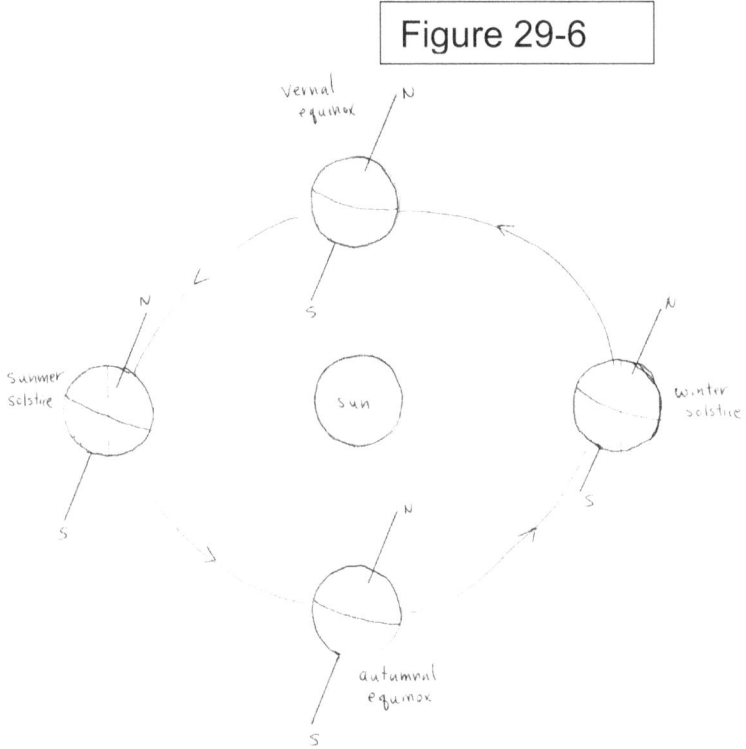

vernal
equinox
N

N

L

N
S

Summer
solstice

sun

winter
solstice

N

S

S

N

autumnal
equinox

S

Here we see the earth's position at different times of the year. Since the earth's axis always points to the north celestial pole, the sun's declination is constantly changing as the earth's moves in orbit. This is not due to movement of the sun but to the earth's movement around the sun.

Copernicus attributed another movement to the earth to account for the precession of the equinoxes. Today we diagram the movement of precession as shown in Figure 29-7.

Figure 29-7

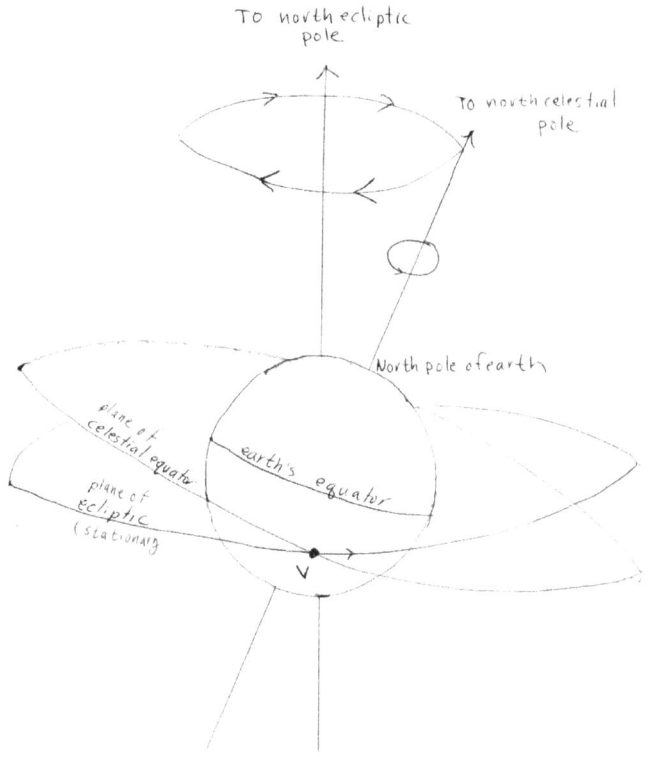

To north ecliptic pole.

To north celestial pole.

North pole of earth

plane of celestial equator

plane of ecliptic (stationary)

earth's equator

V

The north south axis of the ecliptic is considered as stationary as is the plane of the ecliptic. The north south axis of the earth is seen to rotate slowly around the north ecliptic pole clockwise as viewed from the north ecliptic pole or from east to west as indicated by the arrow in the figure. As this rotation occurs the first point of Aries, designated by V, must shift its position on the ecliptic eastward as viewed from the earth.

Because Copernicus accepted the theory of trepidation, which meant the rate of precession varies, he had to attribute a more complicated theory of motion for the earth. We will not attempt to explain the complications of this motion. He also

48

included a theory to explain the variation of the obliquity of the ecliptic over long periods.

For his explanation of the earth's orbit around the sun Copernicus adopted the eccentric circle theory which we learned in our study of the sun's motion in Ptolemy's theory. He found the eccentricity to be .0323, which contrasts with Ptolemy's value of .04167. He found the angle for the line of apsides to be 96° 40' in contrast to Ptolemy's value of 65° 30'.

Because these values are so different from Ptolemy's, Copernicus developed a theory of interacting circles to explain why these values are changing over the centuries. We show his basic idea in Figure 29-8.

Figure 29-8

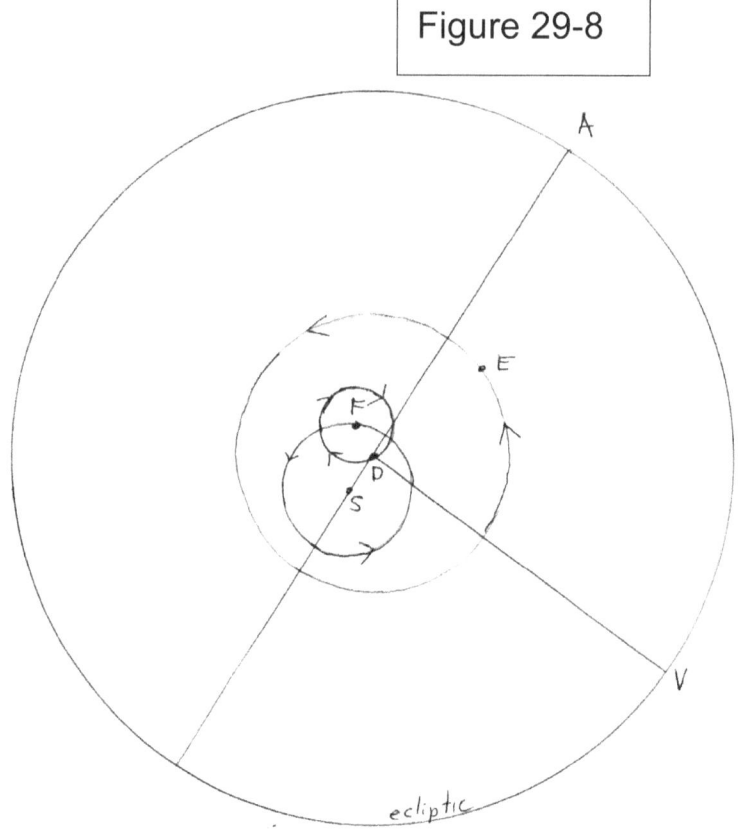

We are viewing from the north celestial pole. The sun is stationary at S. The center of the earth's orbit is at D. However, this point D is not stationary but moves uniformly around a circle with center at F. The earth moves at uniform angular velocity around its center at D. Furthermore, F is in motion around the sun at uniform angular velocity. At the time of Copernicus D was at the position shown in the figure so that the angle VDA was 96° 40' which is the angle of the line of apsides, SDA, from line DV to the first point of Aries at V. Copernicus calculated it took 3,434 years for point D to move around its circle in the clockwise direction as seen from the north celestial pole.

Meanwhile point F is slowly moving around the sun in a counterclockwise rotation as seen from the north celestial pole such that it takes 53,000 years for just one revolution.

If the radius of the earth's orbit, DE, is designated as 1 unit then the eccentricity, SD, is .0368 units, and the radius of the circle with F as the center, FD, is .0047 units. Copernicus felt D was at a position farthest from the sun at 64 B.C., so the eccentricity at that time was at its largest value. 100 years after Copernicus D would be at its closest position to the sun so the eccentricity would be at its smallest value.

This complicated system of interacting circles each with uniform motion was necessary for Copernicus to explain the changing values of the solar parameters as determined by Ptolemy in "The Almagest". It also shows us that Copernicus was determined to resolve all apparent irregularities of motion in the heavens by a combination of circular motions with points moving at uniform angular velocities around their centers.

We come now to Copernicus' planetary theory as seen from the moving earth. Copernicus' theory had the great advantage over Ptolemy's theory in that the theory of Copernicus gave reasons why the retrograde arcs of the superior planets always occur when the planet is in opposition to the sun.

Copernicus also explained why the radius of the epicycles calculated by Ptolemy had the largest value for Mars followed by Jupiter and Saturn.

For Ptolemy these parameters were what was found by observations, but there was no logic in why the retrograde arcs and hence the radii of the epicycles diminished as one went from Mars to Jupiter to Saturn. Copernicus system made it clear why this had to occur. Likewise, Copernicus explained why the retrograde arcs of Venus and Mercury had to occur in conjunction with the sun, and why the sun and the moon never moved with retrograde arcs. Although Copernicus' final theory did not allow more accurate predictions of planetary positions than did Ptolemy's system with the correct parameters, Copernicus' system made sense out of the retrograde arcs and why they occurred when they did. This reasonableness greatly influenced Kepler and Galileo the two most important advocates of Copernicus' system.

Let us look first into the reasons retrograde arcs appear for the planet Mars and why they must occur in opposition to the sun according to Copernicus. See Figure 29-9.

Figure 29-9

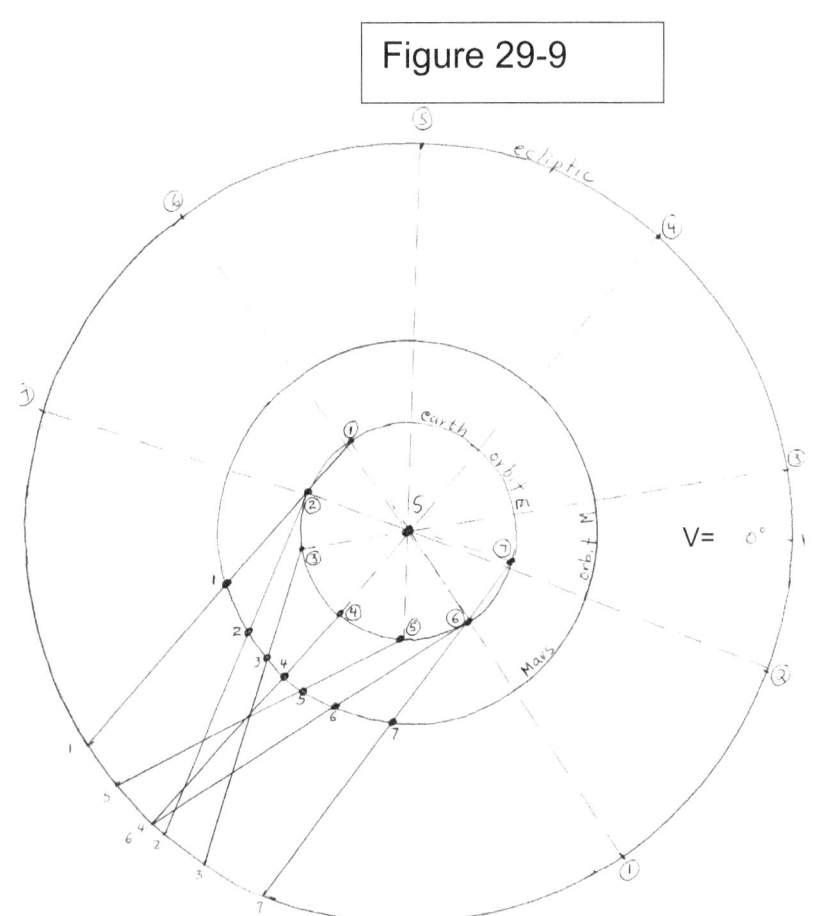

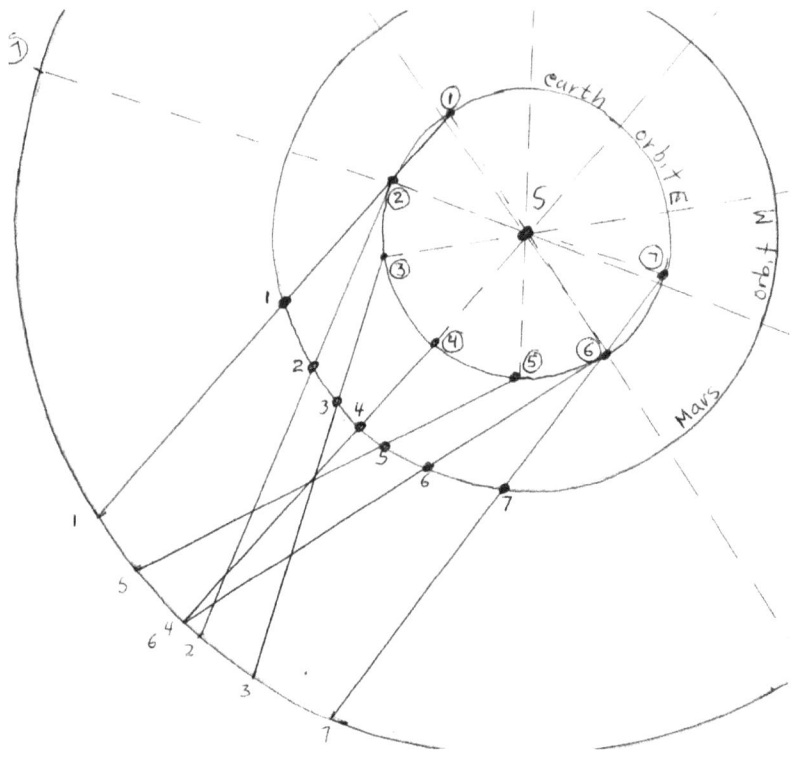

In this figure we see the celestial sphere in cross section viewing from the north ecliptic pole. The largest outer circle represents the ecliptic, and the point 0° is the first point of Aries. The sun is at S, the center of the circles and is stationary as is the ecliptic. The orbit of the earth is circle E, and the orbit of Mars is circle M. Although we know Copernicus did not have the sun at the center of the earth's orbit, for the purpose of explaining retrograde arcs we draw the orbits of Mars and the earth as centered at S. We plot the positions of Mars and the earth on 7 different dates based on the data for 1984 in chapter 22 table 3, for times before during and after Mars moves through a retrograde arc as seen from earth. The celestial longitude of Mars is marked by lines from the earth through Mars extended to the ecliptic. The position of the sun for each of the dates is marked by broken lines from the earth through the sun to the ecliptic. This allows us to follow visually why the retrograde motion occurs when the sun is opposite Mars.

We list the seven positions in the following table from the Nautical Almanac. Note two of them are the stationary points for the retrograde arc of 1984.

Celestial Longitudes

Date (1984)

	Sun	Mars
1 . Jan 21	300.65°	215.83°
2. Mar 2	342.12°	232.80°
3 . Apr 1	11.93°	238.53°(stationary point)
4 . May 11	50.96°	230.7 (retrograde motion)
5 . Jun 20	88.37°	220.09°(stationary point)
6. Jul 30	127.50°	230.62°

7. Sep 8   166.02°   252.17°

As we study the figure we see Mars moving prograde and increasing its celestial longitude through position 3 marked on the figure. Because the earth has a more rapid angular velocity around the sun 360°/365 d = .99°/d) compared to Mars ($\cong$ 360°/687 d = .52°/d), the earth 'passes by' Mars on its orbit which is inside or closer to the sun than Mars' orbit. When we pass a car on the highway the car we pass appears to be moving backward against the background of distant hills and trees. In this manner Mars appears to the earthbound observer to be moving backward as the earth passes Mars against the background of the stars of the ecliptic. This phenomenon which involves the motion of the earth and Mars against a stationary ecliptic accounts for the periodic retrograde arcs of Mars as seen from earth. The spacing of the arcs at different positions of the ecliptic is a consequence of the motion of both planets relative to the ecliptic.

Note the retrograde motion has to occur when the earth is between the sun and Mars (when the sun is in opposition to Mars), since this is the only time in the orbital movements of the planets when the earth 'passes by' Mars.

The pictorial approach shown in Figure 29-9 is not found in *De Revolutionibus*, but was first shown in Galileo's book of 1632, "Dialogue Concerning the

Two Chief World Systems".

Copernicus explains the phenomenon in Book V of De Revolutionibus.

The width of the retrograde arc in Figure 29-9 is about 18°. We know that the widths of the retrograde arcs for Mars vary between about 10° to 20°, depending on how close Mars is to the earth when it is in opposition to the sun. The closer Mars is to earth at the time of its retrograde motion the wider will its arc be. We can imagine varying the distance and hence the radius of the circle M in Figure 29-9 and how it would affect the width of the retrograde arc.

Because Jupiter and Saturn are much farther from the sun in Copernicus' system, the width of their retrograde arcs when the earth 'passes by' these planets will be much smaller than the retrograde arcs of Mars. Because their angular velocities are much less than the earth's more retrograde arcs will occur per revolution of these planets around the sun. These phenomena account for the smaller epicycle radii for Jupiter and Saturn compared to Mars.

When we examine Figure 29-9 from positions 3, 4, and 5, we see why Ptolemy and the Greeks observed Mars, Jupiter, and Saturn appeared brighter during the period of retrograde motion. This is because when the earth 'passes by' the superior planets, the earth is closer to the planets than at

57

other times in its orbit. By the time of position 7 when prograde motion has fully resumed the earth has moved farther away from the superior planet than at the period of retrograde motion so the planet will appear less bright.

Copernicus' theory gives a reasonable explanation of the phenomena of retrograde arcs in opposition to the sun for the superior planets which Ptolemy's system did not give. Let us now look at Copernicus' system for the inferior planets Venus and Mercury.

Figure 29-10 shows Copernicus' explanation of the retrograde arcs for the inferior planets, Venus and Mercury.

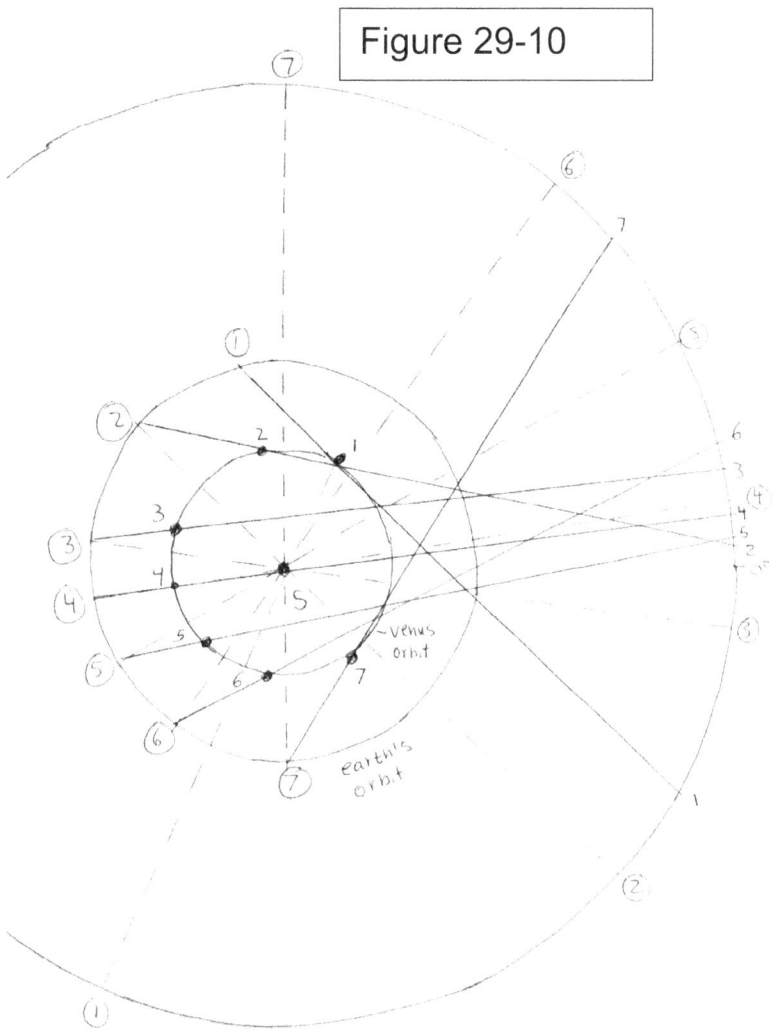

Figure 29-10

We have chosen to plot positions of Venus relative to the earth's motion for seven dates in 1985 from data in the Nautical Almanac. We will give a table of data with the celestial longitudes of Venus and the sun as seen from earth.

Date (1985)	Celestial Longitude Sun	Celestial Longitude Venus
1. Jan 6	332.97°	286.15°
2. Feb	2.11°	316.65°
3. Mar 12	18.90°	346.87° (stationary point)
4. Apr 6	9.30°	16.62° (retrograde arc)
5. Apr 26	4.70°	36.16° (stationary point)
6 . May 25	20.35°	64.26°
7. Jun 24	48.15°	92.65°

In Figure 29-10 the sun and the ecliptic are stationary and with the earth's orbit and venus' orbit as marked. Stationary points occur at positions 3 and 5. The broken lines trace the celestial longitudes of the sun for the corresponding positions of earth, and the solid lines trace the celestial longitudes of Venus for the corresponding positions of the earth. Position 4 is approximately at the middle position of the retrograde arc and is in conjunction with the sun, that is the celestial longitudes of the sun and Venus are nearly identical on that date.

Because the orbit of Venus is inside the orbit

of the earth in Copernicus' system we see from positions 1 and 7 that Venus cannot move more than about 48° maximum elongations from the sun. For position 1 the elongation is 332.97° - 286.15° = 46.82°, and for position 7 the elongation is 92.65° - 48.15° = 44.50°. In these positions the line of sight from the earth to Venus is essentially a tangent line to the orbit of Venus, and thus these positions represent points that are very near the maximum elongations of Venus for this period of observations.

If we imagine the orbit of Mercury further inside of the orbit of Venus we can see that the maximum elongations for Mercury will be of less angular magnitude than the maximum elongations of Venus.

In Ptolemy's theory for Venus and Mercury the radii of the epicycles are determined by the maximum elongations so that the epicylce for Venus is larger than the epicycle of Mercury. In Copernicus' theory the elongations of Venus and Mercury are a result of the earth being the third planet from the sun. The angular velocity of Venus is greater than that of the earth (we shall see shortly that one orbit around the sun for Venus is 225 d) so that Venus 'passes by' the earth from positions 3 through 5 in Figure 29-10. Because of the circularity of the orbits of Venus and the earth and the orbit of Venus being inside that of the earth's, we see that as Venus 'passes by' the earth observers on the earth see Venus in retrograde motion as we have indicated by the sequence of solid lines in Figure 29-10. The conjunction that occurs with retrograde motion, we recall, is the inferior conjunction for Venus. This is very close to position 4 where Venus is closest to the earth.

At a time later than position 7 Venus will again be in conjunction with the sun, but this will occur when Venus is on the other side of the sun than the earth, and this conjunction is the superior conjunction which occurs during prograde motion of Venus as seen from the earth.

Of course Copernicus and astronomers of his era before the invention of the telescope could not observe the conjunctions of Venus with the sun because of the sun's intense brightness. However, after the telescope was invented and filters could be used to dim the sun's light, astronomers could observe the inferior conjunction of Venus when it is in retrograde motion moving in front of the sun. They also confirmed that the superior conjunction could not always be observed because Venus may be eclipsed by the sun which confirms Copernican theory. (Because Venus' orbit is tilted to the ecliptic some superior conjunctions can be observed). In the Ptolemaic theory Venus is always in front of the sun during conjunctions as we showed in Figure 28-3.

Figure 28-3

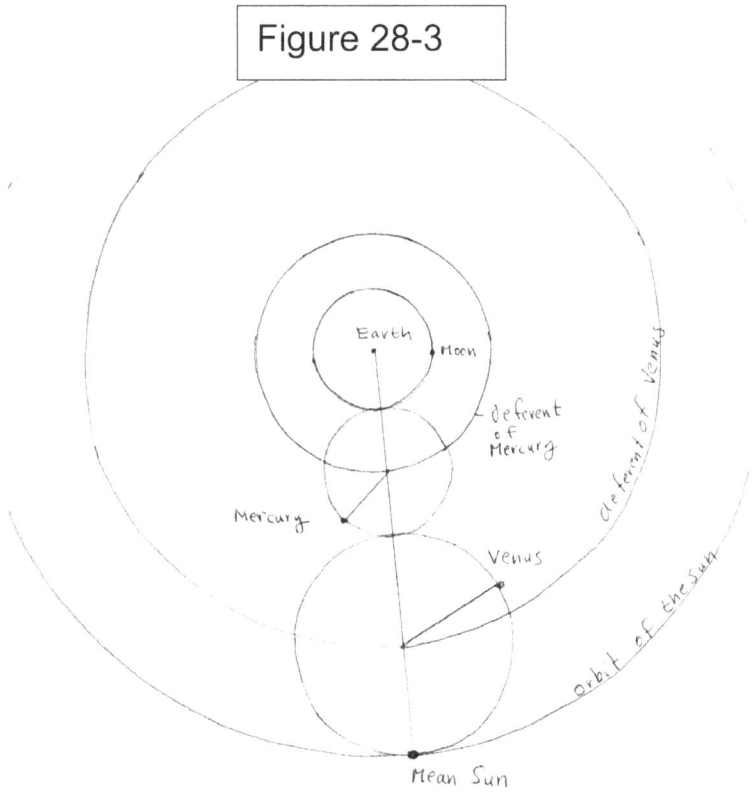

Later astronomers showed that the angular diameter of Venus at inferior conjunction is about 60" of arc. At superior conjunction when Venus is on the other side of the sun than the earth its angular conjunction when observed is about 10" of arc. This does not contradict Ptolemy's theory since Venus will move closer and then farther from the earth as it moves on its epicycle.

Before the telescope astronomers knew the brightness of Venus seemed the same before both inferior and superior conjunctions. This means, for example, in Figure 29-10 that to the unaided eye Venus would be just as bright at position 3 as at position 7 even though according to Copernicus,

Venus is farther from the earth at position 7 and should be slightly dimmer. No explanation of this phenomena was found until Galileo observed Venus with the telescope. The heliocentric theory was criticized because Venus did not change in brightness. Of course, in Ptolemy's theory for Venus in Figure 26-5 Venus should also change in brightness. Galileo after 1609 found the reason Venus maintained the same brightness was that just before and after its inferior conjunctions Venus was in a crescent phase like the moon and reflected less of the sun's light. Just before and after the superior conjunctions Venus is in a gibbous phase and thus reflects more of the sun's light, and thus even though farther away, Venus appears to have the same brightness as it does before and after inferior conjunctions. We will discuss more of the significance of the phases of Venus when we come to the work of Galileo in a subsequent chapter.

In the discussions involving both Figures 29-9 and 29-10 we have been explaining Copernicus' solution of the problem of the second inequality of planetary motion, the inequality with respect to the sun, which accounts for the retrograde arcs. To do this we approximated Copernicus' planetary theory by assuming the orbits of the earth and the planets had their centers of uniform motion located at the sun. Copernicus knew this could not be the situation because of the first inequality of planetary motion, the inequality with respect to the ecliptic. The observations that the retrograde arcs for the planets are irregularly spaced required centers for their orbits at a point along a line of apsides just as in Ptolemy's theory. We have already shown the Copernicus' model for the earth's orbit and how the center of the earth's orbit changes with time, and is not located at the position of the sun. Copernicus developed a theory for the first inequality which was just as complicated as Ptolemy's also using epicycles. His goal was to maintain uniform motion around the centers of the circles, avoiding the use of the equant point.

Before looking into the basic way Copernicus accounted for the first inequality, we would like to mention why the moon and sun do not have retrograde arcs like the planets in the heliocentric theory. In Ptolemy's theory the sun and moon orbit the earth, but were never observed to have retrograde motion. Ptolemy had no explanation why this was so, and the why the planets did have retrograde motion. We recall Ptolemy chose the eccentric circle theory for the sun, and that it gave the same results as an epicycle deferent theory. Both theories predicted the same observational results for the sun's position on the ecliptic. In Copernicus' heliocentric theory the sun is stationary while the earth orbits the sun with continuous motion,

although requiring an eccentric center to make the motion uniform with respect to the sun. Because the sun and the ecliptic are stationary, the sun can only appear to have prograde motion for earth observers. The earth never 'passes by' the sun in the sense the earth 'passes by' the superior planets as in Figure 29-9. Likewise, the sun never 'passes by' the earth in the sense the inferior planets 'pass by' the earth as in Figure 29-10. Thus the sun cannot have retrograde motion in the Copernican or heliocentric system. In Ptolemy's system the sun could have had retrograde motion just as did the planets, in the Copernican system the sun cannot have retrograde motion because of the geometry of the system.

In the case of the moon Ptolemy's system could have resulted in retrograde motion for the moon, that is, his system allows for it as a possibility depending on the parameters. In Copernicus' system he too regards the earth and the ecliptic as stationary as regards the moon. In fact he does use epicycles to account for the inequality with respect to the ecliptic seen in lunar motion. But because retrograde motion in Copernicus is due to the simultaneous motion of the earth and a planet relative to a stationary sun such that either the earth 'passes by' the planet or the planet 'passes by' the earth, and this situation does not apply to the moon and earth,   the moon does not have retrograde motion.

In book V of *De Revolutionibus* Copernicus calls the inequality with respect to the sun the motion in parallax. We will quote Copernicus in translation.

"Their double motion [refering to the planets] in longitude produce very different appearances in them. One is that due to the motion of the earth it has not undeservedly been decided to call the motion in parallax, since it is that which makes all of them appear to have stations, progressions and regressions... through the agency of the parallax produced by the motion of the earth in proportion to the difference in size of their orbits."

Copernicus explanation for the first inequality, the inequality with the respect to the ecliptic, is called the particular motion of each planet. In constructing a theory to explain this motion the period of revolution is the time for the planet to rotate once around the sun. Because the motion is irregular as viewed from the sun an eccentric circle around which uniform motion occurs is used. See Figure 29-11.

Figure 29-11

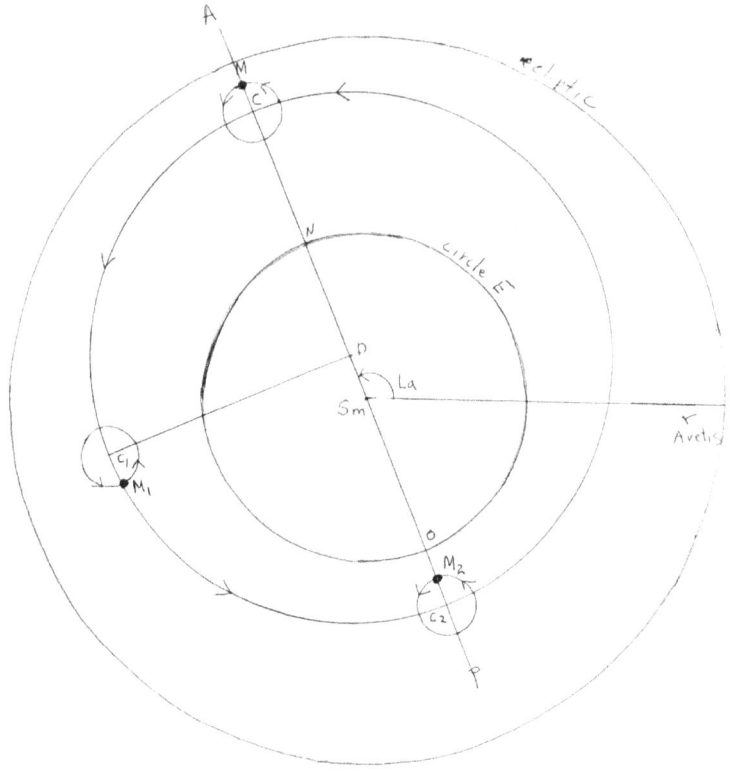

We will first discuss the theory for the superior planets. We again are looking at the situation from the north ecliptic pole. The key point in the figure is point Sm which is the center of the earth's orbit, the innermost circle marked circle E.   We can also consider this to be the point of the mean sun. Copernicus referred all planetary motion to this point and the line from Sm to the westernmost star in the constellation Aries which is gamma (γ) Aretis. This is a fixed star and does not change its position as does the first point of Aries. Of course the

68

celestial longitude of this star does vary with the precession of the equinox. The diameter of the earth's orbit is NO. The theory to explain the relationship between the earth and the sun is given in Figure 29-8.

In Figure 29-11 we see Copernicus does use an epicycle deferent system, but it is designed so the epicycle's center, C, moves with uniform angular velocity around the center of the deferent circle, D. There is not an equant point nor a *circulus equans* as in Ptolemy's system. The line that connects Sm to D when extended, is the line of apsides, AP, for the planet. Angle La is the angle the line of apsides makes with the line from Sm to γ Aretis. Note that Copernicus as Ptolemy has the center of the deferent D eccentric to Sm, the center of the earth's orbit.

In Ptolemy's system the epicycle was designed to result in the retrograde arcs for the planets while the equant was designed to explain the irregular motion of the planet with respect to the ecliptic. In Copernicus' system the motion of the earth in its orbit combined with the eccentric deferent explains the retrograde arcs while the epicycle with its small radius, CM, is designed to account for the inequality with respect to the ecliptic. In Copernicus it is the irregular motion as seen from point Sm which is explained by the use of the epicycle and eccentric deferent. Copernicus has CM = 1/3 the distance DSm

We call the epicycle of Copernicus a minor epicycle because of its small radius relative to the radius DC for the deferent. In many elementary textbooks of science or physics the authors present Copernicus' system as having the center as the true sun with circular orbits for the planets centered at the true sun moving presumably with uniform angular velocity. In part the fault for such simplification of Copernicus lies with a figure in *De Revolutionibus* showing such a system. We know, however, planetary observations for centuries indicate that combinations of circles are required to get a close approximation of what astronomers observe to be the motion of the planets.

Because both Ptolemy and Copernicus required such complicated interactions of circles, future mathematicians such as Kepler and Newton sought better explanations for the correct explanation of planetary motion.

As we mentioned there is no equant in Copernicus' system. DC rotates with uniform angular velocity, ω_d, around point D, and CM rotates with uniform angular velocity equal to ω_d, around C. Both rotations are counterclockwise as viewed from the north ecliptic pole. The system is such that the planet is at its maximum distance from Sm when it is at point M on the line of apsides. This is aphelion for the planet.   If the earth is at point O when the planet is at M, then the earth is at its maximum distance from the planet in conjunction with the true sun which would be very close to position Sm. When DC has rotated 90° to position DC_1 then CM has rotated 90° to position C_1M_1 with the planet at M_1.

After another 90° rotation DC_1 moves to DC_2 while CM_1 moves to C_2M_2. The planet is at M_2 along the line of apsides. If the earth is at point O on its orbit then the planet is at its closest point to the earth and this position of the planet is called perihelion.

We see how Copernicus has designed his system for the superior planets so that there is uniform angular velocity around the centers of the deferent and epicycle centers. He does place the deferent center eccentric to the center of his system, which is Sm, the position of the mean sun which is the center of the earth's orbit. He has eliminated an equant point and a *circulus equans* around which uniform angular velocity takes place. His next task is to determine the parameters for each of the superior planets. His method of parameter determination uses the same mathematical approach we learned in Ptolemy's system. Furthermore, Copernicus uses many of the observations given in "The Almagest" to determine the planetary parameters. He uses very few of his own observations or the observations of other astronomers of the16th century. We recall for each of the superior planets CM = 1/3 DSm. Copernicus also determined his parameters so that CM + DSm would equal the same value Ptolemy had determined for the distance QE, the distance from the equant point to the earth. This is another example of Copernicus' great dependence on Ptolemy's work. Nonetheless Copernicus did develop a complete system of astronomy with fully determined parameters based on observations, although many of them ancient, and which eliminated the equant point so long considered necessary to explain the inequality of planetary motion with respect to the ecliptic. He also made use of the principles of trigonometry and geometry so his system was mathematically based.

When we explained the Ptolemaic system we worked many mathematical demonstrations and calculations so the reader could see the mathematical basis for the system. But we will not give the mathematical demonstrations for the determinations of the parameters used by Copernicus. We feel the reader is now sophisticated in solving triangles, arc lengths, and chords so that he could follow Copernicus' own work in Book V of *De Revolutionibus*, or the book by Swerdlow and Neugebauer in the bibliography for this chapter.

We turn now to Copernicus' theory for the

inferior planets Venus and Mercury. The theory for these planets is even more complicated than the theory for the superior planets. See Figure 29-12.

Figure 29-12

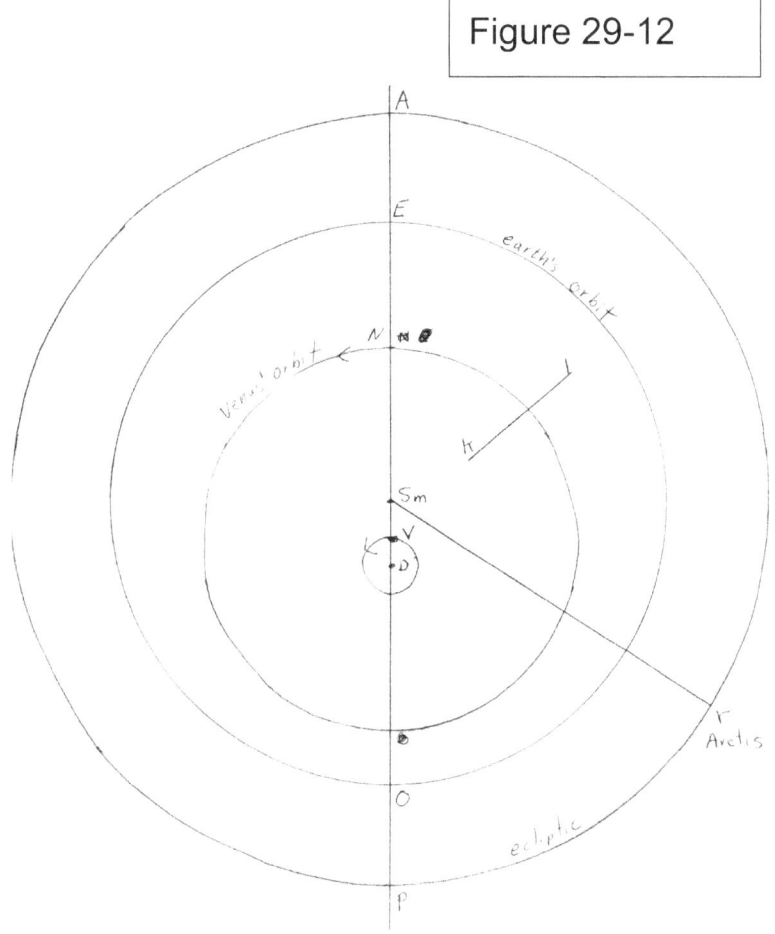

We start with the system for Venus. The reference point for the system is again Sm, the center of the earth's orbit and the position of the mean sun. The line from Sm to γ Aretis is indicated by line Sγ. The earth's orbit is the circle with radius SmO , which is the line of apsides for Venus.

To account for the observations that the

greatest elongation of Venus from the sun either east or west of the sun varies from elongation to elongation, Copernicus has a moving center for the circular orbit of Venus. At the position shown in the figure the center of Venus' orbit is at V, and the radius for the orbit is VN. Venus moves around its orbit at uniform angular velocity as seen from the center of its orbit V. At the same time point V moves around the small circle with center at D and radius DV. The uniform angular velocity of V around D is twice the angular velocity of Venus around its center and in the same counterclockwise direction as seen from the north ecliptic pole.

The system is such that when the earth is at position E or O on its orbit then point V is at the position in the figure on the line of apsides. Radius DV = 1/3 of the distance from Sm to D. The moving center and the eccentric center for Venus will result in the variations for the greatest elongations of the planet east and west of the sun.

The data Copernicus had for Mercury indicated even greater irregularities than could be accounted for by the system for Venus. Copernicus then added line kl to the model on which Mercury oscillates as it moves around its circular orbit. We see the great lengths Copernicus used to preserve his fundamental dictum that motion around the centers had to be of uniform angular velocity.

As for the superior planets Copernicus used

observations in "The Almagest" plus one of his own to determine the values for the parameters of his model for Venus. He had to borrow data on Mercury from other astronomers to find his parameters for Mercury as he was unable to make any satisfactory observations of Mercury from his observatory at Frauenberg due to atomospheric conditions. Some writers have indicated Copernicus never saw Mercury at all, however, what Copernicus tells us is that he could never ascertain its celestial longitude by his own observations.

Once Copernicus had worked out the values for the parameters of all the planets he was in a position to do calculations which establish the mean distances of each planet from the sun relative to the distance of the earth from the sun which is designated as 1 unit. Ptolemy's theory did not really allow this without making the assumption that the outermost point epicycle of one planet was at the same distance as the innermost point of the epicycle of the next planet as shown in Figures 28-2 and 28-3. The assumption is not warranted by observations made by Ptolemy or the ancients as explained in chapter 28.

In Copernicus' system observations can be used to calculate the distances of the planets from the sun in reference to the radius of the earth's orbit. We will show how this follows from the Copernican system. See Figure 29-13.

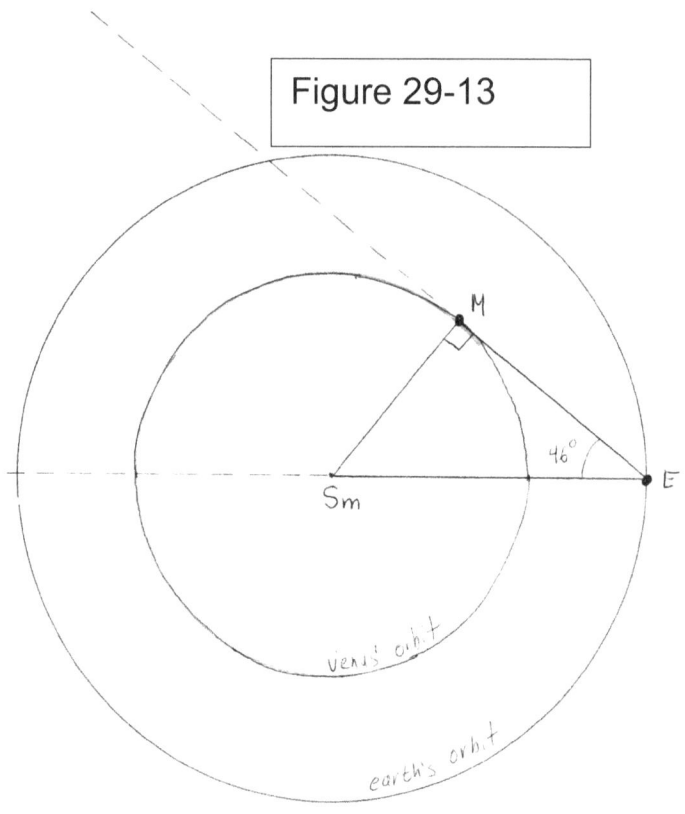

Figure 29-13

M

46°

Sm

E

venus orbit

earth's orbit

In Figure 29-13 when an inferior planet, M, is at greatest elongation from the sun the line EM from the earth to the planet is tangent to the circular orbit of the planet. If we assume the sun is at Sm and that Sm is the center of the planet's orbit (a close approximation), then triangle SmME is a right triangle, this means,

sin E = MSm/ESm. If ESm = 1 unit, MSm = sin E

If we use 46° for angle E as a close approximation for the mean value of the greatest elongation for Venus as an example then, MSm = sin 46°= .7193.   Thus the distance of Venus from the sun is .7193 of the distance of the earth from the sun as a mean value. This happens to be the value Copernicus found for this value.

A more involved method is required to approximate the distances of the superior planets to the sun relative to the earth's distance to the sun. See Figure 29-14.

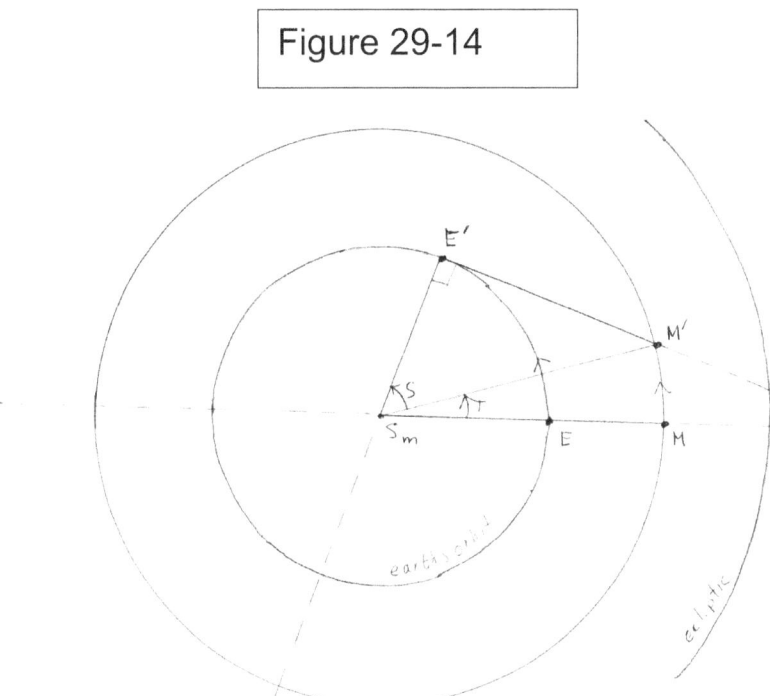

Figure 29-14

The figure shows the earth's orbit inside that of the superior planet so that the celestial longitude of the planet is determined at two positions for the planet, M and M'. In the interval between the two positions the earth has moved on its orbit from E to E'. The first celestial longitude determined is when the planet is at M, which is the time when the planet is in direct opposition to the sun, Sm, which is taken to be the center of both the earth and the planet's circular orbits. To make this approximation we disregard the details of Copernicus' theory presented in Figure 29-11.

The observer follows the planets position each night until prograde movement takes the planet to position M'. This position is the point where the elongation of the superior planet is 90° east of the sun. As shown in the figure Sm EM is a straight line, and SE'M' is a right angle. Note that in the time interval for the superior planet to move to M' from the position at M, the earth has moved from E to E'. The observer on the earth determines that the angle SmE'M' has reached 90^0 by noting the angular distance from E'Sm on the ecliptic to E'M' on the ecliptic is 90^0.

Next the angle ESmE' is calculated by the time in days for the earth to move from E to E' multiplied by the earth's mean motion of 360°/ 365.25 d. For example, let us say the planet is Mars and that it takes 106 d between the two positions. Then it takes   106 d for the earth to move through angle ESmE'. Then,

 angle ESm E'   = angle (S + T) = 106d x 360^0/365.25d = 104.47°.

We use the value of 365.25 d for the time it takes the earth to move 360° around Sm. From the theory of Mars in chapter 23 we learned the tropical period for Mars is 1.88 y = 687 d. This value is the sidereal period for Mars for it is the mean time for Mars to circle the earth in Ptolemy's system, or the sun in Copernicus' system. In 106 d Mars has moved from M to M'. This means,

angle MSm M' = angle T = 106d x 360^0/687d = 55.55°.

We see angle S = angle (S + T) - angle T.

Angle S = 104.47° - 55.55° = 48.92°.

Triangle M'E'Sm is right triangle so

cos S = E'Sm/SmM', and since ESm = E'Sm =1 unit, then,

SmM' = 1/cos S = 1/cos 48.92^0 = 1/.6571 = 1.5218.

 This means Mars is 1.52 times as far away from the sun as is the earth from the sun. Of course Copernicus in his calculations used his own table of sines and took into account the details of his planetary theory to derive his results, but the mathematical approach is similar to what we have shown. We will give a table of Copernicus' determinations and compare it to the modern values. The distance from the earth to the sun is assigned the value of 1.000 unit, which we recall is about 93,000,000 miles.

	Copernicus' Value	Modern Value
Mercury	.3763	.3871
Venus	.7193	.7233
Earth	1.000	1.000
Mars	1.5198	1.5237
Jupiter	5.2192	5.2028

Saturn   9.1743   9.5389

We have seen from Ptolemy's work, (chapter 23), how it is possible to determine the sidereal periods by comparing the tropical and synodic periods as seen from earth. As mentioned this gives the value of 1.88 y = 687 d for the time it takes Mars to orbit the earth in Ptolemy's theory, or the sun in Copernicus' system. The value for the tropical period of Jupiter is 11.86 y, and for Saturn is 29.49 y. However, the tropical period for both Venus and Mercury were equal to 1 year. Ptolemy, of course, did not consider them to be rotating around the sun, and instead considered their motions as linked to the sun as we have shown in chapter 26. Ptolemy did determine the uniform angular velocity of these planets around their epicycles.

For example, in the case of Venus in the table in chapter 26, ω_e = .6165°/d. If we assume that the center of the epicycle is the mean sun, and if we start motion around the epicycle at a greatest eastward elongation the time to the next greatest eastward elongation will be,

$360° \times 1d/.6165° = 583.94$ d.

Copernicus could have used this value in his theory, (he actually determined this value from long times intervals between many greatest elongations), to determine the sidereal period of

Venus (the time it takes for Venus to orbit once around the mean sun as seen from the mean sun). To visualize how this can be done see Figure 29-15.

Figure 29-15

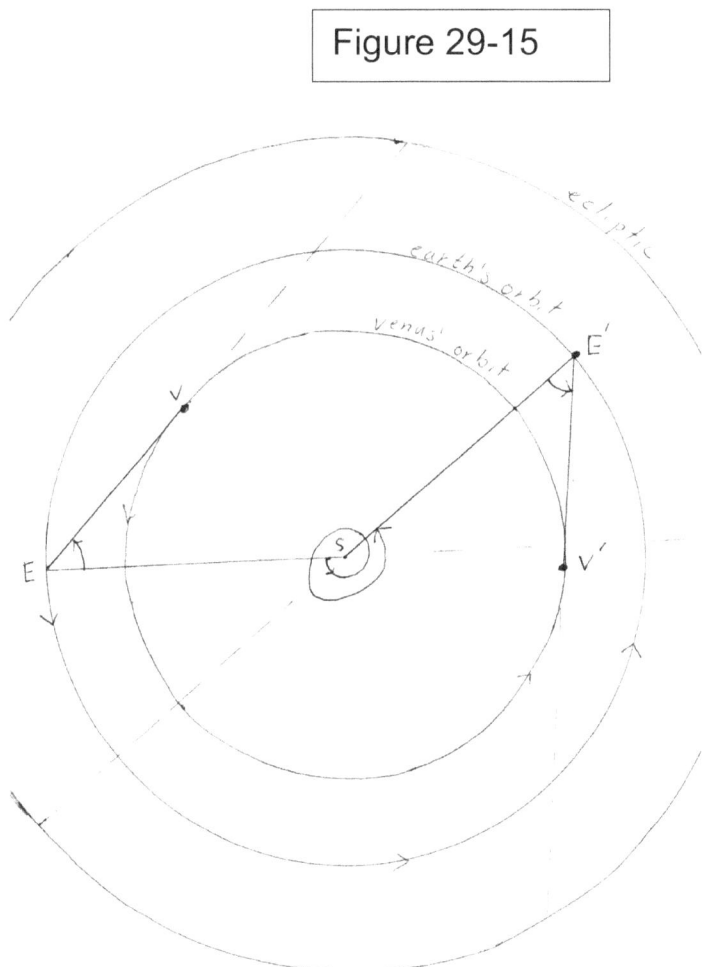

In the figure we show the relative positions of the earth, E, and Venus, V, at a greatest elongation when Venus is seen from earth to be east of the sun as an evening 'star'. After 584 d the next greatest elongation when Venus is east of the sun occurs. The earth has moved from E to E'. We can calculate the number of degrees the earth has rotated by the following.

584d x 1 orbit/365.25d = 1.60 orbits.

1.60 orbits x 360^0/1 orbit = 575.6°.

This is, 575.6° - 360° = 215.6°, more than one revolution of the earth around the sun, S. In Figure 29-15 angle ESE' measured counterclockwise equals 215.6°.

In the same 584 d Venus is moving at a more rapid angular velocity and 'passes by' the earth so that in 584 d Venus has orbited the sun more than 2 revolutions as shown in Figure 29-16.

Figure 29-16

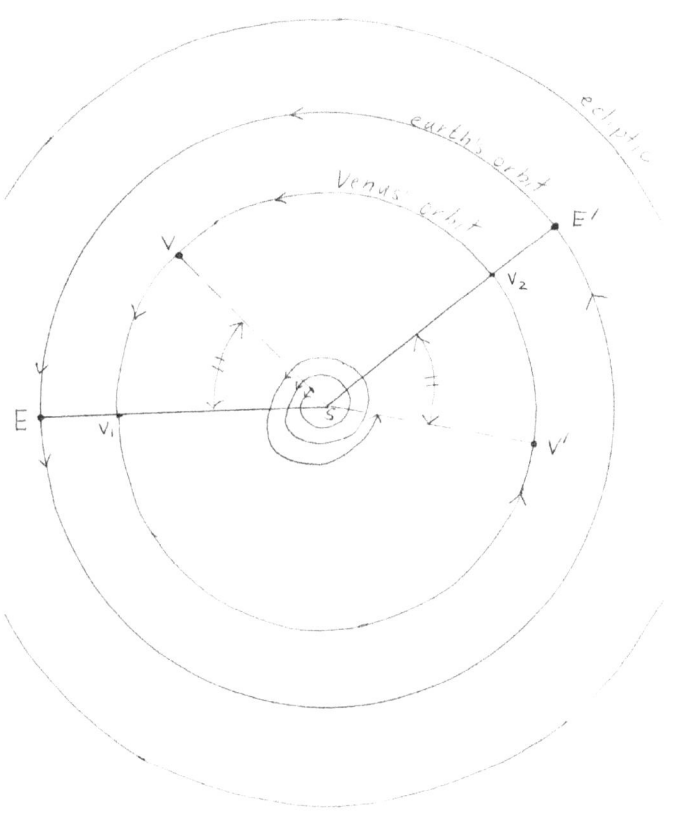

We see Venus has moved to position V'. Since we can assume for this calculation that the angles of greatest elongation are the same, then angle ESV = angle E'SV' in both Figures 29-15 and 29-16.

We can argue that if Venus was at point V_1 in Figure 29-16 then after 584 d it would be at position V2, since angles ESV and E'SV' are equal. Thus in 584 d Venus moves 2 revolutions (720°), plus an angle equal to angle ESE' = 215.6°. This means in 584 d Venus rotates 720° + 215.6° = 935.6° around the sun, S. Thus,

584d/935.6^0 x 360°/1 orbit = 224.7 d per orbit.

This means the sidereal period for Venus in Copernicus' system is about 225 d.

Ptolemy's data in "The Almagest" can also be used to find the sidereal period for Mercury. From chapter 25, ω_e, approximately 3.107°per day for Mercury. If we assume the epicycle for Mercury has its center at the sun as a close approximation of Copernicus' theory, then for Mercury to move once around the epicycle from one greatest elongation east of the sun to the next greatest elongation, angle MSM' in Figure 29-17, east of the sun it will take,

360° x 1d/3.107^0= 115.9 d.

In this time the earth moves,

115.9d x 1 orbit/ 365.25d = .317 orbits.

.317 orbits x 360°/1 orbit = 114.2°.

As the earth moves 114.2° Mercury moves 1 revolution around the sun plus 114.2°.

See Figure 29-17.

Figure 29-17

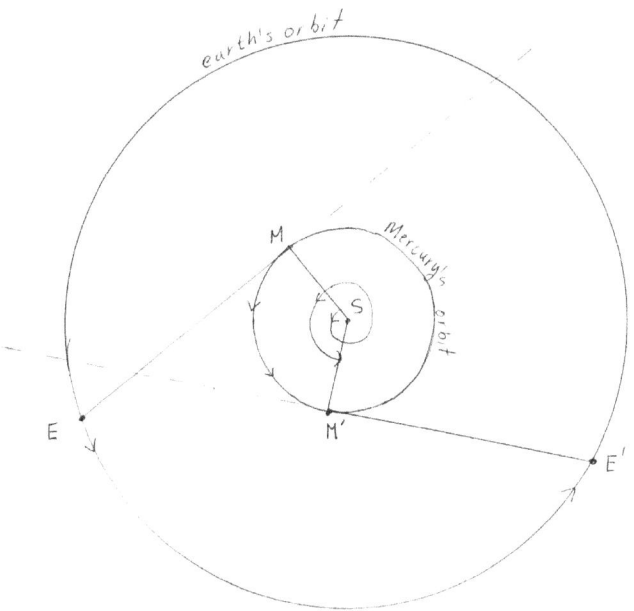

We then calculate the time for 1 revolution for Mercury.

115.9 d/474.20^0 x 360°/1 orbit = 88 d per revolution.

We see the Copernican system allows calculation of not only the relative distances of the planets from the sun, but also allows the calculation of the sidereal periods for all of the planets. We have seen how Ptolemy's data is used throughout Copernicus' work, and we could say that Copernicus is incomprehensible without a familiarity with "The Almagest".

Copernicus' system is not more accurate in predicting planetary positions than is Ptolemy's system, nor is it a simpler system when studied in its detail. In our brief survey we have left out the details of many determinations, which if read in *De Revoulutionibus*, would convince the reader of the complexity of Copernicus' system. Copernicus calculated all the parameters by mathematical methods quite similar to the work of Ptolemy. The Copernican system is also not a true heliocentric system, since the true position of the sun is not used as the key reference point of planetary theory. Rather it is the center of the earth's orbit that is the key point of reference.

There are certain features of Copernicus' system that are quite appealing to the mathematical mind. The system explains why the planets have their retrograde arcs related to the sun's opposition in the case of the superior planets, and related to conjunctions of the sun in the case of the inferior planets. The system also explains why the inferior planets have limited elongations from the sun (being inside the moving earth's orbit), while the superior planets have both conjunctions and oppositions with the sun (being outside the orbit of the moving earth). Further, Copernicus can calculate the distances of the planets from the sun compared to the earth's distance from the sun and determine their sidereal periods.

It was these features and new astronomical observations in the period following publication of *De Revolutionibus* that would lead both Kepler and

Galileo to become ardent supporters of the Copernican hypothesis that the earth moves and the sun is stationary.

With the data from the new work on planetary observations made by Tycho Brahe, Kepler put together a truly new astronomy abandoning the long held belief that celestial motion had to be circular based on deferents and epicycles. It is perhaps somewhat ironic that the new astronomy, using elliptic orbits as developed by Kepler, was based on the mathematics of the conic sections. It was none other than Apollonius who had written the master treatise on conic sections to which Kepler turned to develop the theory of elliptical motion. Apollonius, who was one of the pioneers in the idea that planetary motion could best be explained by the use of deferents and epicycles, was the primary source for the mathematics of the ellipse.

In order for us to understand properly the mathematics of the new astronomy of the 17th century, we need to extend our knowledge of mathematics making use of the fruits of 17th century mathematical development as well as the Greek work on conic sections. In the following chapters we will take up the subjects of coordinate geometry and the conic sections. This preparation will allow us to proceed to our study of Kepler, Galileo, and Newton.

Bibliography for Chapter 29

1.Copernicus, "On the Revolutions of the Heavenly spheres", translated from the Latin by A.M. Duncan, Barnes and Noble Books (a division of Harper and Row Publishers,   Inc.),   New York, 1976.
2.Kuhn, Thomas S., "The Copernican Revolution", Harvard   University Press, Cambridge, Massachusetts, and London, England, 1 957.
3.Dreyer, J.L.E., "A History of Astronomy from Thales to Kepler", Dover Publications,   Inc., 1953.
4.Koyre, Alexandre, "The Astronomical Revolution", translated by Dr.   R.E.W. Maddison, Cornell University Press, Ithaca, New York, 1973.
5.Swerdlow, N.M., Neugebauer, 0., "Mathematical astronomy in Copernicus's De Revolutionibus, Springer-Verlag, New York, Berlin, Heidelberg, Tokyo, This is Vol.   10 in "Studies in the History of Mathematics and Physical Sciences", 1984.
6.Berry, Arthur, "A Short History of Astronomy", Dover Publications, Inc., New York, 1961.

Chapter 30

Coordinate Geometry

As we have learned the Greek mathematicians represented triangles, circles, and their relationships by figures constructed on a plane or with the concepts of spherical trigonometry.   The mathematics of ratio and proportion were used to demonstrate the theorems concerning geometric properties. We learned that we still depend on ratios and proportions in modern mathematics, but we use the modern notation of algebraic equations. For example, the sine of an angle is the ratio of the side opposite the angle in a right triangle to the hypotenuse of the triangle.   We can also write the equation, $y = \sin A$, where we indicate the sine function of two variables, the independent variable being the angle and the dependent variable being the value of sine A. We have learned that we can also represent, $y = \sin A$, as a graph or curve relating the two variables.

In relating the two variables by a graph or geometric curve we are using the method of coordinate geometry. The development of this method began in the 17th century by two mathematicians working independently of each other.

The philosopher and mathematician Rene Descartes has received most of the credit for the creation of coordinate geometry since his book, "The Geometry", was the first book on the subject published.   This occurred as an appendix to his philosophical text, "Discourse on Method", which was published in 1637.

But Pierre Fermat (1601-1665) also developed the principles of coordinate geometry, and in fact wrote them in a system similar to our modern system of notation. However, Fermat's work was not published until after his death in the 1670's, and thus he did not receive as much credit as has Descartes.

Fermat was a lawyer by profession and served as King's Councilor for the parliament of Toulouse in France. Mathematics was a hobby for him, yet he contributed fundamental work in many branches of mathematics.

Descartes (1596-1650) devoted much of his adult life to a study of knowledge and philosophy, and he used mathematics as a major source of his thinking.   The book, "Discourse on Method", used a new approach for those who wished to learn to acquire valid truths.   As his starting point for the acquisition of true knowledge he rejected all the previous authorities, dogmas, and previous philosophical works, in particular the works of Aristotle.   He wished to establish truth by reasoning in such a way that all his conclusions were certain and irrefutable. He uses reasoning, intuition, and deduction to establish his truths. We can agree this is very much the method of mathematical development, and Descartes approved of mathematical demonstrations because they are certain and irrefutable. He wrote,

"There are two things that I distinguish in the geometrical mode of writing, viz., the order and method of proof. The order consists merely in putting forward those things first that should be known without the aid of what comes subsequently".

This is the idea of starting with the fundamental axioms as Euclid had done in the "Elements", and then using reasoning and deduction to establish further truths. Descartes went on,

"The method of proof is two fold, one being analytic, the other synthetic. Analysis shows the true way by which a thing was methodically discovered and derived, as it were effect from cause... Synthesis contrariwise employs an opposite procedure [to analysis], one in which the search goes as it were from effect to cause (though often here the proof itself is from cause to effect to a greater extent than in the former case). It does indeed clearly demonstrate its conclusions, and it employs a long series of definitions, postulates, axioms, theorems and problems, so that if one of the conclusions that follow is denied, it may at once be shown to be contained in what has gone before".

Descartes says both analysis and synthesis may be needed in a proof. Analysis is how a relation is discovered. In the case of the sine function, for example, we showed how this function was discovered by finding the values for the lengths of chords in a circle if we knew the

91

values of the arcs for the chord and the value of the diameter of the circle.

Synthesis is the chain of reasoning that the philosopher or mathematician uses which convinces the reader that the results claimed, such as the results in the table of sines, are the true values. We recall the many steps in the chain of reasoning required to establish the table of sines. Descartes felt mathematical results were truths that could be used to study nature and increase mankind's understanding of the world and universe so that the conditions of life might be improved.

Fermat on the other hand studied mathematics as an intrinsically beautiful creation of the human intellect full of harmony and symmetry. Both Fermat and Descartes wanted to use mathematics to study more complicated curves than the circle, and wished to develop algebraic expressions for geometric curves. They felt working with algebraic expressions was easier and made the chain of reasoning more clear than working with geometric constructions alone.

Descartes had studied Euclid and had found as we have that many of the arguments are so tied to the geometric constructions that the construction is essential to understand the argument. He wrote that Greek geometry,

"can exercise the understanding only on condition of greatly fatiguing the imagination".

Unfortunately to read Descartes' book on coordinate geometry and Fermat's work is very difficult since the notation is not what is used today.   Furthermore, Descartes wrote in such a way, that many of the steps in his arguments were omitted so that the reader has to do the steps for himself if he is to be convinced.   Thus the book is written for the sophisticated mathematicians of his age.

Fermat's treatise on coordinate geometry is a short work called, *'Ad Locos Pianos et Solidos Isagoge'*.   It is only available in Latin or French so that English readers must depend on commentaries such as that by Carl Boyer whose work, "The History of Analytic Geometry", is listed in the bibliography for this chapter.   Boyer reports that Fermat stated,   *"Whenever in a final equation two unknown quantities are found, we have a locus, the extremity of one of these describing a line, straight or curved."*

Fermat began with algebraic equation with independent and dependent variables, and showed how the equation could represent curved or straight lines. The term 'locus' refers to all the points on the geometric line given by the equation.

Fermat did not use the term 'coordinate geometry', nor did he set up reference axes in the manner we do today.   In our development of coordinate geometry we wish to emphasize those principles that will be of use to us in understanding the work of Kepler, Galileo, and Newton. We will be using modern notation and restrict ourselves to a limited number of the many demonstrations found in modern mathematics books dealing with coordinate geometry.

The first task in modern coordinate geometry is to establish the two coordinate axes. Two mutually perpendicular lines are drawn on a plane as in Figure 30-1.

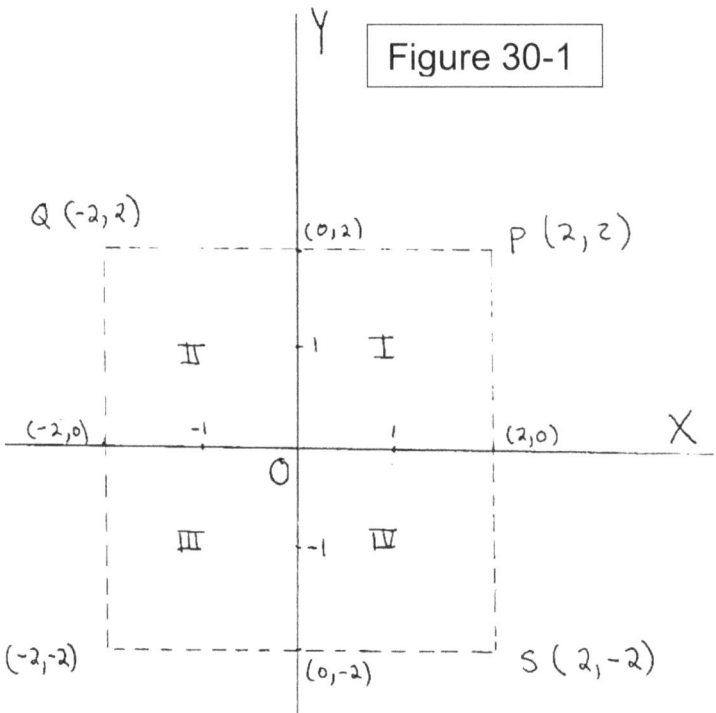

Figure 30-1

Line X is chosen as the horizontal axis, and line Y as the vertical axis. The point of intersection is called the origin, 0. X is called the axis of abscissas or the X-axis. Y is called the axis of ordinates or the Y-axis. Distance is measured from the origin as positive distance along the X-axis to the right of the origin, and along the Y-axis in the upward vertical direction. Distances measured to the left of the origin along the X-axis and downward vertically along the Y-axis are negative distances. Each point on the plane can be related to the XY axes by the perpendicular distances from the point to the two axes as indicated in the figure.

We have shown 4 points on the plane each point being 2 units from the axes. Point P is 2 units of positive distance from both the X and Y axes. The point is assigned 2 coordinates called the x and y coordinates. The x coordinate is +2, and the y coordinate is +2. The coordinates are written P (x, y), or P (2, 2), where the x coordinate is given first.

Point Q is 2 units in the negative direction from the Y axis, and 2 units in the positive direction from the X axis. The coordinates for point Q are, Q (-2, 2). Note again the x coordinate is the perpendicular distance either negative or positive from the Y-axis, and the y coordinate is the perpendicular distance either negative or positive from the X-axis.

Just as in our study of trigonometry we divide the plane into 4 quadrants as noted in the figure by Roman numerals. For example, a point with coordinates (7, -5) has an abscissa of + 7, and an ordinate of -5 which place the point in the fourth quadrant. In general we speak of any point on the plane as having coordinates (x, y) where x is the abscissa, and y is the ordinate, and either coordinate may have values negative, positive, or zero.

The modern system of dividing the plane into 4 quadrants using mutually perpendicular X and Y axes is also called a rectangular Cartesian coordinate system in honor of Descartes. The term abscissa comes from the Latin word 'abscindere' meaning to cut off, and the term ordinate comes from the Latin word 'ordinare' meaning to order.

We will next develop the method of finding the distance between two points using coordinate geometry. See Figure 30-2.

Figure 30-2

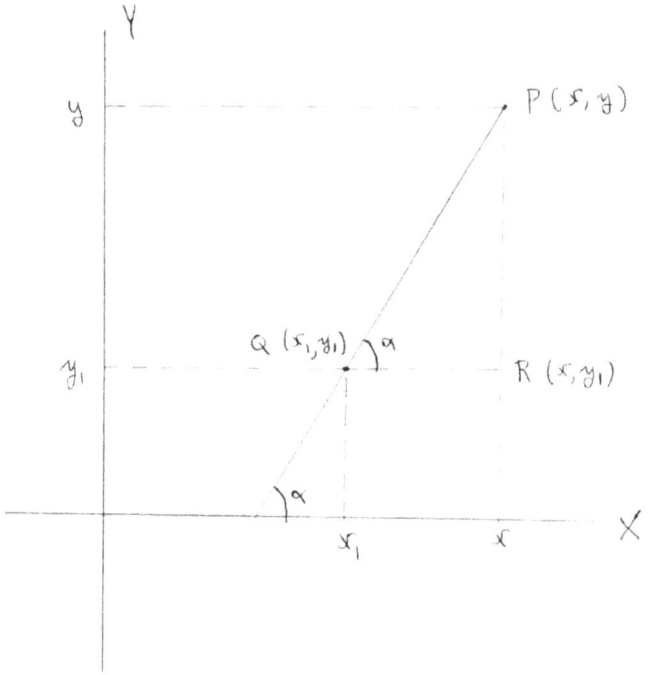

In the figure two points, P (x, y), Q (x_1, y_1), are marked, and the line connecting the points is drawn and extended to the X-axis where it forms the angle α (Greek letter alpha). The perpendicular distances to the axes from the two points are marked with broken lines.

We see in the figure we have right triangle QRS where R has coordinates (x, y_1). Distance QR = $|x - x_1|$. Distance PR = $|y - y_1|$. (We use absolute values so the distance will be positive values if the points were chosen with different orientation, and/or in other quadrants). Using the Pythagorean Theorem,

1. $(PQ)^2 = (PR)^2 + (QR)^2$,

$(PQ)^2 = |y - y_1|^2 + |x - x_1|^2$,

$$|PQ| = \sqrt{|y - y_1|^2 + |x - x_1|^2} \ .$$

The angle α that a line such as PQ extended makes with the X-axis is called the angle of inclination of the line. It is the smallest angle between the X-axis and the line measured in the counterclockwise direction. Mathematicians define the tangent of the angle of inclination of a line as the slope of the line. We will use this concept in later chapters. We see from right triangle QRS how to calculate tangent α

.   2. $\tan \alpha$ = PR/QR = $\dfrac{y - y_1}{x - x_1}$.

We do not use absolute values in this determination. If α is between 0° and 90°, the tangent will be a positive number. If α is between 90° and 180°, then $\tan \alpha$ will be a negative number. If α = 90°, then the slope or $\tan \alpha$, is not a real number, and lines with angles of inclination of 90° are parallel to the Y axis.

In coordinate geometry straight lines have an algebraic equation which states the relationship between the variables x and y.

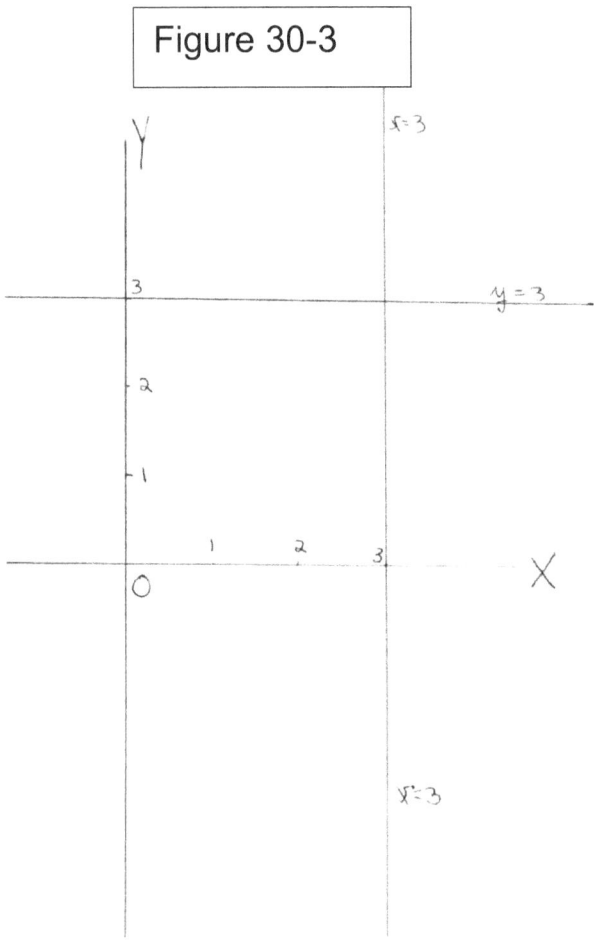

Figure 30-3

In Figure 30-3 we see two straight lines in addition to the lines marking the X and Y axes. The line marked x = 3 is a line parallel to the Y-axis always being 3 units to the right of the Y-axis. For every point on this line the x coordinate is 3. The y coordinates can vary from any positive number to any negative number. This condition is stated in the equation x = 3. Likewise the equation, y = 3, gives us the equation for a straight line such that the y coordinate for any point on the line is 3 while the x coordinates may take on any number. The graph of this line is marked,

y = 3.

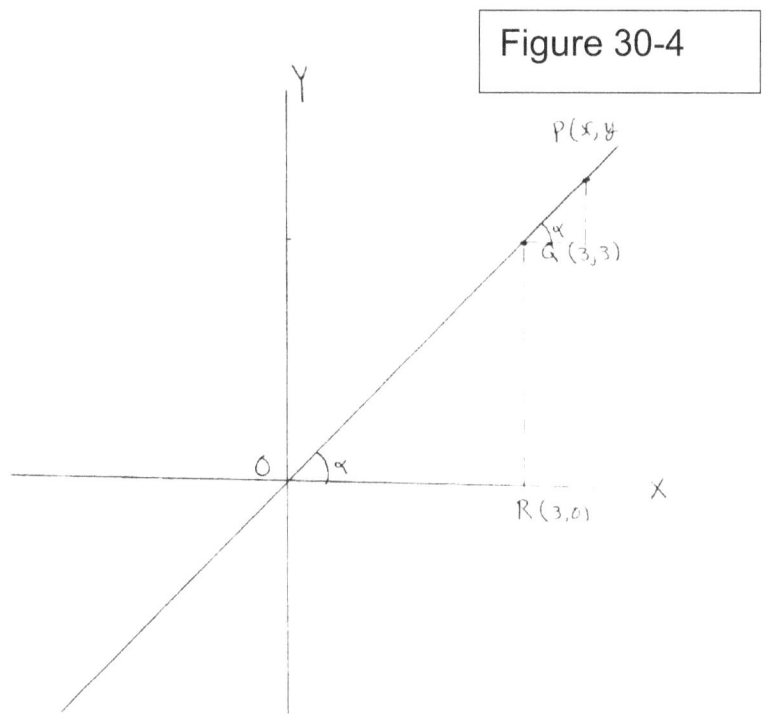

Figure 30-4

In Figure 30-4 we show a straight line which contains the two points O (0, 0), and Q (3, 3).   We ask: what is the equation for this line?

To answer this question we first determine the slope of the line by completing the right triangle ORQ as indicated. We designate the slope of the line by the letter m.

3. tan α = QR/OR = 3-0/3-0 = 1.

Now let P (x, y), be any other point on the line other than O or Q. We then complete the right triangle QSP and determine the slope.

4. $m = \tan \alpha = PS/QS = \dfrac{y-3}{x-3}$.

We can equate the results of step 4 and step 3.

5. $\dfrac{y-3}{x-3}$ =1 , or, **y − 3 = x - 3, or, y = x.**

This represents the equation of the line which connects O and Q.

In general if we have a line with slope m and which crosses the Y-axis at a point Q with coordinates (0, b) as shown in Figure 30-5, then we can develop the formula for the line as follows.

Figure 30-5

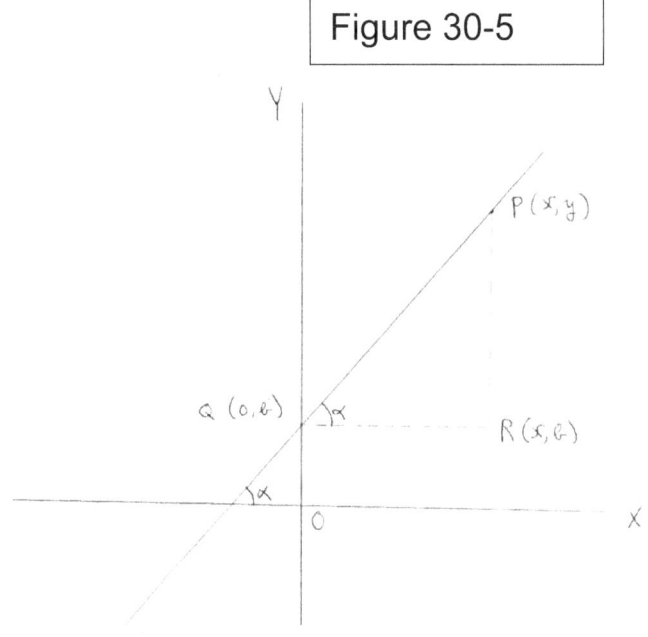

100

We construct the right triangle QRP from any point P (x, y) on the line other than Q as shown in the figure. We see the line has the angle of inclination, α , and because QR is parallel to the X axis, α is equal to angle PQR. Thus,

6. m = tan α = PQ/QR = $\dfrac{y-b}{x-0}$.

m = $\dfrac{y-b}{x}$, or y – b = mx, or,

y = mx + b.

This is the equation of a straight line which has a slope m, and a y coordinate b at the point the line intercepts the Y axis. Of course, if α = 90° this equation does not apply, since such lines do not intercept the Y-axis and are parallel to the Y-axis.   We note the equation is of the form,

Ax + By + C = 0, where A, B, and C are constant terms.

We see if A, B, and C are not 0 we can write,

7. Ax + By + C = 0, By = - Ax - C

y = $-\dfrac{A}{B}$ x - C .

Here - A/B is the slope m of the line, and, - C is the y coordinate of the point where the line intercepts the Y axis.

We will next derive the equation in coordinate geometry for a circle.   The equation has the simplest form if we choose the origin of the rectangular Cartesian coordinate system to be the center of the circle as in Figure 30-6.

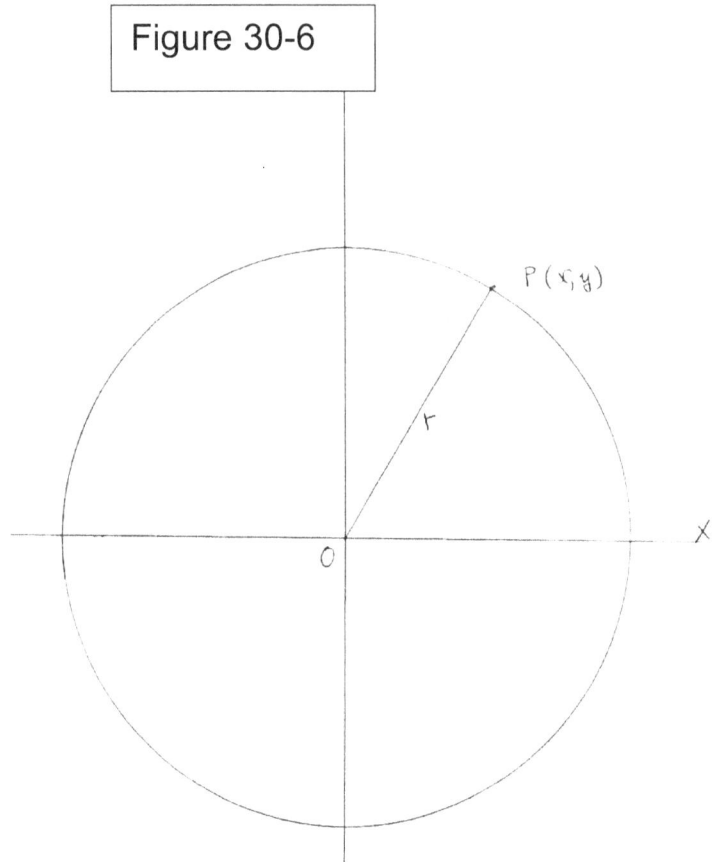

Figure 30-6

The circle has radius r.   Let any point on the circle, P, have coordinates (x, y).   The distance OP = r, is, by the distance formula of step 1,

8. $|OP| = \sqrt{|x - 0|^2 + |y - 0|^2}$,

$(OP)^2 = x^2 + y^2$, thus, $r^2 = x^2 + y^2$, or,

$x^2 + y^2 = r^2$.

This represents the equation for the circle with radius r, and center at the origin.   We can see now that a geometric figure can be represented by an algebraic equation. In the following chapter we will learn about more complicated geometric figures called the conic sections.   We shall see how they were originally defined by the ancient Greek geometers, and how their definitions can be transformed into equations based on the rectangular Cartesian coordinate system.   We will need to understand the conic sections since one of them, the ellipse, is the path of the planets derived by Kepler in his work. Gallileo was to show that objects projected from the earth's surface would follow the path of a conic section known as the parabola. Later Newton showed that celestial objects could follow the paths of any of the conic sections in their motions.

Bibliography for Chapter 30

1. Morrill, W.K., "Analytic Geometry", International Textbook Company, 1964.

2. Descartes, Rene, "Discourse on Method, Optics, Geometry, and Meteorology, Translated with an Introduction by Paul J. Olscamp, The Library of Liberal Arts, Bobbs-Merril Company, Inc., 1965.

3. Boyer, Carl B., "History of Analytic Geometry", The Scholar Bookshelf,   1956, Published 1988 by The Scholar's Bookshelf, 51 Everett Drive, Princeton, NJ.

Chapter 3 1

Geometry of the Conic Sections

The conic sections became very important in science and astronomy with the work of Kepler, Galileo, and Newton. These mathematicians found that the curves derived from the conic sections were found in natural motions. Although these curves were known and studied by the ancient Greek mathematicians, they were not felt to be part of celestial motions. As we have learned the motions of the planets, stars, and the apparent motion of the sun were to be resolved with circular motions or combinations of circles. We will take up a study of the curves made from conic sections so we will be familiar with their properties when we come to the work of Kepler, Galileo, and Newton.

Our work will involve two lengthy chapters, first the geometry of the conics, then the use of coordinate geometry for the conics. The conic sections for the ancient Greeks were originally derived by constructing what is called a right cone.

Figure 31-1

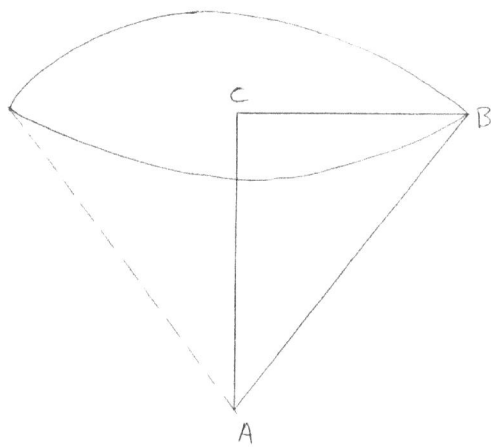

See Figure 3 1 - 1 . The right cone is formed by taking a right triangle in a plane, triangle ACB, and then rotating side AB around AC as a stationary axis. The base of the cone is the circle with radius CB, and the vertex of the cone is point A. If a plane sections the cone through radius CB and axis CA we can call AB the side of the cone. A cone constructed in this manner from a right triangle is called a right circular cone.

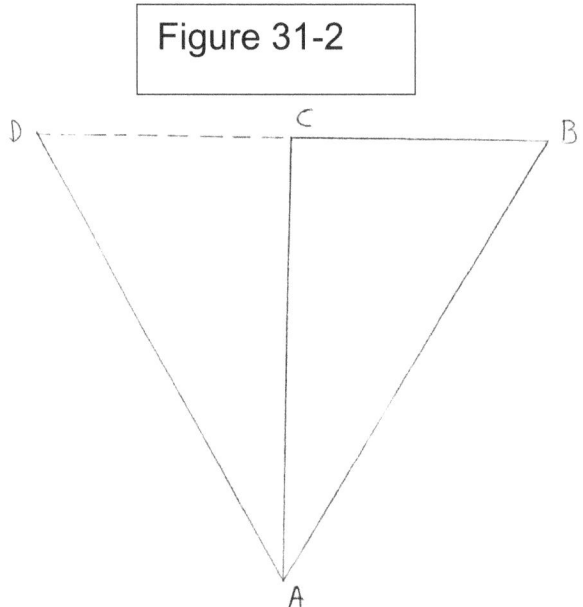

Figure 31-2

In Figure 3 1 - 2 we show the plane section of the right cone including right triangle ACB, side AB, right triangle ACD, and side AD. The vertical angle of this cone is angle DAB. The ancient Greek mathematicians considered three kinds of right circular cones:

A. An acute angled right cone. Here the vertex angle DAB is less than 90°.

B. A right angled right cone. Here the vertex angle DAB equals 90°.

C. An obtuse angled right cone. Here the vertex angle DAB is greater than 90°, but less than 180°.

The conic sections were then obtained by intersecting each of these three right cones by a plane such that the intersecting plane was perpendicular to side AB of the generating triangle for the cone.

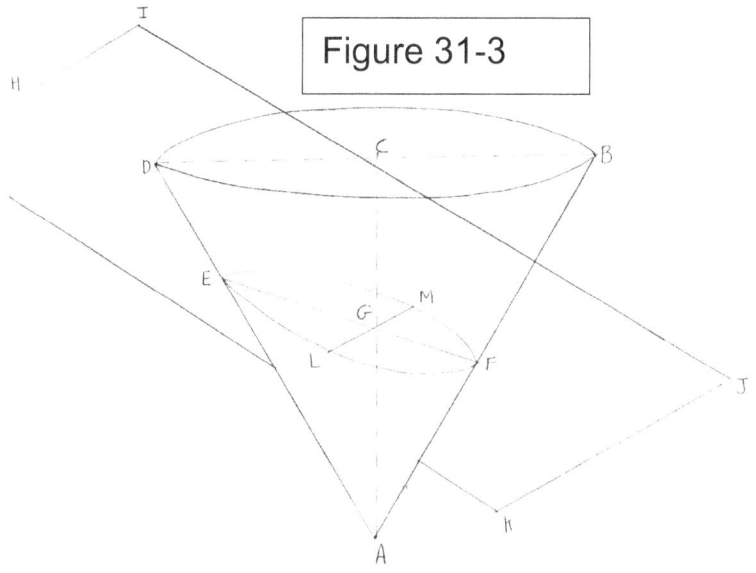

Figure 31-3

In Figure 31-3 we show the result of sectioning an **acute** angled right cone by a plane perpendicular to side AB of triangle ACB which is generating triangle for the cone. Angle DAB < 90^0.

Plane HIJK sections the cone as shown, meeting side AB at F, axis AC at G, and side AD at E, such that line EGF on the plane is perpendicular to AB.   In addition line LGM on the plane intersects axis AC such that angles LGF and MGF are also right angles. The curve of intersection of the plane and the outer surface of the cone, curve ELFM, is called an ellipse. When Euclid discussed such a section he called it a curve similar to the Greek word for shield. The modern names for the conic sections were given by Apollonius of Perga in his classic treatise on the conic sections.

If a **right** angled right cone is sectioned by a plane perpendicular to side AB of the generating right triangle, the resulting curve of intersection is called a parabola. This is shown in Figure 31-4.

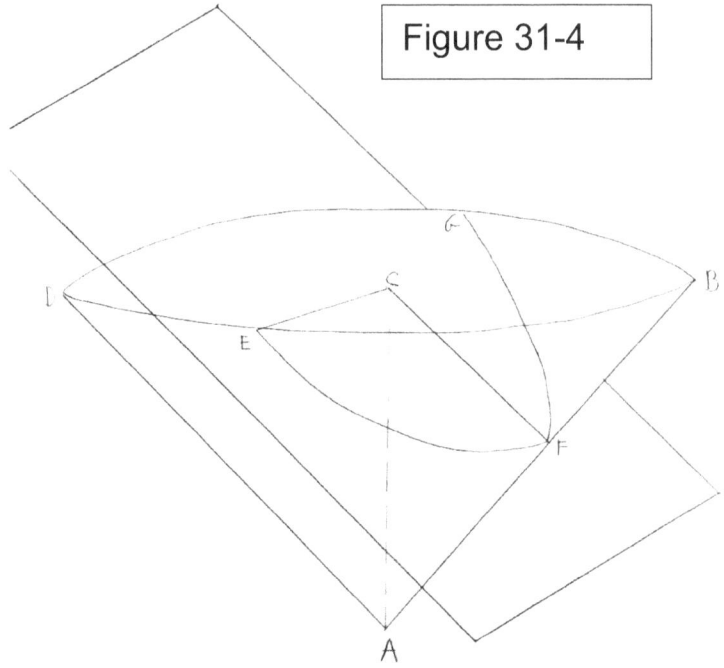

Figure 31-4

In the figure angle DAB is a right angle, and the plane intersects the cone such that angles AFC and BFC are right angles. The intersection creates the curve EFG which is the parabola.

If a right cone with an **oblique** angle at the vertex is sectioned by a plane perpendicular to one side, the curve of intersection is a hyperbola. This is shown in Figure 31-5.

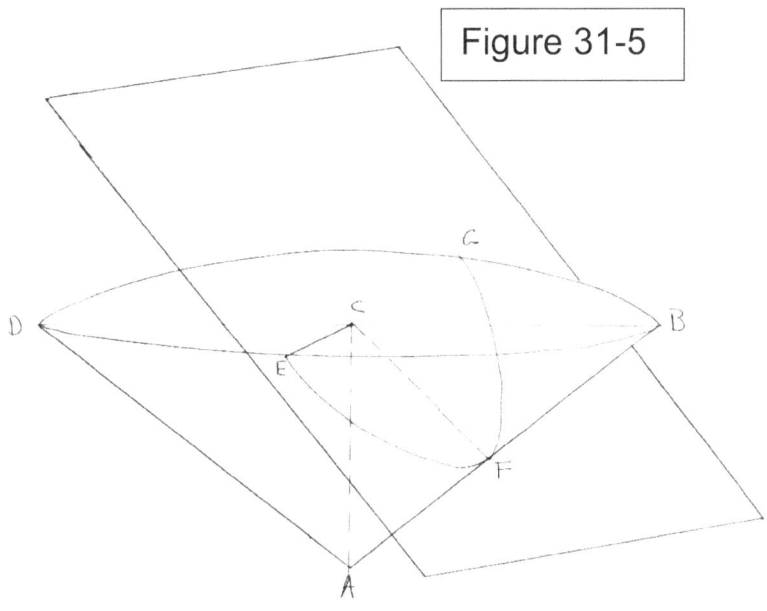

Figure 31-5

Here angle DAB > 90°, and the plane of intersection forms right angles AFC and BFC with side AB of the right cone. The curve of intersection of the plane and curve is curve EFG. Although superficially it may seem similar to the parabola we shall see it has different properties.

The original mathematical work on the conic sections is believed to have been made by the Greek mathematician Menaechmus around 360 B.C. The study of these curves was carried on by other Greek mathematicians including Euclid and Archimedes. The great treatise on conics was made, as mentioned, by Apollonius around 220 B.C. We will first take up the subject as developed by Menaechmus who was a pupil of Eudoxus. Since no extant work of Menaechmus remains to be studied, we will use the work of mathematical historian Sir Thomas Heath who describes in his book, "A History of Greek Mathematics", the most likely methods of Manaechmus.

We construct Figure 31-6.

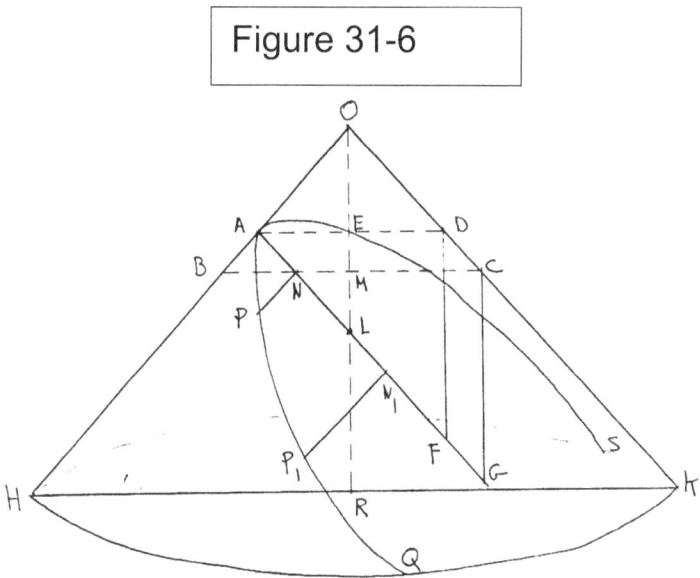

Figure 31-6

This figure represents the parabola formed by the intersection of a plane with a right cone with a vertex angle HOK of 90°. The parabola is the curve QPAS. The plane is perpendicular to the side OH of the right triangle ORH which generates the right cone. AG, the axis of the parabola, is perpendicular to OH so angles HAG and OAG are right angles.

We let P be any point on the parabola formed by intersection of the cone and plane. PN is drawn perpendicular to AG. Once N is located on AG we draw BC through N such that BC is perpendicular to the axis of the cone, OLR. Note L is a point on AG and on the axis of the cone.

AD is drawn parallel to BC. DF and CG are drawn parallel to the axis of the cone, OLR. Now BC, AD, OLR, DF, and CG are all in the same plane. Note AD and BC are perpendicular to the axis of the cone OLR. RLO, the axis of the cone, bisects angle HOK and thus RLO bisects both AD and BC. From the geometry of the figure we will derive the relationship,

$(PN)^2 = 2(AL)(AN)$, for any point P on the parabola.

1.　　We first wish to show that AL = LF, or 2(AL) = AF.

Consider right triangle ADF. Sides AD and AF are both cut by OLR which is parallel to the base of the triangle, DF. From our earlier work this means we can set up the proportion,

2.　　AE/ED = AL/LF. However, AE = ED because OLR bisects AD so,

AL/ LF = 1/1, or AL = LF, and thus 2(AL) = AF.

Next, we wish to show that $(PN)^2 = (BN)(CN)$.　In order to demonstrate this relation, we think of lines BC and NP as being on a plane which sections the cone such that the plane is perpendicular to axis OLR. The intersection of this plane with the cone is a circle with radius MB since M is the midpoint of diameter BC. We draw this circle as Figure 31-7.

Figure 31-7

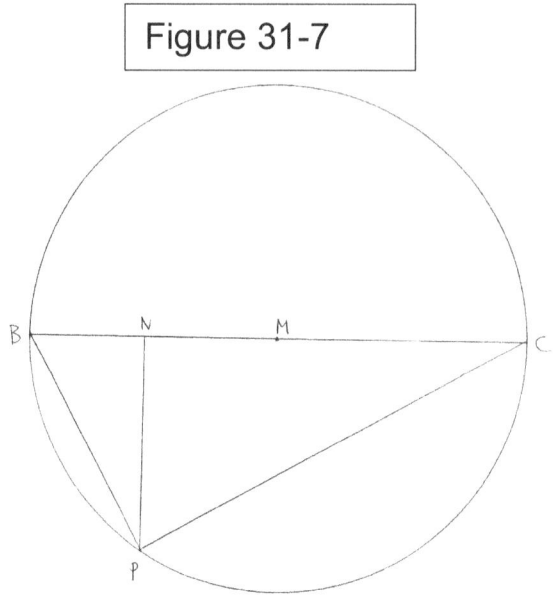

Since PN is perpendicular to BC, we have right triangle BNP and PNC. Note also triangle BPC is a right triangle because angle BPC is an inscribed angle in a circle which intersects the circle at a diameter. We then have angles BPN and CPN as complementary angles, which means angle PBN = angle CPN, since they are both complements to angle BPN. Thus triangles BNP and PNC are similar, and we can set up the proportion between corresponding sides:

3.   CN / PN = PN /BN, or, $(PN)^2 = (BN)(CN)$.

We next wish to show $(BN)(CN) = (AN)(NG)$ in Figure 31-6. Consider triangles BAN and NCG. Both are right triangles, and vertical angles BNA and GNC are equal. These triangles are then similar so corresponding sides are in proportion:

112

4.   BN/ AN = NG/CN, or, (BN)(CN) = (AN)(NG).

Next we show that NG = AF. To see this note FDCG is a parallelogram, because side OK of the plane which contains DC, is parallel to AG which contains FG. They are parallel because the plane of intersection with the cone is drawn perpendicular to OH with OAG as a right angle, and the entire intersecting plane is perpendicular to OAH. The vertex angle, HOK, is also a right angle in Figure 31-6 thus making AG parallel to OK. DF = CG, because they are opposite side of the parallelogram.   Then consider triangles AFD and NGC. They are similar triangles.   The reader should be able to establish this since AD parallel to NC, NG, and the parallel lines, DC, AG make corresponding angles equal in these right triangles. Similar triangles have corresponding sides in proportion:

NG/AF = CG/DF, and since DF = CG, NG/AF = 1/1 , or NG = AF.

We substitute AF for NG in step 4 to get,

(BN)(NC) = (AN)(AF).

We recall we found AF = 2(AL). Thus,

5.   (BN)(NC) = (AN) x 2(AL), and since.

(BN)(CN) = $(PN)^2$,

6. $(PN)^2 = 2(AL)(AN)$, which was our goal before step 1.

From Figure 31-6 we see AL is a fixed value on line AG and has the same length no matter where point P is chosen on the curve.   Whereas, quantities PN and AN vary in length depending on where we choose the position of P.   We show, for example, position P_1 and N_1 for another location of point P.

We now draw the parabola of Figure 31-6 on a plane as shown in Figure 31-8.

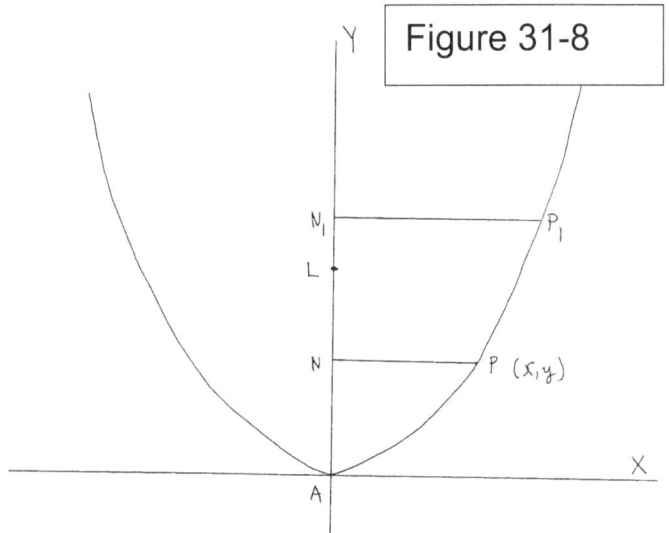

Figure 31-8

Point A is placed at the intersection of two perpendicular coordinate axes marked X and Y as shown. We indicate length AL along the axis of the parabola which coincides with the Y axis. We also indicate point P and N. Points P_1 and N_1 are also shown as representing another position of point P and the resultant change in length PN to length $P1N_1$.

Point P will then have the coordinates (x, y).   Since PN is length x, and AN is length y, the geometric equation for the parabola, $(PN)^2 = 2(AL)(AN)$, becomes,

7. $x^2 = 2(AL)y$, or,

$$y = \frac{1}{AL}\,x^2.$$ If we indicate the constant

1/AL by **a**, then ,

$Y = ax^2$ is the equation for the parabola. The constant factor, a, depends on the point A in Figure 31-6 where the plane intersects the cone.

Sir Thomas Heath feels Menaechmas produced the geometric representation of the hyperbola by using a right cone with a right angle at the vertex as in the case of the parabola, but having the plane intersect the cone parallel to the axis of the cone as shown in Figure 31-9.

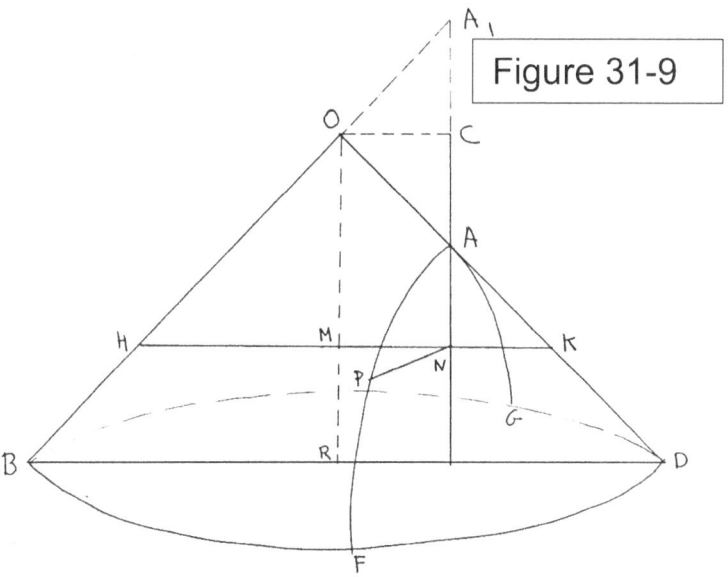

Figure 31-9

The cone has a right angle at BOD with the axis of the cone being OR which is perpendicular to the diameter of the base of the cone, BRD. The plane sections the cone at FAG such that line CAN on the plane is parallel to the cones's axis OR. A plane containing HMK, sides BHO and OKD meets the plane with CAN at right angles, so angles HNA and KNA are right angles.

The curve of intersection is FAG and when formed in this manner is a hyperbola (later Apollonius showed the hyperbola had two branches which we will discuss subsequently).

Point P is considered to be any point on the hyperbola with PN perpendicular to the axis of the hyperbola AN. Depending on where P is located line HMNK moves along axis AN but always perpendicular to AN.

We envision the circle formed by a plane perpendicular to OR and containing line HMK. This circle is drawn in Figure 31-10.

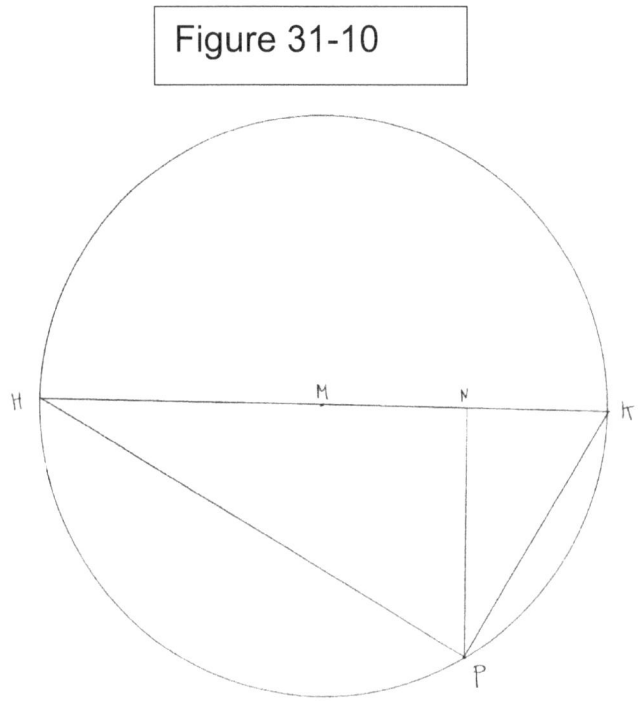

Figure 31-10

HK is a diameter, so angle HPK is a right angle, as are angles HNP and PNK.   Right triangles HNP and PNK are similar so,

8.   HN/PN = PN/NK, or, $(PN)^2 = (HN)(NK)$.

We wish to show, $(HN)(NK) = (MK)^2 - (MN)^2$. To do this note that M is the center of the circle as it is on axis OR of the cone in Figure 31-9.   We see,

10.   HN = HM + MN, or since HM = MK,

       HN = MK + MN

11.   MK = MN + NK, or, NK = MK - MN.

We multiply the left side of step 11, NK, by the left side of step 10, HN, and the right side of step 11, (MK) - (MN), by the right side of step 10, (MK) + (MN).   We are thus multiplying equals by equals, so the results are equal.

12.   (HN)(NK) = (MK - MN)(MK + MN)

       $(HN)(NK) = (MK)^2 + (MN)(MK) - (MN)(MK) - (MN)^2$,

       $(HN)(NK) = (MK)^2 - (MN)^2$, as we wished to show.

13. We use this result in step 9 to get,

       $(PN)^2 = (MK)^2 - (MN)^2$.

We will use this result in a subsequent step. We now redraw part of Figure 31-9 as Figure 31-11.

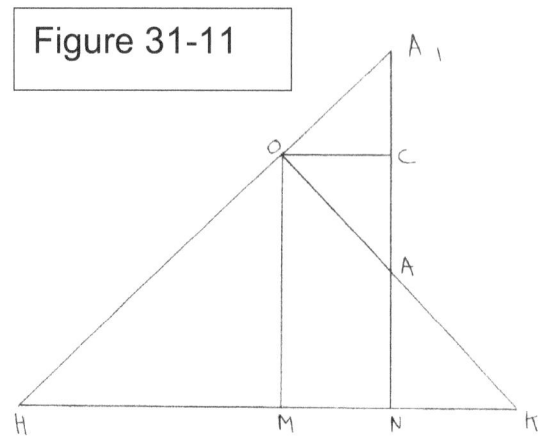

Figure 31-11

Here we have shown the plane containing HMNK and the sides of the cone HO and OK. We extend HO to meet AN extended at A_1. OC is drawn perpendicular to OM and parallel to HMNK. We argue first that, OM = MK. This is true because angle HOK is a right angle, and OM bisects angle HOK so that angle MOK equals 45°. Also, since angle OMK = 90°, angle MKO equals 45°. Thus, sides OM and MK being opposite equal angles in a triangle are equal. We see OM = CN, so MK = CN. We use this in step 13:

14. $(PN)^2 = (CN)^2 - (MN)^2$.

We also see, MN = OC. We argue, OC = CA. Angle A_1 OA is a right angle, and OC is perpendicular to OM and A_1 N. This means triangle OCA is a right triangle. Since angle MOC is a right angle, and angle MOK = 45°, then angle AOC = 45°, and also angle CAO =45°. Thus, the opposite sides OC and CA are equal. We then substitute CA for MN in step 14.

15. $(PN)^2 = (CN)^2 - (CA)^2$, or, $(CN)^2 - (PN)^2 = (CA)^2$, or

$$\frac{(CN)^2}{(CA)^2} - \frac{(PN)^2}{(CA)^2} = 1.$$

Returning to Figure 31-9, we see that CN and PN vary for each point on the curve, the hyperbola, while CA remains constant. The value of CA is determined by the position of intersection of the plane and the cone.

We next find the equation for the hyperbola in the XY coordinate system.   We construct Figure 31-12, which is the plane of Figure 31-9 containing the hyperbola.

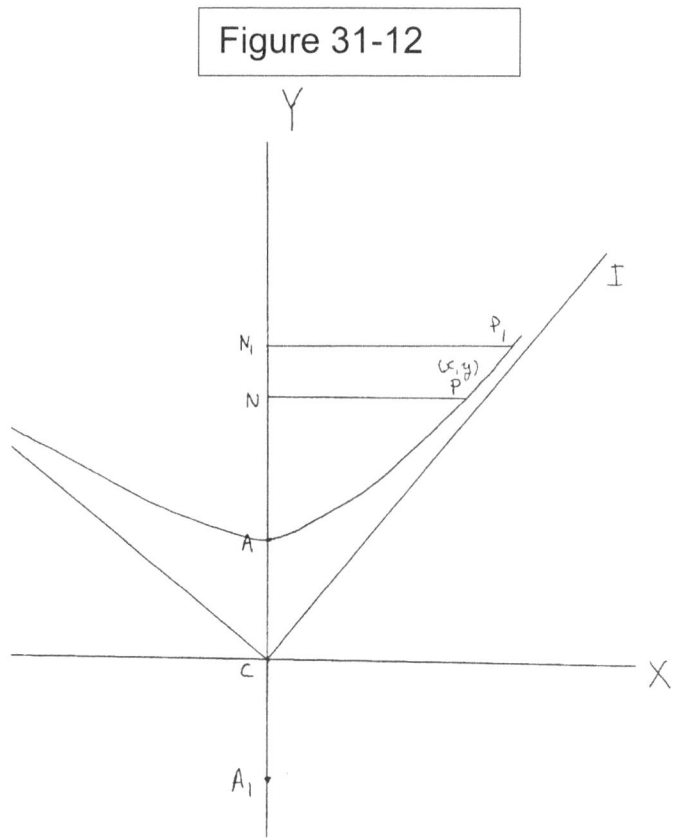

Figure 31-12

We place point C at the origin with points A_1 and A equally spaced from the origin on the Y axis. Thus the hyperbola branches upward in both the negative and

119

positive X directions. We show two positions for point P, at P and P_1, and show that as P changes position from P to P_1, so also does NP, moving from NP to N_1P_1, so that NP is always perpendicular to the Y axis containing A_1CA.

The coordinates for a point P on the hyperbola are (x, y). Thus CN = y, and PN = x. The formula for the hyperbola in step 15 then becomes,

16.

$$\frac{y^2}{(CA)^2} - \frac{x^2}{(CA)^2} = 1.$$

We now add lines CH and CI to Figure 31-12, such that angle HCI equals 90°, and is bisected by the Y axis. These lines, CH and CI, are in the plane of the hyperbola of Figure 31-9.   Since they are perpendicular they are also parallel to the sides of the cone. The hyperbola approaches these lines for larger values of positive or negative x, but never reaches these lines so as to intersect them.   Because angle HCI is a right angle the hyperbola developed by the methods of Menaechmus in Figure 31-9 is called a rectangular hyperbola.

Because the formula for the hyperbola contains both x and y as squared quantities, then for every value of x assigned there are both positive and negative values for y. In Figure 31-12 we have shown the graph for only the positive values of y. In Figure 31-13 we show both branches of the hyperbola.

120

Figure 31-13

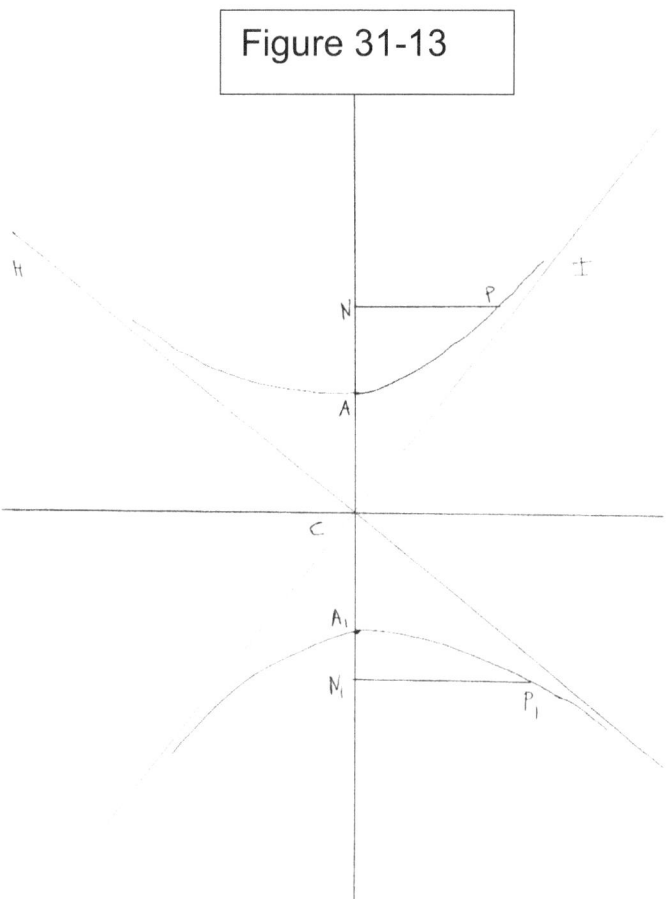

Apollonius of Perga showed that a cone can be considered to have two parts called nappes.

Figure 31-14

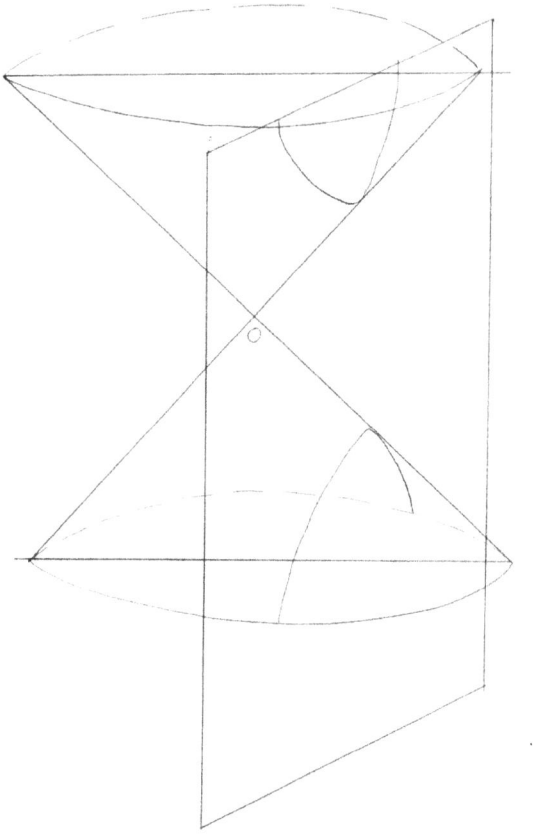

This is shown in Figure 31-14 where a plane
sections both nappes of the cone the resulting curves of
intersection are both branches of the hyperbola. Note
that the cone of Figure 31-14 with two nappes is a right
cone with a right angle at the vertex, 0, and that the
plane of intersection is parallel to the axis of the cone
just as in the case of Figure 31-9.

Going back to Figure 31-13 we see lines CH and CI
contain or limit both branches of the hyperbola. These

lines are called the asymptotes of the hyperbola. There is a geometric property of the asymptotes which can be demonstrated for the rectangular hyperbola of Menaechmus.

To demonstrate this property we will work with one branch of the hyperbola shown in Figure 31-15.

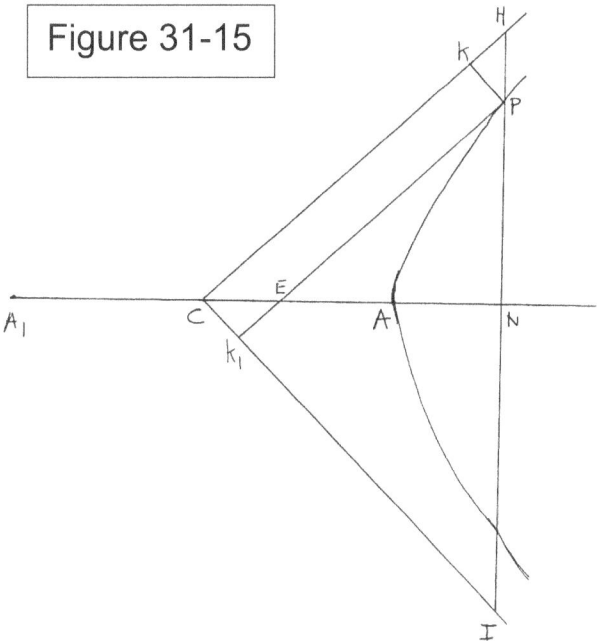

Figure 31-15

The axis of the hyperbola is CA, and P represents any point on the hyperbola. For a rectangular hyperbola the asymptototes are drawn as CH and CI perpendicular to each other at C.

From CH and CI we draw perpendiculars PK and PK_1 as shown. We then extend PN to meet the two asymptotes as shown.

17. $(PK)(PK_1)$ = area rectangle $CKPK_1$.

18. area rectangle CKPK1 = area quadrilateral CHPE.

This is true because triangles CEK$_1$ and PKH are equal in area. (The reader should now be able to show this).

19. Area quadrilateral CHPE = area triangle HNC - area triangle PEN.

Area triangle HNC = ½(CN)(HN), and because CN bisects angle HCI, then CN = HN as is readily shown, and

area triangle HNC = ½(CN)^2.

Area triangle PEN = ½(PN)(EN), and since,

PN = EN,

area triangle PEN = (PN)^2. Thus,

area quadrilateral CHPE = ½(CN)^2 - ½(PN)^2.

Recall,

(PK)(PK$_1$) = CHPE, so,

(PK) (PK$_1$) = ½[(CN)^2 - (PN)^2].

We found above, as part of step 15, that,

(CN)^2 - (PN)^2 = (CA)^2. Thus,

20. (PK)(PK$_1$) = ½(CA)^2.

The value of ½(CA)^2 remains a constant no matter where point P is on the hyperbola.   We will now use this information to graph the hyperbola on the XY coordinate axes. See Figure 31-16.

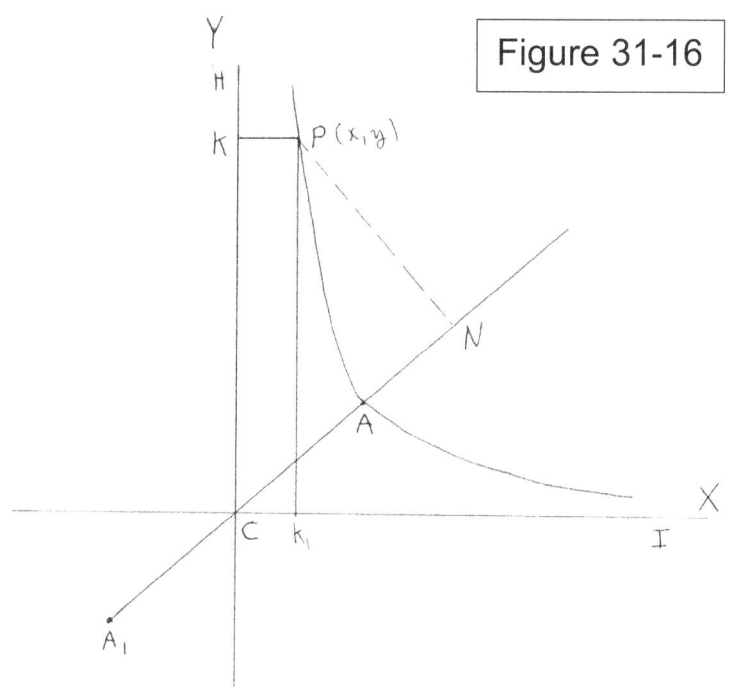

Figure 31-16

In this figure we take the asymptotes as the X and Y axes. The Y axis lies along CH, and the X axis along CI with the origin at C. In the case of the rectangular hyperbola the asymptotes are perpendicular, and thus can be taken as the coordinate axes. We then can represent the x coordinate of a point P on the hyperbola by the distance PK, and the Y coordinated by the distance PK_1. Then step 20 becomes,

21. $xy = \frac{1}{2}(CA)^2$.

Since the value of CA is constant for all points on the hyperbola, we can represent the term, $\frac{1}{2}(CA)^2$ by the constant k, and the equation for the rectangular hyperbola is,

$xy = k$, or,

125

$$y = k \frac{1}{x}.$$

We have been working with just one branch of the hyperbola. The other branch has its vertex at A_1, but has the same formula. When the formula for the rectangular hyperbola is given in this way we have a functional relationship between the variable x and y. Note neither x nor y can take on the value of 0, since if we assign the value of 0 to either x or y we get a mathematically meaningless expression.

We have then shown in the case of the hyperbola studied by Menaechmus that the hyperbola can be expressed by a relation in terms of its axis, $(CN)^2 - (PN)^2 = (CA)^2$, or in terms of relations of distance from its asymptotes, $(PK)(PK_1) = \frac{1}{2}(CA)^2$.

Menaechmus used his work on conic sections to solve a famous Greek mathematical problem known as the problem of the duplication of the cube. The origin of this problem came from a story in Greek mythology. Minos was the king of the island of Crete and had a son, Glaucus, who died as a boy. He had fallen into a cask of honey and was smothered.

Minos erected a tomb for Glaucus which was in the shape of a cube. The length of each side was 100 units. Minos was not satisfied with the tomb after it was constructed, and wished for a tomb with double the volume. Geometers heard of this problem of doubling the cube from this myth and sought a solution.

126

A mathematician named Hippocrates of Chios discovered that if two mean proportionals in continued Proportion can be found between two known lines of length a and b such that b = 2a, then the values of the mean proportional lengths are such that twice the cube of the lesser mean proportional equals the cube of the greater mean proportional.   Let's put this into mathematical symbols to make the statement clear. We call the lesser mean proportional x, and the greater mean proportional y.   Since x and y are mean proportionals in continued proportion between lengths a and b we can write,

22.

$$\frac{a}{x} = \frac{x}{y} = \frac{y}{b}.$$ We wish to show, **$2x^3 = y^3$**, meaning a

twice the cube with side x equals a cube with side y.

23.Using, $\frac{a}{x} = \frac{x}{y}$, we have **ay = x²**, or

y = x²/a, or, **a = x²/y**.

24.Using $\frac{x}{y} = \frac{y}{b}$, we have **bx = y²**, or, **x = y²/b** , or

since, **b = 2a, x = y²/2a**.   We substitute for **a, x²/y** from step 23.

25.

$$X = \frac{y^2}{2\frac{x^2}{y}}, \; \text{or,} \; X = \frac{y^2}{2} \cdot \frac{y}{x^2}. \; \text{Then,}$$

$$x^3 = \frac{y^3}{2}, or, 2x^3 = y^3.$$

Thus, when 2a = b, and when x and y are the two mean proportionals in continuous proportion between a and b, then a cube with sides of length y is twice the volume of a cube with sides of length x. In the case where the original cube has a side of 100 units, we have x = 100, and x^3 = 1,000,000 cubic units. y^3 = 2,000,000 cubic units if the cube with side y units is twice the volume of the cube with sides of 100 units. The cubic root of 2,000,000 is about 125.992.

26.

$$a = \frac{x^2}{y} = \frac{10,000}{125.992} = 79.37.$$

b = 2a = 158.74.

This is the solution of the problem of doubling the cube by Hippocrates of Chios. Menaechmus said the solution could be found by obtaining the values of the point of intersection of a hyperbola and a parabola when the two curves are properly constructed. To see Menaechmus solution we construct Figure 31-17.

Figure 31-17

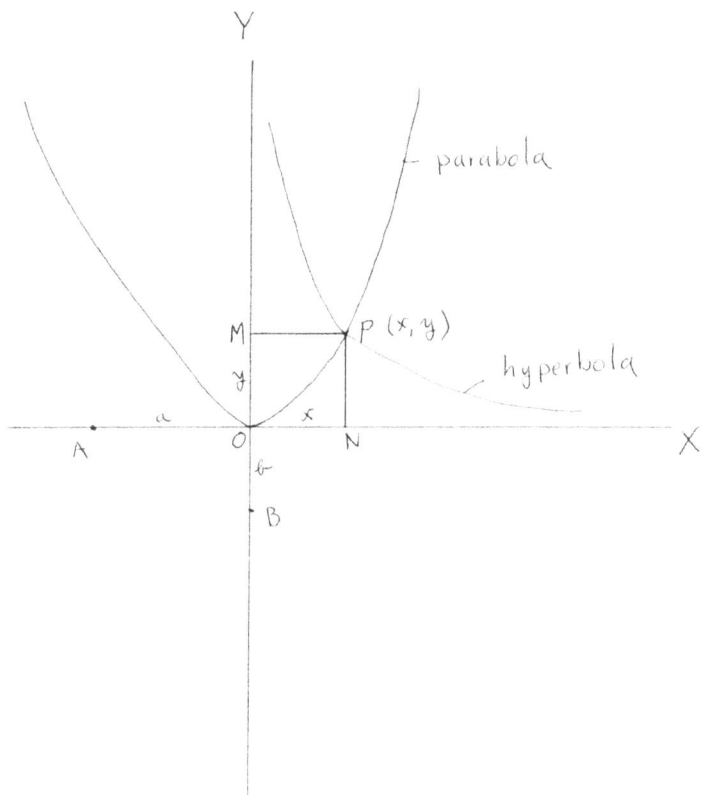

27.   We draw axis AON perpendicular to axis BOM.
We think of AON as the X axis and BOM as the Y axis.
Length AO = a, and length OB = b, where O is the
intersection of the two axes.

28.   We select a length ON = x, and a length OM = y
such that y and x are the the mean proportionals in
continuous proportion between a and b, where 2b = a.
Thus,

$$\frac{a}{y} = \frac{y}{x} = \frac{x}{b}, \text{ or, } \frac{b}{x} = \frac{x}{y} = \frac{y}{a}.$$

Since 2b = a, then x is the lesser mean proportional and y the greater, we demonstrate $2x^3 = y^3$ by a similar process from steps 22-25. Note the change in symbols from the demonstration concluded in step 25 with 2b = a instead of, 2a = b, and the proportion sequence, b/x = x/y = y/a instead of a/x = x/y = y/b.

For the proportion in step 28 the steps are: by = x^2, y = x^2/b, b = x^2/y, and ax = y^2, x = y^2/a, x = y^2/2b, 2x = y^2/(x^2/y) and finally, $2x^3 = y^3$.

What Menaechmus shows is that MP, which is perpendicular to BOM, and NP, which is perpendicular to AON, meet at P such that P is the point of intersection of a parabola and a hyperbola in Figure 31-17. The parabola is as shown in the figure has its vertex at 0 and has the formula, $(PM)^2$ = (OB)(OM), or x^2 = by, or y = (1/b)x^2. The hyperbola has the formula given in terms of distances from the asymtotes of, (ON)(OM) = (AO)(OB), or xy = ab.

We note that if, a/y = y/x = x/b, then using , a/y = x/b, xy = ab, or, (ON)(OM) = (AO)(OB). We recall from Figure 31-16 the equation for the hyperbola oriented as shown with the XY axes as asymptotes has the formula, xy = ½$(CA)^2$. Thus, in Figure 31-17, ½$(CA)^2$ = ab. This means we can consider the relation, (ON)(OM) = (AO)(OB), as representing a hyperbola.

We know, y/x = x/b, or, x^2 = by. Since x = ON = PM, and y = OM, then $(PM)^2$ = (OB)(OM). In Figure 31-8 we

130

learned the formula for a parabola was $(PN)^2 = 2(AL)(AN)$.
In Figure 31-17, (PM) represents (PN), (OM) represents
(AN),and(OB) represents 2(AL). Thus, $(PM)^2 = OB)(OM)$,
which we can also write as $x^2 = by$, or $y = (l/b)x^2$, is a
parabola. Since OM represents AN, then 0 is the vertex
of the parabola.

We can now see that point P is on both the hyperbola
and the parabola since both curves share point (x, y)
when the continuous proportion is true, $a/y = y/x = x/b$,
and if a and b are selected so that 2b = a. As noted above
this means $2x^3 = y^3$ and the point P represents the values
for the problem of doubling the cube.

Menaechmus also studied properties of the ellipse.
The ellipse for Menaechmus was the curve resulting from
the intersection of a right cone with an acute vertex angle,
and a plane where the plane is perpendicular to a side of
the cone as shown in Figure 31-18.

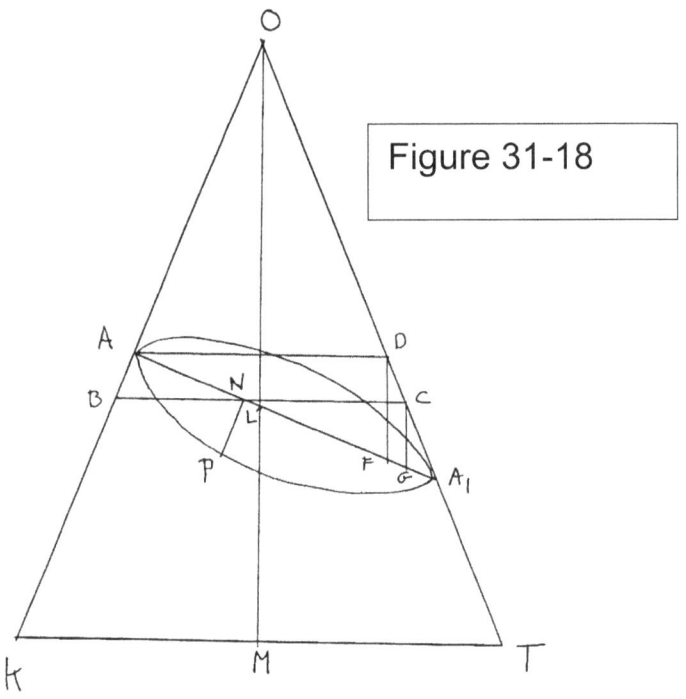

Figure 31-18

Note the vertex angle of the cone, angle KOT is < 90°, and the plane of intersection, which contains line ANLFGA$_1$, and line NP, is perpendicular to side KBAO of the plane. The axis of the ellipse is ALA$_1$, where L is a point on the axis of the cone, and which intersects AA$_1$. P represents a point on the ellipse with perpendicular distance PN to the axis of the ellipse.

BC is drawn perpendicular to the axis of the cone, OLM, and through point N. Thus, as we select different points P on the ellipse the position of line BNC varies. DF is drawn perpendicular to AD which is in turn perpendicular to OLM. Note F is on the axis of the ellipse. CG is drawn perpendicular to BC, and G is on the axis of the ellipse. Thus, DF is parallel to CG.

132

We now consider BC as a line on a plane intersecting the cone perpendicular to the axis of the cone, OLM. We form the plane with BNC including NP, and this will make a circle when it intersects the cone as shown in Figure 31-19.

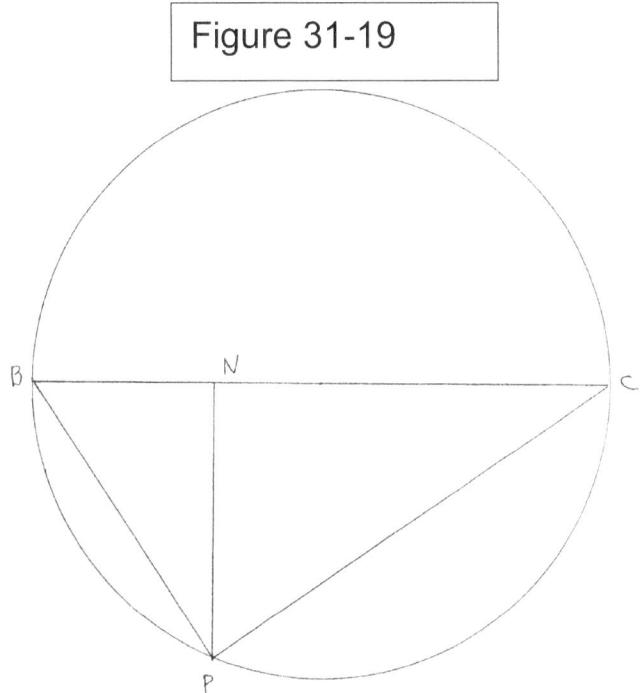

Figure 31-19

In this figure we connect PB and PC, and from our work on the parabola we know triangles BNP and PNC are similar. Thus,

29. CN/PN = PN/BN, or, $(PN)^2 = (BN)(CN)$

In Figure 31-18 we see triangles NAB and NGC are similar right triangles. Thus,

30. AN/BN = CN/NG, or, $(BN)(CN) = (AN)(NG)$

We use this result in step 29:

31. $(PN)^2 = (AN)(NG)$.

In Figure 31-18 right triangles NCG and ADF are similar since DF is parallel to CG. Thus corresponding sides are in proportion.

32. NG/AF = NC/ AD.

Right triangles A_1N C and A_1A D are similar so,

33. NC/AD = A_1N/ AA_1, and we know,

NG/ AF = A_1N/A A_1, or

NG = (A_1N/AA_1) (AF) .

We use this value of NG in step 31 to get,

34.

$$(PN)^2 = (AN) \cdot \frac{A_1N}{AA_1} \cdot AF$$, or

$$(PN)^2 = \frac{AF}{AA_1} \cdot (AN)(A_1N).$$

By inspecting Figure 31-18 we see, AF < AA_1, or AF/AA_1 < 1.

Step 34 gives the fundamental relationship for the ellipse as constructed by Menaechmus. We see that quantities, (PN),(AN), and (A_1N) vary depending on the location of P on the ellipse, while (AF)/(AA_1) remains a constant term less than 1. We will now place the ellipse of Menaechmus on a rectangular coordinate system to find its equation in modern terms.

The equation we derive depends on the location of the origin for the XY axes. The simplest formula occurs when the origin of the XY axes is chosen to be the midpoint of the axis of the ellipse in Figure 31-18, that is ½ the distance between A and A_1. See Figure 31-20.

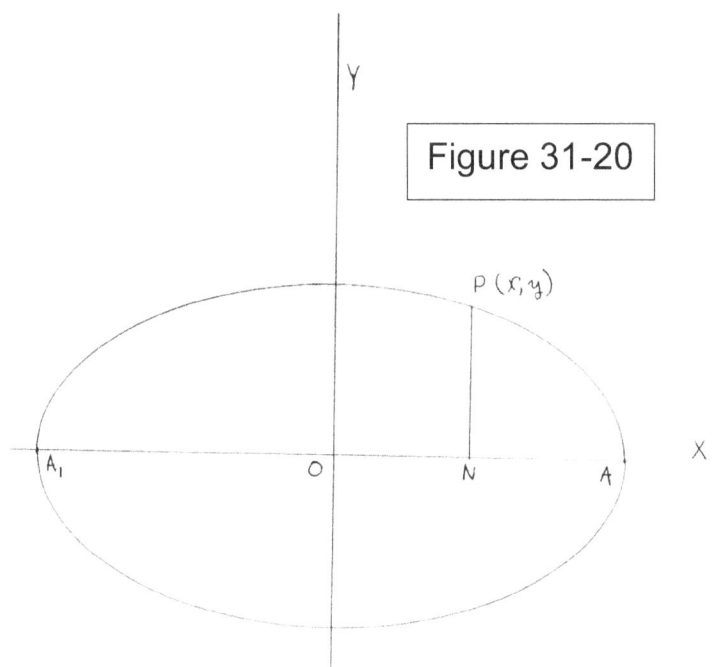

Figure 31-20

Here AA₁ lies along the X axis so that $OA = OA_1$. Point P in the XY coordinate system has coordinates (x, y). PN is equal to y, and ON is equal to x.

35. We know $(PN)^2 = (AF/AA_1)(AN)(A_1N)$.

$A_1N = A_10 + ON$, or, $A_1N = OA + ON$, since $OA = A_10$.

$OA = ON + AN$, or,

$AN = (OA - ON)$. Then,

36. $(PN)^2 = AF/AA_1 \times (OA + ON)(OA - ON)$,

$(PN)^2 = AF/AA_1 \times [(0A)^2 - (0N)^2]$.

$(PN)^2 = (OA)^2 \times AF/AA_1 - (ON)^2 \times AF/AA_1$.

Since $AA_1 = 2(OA)$,

$$(PN)^2 = \frac{(OA)(AF)}{2} - \frac{(ON)^2(AF)}{2OA}.$$

135

We divide each term by (AF), and add the last term to both sides of the equation.

37.

$$\frac{(ON)^2}{2(OA)} + \frac{(PN)^2}{AF} = \frac{OA}{2}.$$

We divide each term by ½(OA).

$$\frac{(ON)^2}{(OA)^2} + \frac{(PN)^2}{\frac{1}{2}(OA)(AF)} = 1.$$

Since ON = x, and PN = y,

38.

$$\frac{x^2}{(OA)^2} + \frac{y^2}{\frac{1}{2}(OA)(AF)} = 1.$$

OA and AF are determined in Figure 31-18 by the angle of the vertex of the cone, angle KOT, and by the distance AO where the plane intersects the cone.

The Greek mathematicians Euclid and Archimedes called the ellipse by the Greek word $\Theta\upsilon\rho\varepsilon o\sigma$. In English letters this word is Thureos. It is the Greek word for shield as used by a Greek soldier, which is elliptical in shape. Although Euclid's book on conics is lost, Euclid knew an ellipse could be obtained by sectioning a right cylinder by a plane at angle between 0° and 90° as shown in Figure 31-21.

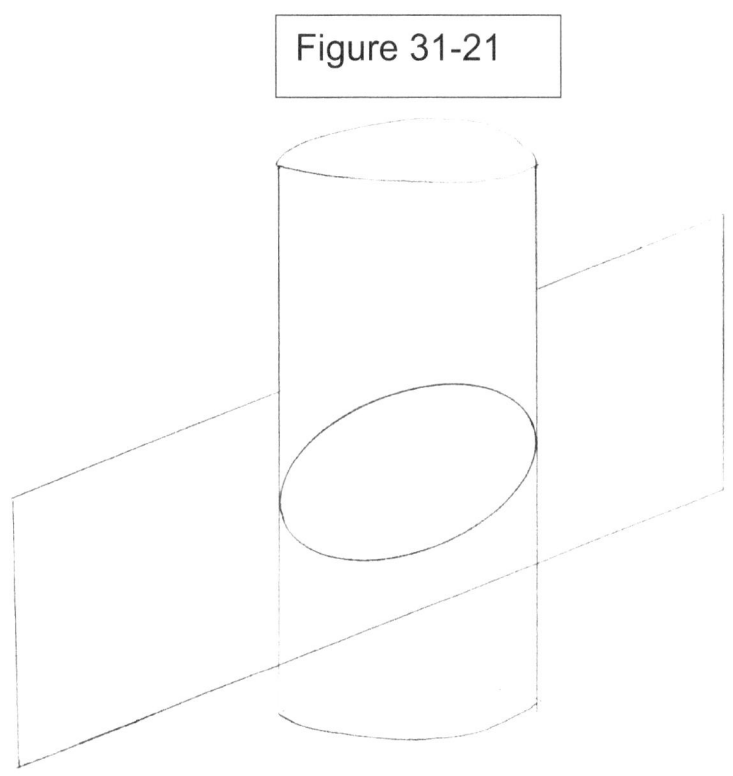

Figure 31-21

We have established the probable means whereby Menaechmus discovered the fundamental properties of the conic sections. We used the work of Sir Thomas Heath. In deriving the relationship of the hyperbola we used Figure 31-9, which was not constructed by a plane intersecting the side of a right cone at a 90° angle. Heath points out in the book, "Apollonius of Perga, Treatise on Conic Sections", which he edited, that it is possible to derive the relations for the hyperbola by a construction where a plane intersects a right cone with a vertex angle of > 90°, such that the plane makes a 90° angle with one side of the cone at the point of intersection. This construction is shown in Figure 31-22, where angle GAO = 90°, and angle KOT > 90°.

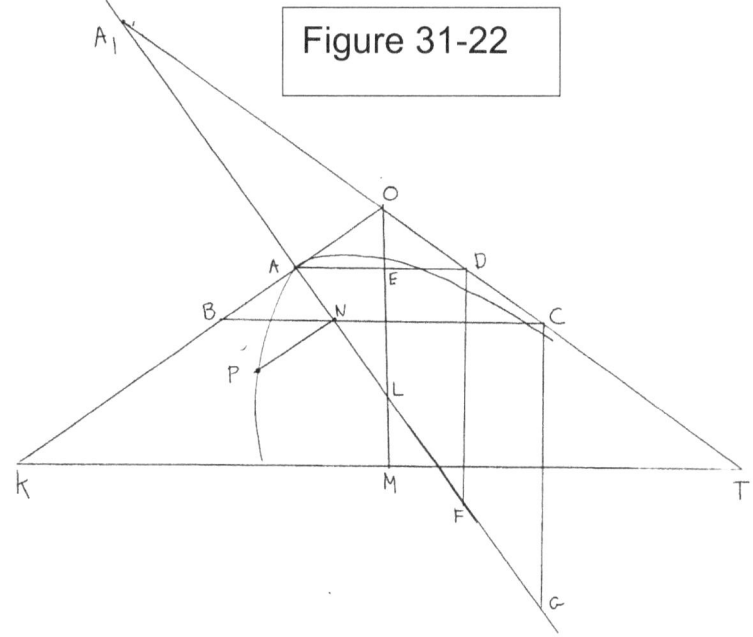

Figure 31-22

Heath is not certain that Menaechmus used such a construction to determine the formula for a hyperbola. However, we will give the derivation of the formula for a hyperbola from the construction in Figure 31-22, since it will reinforce our knowledge about the conic sections as well as show us there is more than one construction that can be used to derive the fundamental relationship for the hyperbola.

The construction in Figure 31-22 has the same basic labels as used in the construction for the ellipse in Figure 31-20. The positions of the points are slightly different, since the vertex angle of the right cone, angle KOT, is greater than 90°. We see the axis of the hyperbola is extended beyond the cone in the plane of intersection to A_1. The lines AD and BC are parallel and perpendicular to the axis OM, and DF and CG are perpendicular to AD and BC respectively as in the case of the construction for the ellipse. Using the same reasoning as in the derivation for the ellipse we have the following steps.

39. CN/PN = PN/BN, or,

   $(PN)^2 = (BN)(CN)$.

40. AN/BN = CN/NG, or,

   $(BN)(CN) = (AN)(NG)$.

41. Thus, $(PN)^2 = (AN)(NG)$.

42. NG/AF = NC/AD.

43. Triangles A_1N C and A_1AD are similar so,

   $NC/AD = A_1N/AA_1$. We use this in step 42:

44. $NG/AF = A_1N/AA_1$, or, $NG = (A_1N/AA_1)(AF)$.

   We use this value for NG in step 41:

45. $(PN)^2 = (AN) (A_1N/AA_1) (AF)$.

   $(PN)^2 = AF/AA_1 \quad x \quad (AN)(A_1N)$.

This is the basic relation for the hyperbola as constructed in Figure 31-22. We see it has the same form as the equation for the ellipse derived from Figure 31-20 in step 36. However, in the case of the ellipse, $AF/AA_1 < 1$, because in the construction for the ellipse the vertex angle of the cone, angle KOT, is less than 90°. When AF/AA_1 is equal to 1, the equation in step 45 gives a rectangular hyperbola. We then have,

46. $(PN)^2 = (AN)(A_1N)$.

Let us place the curve on a rectangular coordinate plane so that the origin, 0, is ½ way between A₁ and A on the Y axis as shown in Figure 31-23.

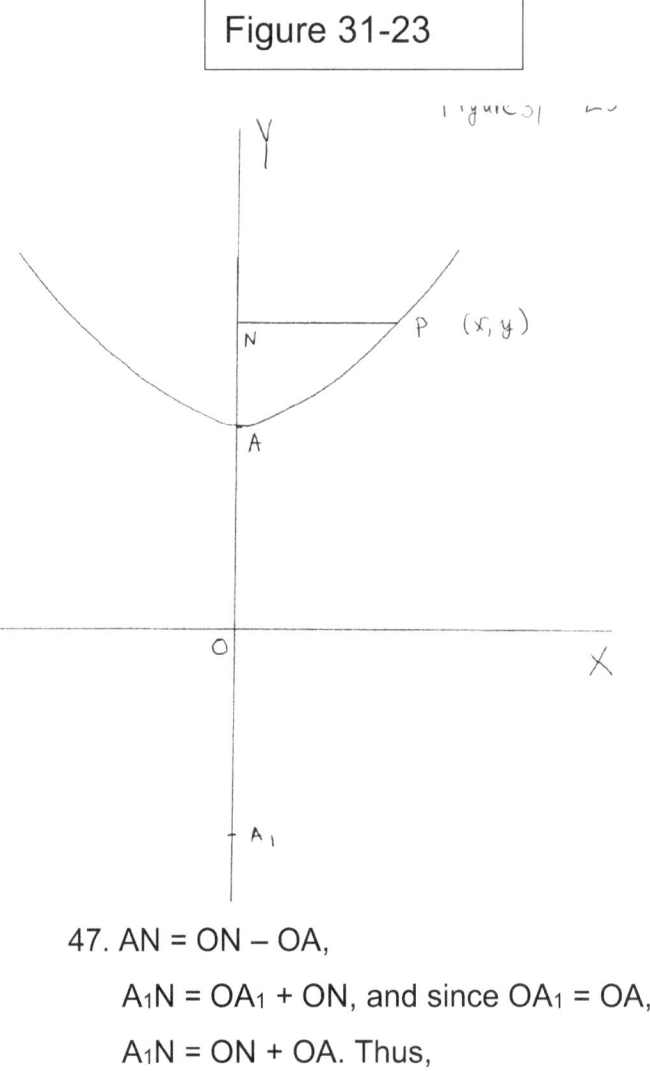

Figure 31-23

47. AN = ON − OA,

   A₁N = OA₁ + ON, and since OA₁ = OA,

   A₁N = ON + OA. Thus,

48. $(PN)^2 = (ON - OA)(ON + OA)$,

   $(PN)^2 = (ON)^2 - (OA)^2$,

   $(ON)^2 - (PN)^2 = (OA)^{2+}$,

140

$$\frac{(ON)^2}{(OA)^2} - \frac{(PN)^2}{(OA)^2} = 1$$

and since ON = y, and PN = x, then,

$$\frac{y^2}{(OA)^2} - \frac{x^2}{(OA)^2} = 1.$$

This is the equation we derived for the rectangular hyperbola in step 16, except that in step 16, C was at the origin instead of O. Thus an equation we derived from Figure 31-22 will be a rectangular hyperbola if,
AF/AA$_1$ = 1.

In Figure 31-22 we can show that AF = 2(AL), by the same reasoning we used in our analysis of the parabola in Figure 31-6. Our next task is to show how Menaechmus may have constructed the rectangular hyperbola given by the equation in step 48. We make use of the relation that 2(AL) = AF, and that, (AF)/(AA$_1$) = 1. This means AF = AA$_1$, so, 2(AL) = AA$_1$ for the rectangular hyperbola. See Figure 31-24.

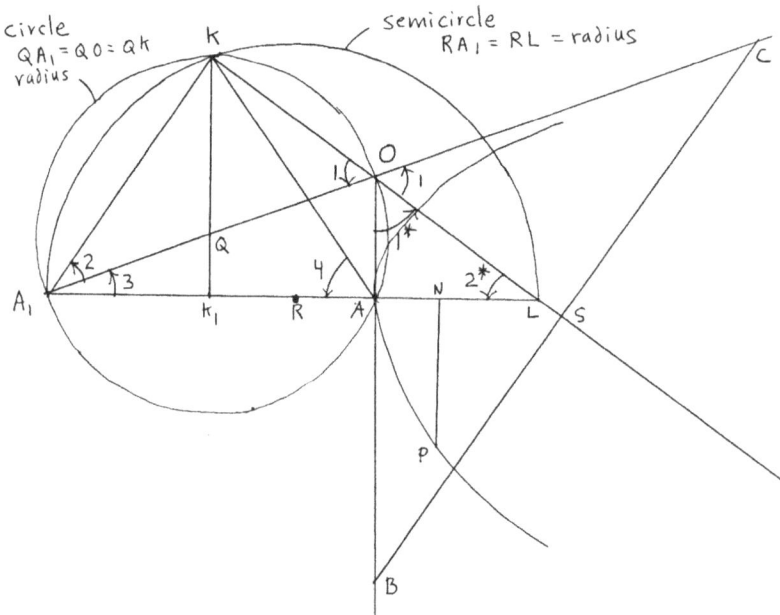

Figure 31-24

We first choose a length for AL which is on the axis for the rectangular hyperbola. We know $2(AL) = AA_1$.

49. $2(AL) + AL = AA_1 + AL$, by adding AL to each

side. $3(AL) = A_1L$, as seen from the figure.

This means in the construction line A_1L is three times the length of AL to obtain the rectangular hyperbola.

Thus in constructing the rectangular hyperbola from a plane intersecting a right cone with a vertex angle > 90°, such that the plane is perpendicular to a side of the cone, the fundamental length that is established first is length AL. Length A_1L is then marked off as three lengths of AL.

We next bisect line A_1L to find point R. Then with $RL = RA_1$ as a radius, we construct the semicircle as shown. We measure off distance AL to the left of A to find point K_1. We can also obtain point K_1 by bisecting segment AA_1. Thus $A_1K_1 = K_1A = AL$. We construct a line perpendicular to A_1A from K_1, and extend it until it meets the semicircle at K. We then connect KL and extend the line beyond L. We connect A_1K so we have an inscribed right triangle in a circle with diameter A_1L, with angle A_1KL being a right angleintersectingthe semicircle at diameter.

Next from point A we construct a line perpendicular to A_1L, and extend the line until it meets KL at point O. Thus, AO is perpendicular to A_1L. The line is also extended downward through point B. From A_1 we connect A_1O, and extend the line through C. We then take a point beyond L on KL and draw a line perpendicular to KL at S to meet OA extended at B, and A_1O extended at C.   We claim that angle BOS = angle COS. Once this is established, we think of triangle OSB as a generating triangle for the right cone with BOC as the vertex angle, and OLS as the axis of the cone.   A line NP on the plane containing A_1L is drawn perpendicular to A_1L and extended to P which is a point of intersection of the plane and the cone. The curve of the hyperbola is shown formed by the intersection of all such points. We note the figure is now analogous to Figure 31-22.

In other words, our construction technique has shown how to determine length OA and angle BOC so a rectangular hyperbola results from the intersection of the cone and plane. Recall the construction begins by knowing length AL. We have yet to demonstrate that

143

angle BOS = angle COS. Once we know that, we are sure that when we rotate triangle BSO around OS as an axis then OB will coincide with OC, since triangles OSB and OSC are congruent.

We bisect line OA_1 at Q and with QA as radius we draw the circle as shown in Figure 31-24 Because triangles A_1AO and A_1KO are right triangles they will be inscribed in the circle formed when A_10 is the diameter. Euclid had shown in the "Elements", Book III, proposition 3, that because chord A_1A is bisected
by perpendicular KK_1, then KK_1 is on a diameter of the circle with Q as center, as well, so the center of the circle, Q, is at the intersection of $0A_1$ and KK1.

We connect KA and label angles 1, 2, 3, 4, 1*, and 2* as shown in the figure.   We wish to demonstrate angle 1 = angle 1*.

50. We will refer to the angles by their numbers. $\angle$4 and $\angle$1 are inscribed angles in a circle intercepting the same arc, arc A_1K. Thus, $\angle$4=$\angle$1.

51. In triangle A_1KO angle A_1KO is a right angle so, $\angle$1 + $\angle$2 = 90°.

52. Since KK_1 is the perpendicular bisector of A_1A, then K is equidistant from A_1, and A, so A_1K = AK. Thus, triangles A_1KK_1 and AKK_1 are similar and,
($\angle$2 + $\angle$3) = $\angle$4.   From step 50, $\angle$4=$\angle$1, so,
($\angle$2 + $\angle$3) = $\angle$1.

53. Triangle A_1KL is a right triangle with the angle at K being 90° so,

($\angle$2 + $\angle$3) + $\angle$2* = 90°. Thus, $\angle$1+ $\angle$2* = 90°.

144

But,   1 + 2 = 90°, so, 2 = 2*.

54. Triangle OAL is a right triangle with the right angle at A, so 1* + 2* = 90°.   Using step 53, 1* + 2 = 90°.

From step 51, 1 + 2 = 90°, so 1* = 1.

Thus we have shown our construction does result in angle BOS (1*) being equal to angle COS (1).   Then it follows triangles BSO and CSO are congruent, so that rotating triangle BSO around OS as an axis will produce a right cone with side OB and OC. We had demonstrated previously if 2(AL) = A_1A, the intersection of plane and cone will result in the rectangular hyperbola of Menaechmus The above construction is found in the book, "Apollonius of Perga", edited by Sir Thomas Heath in his commentary. We changed the argument slightly from that of Heath to demonstrate the geometrical construction technique.   Heath states the actual work of Menaechmus has been lost so the demonstration is given as a possible sequence that Menaechmus may have followed.

Apollonius greatly expanded upon the conic sections and broadened the definition of the conic sections so they could be formed from cones with any vertex angle between 0° and 180°,and sectioned by planes at any angle to the side of the generating triangle.   He also used cones that were not right circular cones. He introduced cones with two parts or nappes as shown in Figure 31-25.

Figure 31-25

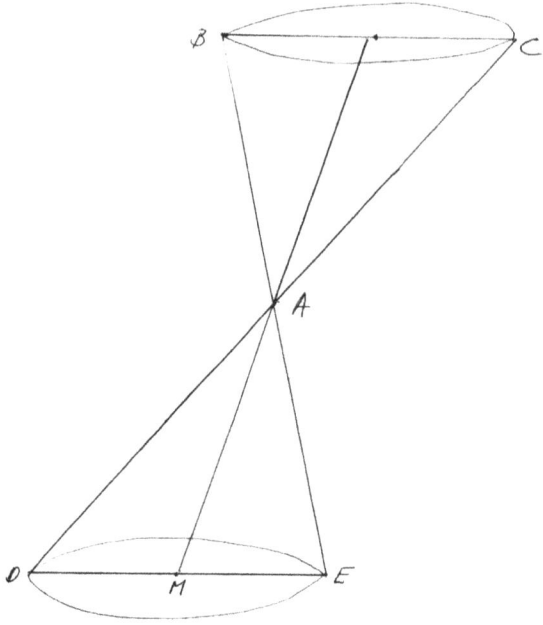

In Apollonius work the axis of the cone did not have to be perpendicular to the circular base DE. Thus angles DMA and EMA do not have to equal 90°. Such cones are called oblique cones.

Modern work in coordinate geometry usually deals with right circular cones with one or two nappes allowing the vertex angle to range between 0° and 180°. We shall see how the conic sections are defined by intersecting such cones with various planes. We consider the parabola first in Figure 31-26.

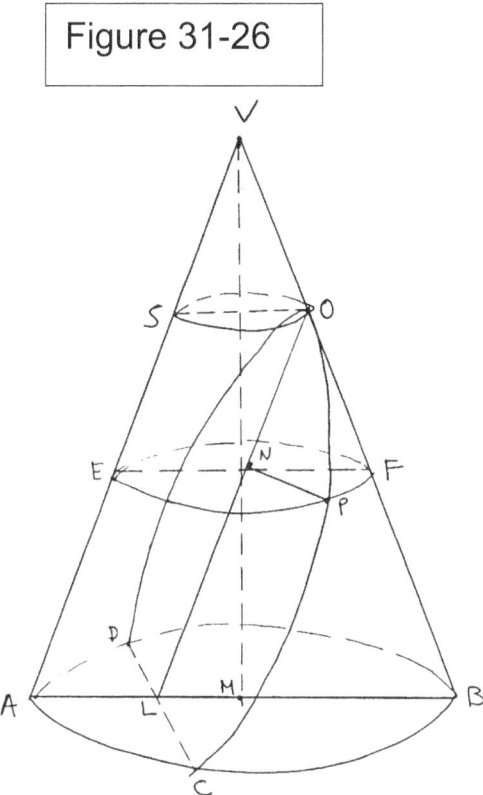

Figure 31-26

For the parabola a cone with one nappe is constructed from the generating right triangle AMV. We can think of triangle AVB as being in a plane sectioning the cone through the vertex, V, and through the diameter of the circular base AB. AB is perpendicular to VM, and angle AVB can be any angle between 0° and 180°. To form the parabola, we section the cone with the plane containing DCO such that the plane is perpendicular to the plane with triangle AVB. Line LO is formed by the intersection of the two planes, and plane DCO must be oriented so LO is parallel to AV. This condition results in the curve of intersection of the plane and the surface of

147

the cone, DOC, which is a parabola. We then can derive the formula for the parabola and show that it is of the same form as found by Menaechmus.

To do this we intersect the cone with planes perpendicular to the axis of the cone, VM, and parallel to the base. These planes will form circles at the intersection with the surface of the cone. Circle SO is formed when the plane goes through point 0, the apex of the parabola. A point P is chosen on the parabola other than point O. P can be any point on the parabola. Circle EF made from a plane perpendicular to VM includes points P and N. N is on the diameter of circle EF such that NP is perpendicular to EF. This circle is shown in Figure 31-27 where we have connected EP and PF.

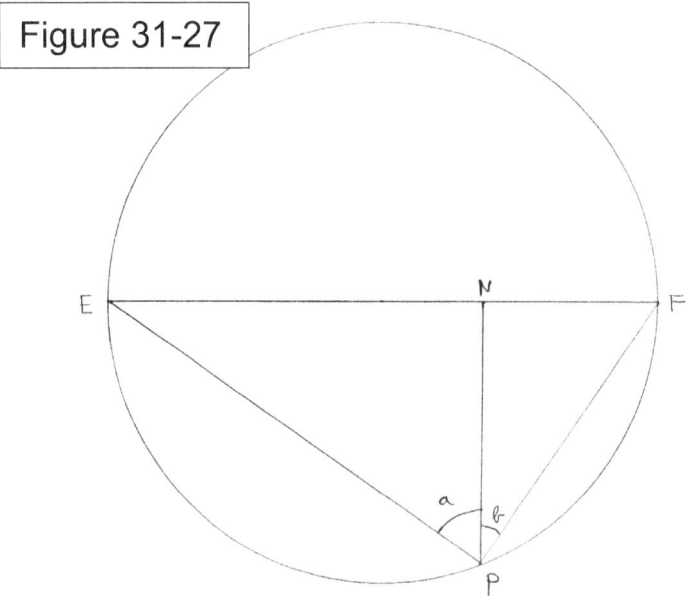

Figure 31-27

We have worked with circles such as this with an inscribed right triangle previously and know angle a + b =

148

90°. Since angles PNF and PNE are also right angles we have similar right triangles PNF and PNE so,

55. NP/EN = NF/NP, or,

$$(NP)^2 = (EN)(NF) .$$

We construct Figure 31-28 which shows triangles VSO and ONF from Figure 31-26 in the plane of triangle AVB.

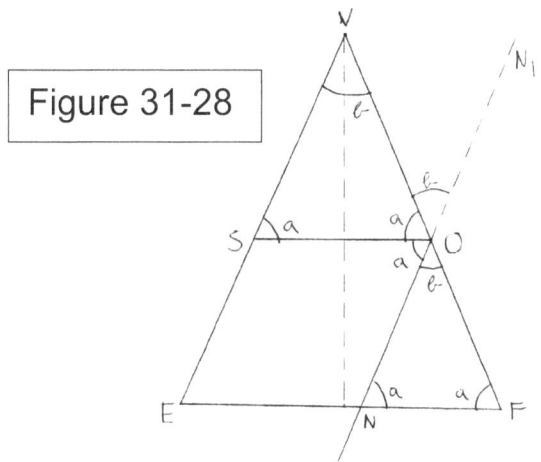

Figure 31-28

We know SV must be parallel to NO. Thus, angle VSO = angle SON, and they are both marked angle a. Since, SV = OV, angle VOS = angle a. Angle SVO is marked as angle b. From parallel lines, angle VON$_1$ = angle b, and thus, angle NOF = angle b, since they are vertical angles.

Angles ONF and OFN must equal angle a, because SO is parallel to EF. Thus triangles SVO and NOF are similar so,

56. NF/ ON = SO/SV, or,

NF = (ON)(SO)/SV, using this in step 55:

57. $(NP)^2 = EN \cdot \dfrac{(ON)(SO)}{SV}$.

Because SENO is a parallelogram,

EN = SO.   Then,

$$(NP)^2 = ON \cdot \dfrac{(SO)^2}{SV}.$$

We now consider point O in Figure 31-26 as being the origin of the XY coordinate axes, and OL as lying on the positive Y axis, while P lies in the first quadrant, the coordinates of point P are then (x, y), where x = NP, and y = ON.   The equation for the parabola then becomes,

58. $x^2 = y \cdot \dfrac{(SO)^2}{SV}$, or, $y = x^2 \cdot \dfrac{SV}{(SO)^2}$.

x and y will vary depending on the position of P on the curve, but SV and SO remain constant. If a = SV/(SO)^2, then,

59. $y = ax^2$.

This is the formula for the parabola in coordinate geometry when the apex of the parabola is at the origin, and the axis of the parabola lies along the positive Y axis. This has the same form as the equation for the parabola we found in our work with Menaechmus.

The ellipse can also be defined for a right cone with any vertex angle between 0° and 180°. See Figure 31-29.

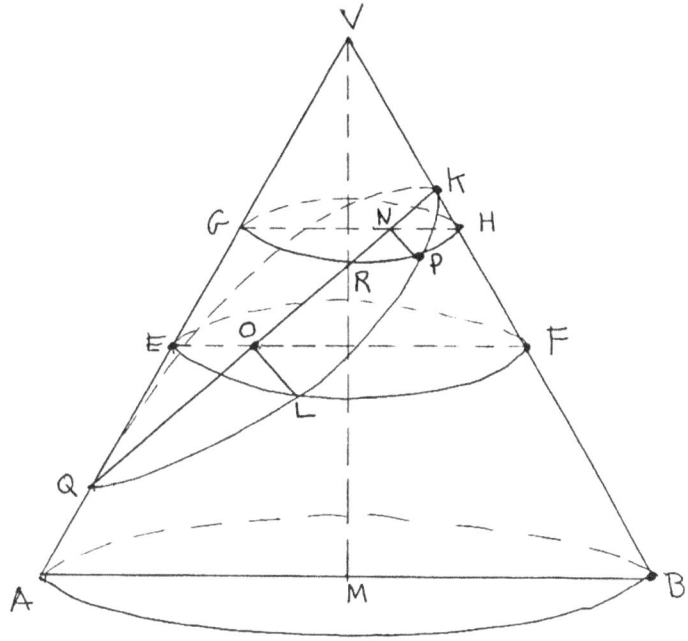

Figure 31-29

VMB is the generating right triangle which we
consider to be in a plane through the cone which
includes the vertex V, and a diameter of the circular
base, AMB.   The cone is intersected by a plane
containing line QK such that the plane is perpendicular
to the plane containing triangle AVB.   QK is the
intersection of these two planes. To form an ellipse the
plane with QLKO must form an angle with the axis of the
cone greater than ½ the vertical angle of the cone.
Thus, angle QRM is greater than angle AVM. We recall
for the parabola these angles would be equal. We shall
see later that if angle QRM were less than angle AVM
the curve of intersection would be a hyperbola.

151

We take any point P on the ellipse other than Q or K and draw line PN to the axis of the ellipse, QK, such that PN is perpendicular to QK. We then intersect the cone with two planes perpendicular to the axis VM. These planes form circles with the intersection of the surface of the cone. The first of these planes contains NP and the circle is GH. The location of the circle depends on where we choose the position of point P.

The second plane which intersects VM perpendicular to VM is a plane through point L of the ellipse and point 0 on the diameter of the circle EF. The plane is chosen to intersect VM so point 0 is the midpoint of QK. Thus QO = OK. 0 and L are such that line OL is perpendicular to the axis of the ellipse, QK, and perpendicular to the diameter, EF, of the circle.

Consider circle GH shown in Figure 31-30.

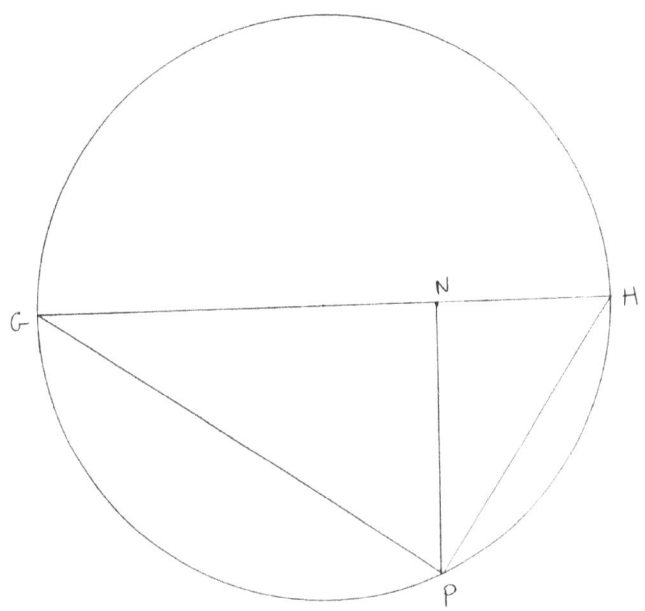

Figure 31-30

When we connect GP and PH we have the same situation as in figure 31-27. It follows from similar right triangles GNP and PNH,

60. NP/GN = NH/NP, or,

$(NP)^2 = (GN)(NH)$.

We consider circle EF. We have again the same situation of two similar right triangles, EOL and FOL. Thus,

61. OL/EO = OF/OL, or,

$(OL)^2 = (EO)(OF)$.

We divide the result of step 60 by the result of step 61.

62.

$$\frac{(NP)^2}{(OL)^2} = \frac{(GN)(NH)}{(EO)(OF)}.$$

Next trace out triangles EQO and GQN. The reader should be able to show they are similar triangles. Then,

63. GN/QN = EO/ QO, and by alternating the means,

GN/EO = QN/ QO .

Now trace out triangles NHK and OFK. They are also similar. Thus,

64. NH/NK = OF/OK, or, NH/OF = NK/OK. We substitute QN/QO for GN/EO, and NK/OK for NH/OF , in step 62.

65.

$$\frac{(NP)^2}{(OL)^2} = \frac{QN}{QO} \cdot \frac{NK}{OK}.$$

We let OL = b.   Since 0 is the midpoint of QK, QO = OK, and we let the value of QO be a.

66.

$$\frac{(NP)^2}{b^2} = \frac{(QN)(NK)}{a^2}.$$

We now transform this equation into coordinate geometry by placing the origin of the XY axes at point 0, and let OK lie on the positive X axis, while OL lies on the positive Y axis. Then OQ will lie on the negative X axis. This is shown in Figure 31-31.

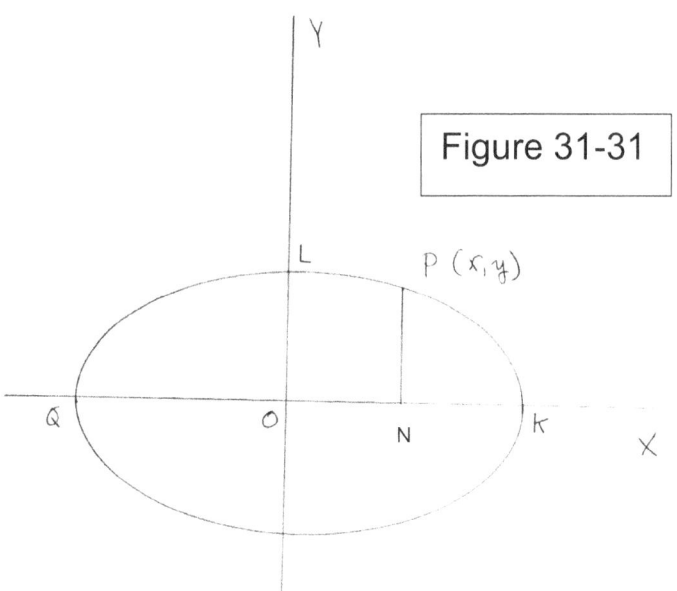

Figure 31-31

Point P on the ellipse has coordinates (x, y). This means ON = x, and NP = y. It follows:

67. QN = Q0 + x = a + x.

NK = OK - x = a - x. The equation in step 66 becomes,

68.

$$\frac{y^2}{b^2} = \frac{(a+x)(a-x)}{a^2}, or, \frac{y^2}{b^2} = \frac{(a^2-x^2)}{a^2}.$$

$$\frac{y^2}{b^2} = 1 - \frac{x^2}{a^2}, or, \frac{x^2}{a^2} + \frac{y^2}{b^2} = 1.$$

We also found the equation for the ellipse to be of this form in our work with Menaechmus.

We next consider how to define a hyperbola as the curve representing the intersection of a plane with any right circular cone with a vertex angle between 0° and 180°. See Figure 31-32.

155

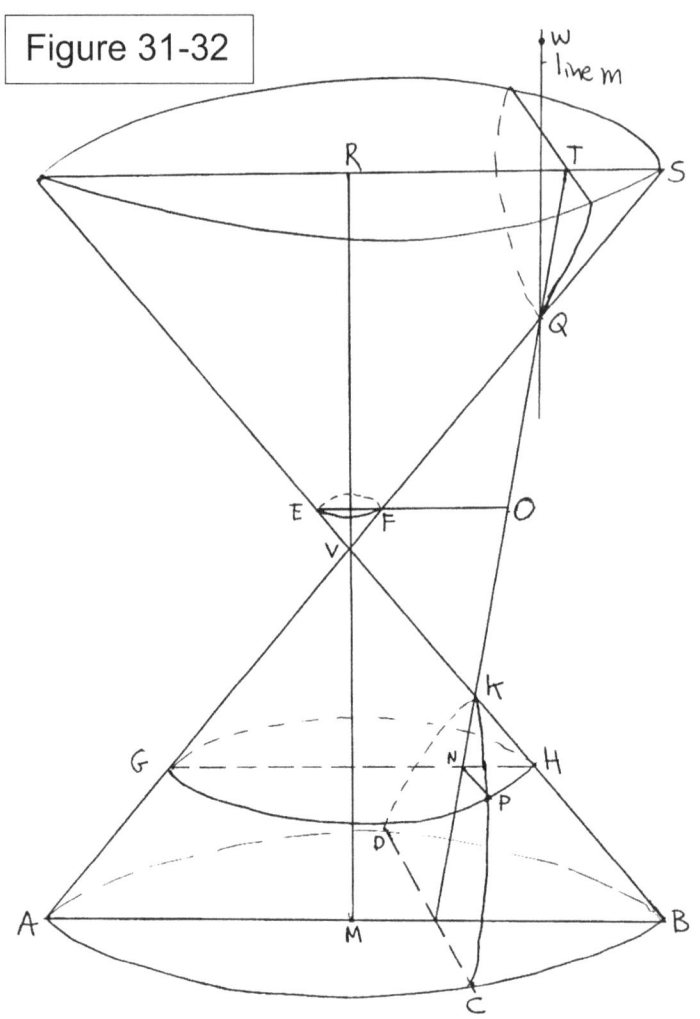

Figure 31-32

The cone is shown with two nappes.   We section the cone with plane AVB which contains a diameter of the circular base, AB, and the axis of the cone, MVR. Angle AVM equals angle BVM. The vertex angle, AVB, can range between 0° and 180°.

We then section the cone with plane QKDC which is perpendicular to plane AVB.   We imagine a line, indicated by line m, parallel to the axis of the cone,

156

MVR, in the plane of AVB which intersects plane QKDC at point Q. Angle TQW must be less than angle SVR if the curve of intersection of plane QKDC with the surface of the cone is to be a hyperbola.

We recall if angle TQW were to equal angle SVR the curve would be a parabola, and if angle TQW were to be greater than angle SVR the curve of intersection would be an ellipse.

We see in Figure 31-32 segment QK of TQK lies between the two nappes of the cone. We call point 0 the midpoint of QK, and then imagine a plane through 0 parallel to the base of the cone and thus perpendicular to the axis of the cone, MVR. This plane will intersect the cone in circle EF.

We let P be any point on the curve of the hyperbola other than Q or K. We again intersect the cone with a plane through P which is parallel to the base of the cone and perpendicular to the axis of the cone. This plane intersects the surface of the cone in circle GH. We connect P with a perpendicular to the diameter of circle at N. NP is also perpendicular to the axis of the hyperbola, QK extended.

We envision the similar right triangles GPN and PNH in circle GH, and use the arguments developed from Figure 31-27. Thus,

69. NP/NG = NH/NP, or,

$$(NP)^2 = (NH)(NG).$$

We inspect triangles GNQ and FOQ and determine they are similar. Thus,

70. NG/NQ = FO/OQ, or,

NG = (NQ)/OQ x (FO).

We inspect triangles NHK and EOK which we reconstruct in Figure 31-33.

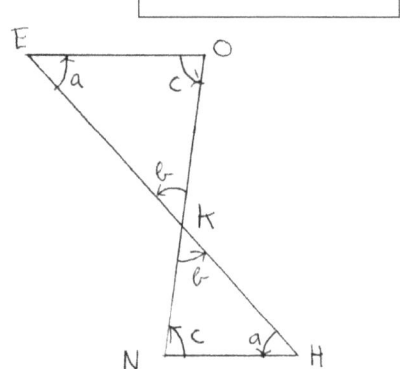

Figure 31-33

We see they are similar triangles since they have equal corresponding angles.   The two angles marked a are equal since they are alternate interior angles between two parallel lines. The angles marked b are equal since they are vertical angles. The corresponding sides of the triangles are thus in proportion.

71. NH(opposite b)/ NK(opposite a) = E0(opposite b)/ KO(opposite a) , or,

NH = (NK/KO)(EO).

From steps 70 and 71 we now have expressions for NG and NH which we can use in step 69.

72.

$$(NP)^2 = \frac{NK}{KO}(EO) \cdot \frac{NQ}{OQ}(FO), \text{ or,}$$

$$(NP)^2 = \frac{NQ}{OQ} \cdot \frac{NK}{KO} \cdot (FO)(EO).$$

Now we choose plane DCKQ in Figure 31-32 for the XY coordinate plane, and choose 0 as the origin. The positive X axis will lie along OK. A line through 0 parallel to NP will be the positive Y axis. We show this arrangement in Figure 31-34.

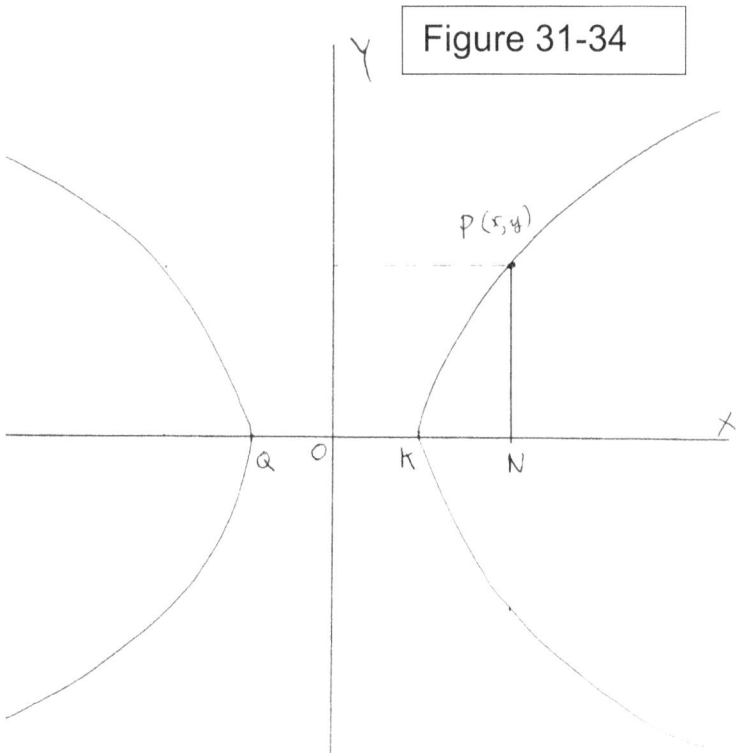

Figure 31-34

The coordinates of point P on the hyperbola are (x, y). We see x = ON, and y = NP.

Because 0 is the midpoint of QK, OQ = OK. We let a equal OQ =OK. No matter where P is located on the hyperbola the term, (FO)(EO) is constant. We let

159

(FO)(EO) = b². If we use these values for a and b in step 74 we have,

73.

$$y^2 = \frac{b^2}{a^2}(NQ)(NK).$$

We see from the figure NQ = QO + ON. Since QO = a, and ON = x, then, NQ = x + a. We see NK = ON - OK, or NK = x - a. Step 73 then becomes,

74.

$$y^2 = \frac{b^2}{a^2}(x - a)(x + a) = \frac{b^2}{a^2}(x^2 - a^2)$$

$$= \frac{b^2 x^2}{a^2} - b^2.$$

Then, $\dfrac{y^2}{b^2} = \dfrac{x^2}{a^2} - 1,\ or,\quad \dfrac{x^2}{a^2} - \dfrac{y^2}{b^2} = 1.$

This is the general formula for a hyperbola when the origin is chosen as in Figure 31-34. If a² = b², then we have the formula for a rectangular hyperbola as we learned in our work with Menaechmus.

In his treatise on the conic sections Apollonius used a more complicated geometrical construction to develop the definitions and properties of the conic sections. In all Apollonius developed about 400 propositions related to the conic sections. We will refer the interested reader to the works in the bibliography on Apollonius' work.

Apollonius did give the conic sections their current names of parabola, ellipse and hyperbola. We will review briefly the concept he used to name them. For each conic section he found a length related to the conic section which he called the parameter of the conic section. See Figure 31-35.

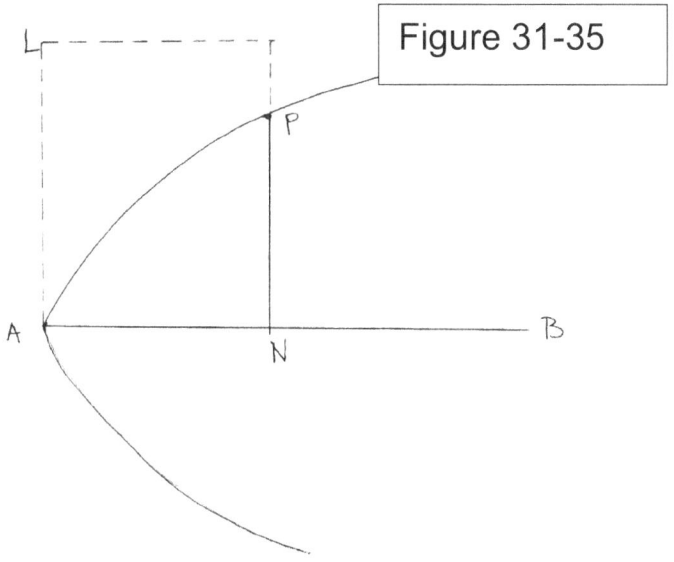

Figure 31-35

The figure shows the curve of the conic section. P represents any point on the curve, and PN is the perpendicular length from P to the axis of the conic section. The parameter of the conic section is length AL. We have drawn AL perpendicular to axis AB. We see that length PN and AN vary depending on the location of point P on the curve. Once the curve is constructed AL remains the same length. The length of AL for the parabola is such that the area of rectangle AL by AN equals the area of the square PN by PN. (In the figure AL is drawn too long, but see the following.)

Apollonius chose the term parabola from the Greek word, $\pi\alpha\rho\alpha\beta o\lambda\eta$. π = pi = p, α = alpha = a, ρ =

rho = r, β = beta = b, o = omicron = o, λ = lambda = l, and η = eta = long e. The Greek word is then, 'parabole', which is modernized to parabola. The word meant 'juxtaposition'. Thus the area $(PN)^2$ exactly equals the area of the rectangle (AL)(AN) in Figure 31-35, and one might loosely say their areas can be juxtaposed.

Our English word parable comes from the Greek word for parabola. A parable is a story which 'fits' a deeper truth.

If the conic section in Figure 31-35 is an ellipse then the area of rectangle AL by AN is less than the area of the square PN by PN. The word ellipse comes from the Greek word $\varepsilon\lambda\lambda\varepsilon\iota\psi\iota\sigma$ used by Apollonius. ε = epsilon = e, ι = iota = i, ψ = psi = ps, and σ = sigma = s. The Greek word transliterated is 'elleipsis'. The Greek word meant 'falls short of'. Thus area (AL)(AN) falls short of area $(PN)^2$.

The English word "ellipsis" which comes from the Greek word for ellipse, means a sentence or phrase where certain words are left out, so the phrase 'falls short', yet still has meaning.

If the conic section in Figure 31-35 is a hyperbola, then the area of AL by AN is greater than the area of square PN by PN. The Greek word chosen by Apollonius for the hyperbola is $\upsilon\pi\varepsilon\rho\beta o\lambda\eta$. (u = upsilon = u). The Greek word transliterates to 'uperbole', which has been modernized to hyperbola.

The Greek word meant 'excess'.   Thus rectangle (AL)(AN) has an excess area when compared to square $(PN)^2$.

The English word hyperbole means speech or writing that exaggerates or goes to excess to make a point or an impression.

We now have a good idea of the origin of the conic sections and their basic formulas.   We went into detail of the geometric constructions of the ancient mathematicians, and we showed how they can be transferred to a rectangular coordinate system.   We used proportions from similar triangles to develop the algebraic equations for the conic sections.   Later mathematicians worked on defining the conic sections without reference to a cone. They sought properties which would allow definitions based on the way the curves behaved in a plane.   In the next chapter we will pursue this approach.

Bibliography for Chapter 31

1. Heath, T. L., "Apollonius of Perga, Treatise on Conic Sections", W. Heffer and Sons Ltd., Cambridge, Originally printed in 1896. Reprinted in 1961.   This book contains an extensive discussion of the work on conic sections before the time of Apollonius plus translating the work of Apollonius into modern symbols so it is more readily understood.

2. Heath, Sir Thomas, "A History of Greek Mathematics", Dover Publications, Inc. New York, 1981.

3. Morrill, W.K., "Analytic Geometry", International Textbook Company, Scranton Pennsylvania, 1964.

Chapter 32
The Focus Directrix Property of Conics

We will next consider a property of the conic sections in the plane which allows them to be defined without reference to the cone. This definition results in the same formulas for the conic sections we found in the last chapter when the location of the XY coordinate axes are chosen appropriately. This property was first given in the work of Pappus of Alexandria in the third century A.D. It is speculated that he found this property in the lost works of Euclid. The idea behind this property is that the parabola, ellipse, and hyperbola can be considered in a plane as loci of points.

The term 'loci', singular, 'locus', refers to all the points drawn on a plane that represent the mathematical relationships for the curve. Today we say the loci of points for a curve are all the points in the XY coordinate system in the plane that satisfy the formula for the curve given as an expression in terms of the variables x and y. For example, the relationship expressed by the formula, $y = x$, can be drawn on the XY coordinate plane as in Figure 32-1 as the straight line such that any point P with coordinates (x, y) lies on the line if the x coordinate equals the y coordinate.

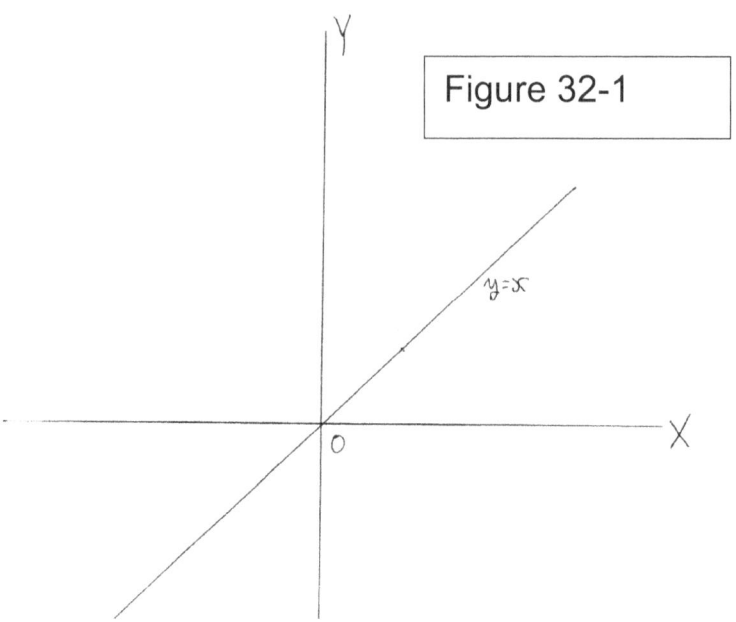

Figure 32-1

The collection of all points that satisfy this requirements are the loci of points for the relationship, y = x.

In Pappus' work he stated conic sections are formed by the loci of points which satisfy the relationship that the ratio of the distance of the points from a fixed line on the plane, called the directrix, to a fixed point, called the focus, is constant. Actually, naming the fixed point the focus did not occur until the work of Kepler in the 17th century.

This definition of the conics can be visualized in Figure 32-2.

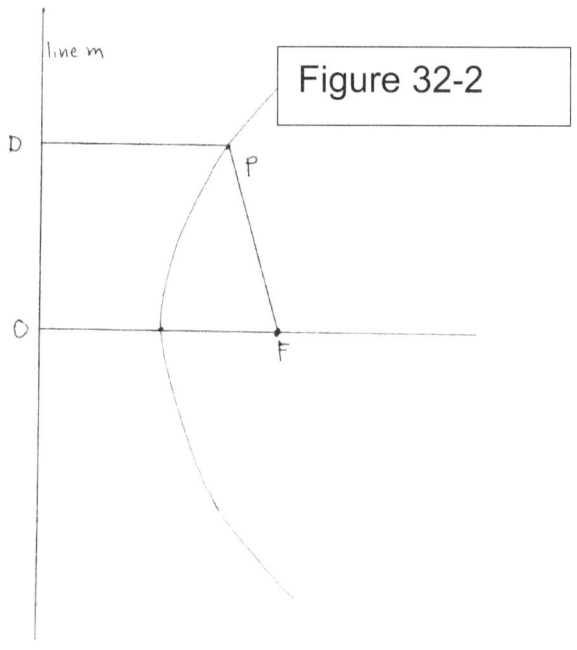

line m

Figure 32-2

D

O

P

F

We first draw the fixed line, line m, and a fixed point, F. We then connect F to the fixed line by a line FO perpendicular to line m. FO extended represents the axis of the conic section. The loci of all points such as point P lie on the conic section. The distance from P to line m is the perpendicular line PD to line m. We then take the ratio, PF/PD, which is fixed for the loci of points on the conic section. If the ratio, PF/PD, = 1 , then the conic is a parabola. If the ratio, PF/PD < 1, the conic is an ellipse. If the ratio, PF/PD > 1, the conic is a hyperbola.

Sir Thomas Heath feels the origin of this idea may have come from the Greek mathematician Aristaeus who lived around 350 B.C. Euclid, around 300 B.C., may have included it in one of his lost works with which Pappus was acquainted.

We will demonstrate this property leads to the same mathematical relationship we found for the ellipse when a right circular cone is sectioned by the proper plane. We will omit the demonstrations for the parabola and the hyperbola.

We recall from our work with Menaechmus in the last chapter the relationship for the ellipse as labeled in Figure 32-3 is,

1. $(PN)^2 = (AF)/AA_1 \times (AN)(A_1N)$, when the ratio $AF/AA_1 < 1$.

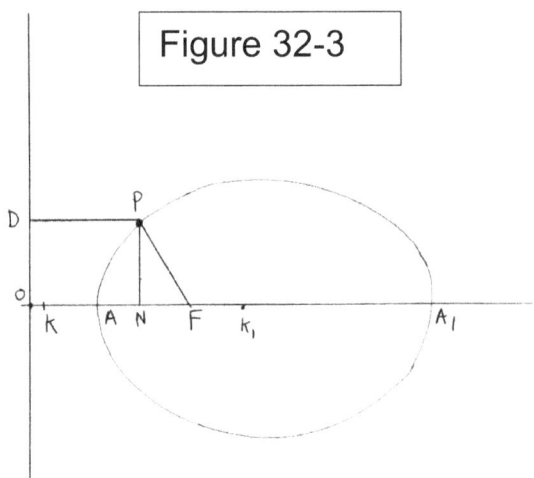

Figure 32-3

We claim the loci of all points P is an ellipse when the ratio PF/PD is a constant < 1. In Figure 32-3 we see A and A_1 are the points where the ellipse crosses line FO, which is perpendicular to the directrix, DO. We see the values of PN, PD, AN, and A_1N vary for different positions of P on the ellipse. If we think of P moving

168

around the curve, we see the position of D on DO moves along the line DO.   We will demonstrate that,

$$(PN)^2 = 1 - \frac{(PF)^2}{(PD)^2} \cdot (AN)(A_1N).$$

The term, $1 - \dfrac{(PF)^2}{(PD)^2}$, will be less than 1 when PF/PD is less than 1, so if we can equate [1 − (PF)^2/(PD)^2]   with AF/AA$_1$ this will establish that the curve by the equation based on the focus directrix property is an ellipse.

2. We let PF/PD = e. Squaring e^2 = (PF)^2/(PD)^2.

3.   For right triangle PNF: (PF)^2 = (PN)^2 + (NF)^2.

We divide both sides by (PD)^2

$$\frac{(PF)^2}{(PD)^2} = \frac{(PN)^2 + (NF)^2}{(PD)^2} = e^2.$$

In this demonstration we use the approach of Pappus.   At this point Pappus gives a new definition for e. He selects a point K on FO in Figure 32-3 such that,

4.   e = NF/NK, or, e^2 = (NF)^2/(NK)^2 Note K is chosen such that

NF/NK = (PF)/PD .

It is not clear why Pappus introduced the new point K in the argument.   We shall see the selection of K to

169

satisfy step 4 does lead to the demonstration that the curve must be an ellipse when e < 1. Perhaps Pappus made various trails to prove the proposition, and found this definition necessary. Often Greek mathematicians gave their work in finished form without explaining the origin of their ideas.

Pappus chooses another point, K_1, in Figure 32-3 on the other side of F so that NK = NK_1. Then,

5.

$$\frac{e^2}{1} = \frac{(PN)^2 + (NF)^2}{(PD)^2} = \frac{(NF)^2}{(NK)^2}.$$

We alternate the means of the second proportion:

$$\frac{(PN)^2 + (NF)^2}{(NF)^2} = \frac{(PD)^2}{(NK)^2},$$

$$\frac{(PN)^2 + (NF)^2}{(NF)^2} - \frac{(NF)^2}{(NF)^2} = \frac{(PD)^2}{(NK)^2} - \frac{(NK)^2}{(NK)^2}.$$

Then we have,

6.

$$\frac{(PN)^2 + (NF)^2 - (NF)^2}{(NF)^2} = \frac{(PD)^2 - (NK)^2}{(NK)^2}.$$

We alternate the means of this proportion.

7.

$$\frac{(PN)^2}{(PD)^2 - (NK)^2} = \frac{(NF)^2}{(NK)^2} = \frac{e^2}{1}.$$

In Figure 32-3 distance PD = NO. Thus,

8. $(PD)^2 - (NK)^2 = (NO)^2 - (NK)^2$.

We factor, $(NO)^2 - (NK)^2$ into, $(NO + NK)(NO - NK)$.

Since, $(NO - NK) = OK$, and $NK = NK_1$, then,

9. $(PD)^2 - (NK)^2 = (NO + NK_1)(OK)$.

We see $(NO + NK_1) = OK_1$, so,

10. $(PD)^2 - (NK)^2 = (OK_1)(OK)$.

We use this in step 7 to get,

11.

$$\frac{(PN)^2}{(OK_1)(OK)} = \frac{e^2}{1}.$$

When P moves to A on the axis AA_1 as seen in Figure 32-4,

Figure 32-4

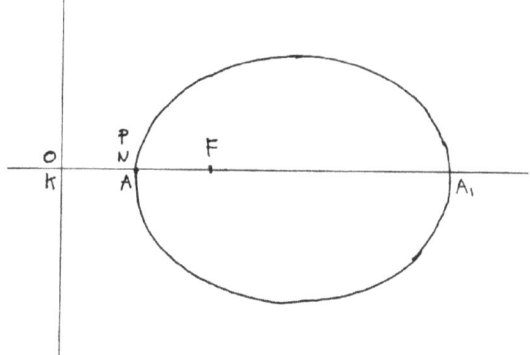

12. e = PF/PD, becomes, e = AF/AO, or, 1/e = AO/AF.

Using Pappus' definition e = NF/NK, or, 1/e = NK/NF, and step 12 we can write by composition,

13.

$$\frac{1+e}{e} = \frac{NK + NF}{NF} = \frac{AO + AF}{AF}.$$

Since NK + NF = FK, and AO + AF = FO, in Figure 32-3,

14.

$$\frac{1+e}{e} = \frac{FK}{NF} = \frac{FO}{AF}.$$ Alternating the means of the last two terms gives,

15.

$$\frac{FO}{FK} = \frac{NF}{AF}, \textbf{\textit{by inversion,}}$$

16.

$$\frac{FO}{FK} = \frac{AF}{NF}.$$ We subtract the equivalent of 1 from both sides,

17. FO/FK – FK/FK = AF/NF – NF/NF, or,

$$\frac{FO - FK}{FK} = \frac{AF - NF}{NF}$$, alternating the means,

18.

$$\frac{FO - FK}{AF - NF} = \frac{FK}{NF}.$$

Since FO - FK = OK, and AF - NF = AN,

19.

$$\frac{OK}{AN} = \frac{FK}{NF}.$$ Using step 14

20.

$$\frac{1 + e}{e} = \frac{OK}{AN}.$$

This holds for points of the curve except when P is at A, as in Figure 32-4, where OK and, AN equal 0.

When P is at position A_1 in Figure 32-3, then N coincides with P. Then, PF/PD = e becomes, $A_1F/A_1O = e$

By Pappus' definition,

21. e = NK/NF, so, $A_1F/A_1O = NF/NK$, and using inversion and the definition that $NK = NK_1$:

22.

$$\frac{A_1O}{A_1F} = \frac{NK_1}{NF} = \frac{1}{e}.$$

By division of this proportion:

23. $\frac{A_1O - A_1F}{A_1F} = \frac{NK_1 - NF}{NF} =$

$$\frac{1-e}{e}.$$

From Figure 32-3, $A_1F + FO = A_1O$, or,

$A_1O - A_1F = FO$. Also we see $NK_1 = NF + FK_1$, or,

$NK_1 - NF = FK_1$. We use these results in step 23.

24.

$$\frac{FO}{A_1F} = \frac{FK_1}{NF} = \frac{1-e}{e}.$$

174

We consider the first two terms of this continuous proportion.   By alternating the means we have,

25. $FO/FK_1 = A_1F/ NF$. By composition of this proportion,

$$\frac{FO + FK_1}{FK_1} = \frac{A_1F + NF}{NF}.$$

From Figure 32-3 we see, $FO + FK_1 = 0K_1$, and, $A_1F + NF= A_1N$.   Using these values we have,

26.

$$\frac{OK_1}{FK_1} = \frac{A_1N}{NF}$$, by alternating the means,

$$\frac{OK_1}{A_1N} = \frac{FK_1}{NF}.$$

From step 24,

$$\frac{FK_1}{NF} = \frac{1-e}{e}$$, so

27.

$$\frac{1-e}{e} = \frac{OK_1}{A_1N}.$$

In step 20 we found, $\dfrac{1+e}{e} = \dfrac{OK}{AN}.$

We multiply the left hand term of step 20 by the left hand term of step 27, and we multiply the right hand term of step 20 by the right hand term of step 27. We are multiplying equals by equals so the results are equal.

28.

$$\frac{(1+e)(1-e)}{e^2} = \frac{(OK)(OK_1)}{(AN)(A_1N)},$$

$$(1-e^2) = e^2 \cdot \frac{(OK)(OK_1)}{(AN)(AN_1)}.$$

From step 11 above we had found,

$$e^2 = \frac{(PN)^2}{(OK)(OK_1)},$$

we substitute this value for e^2 into the result of step 28.

29.

$$(1-e^2) = \frac{(PN)^2}{(AN)(A_1N)}, \quad or,$$

$$(PN)^2 = (1-e^2)(AN)(A_1N).$$

For the ellipse, e < 1, so the term, $(1 - e^2)$ must also be less than 1.   We recall the formula for the ellipse we derived in the previous chapter from the intersection of a plane and a cone, and from step 1, was,

$(PN)^2 = [AF/AA_1](AN)(A_1N)$, where $[AF/AA_1]$ was less than 1. We now see that by starting with the focus directrix property, $e = (PF)/(PD) < 1$, we derived the equivalent formula where $(1 - e^2)$ is equal to $[AF/AA_1]$. Thus we have demonstrated that the focus directrix property can be used to define the ellipse.

As we noted earlier it can be established by the same sort of argument that the focus directrix property can be used to define the parabola and the hyperbola, and then demonstrate that the derived formulas are equivalent to the formulas for these conic sections we obtained by intersecting the appropriate plane with a right circular cone.

We will now derive the equations for the conic sections in terms of coordinate geometry starting with the definitions of the conic sections using the focus directrix property of the conics on a plane. We start with the parabola which is defined as the loci of points on a plane which satisfy the condition that the distance of any point on the parabola from a fixed point is equal to the distance of that point to a fixed line. To have the simplest equation for the parabola we choose the XY axes as in Figure 32-5.

Figure 32-5

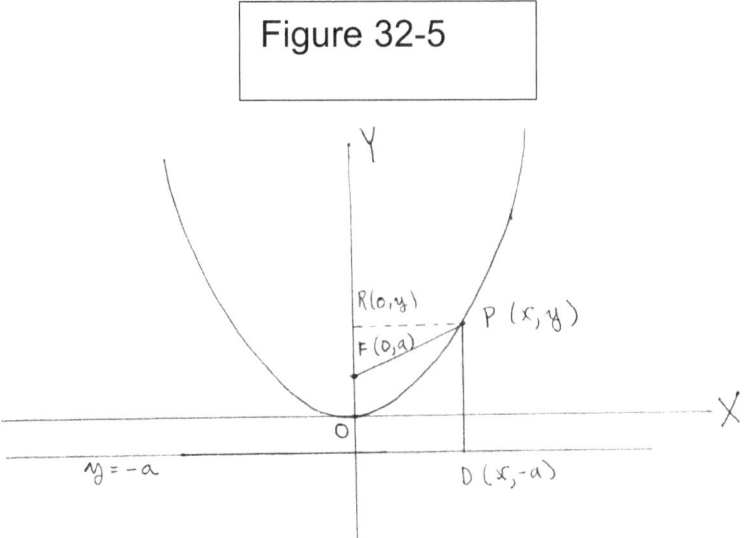

The fixed point is F placed with coordinates (0, a), and is the focus. The fixed line is the line with equation y = - a. The condition for the loci to be a parabola is,

30. |FP| = |DP|, or, |FP|/|DP| = 1 = e.   Note we use absolute values for FP and DP so they are positive quantities.

As noted above in the case of the parabola e, the eccentricity, is equal to 1. We see that when the point P is at the origin of the coordinate system then, |PD| = a, and, |PF| = a, so the origin is ½ the distance between the focus and the fixed line, the directrix.   We complete right triangle PRF as indicated.   Using the distance formula from coordinate geometry,

31. $|PF| = \sqrt{|x - 0|^2 + |y - a|^2}$

|PD| = |y - (-a)| = |y + a|.

Since, |PF| = |PD| then,

178

$$\sqrt{x^2 + |y - a|^2} = |y + a|.$$

We square both sides.

32. $x^2 + y^2 - 2ay + a2 = y^2 + 2ay + a^2$.

We subtract the term, $y^2 + a^2$, from both sides of the equation.

$x^2 - 2ay = 2ay$,

$4ay = x^2$,

$y = (1/4a) x^2$.

This is the equation for the parabola oriented to the coordinate axes shown in Figure 32-5. Different orientations with the coordinate axes are, of course, possible.   If we start with the equation, $y^2 = 4ax$, where a is a constant, we shall find a different orientation of the parabola with the XY axes.   We first add, $x^2 - 2ax + a^2$, to both sides of the equation.

33. $x^2 - 2ax + a^2 + y^2 = 4ax + x^2 - 2ax + a^2$,

$(x - a)^2 + y^2 = x^2 + 2ax + a^2$,

$(x - a)^2 + y^2 = (x + a)^2$.

We take the positive square root of both sides:

34. $\sqrt{(x - a)^2 + y^2} = (x + a).$

If a point P (x, y) is on the curve given by this equation, we can choose point F to have coordinates (a, 0) as shown in Figure 32-6.

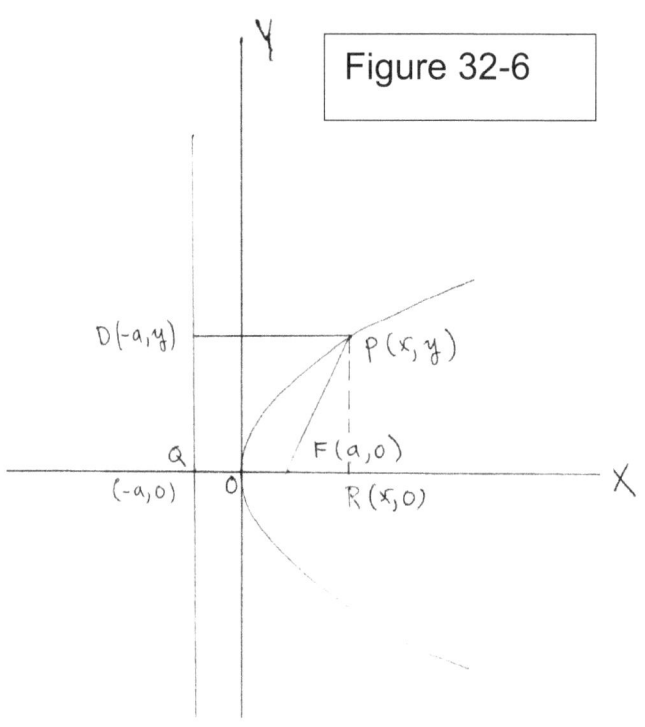

Figure 32-6

We draw the line with the equation, x = - a.

Then,

35. $|PF| = \sqrt{|x-a|^2 + |y-0|^2}$,

$|PD| = |x + a|$.

Using step 34 we see that, |PF| = |PD|, which is the condition for a parabola as drawn in Figure 32-6.

In Figure 32-7 we have drawn four different orientations for the parabola in reference to the XY coordinate axes and the formulas for the 4 parabolas which can be derived for them by methods above.

Figure 32-7

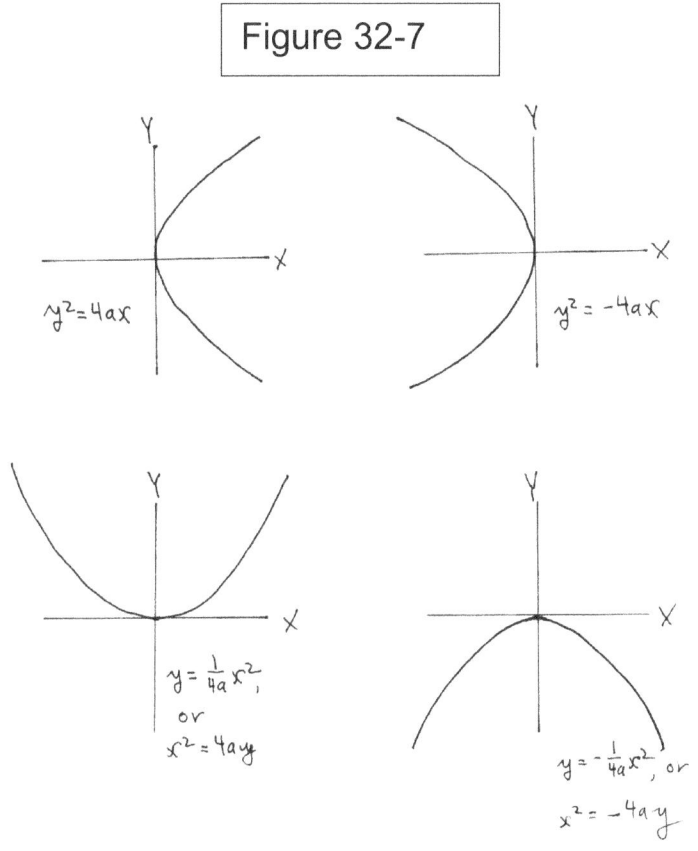

In general if we have an equation of the form, $Ax^2 + Ey = 0$, where A and E are constants, then we can write,

36. $y = - (A/E) x^2$.

We see this is equivalent to $y = (1/4a) x^2$, where, - (A/E) is equal to (1/4a), and we have the equation for the parabola as drawn in Figure 32-5.

Likewise an equation of the form, $Cy^2 + Dx = 0$, can be written as,

$y^2 = - (D/C)x$, where $-(D/C)$ represents $4a$, and we have the equation of a parabola as shown in Figure 32-6.

We turn next to the focus directrix definition of the ellipse in the XY coordinate system. The ellipse is the loci of points on a plane which satisfy the condition that the ratio of the distance from any point on the loci to a fixed point (the focus) taken to the distance from that point to a fixed line (the directrix) is equal to a constant e, the eccentricity, which is less than 1. We see an ellipse on a plane with reference to the XY coordinate axes with its major points indicated in Figure 32-8.

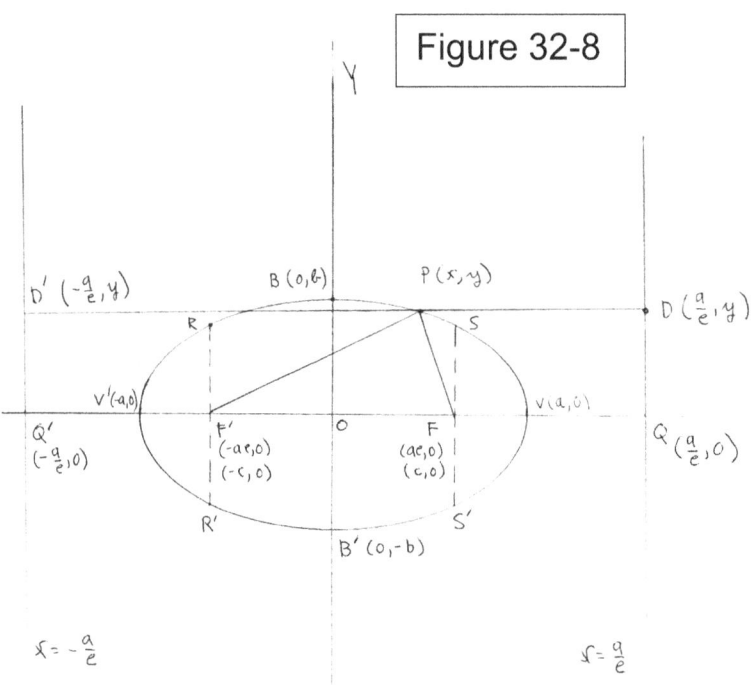

Figure 32-8

We will discuss how the coordinates of these points are determined in the following paragraphs.

We start with point P, which is any point on the ellipse. Its coordinates are (x, y) which are variable. The focus is at F which is placed on the X axis in such a way to produce the simplest formula for the ellipse. The directrix is the line DQ to the right of F. It is oriented to be perpendicular to the X axis. The fundamental condition from the focus directrix definition of the ellipse is that,

$$\frac{|PF|}{|PD|} = e$$

, and where e is a constant value for the particular ellipse with the condition that e < 1.

For the simplest equation for the ellipse the center is at the origin,　(0, 0), with the focus along the X axis. The intersection of the ellipse with the X axis is the point V with coordinates (a, 0), which means the distance from the origin to V is length a. Thus, OV = a.

PD is the perpendicular distance from the directrix to any point P on the ellipse. We designate the coordinates of F as (c, 0). Thus, FV = a - c.

Because, |PF|/|PD| = e, and e < 1, |PF| < |PD|.

This means when P is taken to be on the X axis to the right of the origin then P is coincident with point V.

183

Also point D is coincident with point Q. Thus the relation, |PF| < |PD|, becomes, VF < VQ. This clearly shows that V lies between F and Q.   The lengths of FV and VQ depend, of course, on the numerical value of e, but when e < 1, as it must be for an ellipse then V lies closer to F than it does to Q.

The ellipse also contains another point of intersection with the X axis we have designated as point V'.   There is a geometric method to find point V' on the X axis of the ellipse such that the relation,

$$\frac{FV'}{V'Q} = \frac{FV}{VQ} = e < 1$$

, is demonstrated.

In order to find V' we refer to Figure 32-9.

Figure 32-9

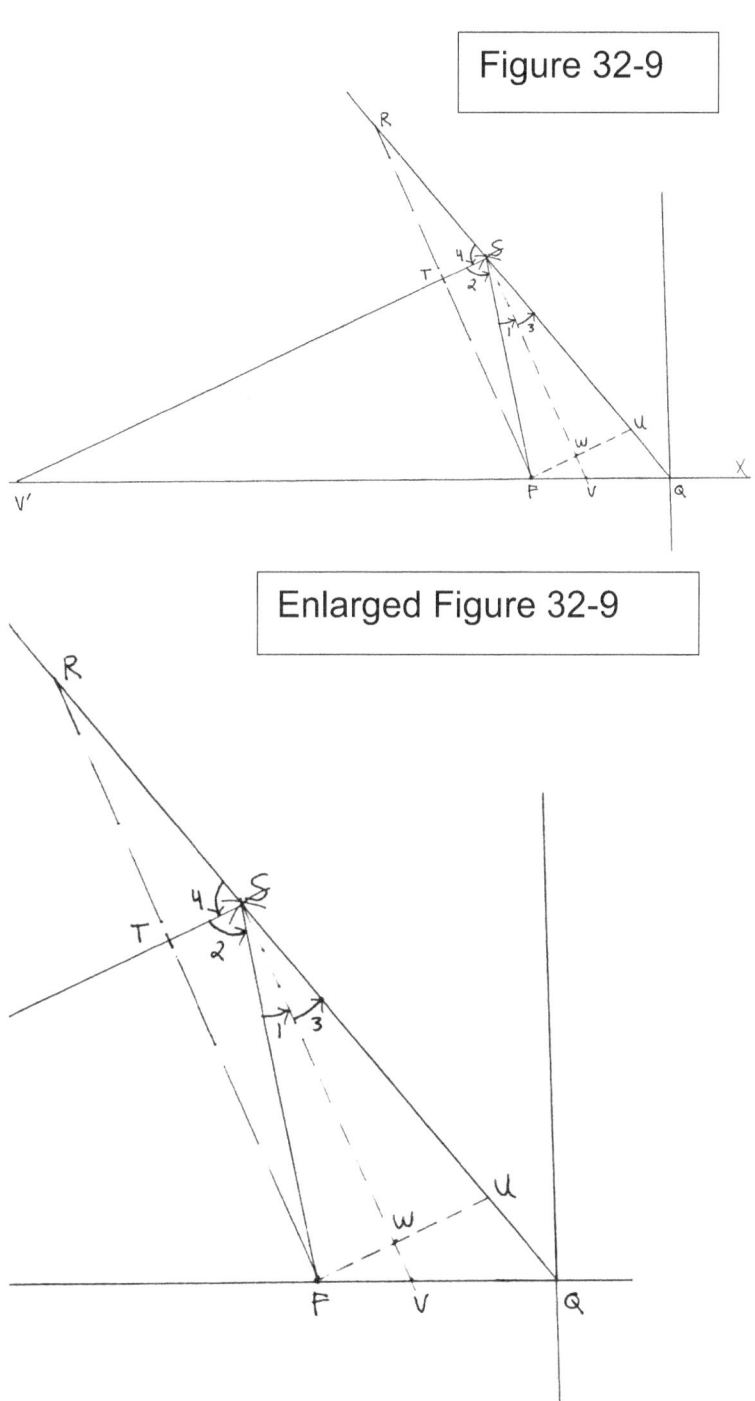

Enlarged Figure 32-9

185

We must have the value of e and the length FQ in figure 32-8.   We could find FQ if we are given e and FV, since VQ = FV/e, and FQ = FV + VQ.   For the sake of illustration in Figure 32-9 we let FQ = 2 inches, and e = .75. We take an arbitrary length of 4 inches and draw a circular arc with center at Q and radius of 4 inches. Lengths other than 4 inches could be used to obtain the final result. We then make a circular arc with F as center and radius of e x 4 inches = (.75) (4) = 3 inches. The intersection of the two arcs is point S in Figure 32-9.

We bisect angle FSQ so angles 1 and 3 are equal.   We connect the line that bisects angle FSQ to line FQ to determine point V. We will shortly demonstrate that FV/VQ = e, as required for point V to be on the ellipse.   Note that point S is not on the ellipse and is constructed for the purpose of determining the locations of points V and V' on the X axis, given the values of e and FQ.

We extend QS towards R, and then bisect angle FSR. The line that bisects angle FSR is extended to the X axis, which it intersects at V'. We note the angles marked 2 and 4 are then equal. We then wish to prove that,

$$\frac{FV}{VQ} = \frac{FV'}{V'Q} = e.$$
Once this is proven we know that we have demonstrated that V and V' are the points on the ellipse with the given values of e and FQ.

37. By the construction of point S,

$$FS/QS = 3/4 = .75 = e.$$

38. Consider triangle FSQ. VS is the bisector of one angle, angle FSQ, of the triangle. Euclid had shown the bisector of an angle of a triangle divides the opposite side of the triangle into segments which are proportional to the adjacent sides. This means,

$$\frac{FV}{VQ} = \frac{FS}{QS}.$$
Since FS/QS = e, then,

$$FV/VQ = e.$$

This means V is on the ellipse, and is the point of the ellipse which intersects the X-axis between V and Q. To demonstrate this proposition of Euclid we note,

angle 1 = angle 3,

angle 2 = angle 4.

We add the respective sides of these equations.

1 + 2 = 3 + 4.

Since the sum of angles 1 + 2 + 3 + 4 = 180°,

and, 1 + 2 = 3 + 4, then,

1 + 2 = 90°, and 3 + 4 = 90°.

39. Since angle 1 + angle 2 = 90°, then triangle VSV' is a right triangle. We draw FR parallel to VS.

Since V'S is perpendicular to VS, FR is also perpendicular to V'S at the point of intersection, point T. This in turn means triangles RTS and FTS are right triangles. Furthermore, they are congruent triangles since they have three equal angles (the right angles, angle 2 = angle 4, and the remaining angles), and share side ST. Thus RT = FT, and RS = FS.

40. Since SV is parallel to RF, SV cuts off proportional segments of the sides in triangle FRQ. The segments in proportion are : QV/VF = QS/SR, or after inverting the proportion and rewriting, FV/VQ = RS/QS. We demonstrated this proposition in steps 16-20 in chapter 7. Since we know RS = FS from step 39, FV/VQ = FS/QS, which demonstrates Euclid's proposition which stated the bisector of angle FSQ in triangle FSQ divides the opposite side of the angle, FQ, into segments, FV, VQ, which are proportional to the adjacent sides FS and QS. Once again, since FS/QS = e, then, FV/VQ =e, and V is on the ellipse. We now wish to show, FV'/V'Q, also equals, FS/QS = e. We draw in Figure 32-9, FU parallel to V'S. Then FU divides the side of triangle

QV'S, lines V'F, QS, into segments that are in proportion.

41. QF/V'F = QU/US, by composition,

$$\frac{QF + V'F}{V'F} = \frac{QU + US}{US}.$$

Since QF + V'F = V'Q, and QU + US = QS,

V'Q/ V'F = QS/US, or by inverting and rewriting

FV'/ V'Q = US/ QS .

42. Because FU is parallel to V'S, it is perpendicular to SV at W.   Triangles SWF and SWU are then right congruent triangles since they have three equal angles and share side SW. Thus US = FS. We use this in the final proportion in step 41.

FV'/ V'Q = FS/QS. Since FS/QS = e,

we then have,

FV'/ V'Q = e = FV/VQ < 1.

We then have shown how to find the second point of intersection of the ellipse with the X axis in Figure 32-8 starting with the values of e and FQ.   The demonstration is an example of the general concept that a given line, such as FQ in Figure 32-9, can be divided into segments, in this case FV and VQ, such that if the ratio of the segments, FV/VQ, is less than 1, then the line, FQ, can be divided by a point on the line extended, in this case point V', into the equivalent ratio.   Thus, FV/VQ = FV'/V'Q, where the ratio is less than 1. This is called dividing the line externally.

The construction of Figure 32-9 is a demonstration that this can be done.   We shall see later that a line that is divided internally by a point so that the ratio of the segments is greater than 1 can also be divided externally by a point with the equivalent ratio.

We note that if a line is divided internally by a point so the ratio of the segments equals 1, then it cannot be divided externally by a point with the same ratio.   For example, if the ratio FV/VQ =1 = e, we have the situation in the case of the parabola, that if Q is on the directrix, F is the focus, and V the apex of the parabola.   If V is on the X axis for the parabola there is no other point of the parabola on the X axis.

If we attempted to find another point on the X axis by the method of Figure 32-9 we would have the construction in Figure 32-10.

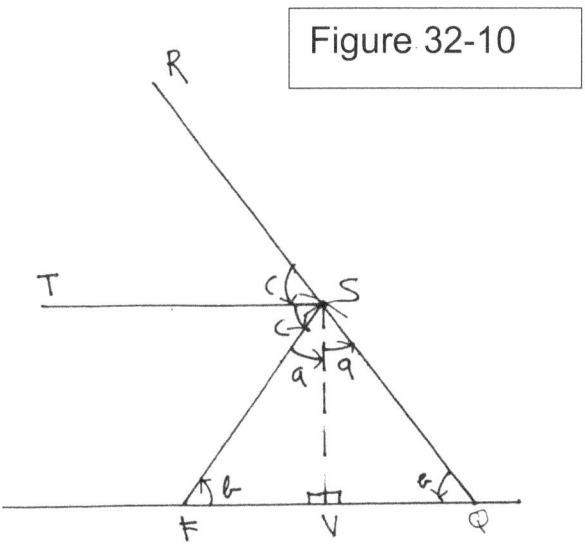

Figure 32-10

Point S is found by the intersection of circular arcs with radii both equal to FS, since FV/VQ = e = 1. Bisecting angle FSQ results in two right triangles, FVS and QVS. Bisecting angle FSR results in the two equal angles marked c.   We see,

43.   a + b = 90°

2a + 2c = 180°, and if we divide both sides by 2:

a + c = 90°.   Thus, c = b.

This means TS is parallel to FQ and thus will never intersect FQ extended at a point V'.   There is no point V', where,

FV/VQ = FV'/VQ = e = 1.

We can now turn to the ellipse in Figure 32-8.

191

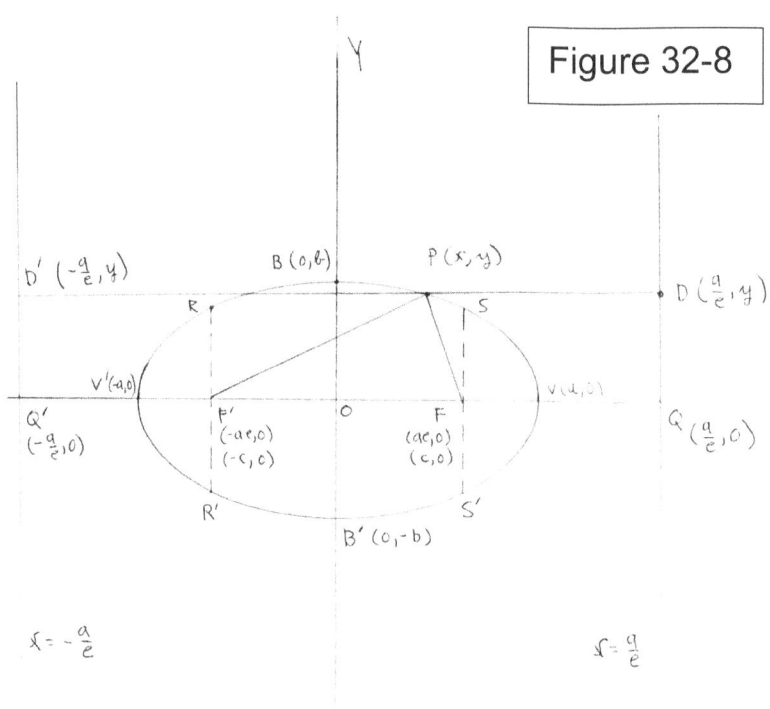

Figure 32-8

We have determined points V and V', and place them equidistant from the origin with coordinates (a, 0), and, (-a, 0), respectively. We let the coordinate of F be (c, 0) as noted. Then,

44. FV = a – c,

FV' = a + c

45. We know FV = e(VQ), and FV' = e(V'Q). From Figure 32-8,

VQ = OQ - a, and, V'Q = OQ + a. Thus,

a - c = e(OQ - a), and

a + c = e(OQ + a).

192

We add the respective sides of the last two equations to get,

46. (a - c) + (a + c) = e(OQ - a) + e(OQ + a)

2a = 2e(OQ),   or OQ = a/e .

This means the coordinates of point Q must be (a/e, 0) in Figure 32-8, and the equation of the directrix must be x = a/e.

If we use a/e for OQ in the equation,   a - c = e(OQ - a), we have,

47.   a - c = e(a/e - a)

a - c = a – ae, Thus, c = ae.

This means the coordinates of F must be (ae, 0). To find the equation of the ellipse in Figure 32-8 we proceed as follows.   Since e < 1, then

48.   ae < a, and a < a/e, or, ae < a < a/e.

We see that this agrees with the relative positions of F, V, and Q on the X axis.

49.   For the ellipse, |PF|/|PD| = e.

$$|PF| = \sqrt{(x - ae)^2 + (y - o)^2}$$, and

|PD| = a/e – x.

These relations follow from the distance formula for two points in Figure 32-8.   Since |PF| = e(|PD|),

$$\sqrt{(x - ae)^2 + y^2} = e(^a\!/_e - x),$$

$$\sqrt{(x - ae)^2 + y^2} = a - ex.$$

We square both sides,

$$(x - ae)^2 + y^2 = (a - ex)^2,$$

$$x^2 - 2aex + a^2e^2 + y^2 = a^2 - 2aex + e^2x^2,$$

$$x^2 + a^2e^2 + y^2 = a^2 + e^2 x^2.$$

We subtract a^2e^2, and e^2x^2 from both sides.

50. $x^2 - e^2x^2 + y^2 = a^2 - a^2e^2,$

$$x^2(1 - e^2) + y^2 = a^2(1 - e^2).$$

We divide both sides by

$$a^2(1 - e^2).$$

51.

$$\frac{x^2}{a^2} + \frac{y^2}{a^2(1-e^2)} = 1.$$

We recall $c = ae$, so $c^2 = a^2e^2$. We subtract a^2 from both sides.

52. $c^2 - a^2 = a^2e^2 - a^2.$

We multiply both sides by - 1.

$$a^2 - c^2 = a^2 - a^2e^2.$$

$$a^2 - c^2 = a^2(1 - e^2).$$

194

In Figure 32-8 when point P is at position B, the intersection of the ellipse with the Y axis, then the x coordinate of P is 0. The equation for the ellipse in step 51 becomes,

53.

$$\frac{0}{a^2} + \frac{y^2}{a^2(1-e^2)} = 1.$$

We note $a^2(1 - e^2) = a^2 - c^2$, from step 52. Mathematicians then let,

$a^2 - c^2 = b^2$, so that step 53 becomes,

54. $y^2 = b^2$, or, $y = \pm b$.

Thus the coordinates of point B are (0, b). There is another point B', where the ellipse crosses the Y axis with coordinates (0, -b).

Because $b^2 = a^2 - c^2 = a^2(1 - e^2)$, the equation for the ellipse in step 51 becomes,

55.

$$\frac{x^2}{a^2} + \frac{y^2}{b^2} = 1.$$

Also note because, $b^2 = a^2 - c^2$, this means, b < a, in Figure 32-8.

We also see that when y = 0, then the equation for the ellipse is,

195

$x^2/a^2 = 1$, or, $x^2 = a^2$, or, x = ± a.

This verifies that the coordinates of V are (a, 0), and the coordinates of V' are (-a, 0), as we had previously determined.

Using the equation, $b^2 = a^2(1 - e^2)$, we have,

56. $b^2 = a^2 - a^2e^2$.

$a^2e^2 = a^2 - b^2$.   Also note, **b =** $a\sqrt{1-e^2}$.

$$e = \sqrt{1 - \frac{b^2}{a^2}}$$.

Thus when we have an equation in the form,

$x^2/a^2 + y^2/b^2 = 1$, we have the equation for an ellipse with the orientation and key coordinates as seen in Figure 32-8.   We can use the values of a^2 and b^2 to determine the value of the eccentricity, e, of the ellipse.

Special names are given to parts of the ellipse. Note in Figure 32-8 that we designate point F', with coordinates (-c, 0), the symmetric point to F on the negative side of the origin. F and F' are called the foci of the ellipse.

We also have drawn the line x = - a/e , the line symmetric to the directrix.   Both lines x = a/e, and x = - a/e, are called the directrices of the ellipse.   The ellipse then has two directrices and two foci, and either pair

196

could be used to derive the equation of the ellipse. We recall in contrast that the parabola has only one directrix and one focus.

The major axis of the ellipse is VV', and is equal in length to 2a. The major axis contains the two foci F and F'. The minor axis is the line BB' and bisects the major axis.

We recall for any point P on the ellipse with coordinates (x, y), that, |PF| = e(|PD|), and from step 49 we found

57. |PD| = a/e - x. Thus,

|PF| = e(a/e - x) = a − ex.

Using the focus F', we can write, |F'P| = e(|PD'|).

Then by the distance formula,

58. |PD'| = $x - (-\frac{a}{e}) = x + \frac{a}{e}$.

Since, |F'P| = e(|PD'|),

then, |F'P| = e(x + a/e) = ex + a.

We add the values for |FP| and |F'P|.

59. |FP| + |F'P| = (a - ex) + (ex + a)

|FP| + |F'P| = 2a.

This is another property of the ellipse, namely that each point on the ellipse satisfies the condition that the sum of the focal radii to the point equals a constant, 2a, which is the length of the major axis, VV'. We used this property in the preface to this book to define the ellipse. It is possible to start with this property as the definition of the ellipse in a plane and then derive the equation for the ellipse from this definition.

Another feature of the ellipse are the two chords of the ellipse RR', and SS' in Figure 32-8.

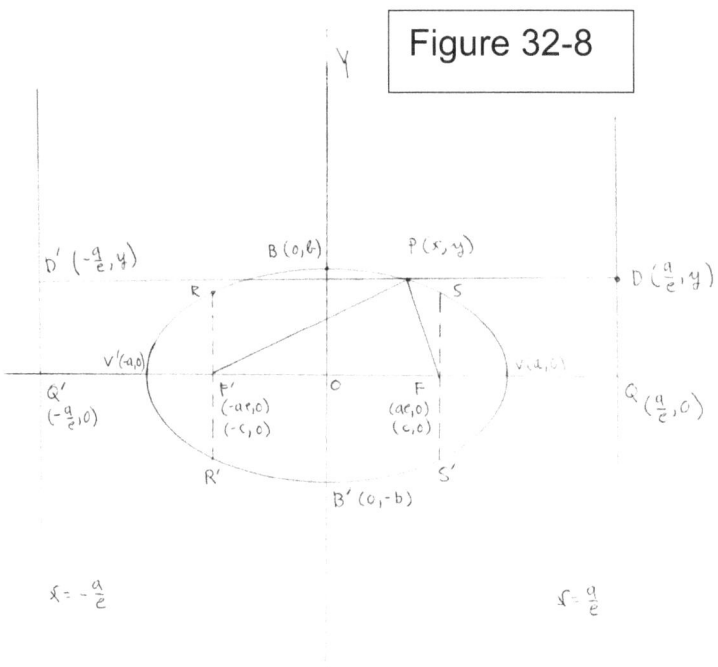

Figure 32-8

These chords are perpendicular to the major axis, VV, and through the respective foci, F and F'. These chords are known as the *latus recti* of the ellipse. We will show how to find the length of the *latus rectum*, SS'. This *latus rectum* lies on the line, x = c, since the

198

latus rectum is parallel to the Y-axis and goes through point F with coordinates, (c, 0).   We then use c for the value of x in the equation of the ellipse.

60.

$$\frac{c^2}{a^2} + \frac{y^2}{b^2} = 1. \ \frac{y^2}{b^2} = 1 - \frac{c^2}{a^2} = \frac{a^2 - c^2}{a^2}.$$

We recall, $b^2 = a^2 - c$.

$$\frac{y^2}{b^2} = \frac{b^2}{a^2}, or, y^2 = \frac{b^4}{a^2}, or,$$

$$y = \pm \frac{b^2}{a}.$$

This result for the y coordinates of the ellipse means that the coordinates for S are (c, b^2/ a), and for S' are (c, - b^2 /a).   We then see from Figure 32-8 that the length of the *latus rectum* , SS' is, 2 · b^2 /a .   It is readily shown that the other *latus rectum,* RR' has the same length.

We now ask the question what happens to the equation for the ellipse if b = a?   As we look at Figure 32-8 it would intuitively appear that the ellipse becomes a circle with radius a. This is correct as we will demonstrate.

61.   The equation $x^2/a^2 + y^2/b^2 = 1$, becomes, $x^2/a^2 + y^2/a^2 = 1$, or, $x^2 + y^2 = a^2$, which we know is the equation of a circle with radius   a.

62. Since $b^2 = a^2 - c^2$, then if b = a,

$a^2 = a^2 - c^2$, or $c^2 = 0$, which means c = 0 for a circle.   This means the circle can be considered as a special case of the ellipse when the foci are coincident at the center of the circle.

63. Recall from step 56, $e = \sqrt{1 - \dfrac{b^2}{a^2}}$.

If b = a, then

$e = \sqrt{1 - 1} = 0$.

When the eccentricity is 0 then the ellipse is a circle. For the ellipse the closer e is to 0, the more the ellipse resembles a circle, or is approximated by a circle with radius a, the major axis.   We shall see later in our work when we consider the orbits of the planets the eccentricity of their elliptical orbit is very close to 0. Of all the planets known to the 17th century astronomers Mars had the highest eccentricity, and fortunately it was the orbit of Mars which Kepler studied in detail, and found the positions of Mars could not be explained by assuming a circular orbit.   He found that an elliptical orbit would resolve his difficulties, and thus he discovered his law that the planets move in elliptical orbits about the sun where the sun occupies the position of one of the foci of the elliptical orbit.

We see that since, $e = \sqrt{1 - b^2/a^2}$, the difference between a and b (b always < a) determines the eccentricity of the ellipse. As b becomes closer to a, we see e becomes closer to 0, the value of e for a circle. Mathematicians say the circle has a single degree of geometric freedom determined by the value of a, and that the circle is perfectly symmetric.

The ellipse is said to have two degrees of geometric freedom. One is represented by the value of a, which determines the size of the ellipse and the other represented by, b which determines the shape of the ellipse.

The ellipse is symmetric about both X and Y axes as constructed in Figure 32-8. This means if we know the shape in any of the quadrants in the rectangular Cartesian coordinate system we have the shape in the other quadrants by reflecting the curve about the X and Y axes. If we had an equation of the form, $x^2/b^2 + y^2/a^2 = 1$, we should recognize that this equation represents an ellipse oriented with its center at the origin with the major axis along the Y axis, and the minor axis along the X axis.

We will introduce more on the mathematics of the ellipse and parabola as we progress in our study of the work of Kepler, Galileo, and Newton.

The last conic we shall discuss in reference to the focus directrix property is the hyperbola. The definition of the hyperbola using the focus directrix property is that the hyperbola is the loci of points on a plane which satisfy the condition the ratio of the distance from any point on the loci to a fixed point (the focus), taken to the distance of the point to a fixed line (the directrix), is equal to a constant, e, the eccentricity, which is greater than 1.

We orient the hyperbola for the simplest equation in coordinate geometry as shown in Figure 32-11.

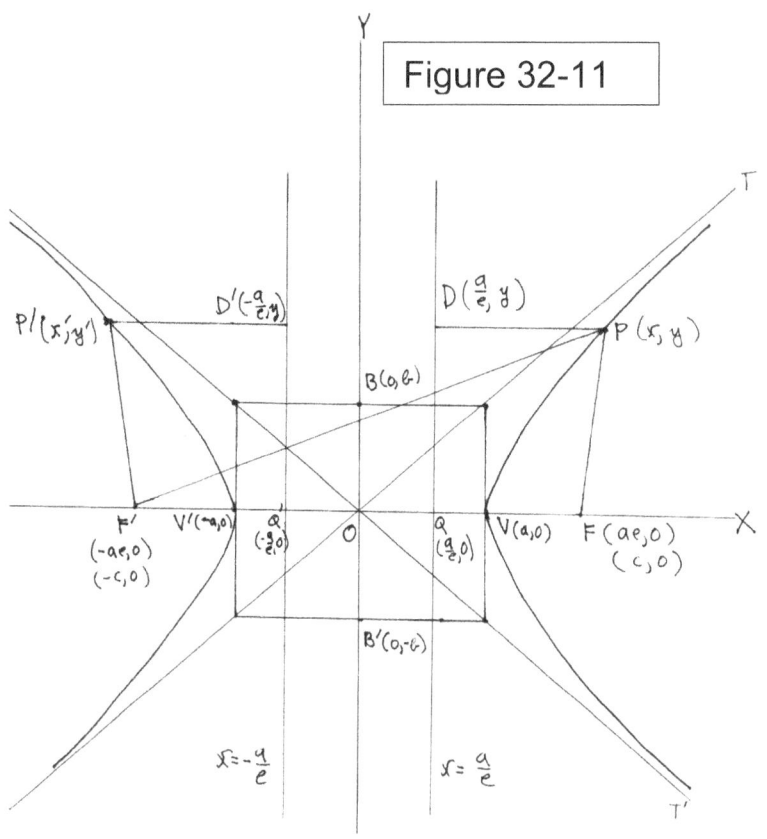

Figure 32-11

The important coordinates for the hyperbola and the lines related to it are shown in the figure, and we will explain how they are obtained in the following discussion. The fundamental property from the focus directrix definition is that for any point P (x, y) on the hyperbola the following ratio holds: $|FP|/|PD| = e > 1$, or,

$|FP| = e(|PD|)$.

When P is on the X axis we designate P as point V with coordinates (a, 0). Point D becomes point Q on the X axis. Thus, $FV/VQ = e > 1$, which means $FV > VQ$.

We recall from the study of the ellipse that when the ratio, FV/VQ = e < 1, that line segment FQ is divided internally by the point V such that the ratio holds, and also that the segment can be divided externally by the point V' in Figure 32-9 such that FV'/V'Q = e < 1. We can use the same reasoning we used in Figure 32-9 to show there is a point V' that divides FQ externally in the same ratio, FV/VQ = e, when e is greater than 1. All we have to do is invert the ratio, FV/VQ = e > 1, to, VQ/FV = 1/e < 1. We then find point V' by the same method we used in constructing Figure 32-9. We then have,

VQ/FV = V'Q/FV' = 1/e < 1. By inverting the ratio and the inequality, FV/VQ = FV'/V'Q = e > 1.

We see in Figure 32-11 where V' is located when FQ is divided externally by V'. We then bisect segment V'V, and place the origin of the coordinate system at the midpoint of V'V. Since the coordinates of V are, (a, 0), then the coordinates of V' are, (-a, 0).

We designate the coordinates of F (c, 0). From Figure 32-11 we see,

64. FV = c − a, FV' = c + a, VQ = a − OQ,

V'Q = a + OQ.

Since FV = e(VQ), and FV' = e(V'Q), then

65. c - a = e(a - OQ), and, c + a=e(a + 0Q).

204

We subtract the second equation from the first in step 65.

66. (c-a) - (c + a) = e(a - OQ) - e(a + OQ),

c − a - c − a = - 2a = ea - e(OQ) - ea- e(OQ),

- 2a = - 2e(OQ), so, OQ = a/e.

Since the origin is point O, this means the coordinates of point Q are, (a/e, 0). The directrix parallel to the Y-axis through Q then has the equation x = a/e. We use the value, a/e, for OQ in the equation,

c - a = e(a - OQ), to obtain,

67. c - a = e(a − a/e), c −a = ea −a, or, c = ae.

Since we designated the coordinates of F as (c, 0) we now know the coordinates of F can also be given as (ae, 0).

We will next derive the equation for the hyperbola in terms of x and y. Since e > 1, we have, c = ae > a > a/e. Figure 32-11 confirms these relations, because F is to the right of V, and V is to the right of Q on the X axis. Compare this to the ellipse in Figure 32-8 where e < 1 .

For the hyperbola for any point P (x, y),

68. |FP| = e(|PD|). Using the distance formula and the coordinates, (ae, 0), for F,

69. **|FP|** = $\sqrt{(x-ae)^2 + (y-0)^2}$,

|PD| = (x – a/e)   . Then,

70.

$$\sqrt{(x-ae)^2 + y^2} = e(x - \frac{a}{e}) = ex - a.$$

Squaring both sides,

71. $x^2 - 2aex + a^2e^2 + y^2 = e^2x^2 - 2aex + a^2$,

$x^2 + a^2e^2 + y^2 = e^2x^2 + a^2$.

We subtract, $e^2x^2 + a^2e^2$, from both sides.

$x^2 - e^2x^2 + y^2 = a^2 - a^2e^2$,

$x^2(1 - e2) + y^2 = a^2(1 - e^2)$.

We divide both sides by $a^2(1 - e^2)$.

72.

$$\frac{x^2}{a^2} + \frac{y^2}{a^2(1-e^2)} = 1.$$

This looks like the equation we found for the ellipse. However, in this equation e > 1, while for the ellipse e < 1. For the hyperbola, e > 1, the term, $\dfrac{y^2}{a^2(1-e^2)}$, must be a negative quantity, since, $(1-e^2)$ will be negative, and both a^2 and y^2 are positive.

Recall c = ae. We square both sides to get $c^2 = a^2e^2$, and then subtract a^2 from both sides.

73. $c^2 - a^2 = a^2e^2 - a^2$.   Multiply by - 1.

$a^2 - c^2 = a^2 - a^2e^2$,

$a^2 - c^2 = a^2(1 - e^2)$.

Knowing that, $a^2(1 - e^2)$, must be a negative number for the hyperbola we let $a^2(1 - e^2) = -b^2$,

(b positive).   Then,

74. $a^2 - c^2 = -b^2$, or, $c^2 = a^2 + b^2$.

We now substitute, $-b^2$, for, $a^2(1 - e^2)$, in the equation for the hyperbola in step 72.

75.

$$\frac{x^2}{a^2} + \frac{y^2}{-b^2} = 1, \, or, \quad \frac{x^2}{a^2} - \frac{y^2}{b^2} = 1.$$

This agrees with the equation we found for the hyperbola in our work with Menaechmus.

76. Note, $-b^2 = a^2(1 - e^2)$, $b^2 = -a^2(1 - e^2)$,

207

$$b^2 = a^2 (-1 + e^2),$$

$$b^2 = a^2(e^2 - 1). \quad b = a\sqrt{e^2 - 1}$$, where e > 1, which makes the quantity inside the square root a positive number.   We recall the value of b for the ellipse when e < 1 was **b = $a\sqrt{1 - e^2}$** .

We could have chosen the focus F'(-ae, 0), and the directrix, x = - a/e, and derived the same equation for the hyperbola. This means the hyperbola like the ellipse has two foci and two directrices, and a center half way between V'V, the two vertices. Because both hyperbolas and ellipses have centers they are called central conics as opposed to the parabola which has no center.

If we let y = 0 in the equation for the hyperbola, we determine the points where the hyperbola intersects the X axis. We get $x^2/a^2 = 1$, or $x = a^2$, where $x = \pm a$ which as we already know are the x coordinate for V and V'.

If we let x = 0 the equation for the hyperbola becomes, $- y^2/b^2 = 1$, or, $y^2 = - b^2$.   This equation has no real number for a solution which means the hyperbola in Figure 32-11 does not intersect the Y axis.

Special names are given for lines associated with the hyperbola. The line through the foci of the hyperbola and the points V'V in Figure 32-11 is called the transverse axis of the hyperbola.

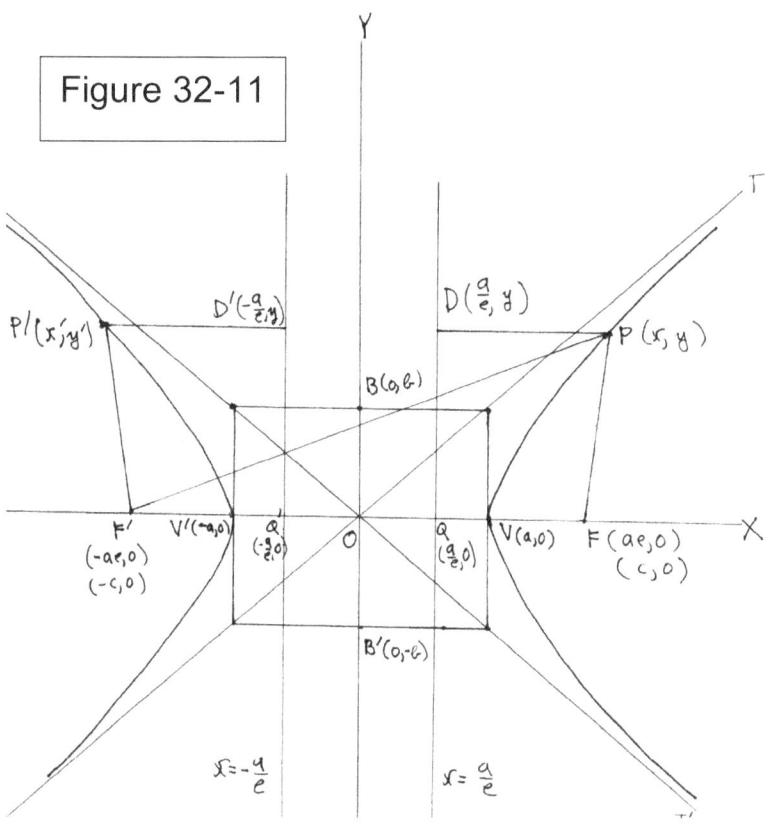

Figure 32-11

The line perpendicular to the transverse axis through the center of the hyperbola is called the conjugate axis.   In Figure 32-11 this is the Y axis. Points V'V where the hyperbola crosses the transverse axis are called the vertices of the hyperbola. The distance between the vertices is 2a, where a is the distance from the center to one vertex.

We designate points B and B' in Figure 32-11, which have coordinates (0, b) and (0, -b) respectively. We then draw perpendiculars to the Y axis through points B and B', and connect them to perpendiculars to the X axis through the vertices. Thus we have the rectangle in Figure 32-11.   We then draw the diagonals through the rectangle and extend them indefinitely. These diagonals go through the center of the hyperbola so we have lines OT and OT'. These lines are the asymptotes of the hyperbola as we shall shortly demonstrate. We first find the equations for these lines.

The Y intercept of line OT is the origin, O, and the slope of the line is, tangent angle TOV = b/a.

The equation of a line with slope m and Y intercept d is y = mx + d. In this case m = b/a, and d = 0. Thus,

y =(b/a) x , or, bx - ay = 0,   is the equation for line OT.

The equation of line OT' by the same reasoning is,

y = - b/a, or, ay + bx = 0.

77. We can multiply the two equations together (right side by right side, and left side by left side).

78. $-a^2y^2 + aybx - aybx + b^2x^2 = 0,$

$b^2x^2 - a^2y^2 = 0.$ We divide by a^2b^2.

$$\frac{x^2}{a^2} - \frac{y^2}{b^2} = 0.$$

This equation is the equation of the two asymptotes, and we note it is similar in appearance to the equation for the hyperbola, $x^2/a^2 - y^2/b^2 = 1.$

210

We wish to demonstrate that the two lines, y = b/a x and, y = - b/a x, meet the condition of asymptotes for the hyperbola with the equation,

$x^2/a^2 - y^2/b^2 = 1$.

The condition for a line to be an asymptote to the hyperbola is that the distance between the line and a point on the hyperbola approaches 0 as the distance of the point from the center of the hyperbola approaches infinitely large values.

Figure 32-12 shows the situation for the first quadrant.

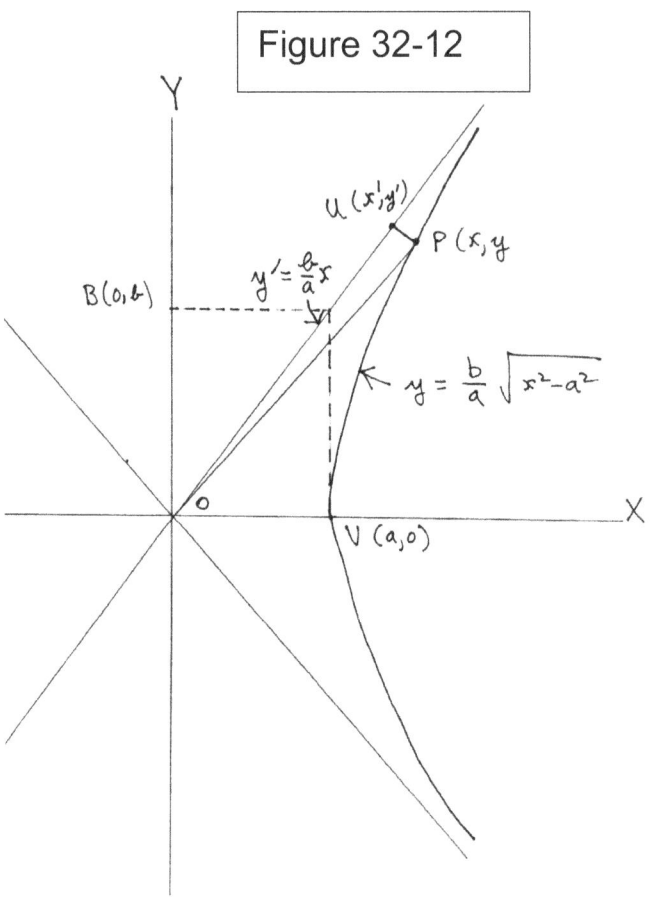

Figure 32-12

Point P(x, y) is a point on the branch of the hyperbola with equation $x^2/a^2 - y^2/b^2 = 1$. The distance from P perpendicular to the line y' = (b/a)x is shown as line PU, where U has coordinates (x', y').

The condition that y = (b/a)x is an asymtote is that the as distance OP increases indefinitely or approaches to the infinitely large, then distance PU approaches to 0 or the infinitely small. To demonstrate this for line y' =

(b/a)x we first solve the equation of the hyperbola for the branch in the first quadrant.

79. $x^2/a^2 - y^2/b^2 = 1$, $\quad y^2/b^2 = x^2/a^2 - 1$,

$y^{2=} = b^2 (x^2/a^2 - 1)$,

$$y^2 = b^2 \frac{(x^2 - a^2)}{a^2},$$

$$y^2 = \frac{b^2}{a^2}(x^2 - a^2).$$

For the first quadrant y and x are positive so,

$$y = \frac{b}{a}\sqrt{x^2 - a^2}.$$

We wish to show that the difference between this value for y and the value of, y' = (b/a)x, approaches 0 as x in both equations approaches the infinitely large. Thus we seek, (y' – y), as x becomes increasingly large.

80.

$$y' - y = \frac{b}{a}x - \frac{b}{a}\sqrt{x^2 - a^2} = \frac{b}{a}(x - \sqrt{x^2 - a^2}).$$

We now use the mathematical trick of multiplying the right side of the equation by,

$$\frac{x + \sqrt{x^2 - a^2}}{x + \sqrt{x^2 - a^2}}$$

, which is equal to 1, and thus does not change the value of the right side.

213

$$y' - y = \frac{b}{a}\{\frac{(x - \sqrt{x^2 - a^2})(x + \sqrt{x^2 + a^2})}{x + \sqrt{x^2 - a^2}}\}$$

$$y' - y =$$
$$\frac{b}{a}\{\frac{x^2 - x\sqrt{x^2 - a^2} + x\sqrt{x^2 - a^2} - (x^2 - a^2)}{x + \sqrt{x^2 - a^2}}\}$$

$$y' - y = \frac{b}{a}\{\frac{x^2 - (x^2 - a^2)}{x + \sqrt{x^2 - a^2}}\}$$

$$y' - y = \frac{b}{a}(\frac{x^2 - x^2 + a^2}{x + \sqrt{x^2 - a^2}})$$

$$y' - y = \frac{ba}{x + \sqrt{x^2 - a^2}}.$$

In examining this result we see as x becomes larger and larger, the denominator on the right side becomes larger and larger.   Since the numerator, ba, is a constant term for the hyperbola, then the quotient on the right side must become smaller and smaller and approach to zero as x approaches toward the infinitely large, and the difference between, y' −y, also becomes smaller and smaller approaching infinitesimally small. This is the condition required such that the line, y' = (b/a)x, is the asymptote in the first quadrant.   The same arguments can be used for the lines, y = (b/a)x, and y =

(-b/a)x, in the second third and fourth quadrants to show these lines are the asymptotes for the hyperbola, $x^2/a^2 - y^2/b^2 = 1$.

Just as the ellipse has a property involving the sum of the focal radii to any point, (|FP| = |F'P| = 2a), so too does the hyperbola. In Figure 32-11 we have connected a point P(x, y) to the two foci.

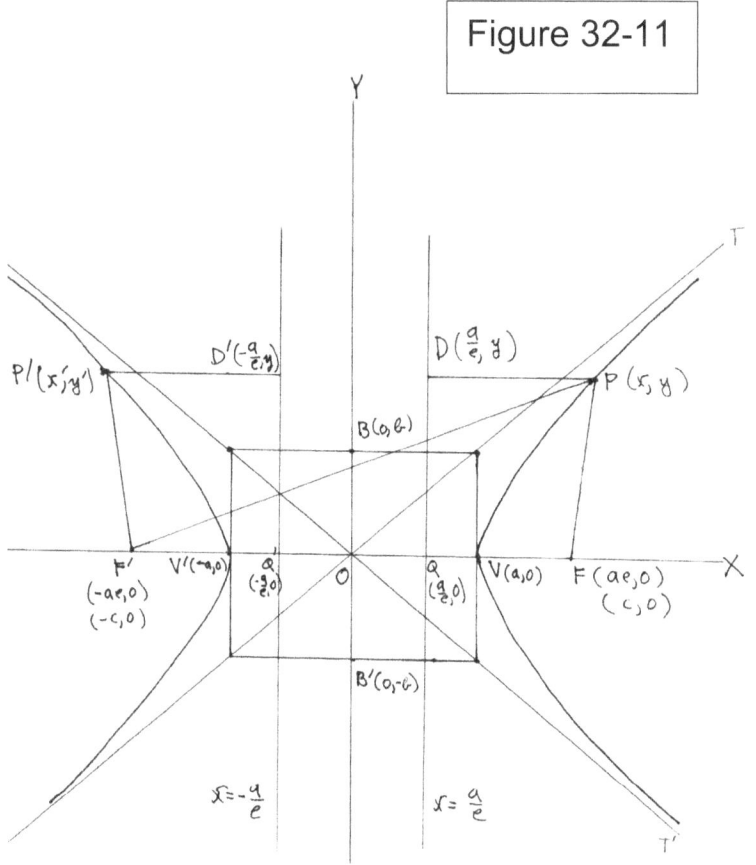

Figure 32-11

We found in our above work, |FP| = ex - a. We know,

82. |F'P| = e|D'P|, which is the fundamental

property for the hyperbola, e > 1.

|D'P| = [x - (-a/e)],   from Figure 32-11.

|D'P| = x + (a/e). Thus,

83. |F' P| = e(x + (a/e))

|F'P| = ex + a.

We now subtract |FP| from |F'P|.

84. |F'P| - |FP| = (ex + a) - (ex - a),

|F'P| - |FP| = ex + a - ex + a = 2a.

Therefore, every point on a hyperbola has the property that the difference in distance between the focal radii from the two foci to any point equals the constant value 2a, which is the distance between the two vertices. This property can also be used to define the hyperbola and derive its equation.

If we have the following two equations,

85. $x^2/a^2 - y^2/b^2 = 1$, and $x^2/a^2 - y^2/b^2 = -1$,

then the last equation can be written as,

$y^2/b^2 - x^2/a^2 = 1$,

so that it represents an hyperbola rotated 90° from the hyperbola in the first equation in Figure 32-11.

The two hyperbolas are called conjugate hyperbolas, and we show them together in Figure 32-13.

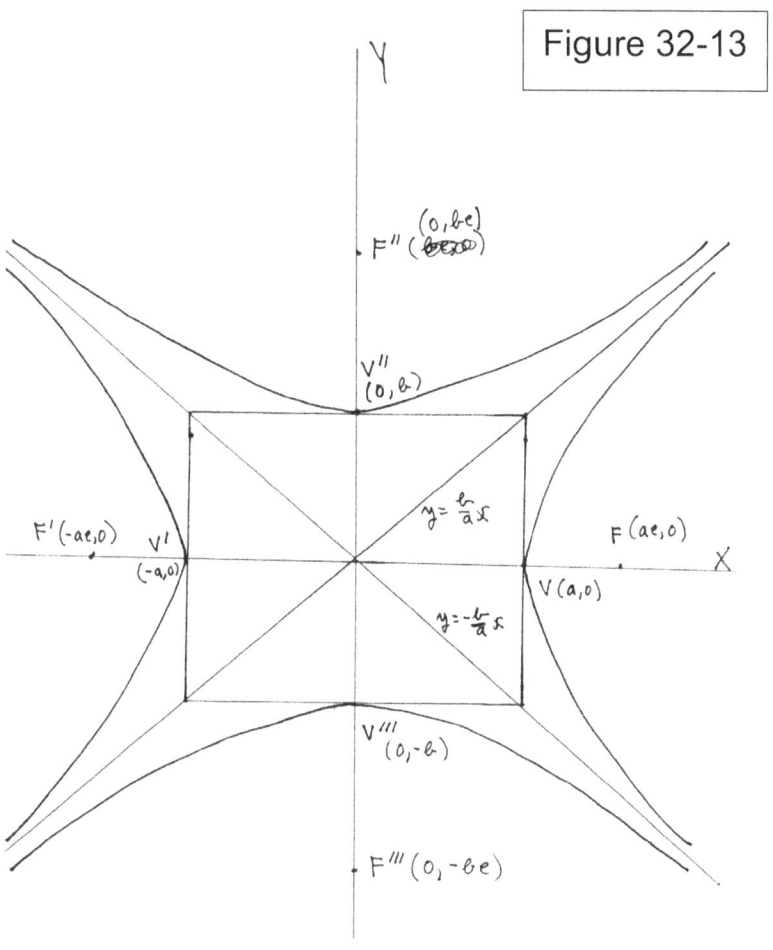

Figure 32-13

We recall also when $a^2 = b^2$, we have hyperbolas that Menaechmus discovered, namely rectangular hyperbolas. The equation for the asymptotes (step 78) becomes,

86. $x^2/a^2 - y^2/a^2 = 0$, $x^2 - y^2 = 0$,

$(x + y)(x - y) = 0$, or $x = -y$, and $x = y$.

217

This means the asymptotes for the rectangular hyperbola are perpendicular, which we also found in our work with Menaechmus. We recall when we derived the equation for the hyperbola from sectioning a cone with a plane with the orientation to the X and Y axes as shown in Figure 32-14, we found a rectangular hyperbola such that, xy = k, where k is a constant.

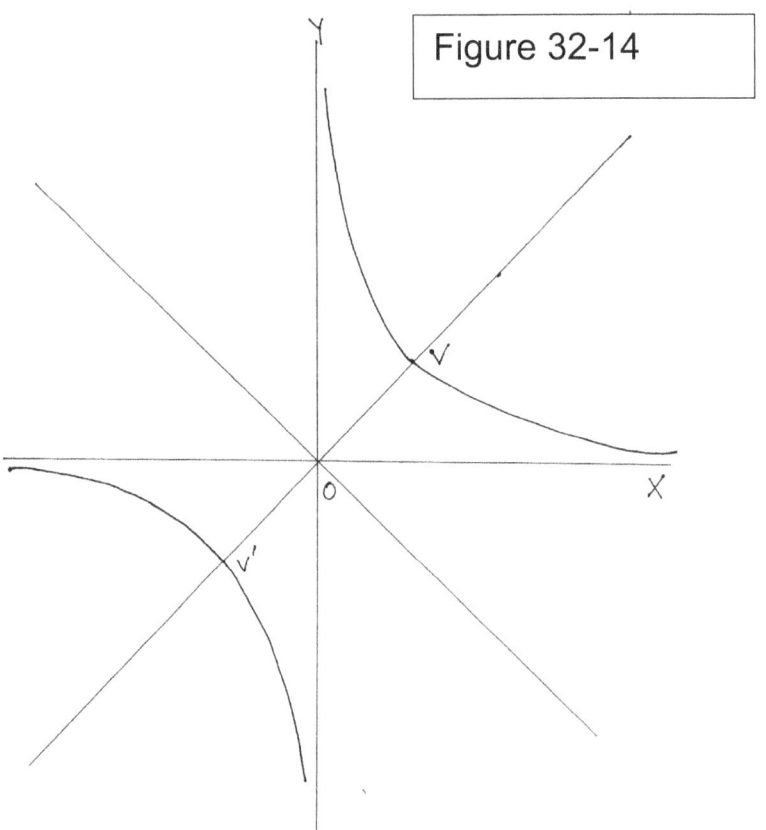

Figure 32-14

In the figure lengths, OV = a, and OV' = a.   It is possible to start with the equation for the rectangular hyperbola, xy = k, and then show by rotation of the transverse and conjugate axes 45° counterclockwise,

218

we will find the equation, $x'^2/a^2 - y'^2/a^2 = 1$, where, k = $a^2/2$, represents the hyperbola on an X'Y' coordinate system rotated 45° counterclockwise from the XY coordinate system as shown in Figure 32-15.

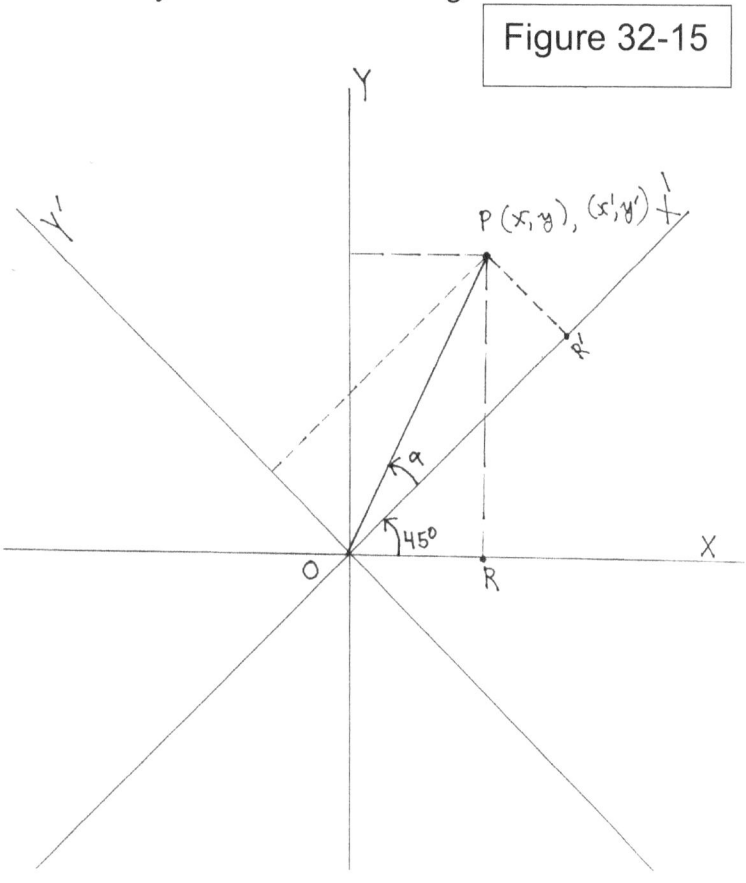

Figure 32-15

In Figure 32-15 we have rotated the XY coordinate axes 45° counterclockwise and indicate the X'Y' coordinate axes with the same origin at 0. Any point P on the plane then has two sets of coordinates; coordinates, (x, y) for the XY axes, and coordinates, (x', y') for the X'Y' axes. We connect OP from the origin to P, and note the angle between OP and the X' is

219

designated by the Greek letter alpha, α. We see from right triangle PRO,

87. cos $(45° + \alpha)$ = x/OP , and

sin $(45° + \alpha)$ = y/OP

x = (OP)cos $(45° + \alpha)$, and,

y = (OP)sin $(45° + \alpha)$.

From right triangle PR'O,

88. cos α = x'/OP, and, sin α = y'/OP.

x' = (OP)cos α , and y' = (OP)sin α .

We recall two trigonometric identities from chapter 13.

89. cos (A + B) = (cos A)(cos B) - (sin A)(sin B),

sin (A + B) = (sin A)(cos B) + (cos A)(sin B).

Thus,

90. x = (OP)(cos α)(cos 45°) - (OP)(sin α)(sin 45°), from step 87.

x = (x')cos 45°) - (y')(sin 45°), using step 88.

y = (OP)(sin a)(cos 45°) + (0P)(cos a)(sin 45°),

y = (y')(cos 45°) + (x')(sin 45°).

We recall from trigonometry that if A and B are complementary angles (A + B = 90°), then sin A = cos B.

Thus, since 45° + 45° = 90°, sin 45° = cos 45°. We recall also from chapter 4 that sin 45° = 1/√2, and thus cos 45° = 1/√2 . We use these values in the results for x and y in step 90.

91. x = x'//2 - y'/√2 , and y = y'/√2 + x'/√2 ,

or, rewriting, y = x'/√2 + y'/√2 . To find the equation of the rectangular hyperbola xy = k in the X'Y' coordinate system of Figure 32-15, we substitute the results of step 91 for x and y.

92. (x'/√2 - y'/√2)(x'/√2 + y'/√2) = k,

$$(\frac{x' - y'}{\sqrt{2}})(\frac{x' + y'}{\sqrt{2}}) = k$$

$$\frac{(x' - y')(x' + y')}{2} = k$$

(x' - y')(x' +y') = 2k,

x'^{2} - y'^{2} = 2k, and if k = a^2/2,

$$\frac{x'^{(2)}}{a^2} - \frac{y'^{(2)}}{a^2} = 1.$$

This is the equation of the rectangular hyperbola in the X'Y' coordinate system, and is of the same form for the orientation of a rectangular hyperbola with respect to the XY axes as the hyperbola in Figure 32-11.

For the hyperbola, $c^2 = a^2 + b^2$, or in the case of the rectangular hyperbola, $c^2 = a^2 + a^2$, or $c^2 = 2a^2$, or $c \pm /2a$.

The coordinates for the foci of the rectangular hyperbola are $(\sqrt{2}a, 0)$ and $(-\sqrt{2}a, 0)$. The equations for the asymptotes are, as we found above, the perpendicular lines, $x' = y'$, and, $x' = -y'$.

When the rectangular hyperbola is oriented so that its equation is $xy = a^2/2$, then the equations for the asymptotes are $x = 0$, and $y = 0$, which are the X and Y axes respectively as seen in Figure 32-14.

We have learned that the same basic equations for the conics can be derived starting with the intersection of cones with planes as in chapter 32, or by starting with the focus directrix property of conics derived in this chapter. We have learned many properties of conics, although there are many we have not discussed. When we reach the work of Kepler, Galileo, and Newton we will introduce other properties.

The ancient Greeks mathematicians expressed their 'equations' in terms geometric areas and lengths of lines. When we write for the ellipse,

$(PN)^2 = (AF)/(AA_1) \cdot (AN)(A_1N)$, with $AF/AA_1) < 1$, for Figure 32-3 the Greek mathematicians thought of $(PN)^2$ as a square area, and the other quantities as lengths of line segments.   In modern coordinate geometry the equation, $x^2/a^2 + y^2/b^2 = 1$, means a curve plotted on the XY coordinate system with x, y, a, and, b representing numerical values assigned by the coordinate axes.

In the next chapter we will show how to construct the conics on a plane using geometric properties of the conics. We will then look into work which suggests the early study of conics by the Greeks may have been inspired not only by the myth of King Minos and the solution of the problem of doubling the cube, but also by phenomena related to the study of sundials as astronomers plotted the shadow of the sun day by day.

Bibliography for Chapter 32

1. Morrill, W. K., "Analytic Geometry", International Textbook Company, Scranton Pennsylvania, 1964.

2. Heath, Sir Thomas, "A History of Greek Mathematics", Dover Publications, Inc., New York, 1981.

Chapter 33

Construction of Conics and the Connection of Conics
with Sundials

In this chapter we will first consider methods of geometric construction for the conic sections.   In geometry the ideal is to construct figures with the use of a straight edge and compass so as to mark lengths or make circles or circular arcs as we have seen in our work with the Greek geometers. Once coordinate geometry was developed, the algebraic equations for curves allowed the calculations of the coordinates of points which are on the curve given by the equation. The points calculated could then be connected by a line, and a sketch of the curve was the result.

For example, if we have the equation for the parabola, $y = (1/4a) (x^2)$, when $a = 1$, then, $y = (1/4)x^2$. We then can make a table of values for x and y which will be coordinates for points on the parabola when plotted on the XY coordinate system.

x	y
-.5	.0625
.5	.0625
-1	.25
1	.25
-2	1

| 2 | 1 |
| -3 | 2.25 |

x	y
-4	4
4	4
-5	6.25
5	6.25

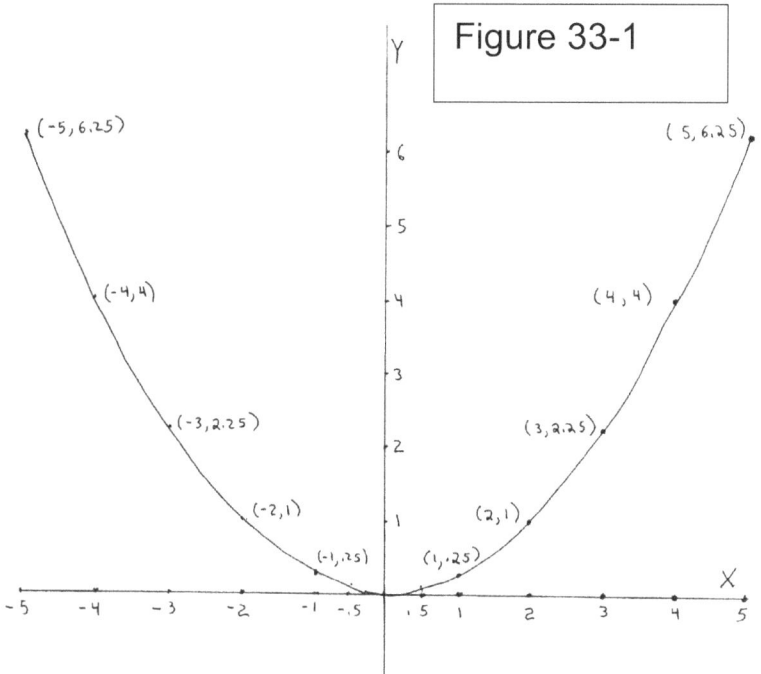

Figure 33-1

We graph the points and connect them with a smooth curve in Figure 33-1, which is the parabola given by the equation, $y = (1/4)x^2$.

If we are given the equations for an ellipse or hyperbola we can proceed in the same manner and

225

make a table of coordinates and then sketch the curve. We wish to show, however, that the conics can be sketched by geometric methods without requiring the calculation of coordinates on the curves. For the parabola this is done as follows making use of the focus directrix property.

1. See Figure 33-2.

Figure 33-2

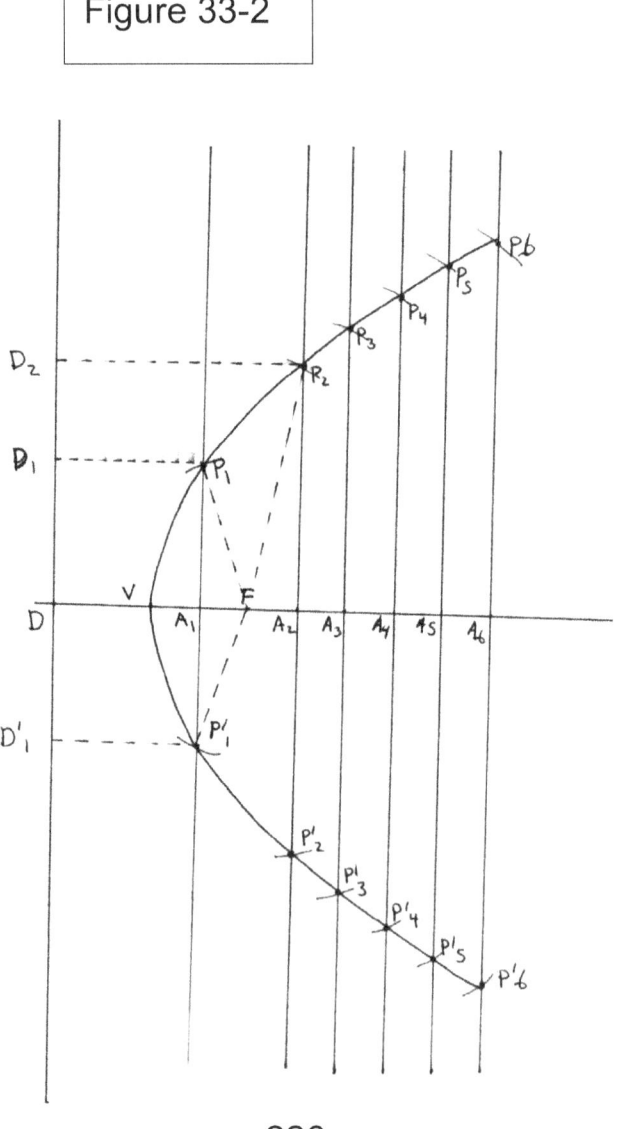

We first draw the directrix and choose a point D on the directrix.   We draw line DVF perpendicular to the directrix.   V and F are chosen so that FV = DV, which is the focus directrix property for the vertex, V, of a parabola. DV is also equal to length a, if the equation of the parabola is of the form $y = (1/4)ax^2$,   or $x = (1/4)ay^2$. The ratio, FV/DV = 1 = e, holds for the parabola.

2. We then choose a series of lengths DA_1, DA_2, DA_3, etc., so that each successive length is greater than the preceding one, and so that $DA_1 > DV$, etc.

3. We mark these lengths off on the axis DVF as shown in Figure 33-2. At each point A_1, A_2, A_3, etc., we draw lines perpendicular to the axis DVF as shown.

4. Starting with length DA_1 measured with a compass, we measure an arc with radius DA_1, and center at F so that the arc intersects the line perpendicular to DVF through A_1. The points of intersection are points P1   and P_1' shown in the figure. These points are on the parabola since they satisfy the property, $FP_1 = D_1 P_1$, and $FP_1' = D_1 P_1'$, which is the focus directrix property for a parabola, $(FP_1/D_1 P_1 = 1 = e)$.

5. This process is repeated again and again for the other lengths DA_2, DA_3, etc., to find more points on the parabola as indicated.   A smooth curved line then connects the points thus determined.   We see this method can be used to construct as many points as one wishes. In this method we did not require setting up the

XY coordinate system.

The geometric construction for the ellipse, which we will demonstrate, makes use of circles and lines as well as the trigonometric functions. See Figure 33-3.

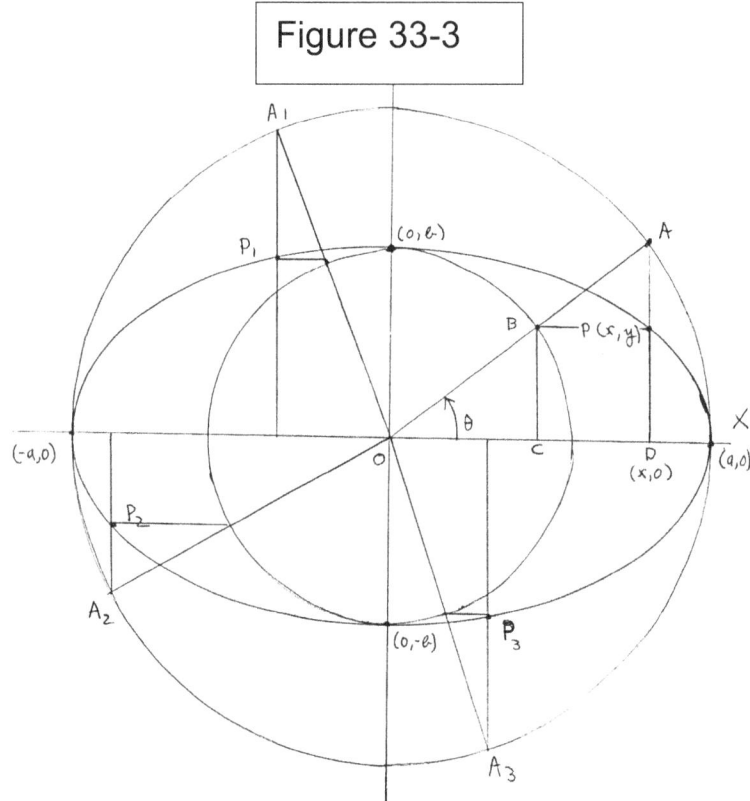

Figure 33-3

For this demonstration we must have the values of a and b in the equation for the ellipse, $x^2/a^2 + y^2/b^2 = 1$, where a > b.

6. We draw the XY coordinate axes, and we draw two circles with centers at the origin. The larger circle has radius equal to length a, and the smaller circle has radius of length b.

7. We draw around the circle a series of radii, OA,

OA$_1$, OA$_2$, OA$_3$, etc., all equal to length a. We will discuss the procedure for finding a point on the ellipse for one of these radii, OA.   Other points on the ellipse are found by using the same procedure.

8. Radius OA = a, intersects the smaller circle at point B. Perpendiculars are drawn to the X axis from A and B which intersect the X axis at C and D respectively.

9. A line is drawn through point B parallel to the X axis which intersects line AD at point P (x, y).   We claim P is a point on the ellipse.   All the other points P$_1$, P$_2$, P$_3$, etc., are found by the same process, and they also are on the ellipse. After enough points are determined the smooth curve connecting all the points, P, P$_1$, P$_2$, P$_3$, etc., is drawn and is the ellipse.

10. To demonstrate that, P (x, y), is on the ellipse we first designate angle DOA by the Greek letter θ, theta. We see from right triangle ADO,

cos θ =OD/OA = x/a, since a is the radius of the larger circle.

Squaring,

11. $\cos^2 \theta = x^2/a^2$.

12. We see from right triangle BCO,

sin θ = BC/OB = y/b.   Squaring,

13. $\sin^2 \theta = y^2/b^2$.

We add the respective sides of steps 11 and 13:

14. $\cos^2 \theta + \sin^2 \theta = x^2/a^2 + y^2/b^2$.

We recall from chapter 13 the trigonometric

229

identity, $\sin^2 \theta + \cos^2 \theta = 1$. Thus,

15. $x^2/a^2 + y^2/b^2 = 1$.   Since P has coordinates, (x, y), which satisfy the equation of the ellipse, then P is on the ellipse. By the same reasoning all the other points P_2, P_3 etc., are on the ellipse also.

To make a geometric construction for the hyperbola we use the property we derived for the hyperbola in the last chapter: $|F'P| - |FP| = 2a$.   We recall F and F' are the two foci, and 2a is the distance between the two vertices.   See Figure 33-4.

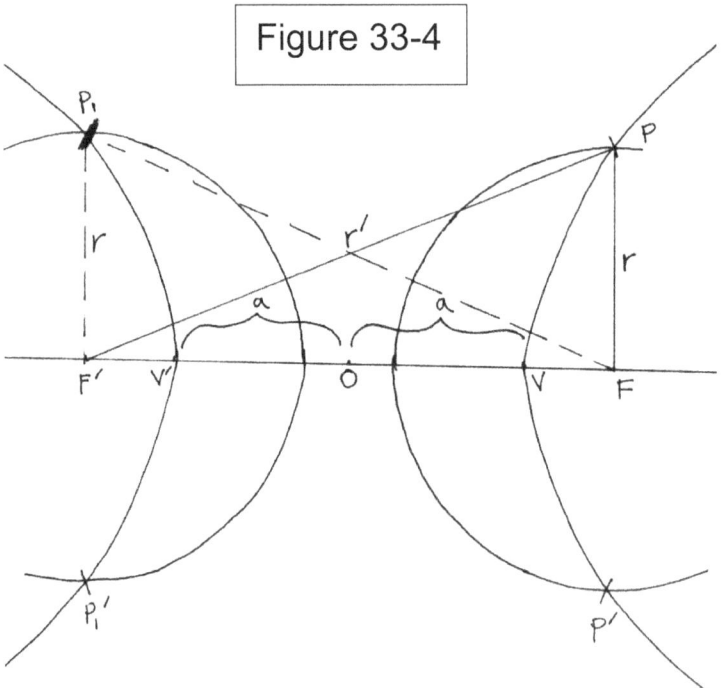

Figure 33-4

16. We draw the transverse axis for the hyperbola. We must be given for this construction the length a, and the value of e, which for the hyperbola is greater than 1. On the transverse axis we mark the center of the hyperbola as point 0.   Then points V and V' are marked

off to the right and left of point 0 such that OV = OV' = a. Next points F and F' are marked off from point 0 such that OF = OF' = ae. For example, if a = 1 inch, and e = 1½, then ae = 1½ inches.

17. We then choose two lengths, r and r' such that the difference, r - r', equals either - 2a or + 2a. In the figure we show the situation where,

r - r' = - 2a. Note this means that, r' = r + 2a. For example, if a = 1 inch then we could choose, r = ½ inch, and thus r" would equal, ½ + 2(1) = 2½ inches.

18. Then using F and F' as centers for radii r and r' respectively we draw circles as shown which are seen to intersect the perpendiculars to the axis from F and F' at P and P1 respectively. (We also show the intersection of circular arcs where radius r' has center at F, and F' making points of intersection P_1 and P_1' on the circle to the left and P and P' on the circle on the right.

19. By varying the lengths of r and r', we can make a series of such arcs and many more points of intersection can be found. All these points, such as P, P', P_1 P_1', etc., will be on the hyperbola because they all satisfy the criteria, |FP| - |FP'| = 2a.

The geometric constructions we choose for the conic sections all follow from the work we did with conics in the last chapter. In the case of the parabola the construction followed directly from the focus directrix property. In the case of the ellipse we needed the values of a and b from the equation of the ellipse. Of course, the equation for the ellipse is a result of the

231

focus directrix definition of the ellipse.

In the case of the hyperbola we used the value of the eccentricity, the value of a from the equation of the hyperbola, and the property of the hyperbola that the difference in lengths from the focal radii to points on the hyperbola equals the constant value 2a. This property was a consequence of a series of steps following from the focus directrix definition of the hyperbola. Knowing these constructions and practicing them with ruler and compass will improve our understanding of the conic sections.

In our study of Menaechmus we learned a motivation for the study of conics came from the Greek problem of doubling the cube, a problem that arose from ancient mythology. We learned Menaechmus only used right circular cones with variable vertex angles, and intersected the cones with planes perpendicular to the side of the generating triangle of the cone. This feature of Menaechmus' work (around 350 B.C.) suggested to the modern historian of mathematics and astronomy, Otto Neugebauer of Brown University, that the study of conic sections by early Greek mathematicians before Apollonius may have been inspired by work on sundials. The art of sundials was studied by the Greeks, and Neugebauer published a paper in 1948 (see the bibliography for this chapter) discussing the connection between sundials and the conic sections made from intersecting right circular cones by planes at right angles to the side of the generating triangles.

We have already shown how to design a sundial which indicates the hours of the day in chapter 19.   In that design the plane of the dial plate is parallel to the plane of the observer's horizon.   We will now discuss the design of Neugebauer's sundial and how it is connected to the conic sections.

In this sundial a gnomen is constructed perpendicular to the dial plate as in Figure 33-5.

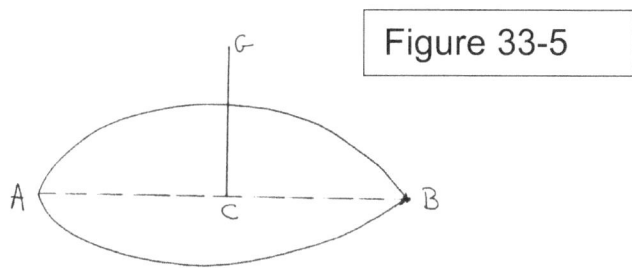

Figure 33-5

Angles ACG and BCG are right angles.   The dial plate is attached to a horizontal plane which can be tilted so that each day the gnomen points directly at the sun at the time of its meridian transit.   This phenomenon is shown in Figure 33-6.

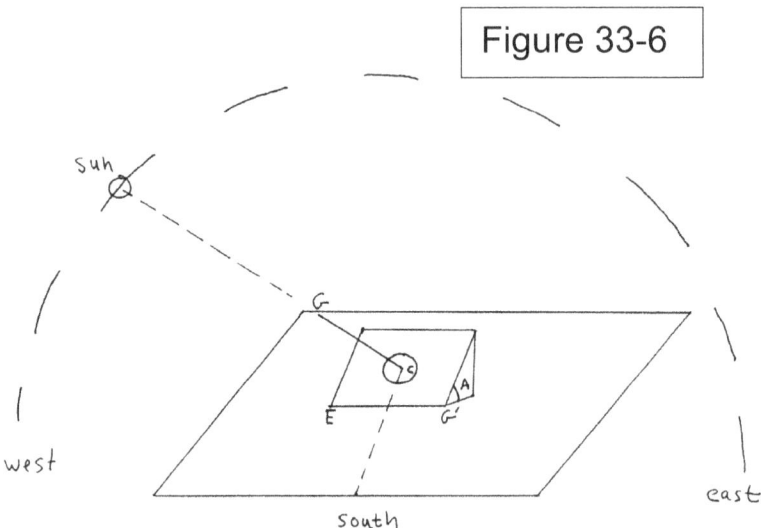

Figure 33-6

In Figure 33-6 angle A is the angle the plane of the dial plate is tilted from the horizon so that the line from the sun through GC is straight when it is 12 noon local solar time. This means angle A is the complementary angle to the altitude of the sun above the horizon at this time. This is shown in Figure 33-7 where we see the dial plate and horizon in profile.

Figure 33-7

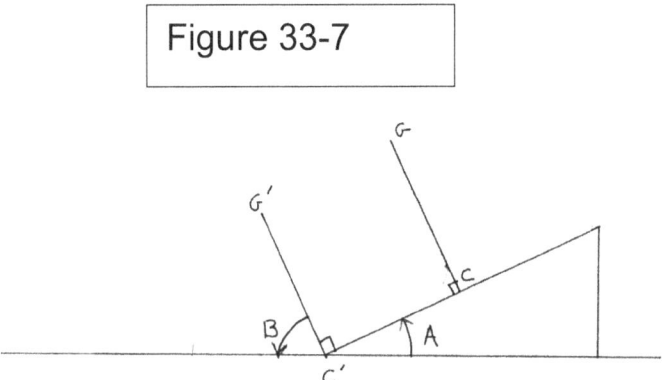

The gnomen, GC, is parallel to G'C', and both are perpendicular to C'C. We see angle B, the altitude of the sun above the horizon, plus 90° plus angle A equals

234

180°.   Thus, angle A + angle B = 90°.

The plane of the dial plate in Figure 33-6 is oriented so that EG' is perpendicular to the local meridian line. Since the altitude of the sun at meridian transit changes every day this sundial needs to have angle A adjusted each day so GC points to the sun at its culmination at meridian transit.

On the days of the equinoxes the sun is located on the celestial equator so that the shadow made by the gnomen traces out a straight line.   See Figure 33-8.

Figure 33-8

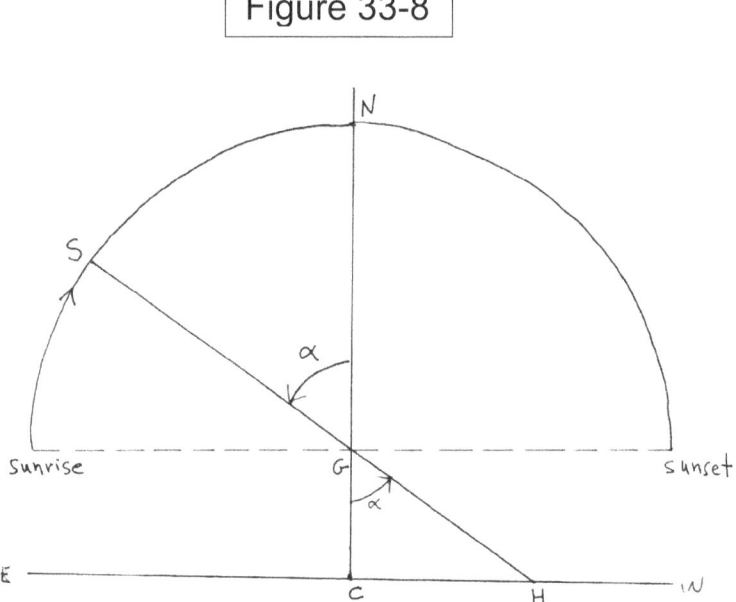

In this figure CG is the gnomen, and the path of the sun is shown as a circular arc across the sky from sunrise to sunset.   Line EW represents the path of the sun's shadow.   S is the position of the sun. Line SGH is a ray of sunlight which travels in a straight line. CH is the length of the gnomen's shadow when the sun is at S. Here the gnomen is kept at 90^0 from th horizon line.

235

We designate angle SGN as angle α (Greek letter alpha). If we let GC equal 1 unit length, then since a = angle CGH, tan α = CH/GC = CH/1 = CH.

This means by measuring CH, the length of the shadow made by a gnomen of 1 unit, we can calculate tan α, and thus α, the angular distance of the sun from its meridian transit, at any time of the day on the equinoxes. We know there are 12 hours from sunrise to sunset on the equinoxes and the length of CH can tell us which hour of the day it is at the time of measurement. But, the question arises, what about other days of the year?

We will look into the situation when the sun's declination is greater than 0°, or the time of year between the vernal and autumnal equinoxes. See Figure 33-9.

Figure 33-9

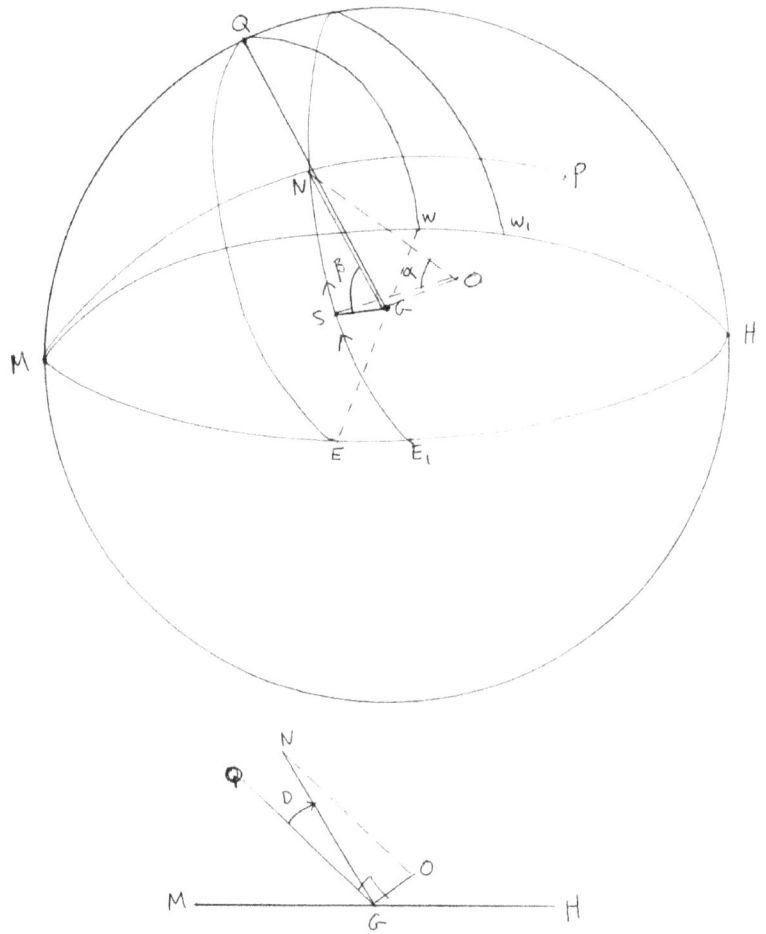

In the figure we are observing the celestial sphere and see the horizon plane of the observer as plane MEHW. The east west direction is EW, and the plane of the celestial equator, 0° declination, is EQW. The tip of the gnomen of the sundial is at G. The sun has a declination greater than 0^0, and it is at position S moving along arc E_1NW_1. The center of arc E_1NW_1 is point O. The meridian line is MNP where P is the north celestial pole, and N is the position of the sun at meridian passage.   When the sun is at position N, the gnomen is positioned such that it points directly at the sun and only has a point for a shadow.

The declination of the sun is angle QGN which we will call angle D and is shown in the lower drawing of the figure.   Now G can be regarded as the center for the circular arc EQW, which is some distance from point 0, the center of arc E_1NW_1, on which the sun travels on the day in question. On the days of the equinoxes point 0 and G do coincide, since on those days the sun moves on arc EQW. This is seen in the lower drawing of the figure.

We again call α the angle between the line from the position of the sun, S, in the figure before meridian transit to its center of apparent motion, 0, and the line from its position at meridian transit, NO. Thus, α = angle SON.

Arcs EQW and E_1NW_1 are on parallel planes. G is considered to be on the plane with arc EQW, and 0 is on the plane with arc E_1NW_1. We connect points G and 0 as line GO.   GO is perpendicular to the two planes since both G and 0 are centers of great circle arcs on the celestial sphere so radii GQ and ON are equal. Thus GO is perpendicular to GQ and ON. We note also ON and GQ are parallel.   This means angles QGN (angle D) and ONG are equal since they are alternate interior angles between parallel lines.

Consider angle SGN.   We denote this angle in the

figure as angle β (Greek letter beta). This angle together with the length of the gnomen determines the length of the shadow made by the gnomen at any time of the day. To make this clearer we redraw parts of Figure 33-9 as Figure 33-10.

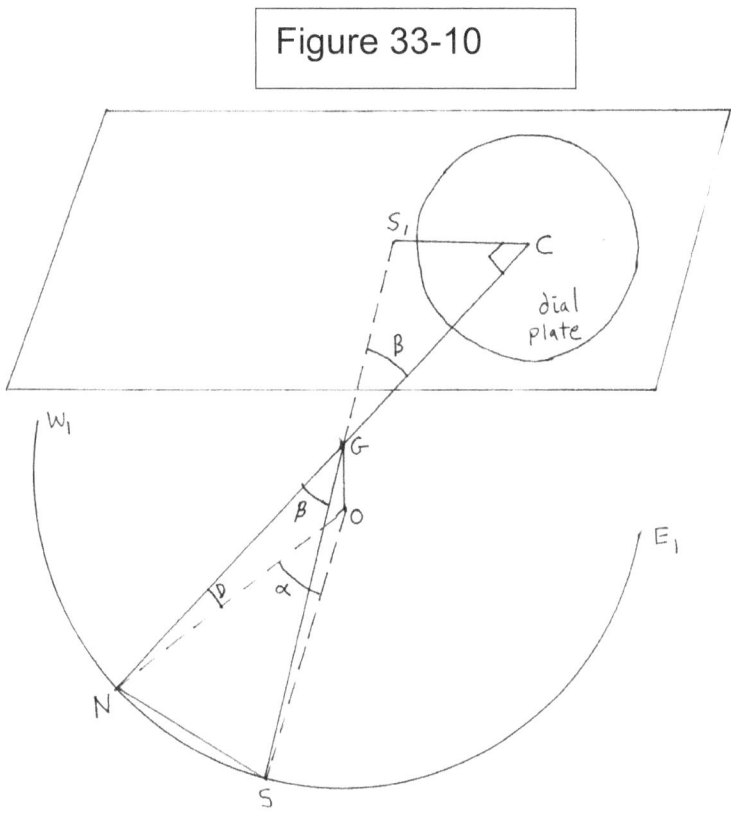

Figure 33-10

In Figure 33-10 we are looking at the plane of the dial plate from above and can see the gnomen GC. The dial plate is represented by the circle. When the sun is at S it sends a ray of light that touches G and reaches the dial plate at S_1. The gnomen blocks other rays of light and casts the shadow, line CS_1 on the plane of the dial plate. We can see the length of CS_1

depends upon length GC and angle β, (angle SGN). In the figure GO, α, and arc E_1NW_1 are the same as in in Figure 33-9.

We connect points S and N as chord SN. We see angle S_1GC equals angle SGN (β), because they are vertical angles. Since the gnomen, GC, is perpendicular to the plane containing S_1C, then angle S_1CG is a right angle. From right triangle GCS_1,

$\tan \beta$ = S_1C/GC, and since GC = 1 unit, $S_1C = \tan \beta$

.

In chapter 5 we learned how to convert chord lengths and angular measures of arcs into angles and their sines by the conversion formula (see Figure 33-11).

Figure 33-11

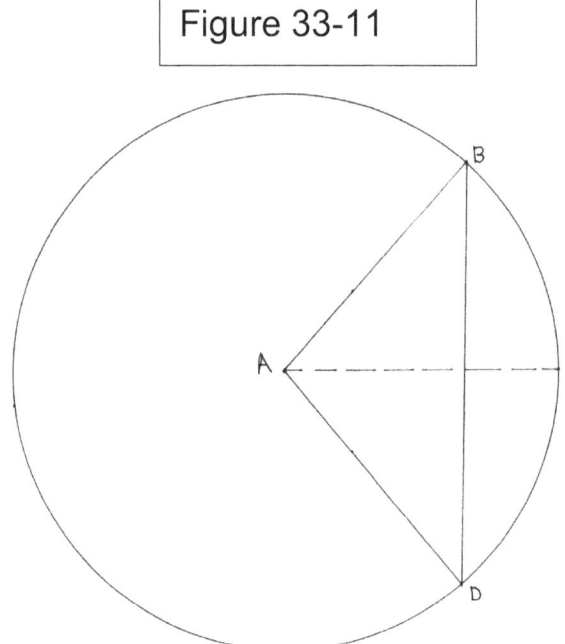

20. $\sin(\text{½ arc }BD) = \dfrac{\frac{1}{2}\,\text{chord } BD}{60\,units}$,

  where angle BAD = arc BD.

  If we let radius AB (60 units) = R, then

21. $\sin(\text{½ angle } BAD) = \text{chord } BD/2R$.

Returning to Figure 33-10 we see the center of the sun's path is O, and angle SON = α. Using the formula of step 21,

22. $\sin \tfrac{1}{2}\alpha = \text{chord } SN/2R$, where R = SO.

  $SN = 2R(\sin \alpha/2)$.

  We note angle ONG = angle D, the declination of the sun. From right triangle NOG,

23. $\cos D = NO/NG$, and since R = SO = NO,

  $NG = R/\cos D$.

We now use the conversion formula of step 21 for the arc SN on the circle with G as the center. In Figure 33-10 we know the center of the arc shown, E_1NW_1, is point O, but there can be drawn a circle with G as the center connecting SN so that chord SN belongs to the circle with G as the center. This is true because GN and GS are equal, and we take either one as the radius for the circle with G as center. Then β is the angle for angle BAD in step 21 : the value for SN in step 22.

26. $2R(\sin \alpha/2) = (2R/\cos D)(\sin \beta/2)$, or,

27. $\sin \alpha/2 = (\sin \beta/2)/\cos D$.

  The purpose of this equation is to find the value of α, the angular distance of the sun in Figure 33-9 from

241

its transit position at N, given the values of the sun's declination, D, for the day in question, and the length of the gnomens shadow, S_1C (figure 33-10). We find the value of β from the previously derived relation, $S_1C =$ tan β. We shall work an example to demonstrate the use of the equation.

Let us say we have a gnomen of length 20 inches which we call 1 unit. It is a day in the spring and we are at 36° north latitude and we know from the theory of the sun's motion of Ptolemy that the sun's declination is 15° north of the celestial equator. Thus angle D = 15°. We wish to know the angular distance of the sun from local noon when we measure the length of the gnomen1s shadow as 18.90 inches.

28. tan β = S_1C,

where S_1C is measured in units.

18.90 inches x1 unit/20 inches = .9450 units.

β = arctan .9450 = 43.38°.

Thus β/2 = 21.69°.

We use this result in step 27 along with the value for D of 15°.

29. sin α/2 = sin 21.69° /cos 15°,

sin α/2 = .3826/.9659

30. sin ½β = chord SN/2NG ,

where R in step 21 equals NG.

SN = 2(NG)(sin β/2).

For NG we use the result of step 23.

31. SN = (2R/cos D) (sin β/2),

where R = S0. We use,

α/2 = arcsin .3826 = 22.50°, or α = 45°.

If we are at 36° north latitude we can use our work from chapter 19 and Ptolemy's theory of the sun to find the length of time the sun is above the horizon.   The celestial longitude of the sun when its declination in the spring is 15° above the celestial equator is found by solving the spherical triangle in Figure 33-12.

Figure 33-12

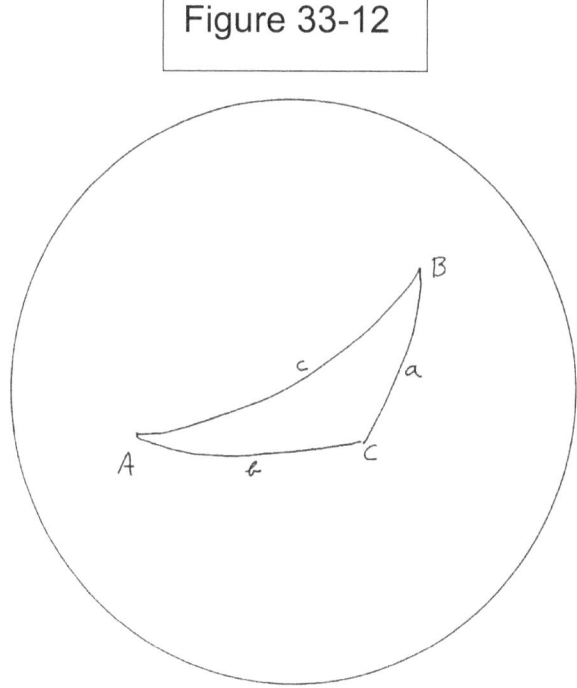

We refer to the methods in chapter 15 and Figure 15-4.

Figure 15-4

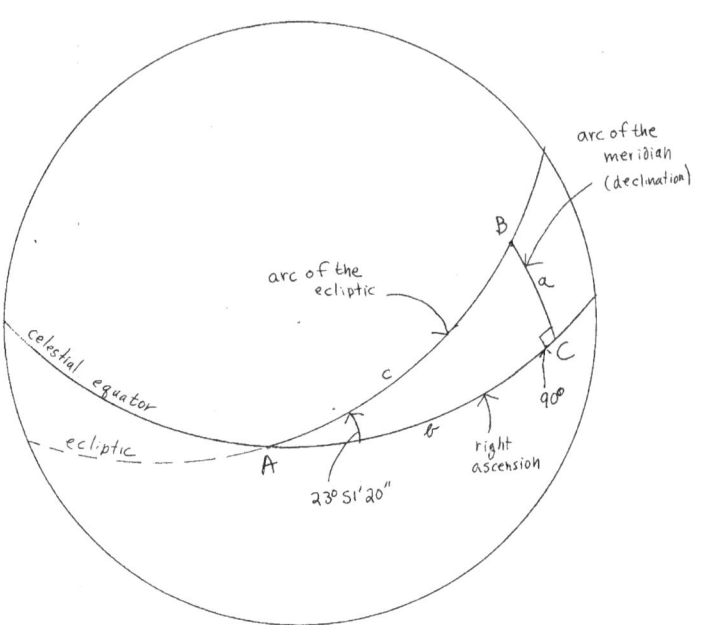

We use the law of sines for spherical triangles.

32. sin c = {(sin a)(sin C)/sin A}.

C = 90⁰, a = D = 15°, and

A = 23° 51' 20", the value for the obliquity of

the ecliptic given by Ptolemy.

33. sin c - {(sin 15°)(sin 90°)}/ sin 23° 51' 20" =

.2588/.4044 =.6400.

c = arcsin .6400 = 39.79°.

This means the sun is 39.79° east of the first point

of Aries, or about Taurus 10°.   From the table of

ascensions for 36° north latitude (Table II, Chapter 19)

when the sun is at 10° Taurus at sunset then Scorpius

10° is rising in the east.

The total time for Scorpius 10° = 228° 47', and total

244

time for Taurus 10° = 26° 13'. The arc of the sun's path in Figure 33-9 above the horizon is arc E_1W_1. This arc equals the difference in total times between Scorpius 10° and Taurus 10°, or 228° 47' - 26° 13' = 202° 34' = 202.57°.

The length of the time the sun is above the horizon is,

34. 202.57° x 1h/15 ° = 13.50 hours = 13 h 30 m.

Because α = 45° the time before local noon when the gnomen's shadow is 18.90 inches is,

35. 45° x 1h/15^0 = 3h. The time from sunrise until local noon is 13h 30m/2 = 6.75 h = 6 h 45 m.

3 h before local noon is 3 h 45 m after sunrise, and sunrise occurred at 12 h - 6.75 h = 5.25 h after local midnight, or 5:15 AM local apparent time.

Thus with Neugebauer's sundial, Ptolemy's theory of the sun and tables of ascensions we, and presumably the ancient Greeks could calculate the position of the sun in the sky on any day such as the one in the example by using a mathematical approach similar to what we have shown.

Thus far in our work with the sundial described by Neugebauer in his paper, we have not shown a connection with the path the shadow of the gnomen traces and conic sections. Our next task is to show this connection. To see this consider Figure 33-13, where we have shown the relations in Figures 33-9 and 33-10 as part of a cone and a plane intersecting a cone.

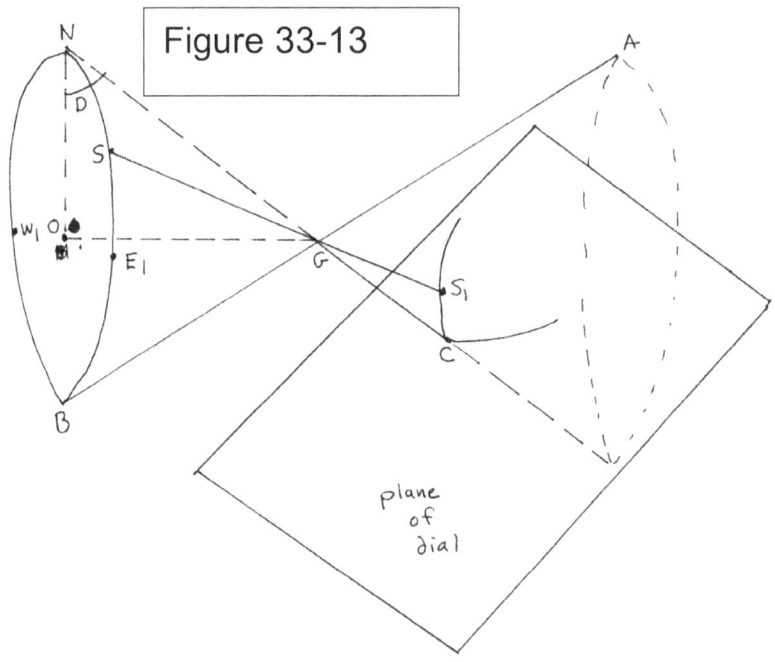

Figure 33-13

Here the gnomen GC is shown pointing to the position of the sun when it reaches its meridian transit at N. Thus, NGC is a straight line. We recall the dial must be adjusted almost daily so this occurs. In practice this means an adjustment is made if the shadow cast by the gnomen becomes more than a dot when the sun is at local noon. Since the declination of the sun changes little in most 24 h periods, morning measurements of the shadow lengths, as in the above example, can be made from the previous transit adjustment.

The arc of the sun's path in Figure 33-13 is E_1NW_1. We have completed the circle for this arc and shown how rays from the sun hitting the tip of the gnomen, G, would make a cone in a 24 h period if the sun did not set. Thus, we can think of the sun's ray SGS_1 as a generating line for the two nappes of the cone in the figure.

Now the earth blocks out the generation of the cone, and in the vicinity of the dial we have a horizontal plane intersecting the cone.   Furthermore, the plane of the dial plate intersects the cone perpendicular to the side of the generating line NGC, because of the requirement for the gnomen to point to point N.   We know from our work with Menaechmus that when a plane intersects a cone at right angles to the side of the generating triangle a conic section is formed.   The conic section formed depends on the vertex angle of the cone. In Figure 33-13 the vertex angle of the cone is angle NGB, and ½ the vertex angle is angle NGO.   We can consider right triangle NOG the generating triangle for one nappe of the cone.

In comparing Figure 33-13 with Figure 33-10 we see angle GNO equals the declination of the sun, D. Angle NGO = ½ angle NGB, the vertex angle of the cone.   In right triangle NOG angle NGO = 90° - angle NGO, or angle NGO = 90° - D.   Then angle NGB, the vertex angle of the cone, = 180° - 2D.

If angle NGB > 90°, the conic section formed by the tracing of the gnomens shadow will be a hyperbola.   If angle NGB = 90°, the conic is a parabola, and if angle NGB < 90°, the conic is an ellipse.

For Neugebauer's sundial the maximum value of D is , according to Ptolemy,   23° 51' 20", so the maximum value for angle NGB = 180° - 2(23° 51' 20") = 132.29°. This situation occurs on the solstice, so the shadow tracing is a hyperbola.

If we use the value of D on the equinoxes, 0°, the minimum value, then angle NGB = 180°, and a cone is not formed. This means the shadow of the gnomen follows a straight line as we discussed above.

The sundial as described by Neugebauer will result in shadow tracings of hyperbolas, and the straight lines on the solstices. However, the sundial we described in chapter 19 where the dial plate is parallel to the observer's horizon can have the gnomen trace out ellipses, parabolas, or hyperbolas depending on the latitude of the observations and the time of year. Above the arctic circle, between the vernal and autumnal equinoxes, for example, the sun is always above the horizon, so the shadow of the gnomen will trace out an ellipse.

We don't have writings from the ancient Greeks describing the relationships between sundials and the conics, but as Neugebauer argues in his paper, the study of sundials may have been very important in the early study of the conic sections, and may be one of the reasons Menaechmus chose right circular cones with planes intersecting the cones perpendicular to the sides of the generating triangles in his work with conic sections.

Bibliography for Chapter 33

1. Morrill, W. K., "Analytic Geometry", International Textbook Company, Scranton Pennsylvania, 1964.

2. Neugebauer, Otto, "The Astronomical Origin of the Theory of Conic sections", Procedings of the American Philosophical Society, Vol. 92, No. 3, July, 1948, Reprinted in: Neugebauer, Otto, "Astronomy and History: Selected Essays", Springer Verlag, New York, 1983.

Chapter 34

Tycho Brahe

In the last chapters we investigated the fundamental properties of the several conic sections as developed both by the Greeks and by the techniques of analytic geometry.   As we have learned neither Ptolemy, nor Copernicus considered celestial motion in terms of paths other than circles or combinations of circles.   Until Johannes Kepler in the early 17th century mathematicians could not resolve all the irregularities of the motion of Mars by combinations or circles.   The planetary tables based on Copernicus could not give accurate future positions for Mars when consulted over long periods.   However, the idea of using conic sections as possible solutions to the problems of celestial motion had not occurred to mathematicians or astronomers until Kepler.   Kepler had started his work as a Copernican and accepted the hypothesis that circular motions were required to explain the appearances of celestial motion. By 1600 he realized that to improve planetary theory he needed more accurate data on the eccentricities of the planetary orbits, data which was in the hands of Tycho Brahe, the great Danish astronomer, who had moved his headquarters to a castle near Prague in central Europe. When Kepler finally worked with Tycho, and after Tycho's death, using Tycho's data on the planet Mars,

Kepler was to discover that an elliptical orbit for Mars was the solution that best fit the data of Tycho Brahe, and to explain the irregular motion of the planet.

In this chapter we shall give a brief presentation of the work and career of Tycho Brahe.   After the publication of 'De Revolutionibus' in 1543 an associate of Rheticus, Eramus Reinhold, from the university at Wittenburg in Germany, used Copernicus' work to produce a table of solar and planetary positions in 1551. This work was done at the expense of the Duke of Prussia and was known as the Prussian Tables or the "Tabulae Prutenia". These tables were more accurate than the older Afonsine Tables.   However, the prediction of planetary positions was still in considerable error, and this was noted by the young Tycho Brahe, who was to become the greatest observational astronomer before the invention of the telescope.

Tycho was born in 1546 in Kundstrup in the Danish province of Scania.   He was of noble birth which later allowed him access to the patronage of the Danish king.   He was well educated at the University of Copenhagen beginning at age 13.   At first he was to follow a political career, but in 1560 he observed a partial eclipse of the sun which resulted in his change of interest to mathematics and astronomy.

In 1563 he observed a conjunction of Jupiter and Saturn and compared the time of its occurrence with the predictions in the Alfonsine and Prussian tables. He found the Alfonsine tables had miscalculated the conjunction by a month, and the Prussian tables, although more accurate, were still off by several days. He became convinced astronomy needed to be improved by new observational records of planetary positions so improved theory could be produced. He traveled greatly around Europe studying mathematics and astronomy and started to design new instruments for astronomical observations. One of his accomplishments was the construction of an enormous quadrant with a radius of 19 feet. This would allow the altitudes of celestial objects, the planets stars, and the sun to be measured within an accuracy of 1 to 2 minutes of arc. Later after he had constructed his observatory, he had another large quadrant built as shown in Figure 34-1.

Figure 34-1

QVADRANS MVRALIS SIVE
TICHONICVS.

The quadrant in the figure was made of brass with a radius of six and three quarter feet. Sights are present which move along the arc of the quadrant and are seen at point F where the observer is standing about 20° below an imaginary horizontal line from a slit in the opposite wall from the quadrant. Tycho is seen at his desk pointing at the slit in the wall.   The quadrant was aligned with the local meridian and the altitude of the star or planet as it crossed the meridian could be read off the scales on the quadrant which are marked to allow readings with accuracy of about 2' arc.   The aperture in the slit could also be adjusted for more accurate readings.   Although clocks are shown in the figure, Tycho did not think the clocks of his day were of enough accuracy and rarely used them in his recordings.   The figure also shows Tycho's dog in the background as well as pictures of other instruments used at the observatory. The dog incidentally had been presented to Tycho by King James I of England on a visit and was highly regarded by Tycho.

Tycho convinced King Frederick and His Queen Sophia of Denmark to be his patron and finance the building of a great observatory known as Uraniborg in 1576 on the small island of Hveen in the sound between Denmark and Sweden.   Tycho organized the observatory like a small city and built many instruments many of which were much larger than those used by the ancients so he could obtain more accurate recordings. He made thousands of observations himself and with the help of trained assistants for the next 21 years.

To determine the angular distances between stars Tycho used either armillary spheres or sextants which were much larger than the instruments used by previous astronomers.   A large equatorial armillary sphere is shown in Figure 34-2, and worked much like the description we gave for the zodiacal armillary sphere described in chapter 14.

Figure 34-2

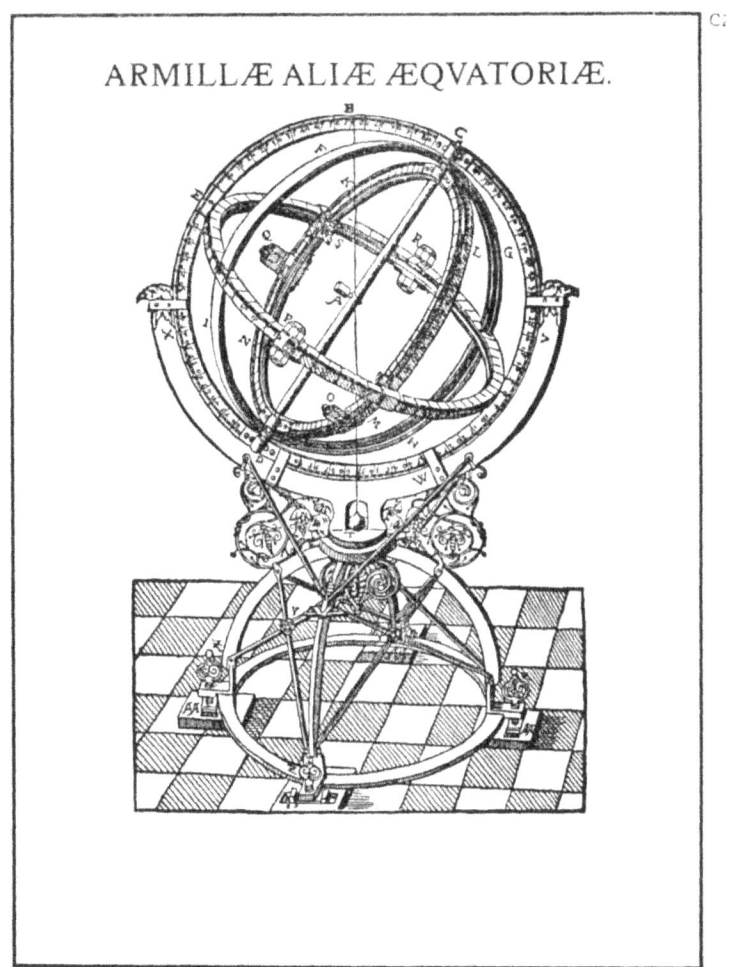

ARMILLÆ ALIÆ ÆQVATORIÆ.

EXPLICATIO

The equatorial sphere would give readings in right ascension and declination instead of celestial longitude and latitude.   Solving spherical triangles would allow Tycho to convert from one system to the other as we have learned.   A giant sextant is shown in Figure 34-3, and could be used by two observers simultaneously as shown in the upper right insert picture behind Tycho in Figure 34-1.

Figure 34-3

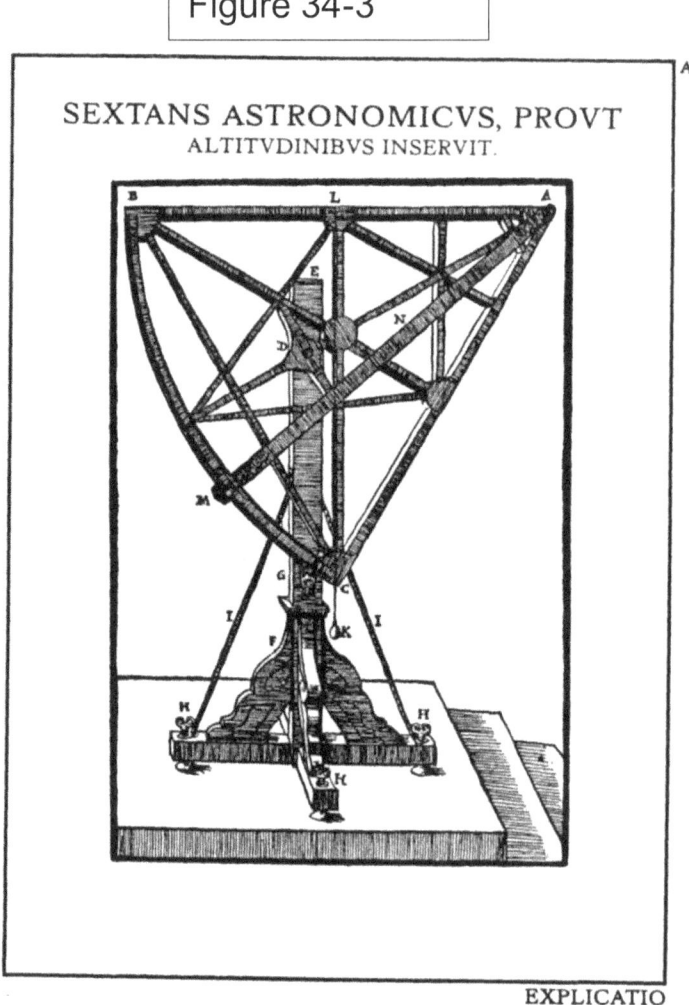

SEXTANS ASTRONOMICVS, PROVT
ALTITVDINIBVS INSERVIT.

EXPLICATIO

Figures 34-1, 34-2, and 34-3 are all taken from a book Tycho publishedin 1598, "*Astronomiae Instauratae Mechanica*".   All these great instruments were lost in

256

the years after Tycho1s death so we must rely on this book for the description of Tycho's instruments.

Tycho made a number of discoveries.   In 1572 before the observatory at Hveen was built, Tycho studied a new star that appeared in the constellation Cassiopeia.   This new star was very bright, brighter than Venus, and could even be seen for a time in the daylight hours.   In the days before telescopes such new stars were not considered to be stars at all by many, and were felt to atmospheric disturbances. Tycho proved, however, that this was a true star as far away as the celestial sphere.   He carefully measured the angular distances between the new star and other stars in the constellation Cassiopeia and found no measurable parallax, which meant the star was farther away from earth than the moon and planets which had detectable parallax.   The demonstration that new stars occurred in the celestial vault proved Aristotle's teaching that the celestial sphere was fixed and unchanging was incorrect.

Tycho also studied the motion of comets very carefully. Comets had been considered atmospheric disturbances since the time of Aristotle.   Tycho demonstrated their variable parallax which argued that comets orbit the sun.   He wrote a book on the comet observed in 1577, and in this book introduced his own planetary theory, the Tychonian system. This is shown in Figure 34-4.

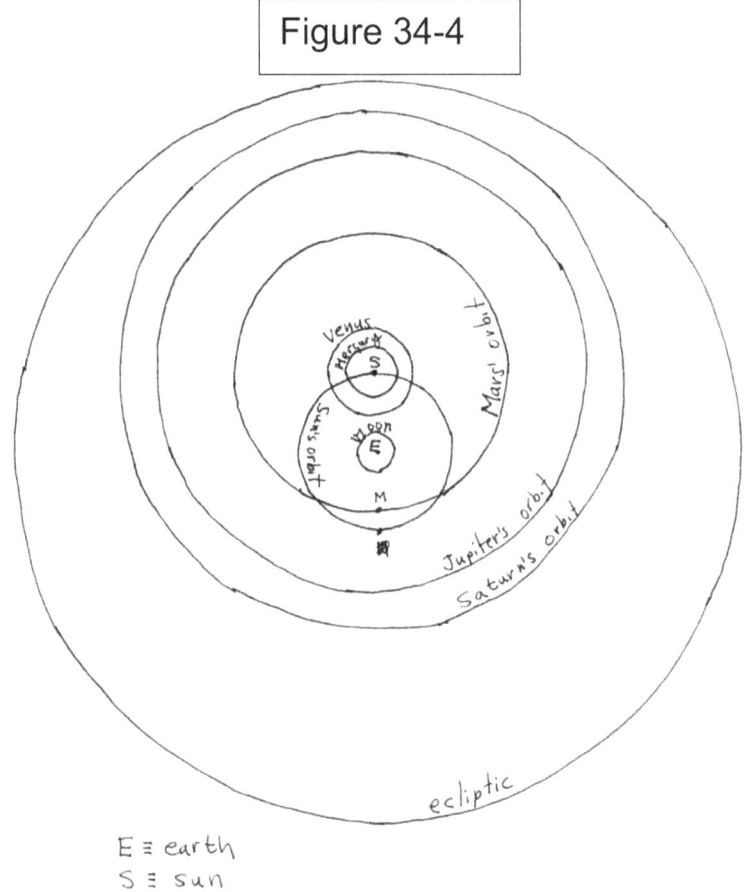

Figure 34-4

E ≡ earth
S ≡ sun

For Tycho the earth was stationary at the center of the universe with the moon orbiting the earth. Also the sun orbited the earth with the planets Mercury, Venus, Mars, Jupiter, and Saturn orbiting the sun. We see from the figure that because Mercury and Venus have orbits closer to the sun than the earth that the elongations of these planets are limited. Mars, Jupiter, and Satrun are so far from the sun and the earth that they can be seen in opposition to the sun.

Note that Mars is the only planet that can be closer to the earth than the sun for part of its orbit around the sun. This occurs when Mars is in opposition to the sun at points near M in the figure. Tycho had made careful observations of Mars for the period before, during, and after the opposition of Mars with the sun from Nov 1582 to Apr 1583. Tycho had accepted the value of 3' for the parallax of the sun used by ancient astronomers, and he came to believe that during opposition the parallax of Mars exceeded this value during oppositions with the sun, which meant Mars had moved closer to the earth than the sun. It appears that Tycho came to this conclusion some years after 1582 after reevaluating his data. Once it was established that Mars could be closer to the earth than the sun, it meant to Tycho the Ptolemaic system was incorrect. He rejected the Copernican system at this time because he could not detect the stellar parallax required by an orbiting earth as we have discussed. We recall in our discussion of the development of Copernicus' theory that Copernicus had rejected a Tychonian system for the outer planets, because to adopt it would mean the spheres to which the earth and Mars were attached would have to intersect. Once again it is the behavior of one planet, Mars, which leads the mathematical astronomers to favor one system over another.

In his book on the comet of 1577, Tycho also discusses the possible orbit of the comet and argues that its orbit is around the sun in an oval path and not a circular path or path made by a combination of circles. This is the first time a major astronomer had suggested abandoning the use of circles to explain the motion of a celestial object.

By 1594 Tycho's personality clashed significantly with numerous people particularly in the Danish court, and he began to lose support. King Frederick had died and eventually it was decided to cut his pension. By 1597 Tycho left Hveen with as many instruments as he could transport seeking a new patron. He found, after some search, support from the Emperor of Bohemia, Rudolph II, who allowed him to relocate at a castle some 20 miles from Prague called Benatek. He was appointed chief mathematician of the court, but he did not have the same grand backing that he had at Hveen.

He sought assistants with mathematical training who could do the calculations necessary to establish the numerical parameters of his planetary system. Johannes Kepler had corresponded with Tycho, and eventually Tycho invited Kepler to Benatek in 1599. We shall see in the following chapter how fortuitous this meeting was for the history of astronomy.

Tycho had also rejected the theory of trepidation, and presented good evidence that the value of the precession of the equinoxes is constant. He determined the value to be 51" of arc per year instead of the value determined by Ptolemy of 36" per year. He pointed out that astronomers had to consider many of the ancient observations in error due to inadequate care and inadequate instruments compared to the work he had done.

Tycho determined many parameters, particularly for the sun. He found the sun's apogee moves relative to the first point of Aries at 45" per year whereas Copernicus believed the value was 24" per year. The modern value is 61" per year. Tycho's solar tables were within 1' of subsequent observations of the sun's position on the ecliptic. The Prussian tables were incorrect by an error of 15' to 20'.

It was because of the extreme care Tycho took in his planetary observations, and the fact that he made observations at all points of the planet's progress around the ecliptic that allowed the new astronomy of Kepler and Newton possible. Kepler came to Tycho believing in the Copernican system and in circular motions. He needed Tycho's data to get better values for the eccentricities of the planets so he could refine and improve the Copernican system, but after working with Tycho's data, he found he had to abandon the long held belief that planetary motion had to be described by circular motions.

Our next task is to learn why Tycho's work led Kepler to the elliptical orbit for the planet Mars.

Bibliography for Chapter 34

1. Berry, Arthur, "A Short History of Astronomy", Dover Publications, Inc., New York, 1961.

2. Lodge, Sir Oliver, "Pioneers of Science", Dover Publications, Inc., New York, 1926.

3. Dreyer, J.L.E., "A History of Astronomy from Thales to Kepler", Dover Publications, Inc., 1953.

4. Dreyer, J.L.E., "Tycho Brahe", Adam and Charles

   Black, Edinburgh, 1890.

5. Brahe, Tychonis, "Astronomiae Instauratae Mechanica", Wandesburg, 1598.

Chapter 35

Kepler and the New Astronomy

Johannes Kepler was born on Dec 27, 1571 in the small town of Weil in the German province of Swabia in southwestern Germany. The area was predominantly Roman Catholic, but the Keplers were Protestants. Kepler was intellectually precocious from his youth. It was felt a career within the church would be suitable for him. He had numerous illnesses in his childhood, but survived them and was sent to monastic schools.   In 1588 he attended the University of Tubingen, a center of Protestant theology.

His career changed under the influence of the professor of mathematics, Michael Maestlin.   Kepler learned in private instruction from Maestlin the elements of geometry, trigonometry, and the conic sections as well as the theories of Ptolemy and Copernicus. Maestlin was a Copernican, and Kepler was duly impressed with the Copernican system.   Still in 1591 he considered the ministry his main career goal until the Protestant Estates of Styria applied to the University of Tubingen in 1593 for a lecturer in mathematics and astronomy for the school at Grantz, the capital of the Austrian province of Styria.   Kepler had two degrees from the university, and with misgivings accepted the appointment recommended to him by the university.   He accepted but with the stipulation that he could return to Tubingen to complete his theological study.

In 1594 he started his work with the young students. Kepler himself was only 23 at the time. In the first year at Gratz he had enough free time to pursue a thorough study of Ptolemy and Copernicus as well as Greek mathematics. As a result of his study bold new ideas and questions formed about the relationship between mathematics and astronomy. Why, he wondered, are the planets arranged as Copernicus asserted? He began to wonder about a physical cause for the arrangement. Further, he felt that God had designed the universe and the solar system according to a mathematical plan which man could learn and understand. Kepler conceived that it was his personal mission to discover the mathematical plan for the six planets in the Copernican theory. In particular he wanted to find the mathematical plan for the distances between the sun and the planets.

His lectures at Gratz were poorly attended partly because he became excited as he spoke so that he constantly digressed from his subject and became incomprehensible to his students.

Another duty given to him was the production of an astrological calendar to be used for forecasts. He considered this work unscientific, yet he needed to do it to earn his income. Also, Kepler had a mystical side to his personality, which led him to wonder that perhaps a true science of astrology might one day be created.

By 1596 his work had progressed to the point that he felt he had discovered the mathematical relationships for the distances of the planetary orbits from the sun. He decided to write a treatise on the subject which became his first book. Its title in English is, "Forerunner of the Dissertations on the Universe, Containing the Mystery of the Universe".   Today it is known as the, *Mysterium Cosmographicum*".   The book shows Kepler's profound faith in the mathematical design of the universe by God. He states he tried many mathematical devices before he found what he believed was the true theory for the distances of the planetary orbits from the sun.   He wrote,

"Finally I came close to the true facts on a quite unimportant occasion. I believe Divine Providence arranged matters in such a way that what I could not obtain with all my efforts was given to me through chance;   I believe all the more that this is so as I have always prayed to God that He should make my plan succeed, if what Copernicus had said was the truth".

We emphasize how differently Kepler proceeds to his discoveries than the modern scientist.   He is convinced his success or failure is in the hands of God and His Providence.   He forms theories first and then tries them on the data at hand using intuition and multiple trials to reach the theory he exposes.   He believed he was on a Divine Mission to reveal the truths that God had designed into the structure of the universe.

The 'true facts' which Kepler discovered and which led to his first theory of the universe came to him on July 19,1595 during a lecture he was giving on the conjunctions of Jupiter and Saturn.   As he thought about the distances of these planets from the sun in the Copernican theory, it occurred to him to circumscribe an equilateral triangle by a circle and then inscribe another circle inside the equilateral triangle. See Figure 35-1.

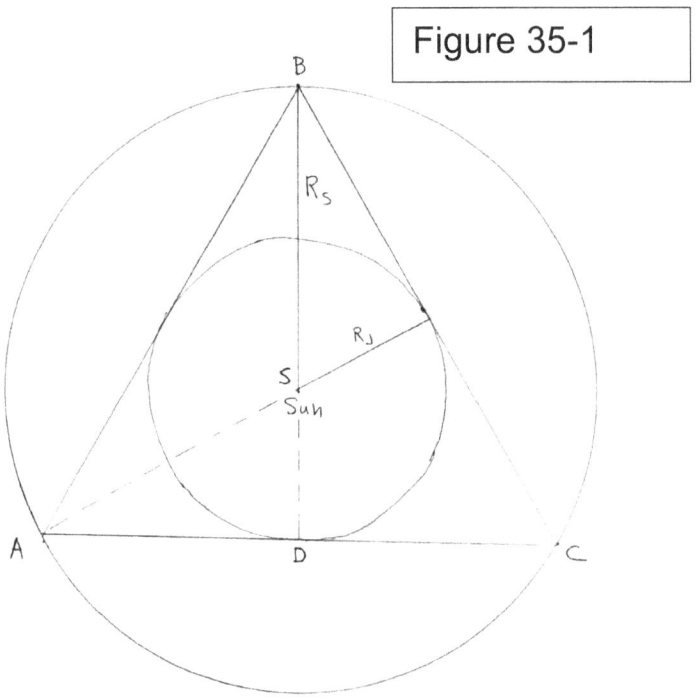

Figure 35-1

The circle which circumscribes the triangle represents the orbit of Saturn and the circle inscribed within the triangle represents the orbit of Jupiter.   The sun is located at the center of the two circles. The radius of Saturn's orbit is R_S, while the radius for Jupiter's orbit is R_J

In Chapter 29 we noted Copernicus found that if the distance from the earth to the sun is assigned the value of 1 unit, then, R_J = 5.22 units, and R_S = 9.19 units. Thus,

R_J/R_S = 5.22/9.19 = .56.

From our work in chapter 5 on equilateral triangles we found the medians of an equilateral triangle meet at a point such that the distance from the meeting point to a vertex of the triangle is 2/3 the length of the median. In Figure 35-1 this means,

BS = 2(BD)/3 . Since BD = R_S + R_J, and BS = R_S, then

$$R_S = \frac{2(R_S + R_J)}{3} = \frac{2R_S}{3} + \frac{2R_J}{3}, or,$$

$$3R_S = 2R_S + 2R_J.$$

$R_S = 2R_J$, $R_J/R_S = \frac{1}{2} = .5$

For Kepler this result was close enough to the value determined from Copernicus' work, .56, that he was sure a geometrical scheme was responsible for the varying distances of the planets from the sun. Kepler reasoned that the triangle is the first figure in geometry followed by the square and the regular pentagon. He then tried to inscribe a square within the orbit of Jupiter and then inscribe a circle within the square, this circle representing the orbit of Mars. Within the orbit of Mars a pentagon is inscribed, and within the pentagon a circle is inscribed representing the orbit of the earth. However, the ratios of the radii of these circles were not close to the ratios of the orbits from Copernicus' work so he knew this was not the correct geometrical scheme for the distances of the planetary orbits.

He then realized there are six planets and that Euclid in the "Elements" had demonstrated that there were only five possible regular three dimensional polyhedra know to us as the five Platonic solids after Plato the Greek philosopher. See Figure 35-2.

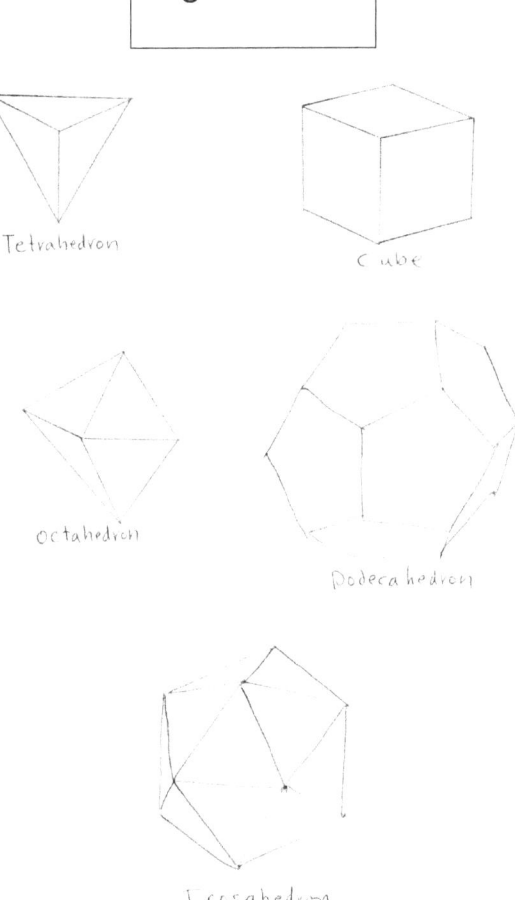

Figure 35-2

Tetrahedron

Cube

Octahedron

Dodecahedron

Icosahedron

The requirement for a three dimensional polyhedron to be a regular polyhedron is that all the edges of the solid be equal in length and that all the faces (two dimensional planes formed by the edges) must be regular polygons.   As a result of these requirements all the angles made by the intersection of the edges are equal.

The first regular polyhedron is called a tetrahedron. It has four faces each an equilateral

269

triangle. Next is a cube with six faces each a square. The third regular polyhedron is called the octahedron. It has eight faces (four are seen in Figure 35-2) each an equilateral triangle. The fourth regular polyhedron is the dodecahedron with twelve faces each a regular pentagon. The fifth and last regular polyhedron is the icosahedron with twenty faces each an equilateral triangle.

The idea occurred to Kepler that it was not a coincidence that there are only six planets and only five regular polyhedra. (The planets Neptune, Uranus, and Pluto and the thousands of asteroids orbiting the sun between Mars and Jupiter were not discovered until several centuries after Kepler). By Sep 14, 1595, Kepler wrote to his teacher Maestlin that he had formed a geometric system relating the distances of the six planets from the sun to the five Platonic solids. It was this scheme which is revealed in the "*Mysterium Cosmographium*" of 1596.

Kepler's geometric plan is shown in the famous figure from the book shown in Figure 35-3.

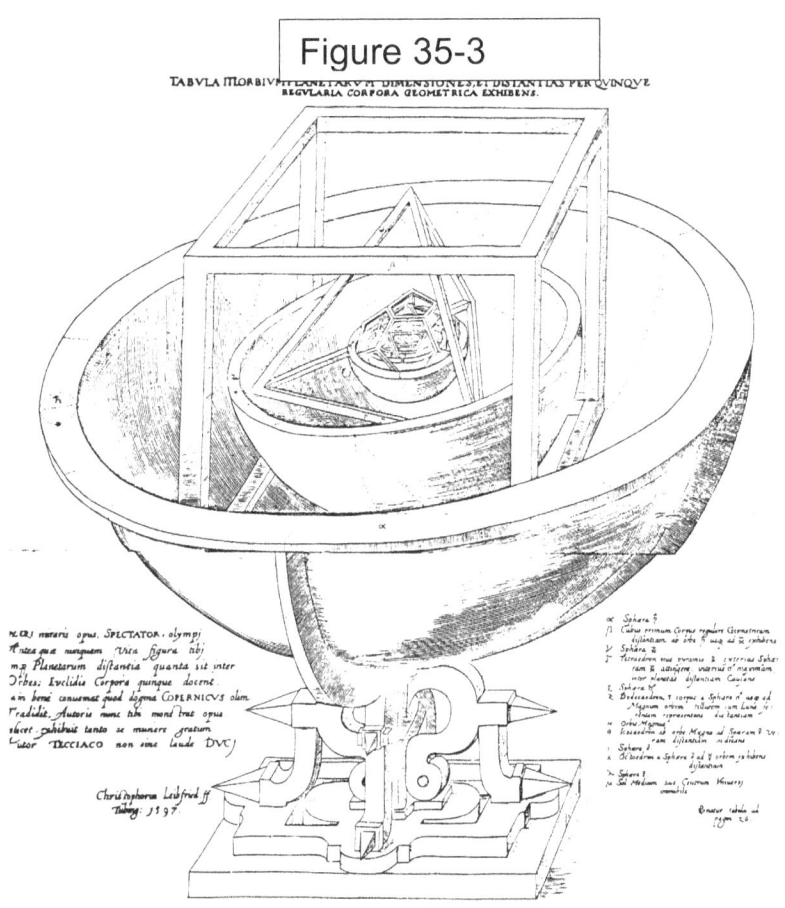

Figure 35-3

TABVLA MORBIV... PLANETARVM DIMENSIONES, ET DISTANTIAS PER QVINQVE
REGVLARIA CORPORA GEOMETRICA EXHIBENS.

Kepler envisions the orbit of Saturn contained within the outer shell of the large outer sphere. The sphere has a thickness determined by the eccentricity of the planet's orbit with respect to the sun. Within the spherical shell containing Saturn's orbit is contained a cube such that the vertices of the cube touch the inner surface of the spherical shell containing Saturn's orbit. Inside of this cube is the next spherical shell which contains the orbit of Jupiter, the second farthest planet from the sun. This spherical shell is arranged so that the outer surface of the shell touches the inner faces of the cube.   The same process is continued with the Platonic solids alternating with spherical shells so the following sequence results:

Saturn

CUBE

Jupiter

TETRAHEDRON

Mars

DODECAHEDRON

Earth

ICOSAHEDRON

Venus

OCTAHEDRON

Mercury.

Thus between each of the planetary spherical shells are five intervals such that the five regular polyhedra fit within them as Kepler conceives in Figure 35-3.   Kepler worked out the details of the mathematical ratios which we shall pass over.   A full discussion is found in the translation of the "*Mysterium Cosmographium*" given in the bibliography for this chapter.   Kepler was so enthralled by his results that he felt he had revealed the mystery of the universe.   He wrote in the preface,

"What delight I have found in this discovery I shall never be able to express in words...I made a vow to almighty God that at the first opportunity I would proclaim among men in public print this wonderful example of His wisdom".

272

Kepler also sought a mathematical relationship between the sidereal periods of the planets and the distances of each orbit from the sun. At this stage of his work he had no success, but the search for such a relationship would drive him on in his future work. Kepler wrote to Maestlin that the sun itself was responsible for the motion of the planets. He postulated an *'anima movens'* in the sun, by which he meant a source of force which causes the planets to move and which diminishes in intensity the farther the planet's orbit is from the sun. He called the effect of the sun's *'anima movens'* the *'virgo motus'*. He did not think of this as having anything to do with the force of gravity which we experience here on earth.

The idea of a physical cause for planetary motion was very important in Kepler's thinking. Unlike Ptolemy and Copernicus before him, Kepler felt a geometric interpretation of planetary motion was not sufficient. For Kepler the geometric theory had to be the result of the physical cause of the motion. This was one of Kepler's major contributions to the new astronomy of the 17th century.

It turned out that Kepler's polyhedral theory for the distances of the planetary orbits from the sun was incorrect.

In 1596, however, Kepler was sure the observational data that Tycho Brahe had collected would confirm his theory, particularly his data on the eccentricities of the planetary orbits, which Kepler would use to determine the thickness of his spherical shells. On Dec 13, 1597 Kepler sent Tycho a copy of his book. Although Tycho did not receive the book until March of 1598, Tycho was impressed with Kepler's mathematical abilities. He did not accept the polyhedral theory, but he wanted Kepler to visit him and work with his data. By 1599 Tycho had left Hveen, and set up his operations at Benatek near Prague, and by 1600 Kepler was able to come to Benatek to work with Tycho. Kepler wrote of the circumstance:

"Tycho Brahe, himself an important part in my destiny, did not cease from then on to urge me to visit him. But since the distance of the two places would have deterred me [that is when Tycho was at Hveen], I ascribe it to Divine Providence that he came to Bohemia, where I arrived just before the beginning of 1600, with the hope of obtaining the correct eccentricities of the planetary orbits".

Kepler was a Copernican in the sense he felt the sun was stationary and the earth moved, and did not accept the Tychonian system. Furthermore, Kepler became convinced that the true sun should be the stationary point from which to calculate all planetary motion rather than the center of the earth's orbit eccentric from the true sun in the Copernican system. The two innovations in Kepler's , which eventually led to the elliptical orbit from Tycho's data, were his hypotheses that the true sun was the stationary point for celestial motion, and that the physical cause of planetary motion arose from the true sun. But Kepler would probably not have found the elliptical orbit if he had not worked on the orbit of Mars. Other than Mercury Mars has the most eccentric orbit and the most irregular motion as seen from earth so that circular orbits show their failure most readily with a study of Mars. Fortunately for science when Kepler arrived at Benatek, Tycho and his assistant Christian Severin, known as Longomontanus, were working on their theory for the motion of Mars in the Tyconic system and had developed a theory for the planet's longitude based on a series of nine accurate observations of the oppositions of Mars from 1580 to 1600. They were about to observe a tenth observation in Jan 1600 soon after Kepler's arrival. Although Tycho claimed his theory could give the longitude of Mars to an accuracy within 2 minutes of arc, his theory did not give accurate values for the celestial latitude of Mars during the series of oppositions. Because of this problem Tycho assigned Kepler the task of studying the oppositions of Mars from Tycho's data. Kepler was to write of this decision:

"That is why I consider it again an act of Divine Providence that I arrived at the time when he was studying Mars; because for us to arrive at the secret knowledge of astronomy [that is the eventual truth that Mars follows an elliptical orbit] it is absolutely necessary

to use the motion of Mars; otherwise it would remain eternally hidden."

Of course this statement gives no hope that future astronomers and mathematicians would not have made the discovery.   At the time Kepler wrote this, Galileo had not yet revealed his work with the telescope which eventually would give even more accurate data on planetary motion.

In 1600 Tycho was 53 years old while Kepler was 29.   Longomontanus left Benatek to return to Denmark leaving Kepler to work out the theory of Mars. Moreover, Tycho needed Kepler's assistance for his major life's work: a new book of tables on the celestial motions of the planets and the sun to replace the inaccurate Alfonsine Tables and the "*Tabulae Prutenicae*".

The problem of an accurate theory for Mars such that it correctly predicted both the longitudes and latitudes for the 10 oppositions of Mars from 1580 to 1600 so that they agreed within 2 minutes of arc with those actually observed by Tycho proved to be a vexing one for Kepler.   Kepler had not resolved the problem by the time of Tycho's death in 1601.   On his death bed it is said Tycho asked Kepler to finish his work honoring his planetary theory.   To aid Kepler Tycho bequeathed to him all his observations.   Tycho repeated the words, "*Let me not seem to have lived in vain*".

After Tycho's death the Emperor of Austria Rudolph II appointed Kepler the Imperial Mathematician to succeed Tycho, a post which he held from 1601 to 1612. Unfortunately, Kepler was paid very little for this position, and had to struggle for some time to get full approval from Tycho's heirs to use Tycho's data in his books.

Kepler's major work as judged by most historians today, came from the work on Mars and occupied him from 1600 to 1606. This book, which was written in Latin, revealed what today we call Kepler's first two laws of planetary motion: (1) The planets move along elliptical orbits with the sun at one focus, (2) If a ray is drawn from the sun to a planet at any point on the orbit then the ray sweeps out equal areas in equal intervals of time. Kepler's third law, which we stated in the preface of this book, was not discovered and published by Kepler until 1614. The importance of Kepler's laws was fully apparent later in the 17th century influencing the work of Sir Isaac Newton. Arthur Koestler in his book, "The Sleepwalkers", wrote of the importance of Kepler's work for Newton,

"The modern version of the universe was shaped more than by any other single discovery, by Newton's law of universal gravitation, which in turn was derived from Kepler's three laws...Although some may never have heard of Kepler's laws, his thinking has been

molded by them without his knowledge, they are the invisible foundations of a whole edifice of thought."

Kepler's book on Mars, although ready by 1606, was not published until 1609, because of the disputes with Tycho's heirs, was titled

"*Astonomia Nova,* ΑΙΤΟΛΟΓΗΤΟΕ *Sev Physica Coelestis tradita Commentariis De Motibus Stellae Martis, Ex observationibus G.V. Tychonis Brahe*".

In English the title translates to "A New Astronomy Based on Causation or A Physics of the Sky derived from Investigations of the Motion of the Star Mars, Founded on Observations of the Noble Tycho Brahe".

The prodigious title is important in telling his readers that Kepler's premise is that a physical cause is responsible for the motion of the planets.   No prior astronomical book based on observations had ever made such a claim.   Furthermore, Kepler gives due credit to the observations of Tycho Brahe which were indispensable for the work.   It is this combination of observations and physical theory which led to the new astronomy.

The book's mathematical parts are difficult and at times obscure.   Yet the book takes up the problem of the irregular behavior of Mars as seen from earth and develops a comprehensive theory for its motion which is accurate with two minutes of arc for both longitude and latitude.   Kepler describes the motion both in terms of Ptolemy's theory, Copernicus' theory, Tycho's theory, and finally his own theory which he argues is the correct interpretation of Tycho's data.

Kepler asserts, as his point of departure from Copernicus and Tycho, that the physical cause of Mars' motion must be found in the true body of the sun, and that the body of the sun is stationary and the reference point of all celestial motion. He recognized as seen from the true sun planetary motion is not uniform, and in the final analysis is not circular nor a combination of circles such as deferent and epicycle. To reach these conclusions Kepler had to work through many laborious steps with theories he formed, but then rejected when the data from Tycho's oppositions did not fit the theory within the accuracy of two minutes of arc for the observations.   It is important to realize that Kepler never doubted that Tycho's observations were this accurate. He knew the work of Ptolemy and Copernicus was only accurate at most to ten minutes of arc, and often in error by many degrees of arc.

Kepler wrote in the preface to his book, which we will hereafter call the "*Astronomia Nova*",
"The first step in tracking down the physical causes of the motions was for me to demonstrate that the point common to the eccentricities is located not at some point near the sun, but at the very center of the solar body contrary to what Copernicus and Brahe believed..I

describe., by my method the positions of Mars in apparent opposition with the Sun then they [Copernicus and Brahe], using their old method, describe the positions of Mars in mean opposition with the Sun...I have proven very firmly that the eccentric of Mars is so placed that its line of apsides intersects the very center of the solar body, not some nearby point... Moreover this holds true not only for the longitude but also for the latitude".

Kepler also believed that the sun was the reference point of planetary theory because the sun was the physical cause of planetary motion including the earth.   He also wrote in the preface,

"The Sun's body is the source of power that drives all the planets.   I have reasoned that although the Sun remains in one position, it still rotates as in a lathe, and indeed emits from itself throughout the extent of the universe an immaterial emanation of its body, analogous to the immaterial emanation of the Sun's light. This emanation rotates with the solar body, like a very rapid whirlwind, covering the whole universe, and it carries around with it the bodies of the planets, with greater or lesser violence depending on whether it is denser or weaker there, according to the rules of its emission."

Notice the 'emission' from the sun is not interpreted by Kepler to be the force of gravity.   Kepler does gives us the feeling that the non-uniform motion of the planets as viewed from the sun is caused by a physical power within the sun, a bold viewpoint for a scientist to adopt.   Regarding the eventual discovery that the planets move in ellipses Kepler wrote in the preface,

"My exhausting task was not finished until I had devised the fourth step toward a physical hypothesis. Often very laborious demonstrations and the analysis of a large number of observations I noticed that the planets route in the sky is not a circle but an oval - perfectly elliptical".

How Kepler reached these conclusions is described in the chapters of the "*Astronomia Nova*". For Kepler the work was a five year struggle, which he called his war with Mars (Mars being the Greek god of war).   Kepler began his study by a careful consideration of the ten oppositions of Mars with the sun observed and recorded by Tycho from 1580 to 1600.   The first eight of these together with the supposed path of the planet as seen from a stationary earth in the theory of Ptolemy and Tycho are shown in Figure 35-4, which we used previously in Figure 23-1, and is from the "*Astronomia Nova*".

Figure 35-4

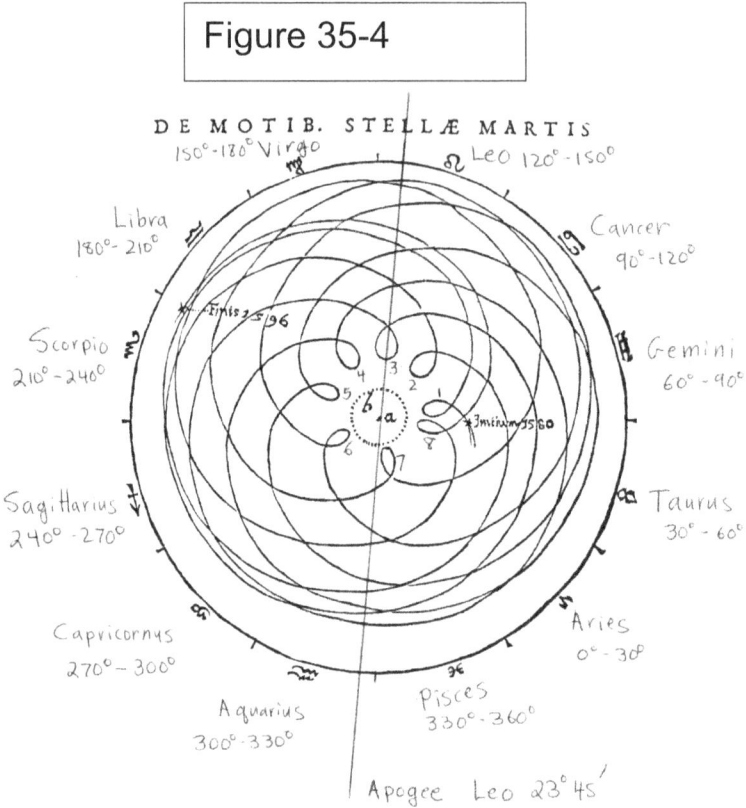

DE MOTIB. STELLÆ MARTIS

We view from the north ecliptic pole with the earth at a, the sun, Venus, Mercury, and the moon move within the circle designated by b, and the oppositions numbered 1 through 8.   The ecliptic is shown as the outer circle marked with the signs of the zodiac. We have written their names with their celestial longitudes. We give in Table 1 the data of Tycho for all 10 oppositions, including the time of occurrence, the celestial longitude as observed from earth, and the celestial latitude as observed from earth.   We number the oppositions 1 through 10.

Table 1

Date	Celestial Longitude	Celestial Latitude
1. 1580 Nov 17 9 h 40 m	66° 46' 10" (Gemini)	1° 40' 0" N
2. 1582 Dec 28 12 h 16 m	106° 46' 10" (Cancer)	4° 6' 0" N
3. 1585 Jan 31 19 h 35 m	141° 10' 26" (Leo)	4° 32' 10" N
4. 1587 Mar 1 17 h 22 m	175° 10' 20" (Virgo)	3°38' 12" N
5. 1589 Apr 15 13 h 34 m	213° 58'10"(Scorpio)	1° 6' 45" N
6. 1591 Jun 8 16h 25	266°32' 0"(Sagittarius)	3° 59"0" S
7. 1593 Aug 24 2 h 30 m	342° 43' 45" (Pisces)	6° 3' 0" S
8. 1595 Oct 29 21 h 22 m	47° 56' 15" (Taurus)	0° 5'15" N
9. 1597 Dec 13 13 h 35 m	92° 28' 0" (Cancer)	3° 33'0" N
10. 1600 Jan 19	128°18'0" (Leo)	4°30'50" N

In Figure 35-4 we have drawn the line of apsides that Tycho had determined for the beginning of 1585 at celestial longitude of 143° 45', or 23° 45' in the sign of Leo.

Kepler worked with the data on the celestial latitude to develop his theory for the celestial latitude of Mars based on the true sun.   To better understand the relationship between the celestial longitude and the celestial latitude for Mars as it moves through its retrograde arcs in opposition to the sun we have constructed Figure 35-5.

Figure 35-5

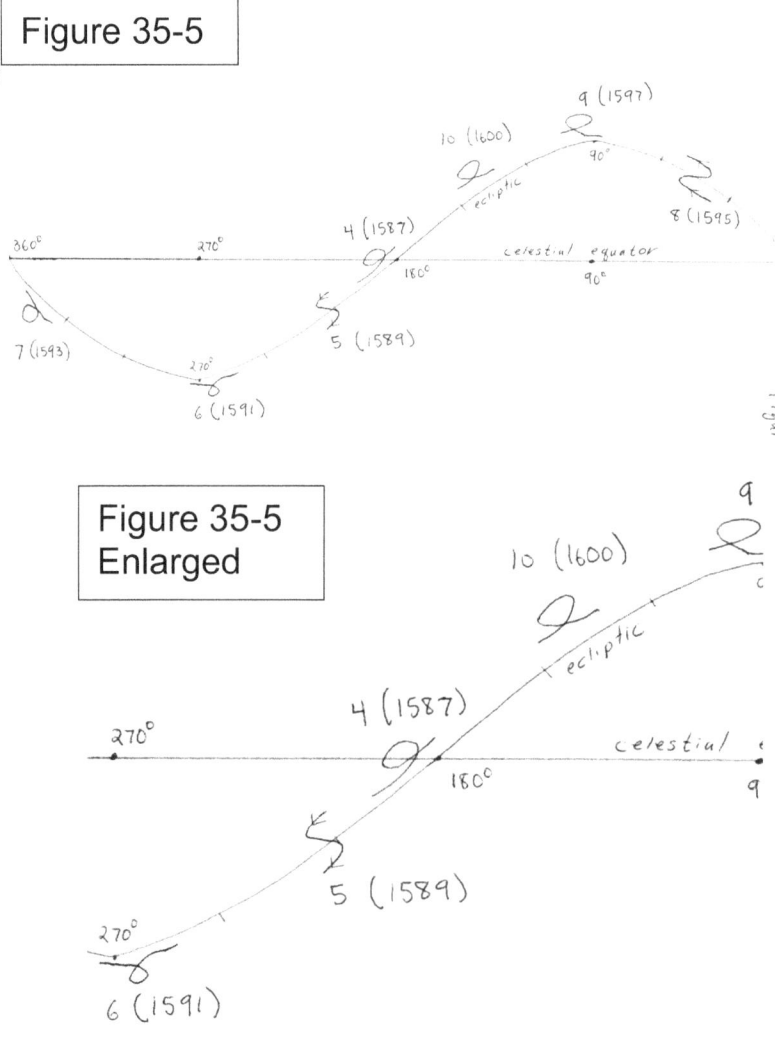

284

In this figure we are viewing the celestial sphere from the earth looking at the entire celestial equator placed on the plane of the paper and running as a straight line from 0° Right Ascension at the right hand end of the line, and 360° Right Ascension at the left hand end of the line.   The ecliptic is the curved line which is above the celestial equator for celestial longitudes between 0° and 180°, and below the celestial equator for celestial longitudes between 180° and 360°. The oppositions of Mars from 4 through 10 are plotted and the years of their occurrence noted.   We did not include the first three observed by Tycho so as not to overlap. This method of drawing the retrograde arcs shows the pattern of the celestial latitudes during the retrograde motion of Mars as observed from earth.

We note first the variable shapes of the arcs. Arcs 6 and 7 appear as loops south of the ecliptic with the retrograde part of the loop pointing away from the ecliptic.   Arcs 9, 10, and 4 are north of the ecliptic with the retrograde part of the loop pointing north of the ecliptic.   Note the two unusual retrograde arcs which appear from the earth as zigzag lines: arcs 8 and 5. Furthermore, these zigzag arcs are essentially 180° apart on the ecliptic, as we see also by inspecting Figure 35-4, where the retrograde arcs are viewed from the north ecliptic pole.

After opposition 5 in Figure 35-5 we see the next two oppositions occur south of the ecliptic, while after opposition 8 the next two oppositions occur north of the

285

ecliptic. It appears that oppositions 5 and 8 are transitions in the orbit of Mars where the planet changes its position in going above and below the ecliptic. Kepler realized these patterns in latitude changes occur because the planes in which Mars and the earth moved are inclined to each other.

Kepler could then envision a situation like that shown in Figure 35-6.

Figure 35-6

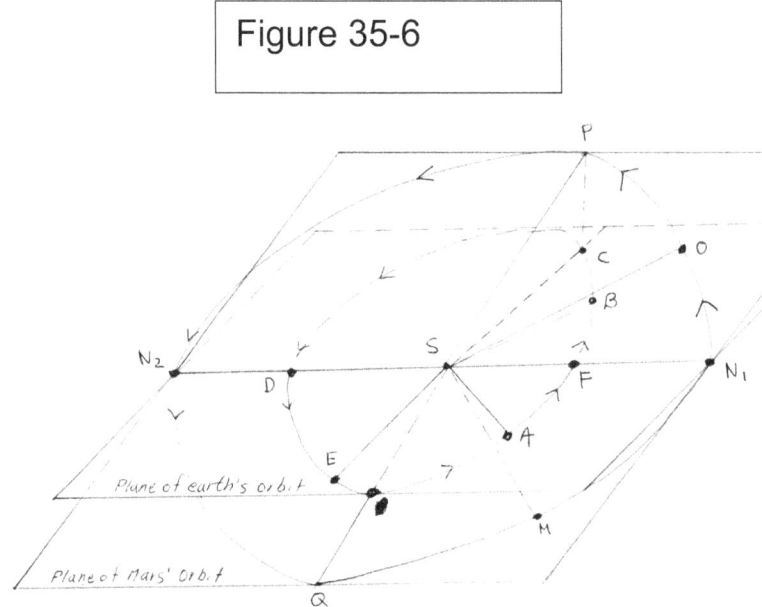

In this figure the true sun is at S which is shown essentially at the center of the earth's and Mars' circular orbit, although Kepler knew that the sun was not at the true center.   The earth's orbit is the circle with points E, A, F, B, C, and D.   Mars' orbit is farther from the sun and is on the circle with points M, N_1, 0, P, N_2, and Q. The earth stays in one plane which includes the true sun, and Mars is on a different plane but which also includes the true sun.   The two planes intersect along

286

line N_2DSFN_1.   In Kepler's early work on the orbit of Mars he sought to find the angle between these two planes.   The angle sought is angle PSC which also equal angle QSE, since they are vertical angles.

When Mars is at position N_1, and at the same time the earth is at position F, then Mars is in opposition to the sun and the retrograde arc produced as Mars moves from before N_1 then through N_1, and after N_1, will appear to the observer on earth just as the zigzag arc observed by Tycho in opposition 8 (Figure 35-5).   We see in Figure 35-6 that after opposition 5, Mars will be north of the ecliptic until it moves to position N_2 on its orbit (Mars and earth are both moving counterclockwise in Figure 35-6 as designated by the arrows).

If Mars is at position N_2, and at the same time the earth is at position D, then the observer on earth will see Mars move through an opposition such as Tycho observed in opposition 5 of 1589.   After Mars moves beyond position N_2, Mars will be seen from earth as south of the ecliptic.   Whenever Mars occupies positions N_1 or N2, it is on the same plane as the earth and the sun and is at 0° celestial latitude.   We see from Tycho's data that the line N_2S N_1, runs between Scorpio and Taurus on the ecliptic (see Figure 35-4).

We should note that if the earth is at position C in Figure 35-6 and Mars is at position N_2 that although the celestial latitude of Mars would be 0°, Mars would not be in opposition to the sun and the line of sight from earth would run from C to N_2, and onto the ecliptic at a position farther to the east of Scorpio such as Libra or Virgo.   The problem Kepler faced is the determination of angle PSC.

What Kepler sought to find in Tycho's data were observations of Mars when the planet is actually at position N1 or N2 in Figure 35-6, and then determine the celestial longitude of these points as seen from the true sun, S.   These would be observations when the celestial latitude of Mars is 0°.   An ascending node is defined as the point where Mars is moving from below the ecliptic to above the ecliptic such as seen in opposition 8, whereas, a descending node is the point where Mars moves from north of the ecliptic to south of the ecliptic.   When we drew the arcs in figure 35-5, we drew the path of Mars as a continuous line as if Tycho had data for each point of its path. Actually Tycho had observations in relationship to the mean sun, and opposition 8 in the table gives the celestial latitude at the time of opposition to be 0° 5' 5" north of the ecliptic instead of exactly 0°. This means Mars is just beyond the actual ascending node (N_1 in Figure 35-6).

Kepler was able to find observations Tycho had made when the celestial latitude was 0° for Mars.   He also found observations in Tycho's data when the planet was at the descending node (N_2 in Figure 35-6).   From these observations and Tycho's theory for the celestial longitude of Mar, he calculated the celestial longitude of

Mars as seen from the true sun when Mars occupies the ascending node or the descending node. The value he found for the ascending node N_1 is celestial longitude 46° 48' (Taurus 16° 48"). The value for the descending node is 225° 44' (Scorpius 15° 44'). The difference between these values is,

225° 44" - 46° 48' = 178° 56'. This is not exactly 180°, and in a later book Kepler gave the values as 47° and 227°, consistent with his theory.

We next construct Figure 35-7 which shows half the orbits of the earth and Mars as viewed from the north ecliptic pole.

Figure 35-7

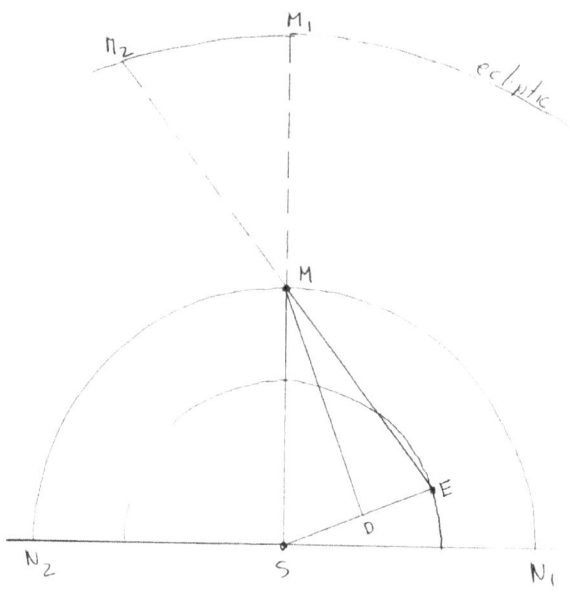

The sun is at S while Mars follows the arc with M, and the earth follows the arc with E.   The line of nodes is N_2SN_1.   When Mars has moved 90° east of N_1, as seen from the sun it is on line SM perpendicular to N_2SN_1.   When Mars is at such a position there will be times when the earth is at a position such as E at an angle N_1SE less than 90°.   If we take the celestial longitude of N_1 as 47°   as seen from the sun, and the celestial longitude of N_2 as 227° as seen from the sun, then N_2SN_1 will be a straight line with respect to the ecliptic, since 227° - 47° = 180°.   Kepler later used this as his value for the line of nodes as we have noted. Thus the celestial longitude of Mars in Figure 35-7 is 137° (47 + 90°) as seen from the sun.

Now there must be a position for point E where angle N_1SE is less than 90°, and where distance EM = distance SM.   At this position triangle SME is an isosceles triangle.   Tycho had calculated the ratio of the distance of Mars from the sun to the distance of earth from the sun when Mars had its greatest celestial latitude north of the ecliptic to be 1.666 to 1.00. When Mars is at M in Figure 35-7 it is at its greatest celestial longitude north of the ecliptic since it is at the equivalent of point P in Figure 35-6.   We then designate length SE in Figure 35-7 as 1 unit, and length SM = EM as 1.666 units.

Next, we bisect angle SME and connect MD. MD will be the perpendicular bisector of SE since SME is an isosceles triangle so SD = DE = .5 units. Then,

cos angle DEM= DE/EM = .5/1.666 = .3012. angle DEM = arcos .3012 = 72.54° = 72° 32'.

Since triangle EDM is a right triangle angle EMD = 90° - 72° 32' = 17° 27'.   Also angle SME = 2(angle EMD), or angle SME = 34° 55'.   Thus when the earth is at E and Mars is at M, the observer on the earth should see Mars at a celestial longitude of 137° + 34° 55' = 171° 55, since angles SME and M_1MM_2 are vertical angles. M_1 and M_2 are the points on the ecliptic made from extending lines SM and EM respectively.

The observer on earth would 'see' the sun at a celestial longitude equal to 171° 55' + angle MES, since line ES extended to the ecliptic is the celestial longitude of the sun. Angle MES is angle DEM.   We found angle DEM = 72° 32' so, the celestial longitude of the sun = 171° 55' + 72° 32' = 244° 27'.

Kepler could now go through Tycho's registry of observations to find times when the celestial longitude of Mars was very near 171° 55', and the celestial longitude of the sun was very near 244° 27'.   We will show how he could use two such observations to estimate the inclination of the plane containing Mars and the sun from the plane containing the earth and the sun.

On Nov 9 1588 at 18h 30 m Tycho observed the celestial longitude of Mars to be 175° 31' with a celestial latitude of 1° 36' 45" north of the ecliptic.   This observation occurred between oppositions 4 and 5 in the table and will illustrate how important is was for Kepler to have accurate observations of Mars at times other than during oppositions with the sun. This observation is not at a time when Mars is exactly 90° from the ascending node in Figure 35-7.   The sun's celestial longitude for this observation was 238°.   Using these values we find angle MES in Figure 35-7 is 238° - 175° 31' = 62° 29' instead of 72° 23' when we know EM = SM.   Thus for this observation EM > SM, and Mars has moved beyond point M in Figure 35-7 where EM = SM.   Thus the celestial latitude of Mars at this observation, 1° 36' 45" must be less than the angle of inclination of the plane containing Mars and the sun to the plane containing earth and the sun.

Kepler found Tycho had observed Mars on 4 Dec 1588 with a celestial longitude of 189° 19' 24", and a celestial latitude of 1° 53' 50" north of the ecliptic.   The sun's celestial longitude at the time was 263°.   In Figure 35-7 angle MES would be 263° - 189° 19' 24" = 73° 40' 36", which is greater than the value of 72° 23' when triangle MES is an equilateral triangle, i.e. when EM = SM.   This means EM <   SM so that the observed celestial latitude of 1° 53'   50" is greater than the inclination angle for the plane containing the sun and Mars to the plane containing the sun and the earth.

Kepler balanced these two observations (9 Nov 1588 and 4 Dec 1588) and concluded that angle of inclination for the two planes must be 1° 50'.

To confirm this value Kepler used another method for determining the angle of inclination for the planes which we will discuss in reference to Figure 35-8.

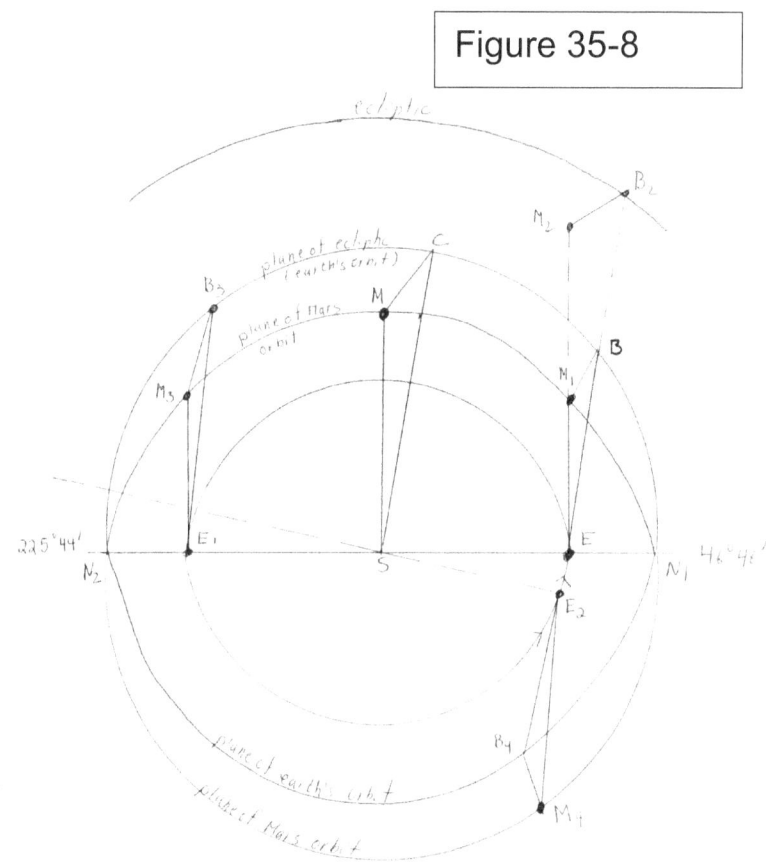

Figure 35-8

We again view the situation from the north ecliptic pole with the sun at S, but this time with the earth at E lying on the line of nodes, N₁ N₂. When the earth is at E the celestial longitude of the sun is 225° 44'.

The plane containing the sun and the orbit of Mars is $N_2 MM_1 N_1$, and the plane containing the sun and the orbit of the earth is $N_2 CBN_1$.   For positions M and M_1 on Mars' orbit, the plane containing the sun and Mars is north or above the plane containing the sun and the earth, and for position M_4 on the orbit of Mars the plane containing the sun and Mars is south or below the plane containing the sun and the earth.

The angle of inclination of the two planes is angle MSC provided MS is perpendicular to $N_2 SN_1$, and CS is perpendicular to $N_2 SN_1$.   Kepler then looked for observations made by Tycho when the earth was along the line of nodes such as position E, and when Mars is at a position M_1, such that angle $M_1 EN_1$ is a right angle. If he could find such an observation then the celestial latitude of Mars would be equal to the angle of inclination of the two planes.   That is angle BEM_1 = angle CSM. This is because the observer on earth will see Mars as north of the ecliptic along line of sight $EM_1 M_2$. $M_2 B_2$ is perpendicular to the ecliptic as seen from E so that line of sight EBB_2 is perpendicular to the line of nodes, $N_2 N_1$.   The same principle can be used when Mars is at a position near M_4 along line $M_1 E$, and at position M_3, when the earth is on the line of nodes such that the celestial longitude of the sun is 46° 48', and line of sight $E_1 M_3$ is perpendicular to the line of nodes.

Observations where the sun and Mars were near the positions in Figure 35-8 were rare in Tycho's data, but Kepler found four with which he could work. We will consider two of them as examples.

Tycho observed Mars on Apr 26, 1585 at 9h 42m to have celestial longitude 141° 26', and celestial latitude 1° 49' 45" north of the ecliptic. The celestial longitude of the sun was 46°. This is very close to the position where the earth is at E_1, and Mars at M_3 in Figure 35-8. The elongation of Mars, angle M_3E_1S is 141° 26' - 46° = 95° 26', beyond 90° so its celestial latitude of 1° 49' 45" is somewhat less than the inclination of the plane of the ecliptic to the plane of Mars' orbit. Kepler argues the figure should be 1° 50' for the inclination from this consideration.

On 16 Oct 1591 at 6h 3 m Tycho observed Mars to have celestial longitude 301° 27' 20", and celestial latitude of 2° 10' 50" south of the ecliptic. The celestial longitude of the sun at this time was 212° 30'. We have represented this observation in Figure 35-8 by placing the earth at E_2 about 13° before it reaches position E on the line of nodes when the celestial longitude of the sun is 225° 44'. Mars is at position M_4 moving counterclockwise in its orbit although at a slower angular velocity than the earth.

In reviewing Tycho's data Kepler found that the celestial latitude of Mars had decreased 27' 40" (moving closer to the ecliptic) from the 2nd to the 16th of Oct, and if its celestial latitude were to continue to move closer to the ecliptic at the same rate over the next 14 d at which point the earth would be at position E, then the celestial latitude of Mars would be: 2°10' 50" - 27' 40" = 1°43'   10"   south of the ecliptic. (Kepler in the "Astronomia Nova" says in 14 d the celestial latitude will decrease 28', and gets the result 1° 45' for the celestial latitude.   Historians have remarked that in the text of the "Astronomia Nova", Kepler does make computational errors.) However, as the earth moves from E_2 to E in 14 d,   Mars moves relatively less from M_4 so that the distance EM_4 > E_2M_4. This means the rate of decrease in celestial latitude for Mars in the 14 d will be less than 28'.   Kepler calculated that in the 14 d after the 16th Oct the celestial latitude would decrease at a lesser rate than 28' for 14 d so that the celestial latitude of Mars when the earth is at E should be 1° 50' south of the ecliptic.   Furthermore, when the earth has moved to E from E_2, the celestial longitude of Mars must increase from 301° 27' 20" to nearly 315° 44', which means angle SEM_4 becomes nearly 90° so the value of angle B_4EM_4 of 1° 50' does represent the angle of inclination for the two planes.

Kepler proceeds to give yet another method to establish the value of 1° 50' for the inclination of the two planes which we will not consider.   Kepler's discussion of the inclination of the plane of the ecliptic to the orbit of

Mars is found in chapter XIII of the second part of the "*Astronomia Nova*".

Thus in his early work with Tycho's data Kepler had succeeded in finding a satisfactory theory for the celestial latitude of Mars.   We note again that to develop his theory Kepler makes use of observations when Mars is not in opposition with the sun unlike his predecessors who did not have access to accurate observations of Mars when the planet was not in opposition with the sun.

It is possible using Kepler's value for the inclination of the planes and the ratio of SM to SE, to calculate the celestial latitude of Mars when the earth is not on the line of nodes and with Mars at points between N_1 and N_2 in its orbit.   Although Kepler does not present such a calculation in the "*Astronomia Nova*" we can show how this is possible using our knowledge of trigonometry.

We will use opposition 3 from Table 1 of oppositions and calculate the celestial latitude using the value of 1° 50' for the inclination of the plane of the ecliptic to the plane of Mars orbit.   The celestial longitude of Mars for this opposition is 141° 10' 26". This is about 94° from the line of nodes, and for our calculation we will assume it lies along line SEM in Figure 35-9 such that SEM is essentially perpendicular to the line of nodes.

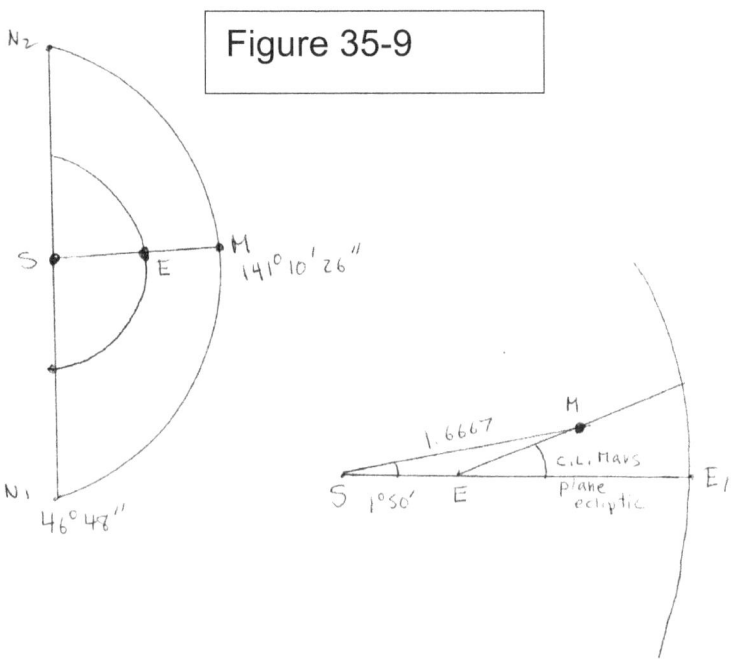

Figure 35-9

In the figure we view the observation from the north ecliptic pole and from N_1 on the line of nodes. In viewing from point N_1, (the drawing on the right) we see the observer on earth will see Mars as having a celestial latitude equal to angle MEE_1. Tycho had found that for these positions of the earth and Mars the ratio of distances SE/SM = 1.0000/1.6667.

We will let SE = 1.0000, SM = 1.6667, and angle ESM = 1° 50'. We then solve triangle ESM for side EM using the law of cosines for plane triangles.

1. $(EM)^2 =$

   $(SE)^2 + (SM)^2 - 2(SE)(SM)(\cos 1°\ 50')$,

   $(EM)^2 = 1 + 1.6667^2 - 2(1)(1.6667)(.9995)$,

   $(EM)^2 = .4462$, or, EM = .6680.

   Then we use the law of sines to find angle

298

SEM .

2. SM /sin SEM = EM/ sin 1.833°

$$\sin SEM = \frac{(\sin 1.8333°)(SM)}{EM} =$$

$$\frac{(.0320)(1.6667)}{.6680} = .0798,$$

angle SEM = arcsin .0798 = 175.4225°,

since angle SEME is an oblique angle. This means,

angle MEE_1 = 180° - 175.4225° = 4.5775°,

or, 4° 35'.

From the table we see the observed celestial latitude by Tycho was 4° 32' .

In the next chapter we will take up the task of examining Kepler's first theories for the celestial longitude of Mars using Tycho's data.

Bibliography for Chapter 35

1. Koestler, Arthur, "The Sleepwalkers", The Universal Library, Grosset & Dunlap, New York,   1963.

2.Kepler, Johannes, "Mysterium Cosmographicum, The Secret of the Universe", translated by A.M. Duncan, introduction and commentary by E.J. Aiton, with a preface by I. Bernard Cohen, New York, 1981.

3. Field, J.V., "Kepler's Geometrical Cosmology", The University of Chicago Press, 1988.

4. Kepler, Johannes,   "Astronomia Nova",   Latin edition

printed by Culture et Civilisation, 115, Avenue, Gabriel
Lebon, Bruxelles, 1968.

5. Kepler, Johannes, "New astronomy", translated by
William H. Donahue, Cambridge University Press, 1992.

6. Small, Robert, "An Account of the Astronomical
Discoveries of Kepler", A reprinting of the 1804 text by
The University of Wisconsin Press, Madison Wisconsin,
1963.

7. Wilson, Curtis, "How Did Kepler Discover His First
Two
Laws?", Scientific American, Vol. 226, no. 3(March
1972), pp. 93-106.

The New Astronomy: The Celestial Longitude of Mars

We have emphasized Kepler's commitment to the Copernican system as he worked with Tycho's data. Kepler believed he would find the correct theory for the celestial longitude of Mars as he had found the theory for the celestial latitude by revising Tycho's table of oppositions so that the table would give the oppositions of Mars in reference to the true sun. Tycho's table, as we shall see, was based on reference to the mean position of the sun, which we recall was the position of the center for the earth's orbit in Copernicus' model. We construct Figure 36-1 to depict Kepler's thinking.

Figure 36-1

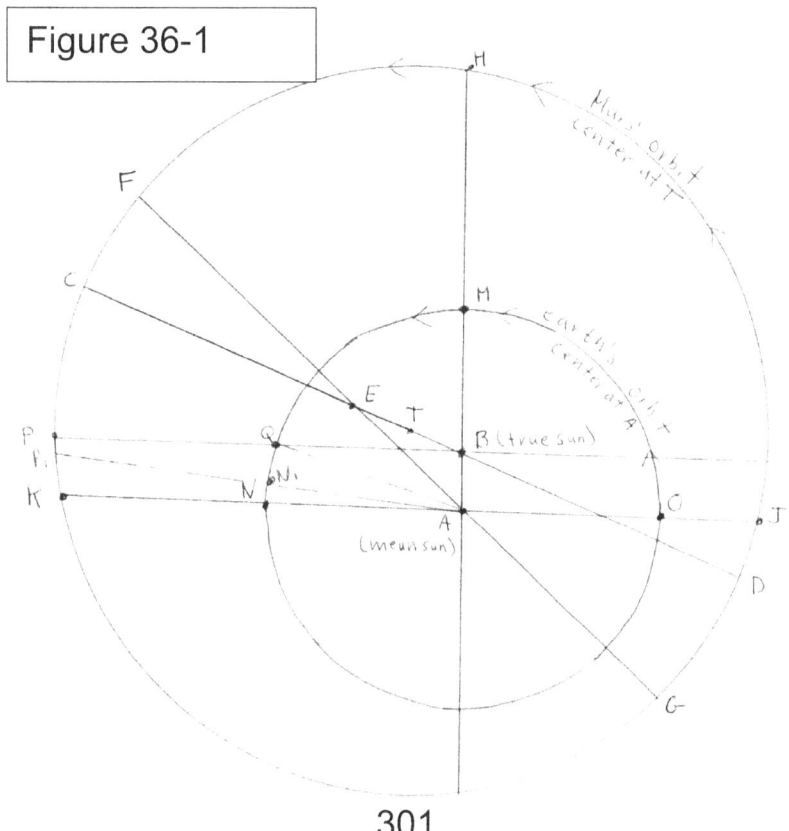

In the figure we view the situation from the north ecliptic pole. Point A is the center of the earth's orbit and the position of the mean sun. The earth's circular orbit is the circle with points 0, M, and N on the circumference. Kepler considered the orbit of Mars to be a circle with a center at T at a distance TB from the true sun at B. Mars orbit is the circle with points J, H, and K on the circumference. We can imagine the ecliptic as a circle so far distant from these circles so that either A or T, or the position of the true sun at B could be the center of the ecliptic. Now for both Copernicus and Tycho an opposition of the mean sun with Mars meant a situation such as when the earth is at position 0 and Mars at position J. We see a straight line runs through AOJ so that the observer on earth would determine the mean sun and Mars have an elongation of 180°.

Kepler did not consider this to be the positions for an opposition. For Kepler an opposition occurs when the earth, Mars, and the true sun are on the same straight line. For example, if the earth is at M, Mars at H, and the true sun at B we would have a case of an opposition of Mars with the true sun. Note in this case we also have an opposition with the mean sun as well, since point A is also on this line. However, if Mars is at P and the earth at Q, Kepler considers this a true opposition, whereas Copernicus would consider P_1 the position of Mars in opposition when the earth is at N_1 so that P_1N_1A is a straight line (opposition the mean sun). Tycho believed the earth was stationary and not moving so N_1 was a stationary point in his theory.

Kepler's next reform of the Copernican system was to use an equant point E, a point from where the motion of Mars is seen to have uniform angular motion, the same principle as used by Ptolemy. The equant point will lie along a line of apsides, ETB.   In this model Kepler then has to find the parameters for Mars: ω_e , the uniform motion around the equant point, E; the values of lengths TE and TB in relation to the length of the radius of Mars orbit TC. We recall Copernicus had developed his theory with the purpose of abolishing the equant point from astronomy.

Once again Kepler relates his system to the true sun at B rather than Copernicus' reference to the mean sun at A.   Kepler believed the force of the sun causing the motion of Mars would be most powerful along line BD since when Mars is at position D, its perigee, it is closest to the sun, and as seen from the true sun it is moving at its maximum angular velocity, although as seen from E, the equant it is moving at uniform angular velocity.

When Mars is at position C on its orbit it is at apogee, and when seen from the true sun it has its slowest motion since it is at its greatest distance from the true sun.   Note again that the line of apsides in Kepler's hypothesis must be through the true sun, B, and not along line FEAG which would be the line of apsides in the Copernican model through the mean sun.

It was clear to Kepler that Mars would, for example, be farther from the true sun, B, when it is at position G than when it is at position D, and thus be moving slower as seen from the true sun, B, than when it is at D, the true perigee.   Whereas Copernicus would say the perigee is at G and Mars would appear to be moving faster as seen from the mean sun, A, than from the true sun, B.

Kepler's ideas meant he had to revise the table of oppositions as presented by Tycho Brahe (table 1, chapter 35) before he could proceed to calculate the parameters for Mars. We will look into some of the reasons why the revision of the table was necessary.

Consider first an opposition of Mars with the true sun as seen from the earth when the earth is at position M and Mars at position H.   Here the line of sight of the mean sun and the true sun coincide as noted above so the celestial longitude of the mean sun, A, and the true sun, B, will be the same and both will have an elongation of 180° from Mars as required for an opposition.   However, if we consider the opposition to the true sun when the earth is at position Q and Mars at position P, then we see that the observer on earth will measure the angle of elongation to the true sun as 180° (angle PQB), whereas the angle of elongation to the mean sun (angle PQA) is not 180°. Tycho would not record such a situation as an opposition.

For Tycho the opposition occurs when the earth is at N_1, and Mars has moved to P_1 from P.   As we learned in our discussion of the Copernican system the angular velocity of Mars is slower than that of the earth so that as the earth, in Kepler's theory, moves from Q toward N_1 the earth 'passes by' Mars so that the observer on earth perceives Mars to be moving on its retrograde arc before the situation of an opposition to the mean sun is realized. We should, therefore, depict the opposition to the mean sun along line P_1N_1A to emphasize that arc QN_1 is greater than arc PP_1. Because the celestial longitudes of Mars as seen from earth for positions P and P_1 are different, Kepler knew he would have to correct the celestial longitudes found in Tychos's table of oppositions which were based on oppositions to the mean sun.

Kepler carefully reviewed Tycho's data for each of the 10 oppositions in table 1 chapter 35, and compared the celestial longitude given for Mars at opposition to the mean sun to the celestial longitude of the true sun as calculated by Kepler. In order to do this he had to find observations by Tycho where Tycho recorded the celestial longitude of Mars at 180^0 from the sun's position at sunrise. He was able to locate in Tycho's data when Tycho saw the true sun at sunrise and Mar's was at 180^0 from that point on the ecliptic, a true opposition. He had to make a correction for the distortion of its position created by atmospheric conditions at sunrise. In other words the position of Mars in the table of Tycho for these observations was changed to reflect where Mars' position would be at exactly 180^0 from true sunrise. The work on learning what correction to make for atmospheric distortion of the sun at sunrise took Kepler about a year to complete. When the sun rises or sets it rays are bent (refracted) by the atmosphere so that the sun appears higher than it actually is. This means at sunrise Kepler had to add some minutes to its celestial longitude to have the value at true sunrise.   If Mars is setting when we see the sun at the horizon at sunrise Kepler subtracts some minutes of from Mars' celestial longitude to make the angle between the true sun and Mars 180^0.

In addition there had to be a correction based on the theory of latitude discussed in the last chapter. This correction added to the celestial longitude for the oppositions in Tycho's table.   Kepler's angle of 1° 50' for the tilt of the orbit of Mars form the orbit of earth was different than Tycho's value.   Kepler found the difference resulted in changes for the celestial longitudes of the oppositions compared to Tycho's table. For example, opposition 6 in the table lists the celestial longitude of Mars at opposition to the mean sun as 266° 32' 0".   Kepler's calculated position for the true sun at the time of the true opposition was 86° 45' 24", a difference of 179° 46' 36", which is13' 24" from 180°. Mars celestial longitude was 13' 24" greater than Tycho's value. This confirmed to Kepler that he needed a table of oppositions based on the true sun.   Tycho's value for the angle between the plane of the ecliptic and the plane of Mar's orbit contributed to the error of the celestial longitudes of Mars as calculated by Tycho.

We can see why Tycho had a different value for
the angle between the plane of the ecliptic and the plane
of Mars' orbit and how this could affect the value for the
celestial longitude of Mars at opposition by studying
Figures 36-2, 36-3 and 36-4. Figure 36-2 is very similar
to Kepler's figure in the "*Astronomia Nova*", chapter XII,
part 2.

Figure 36-2

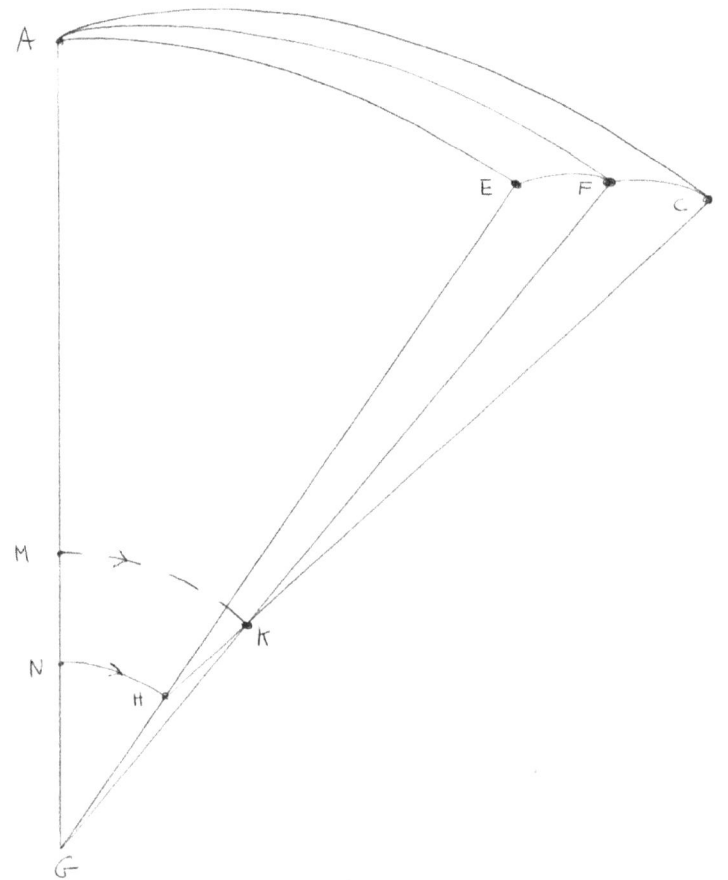

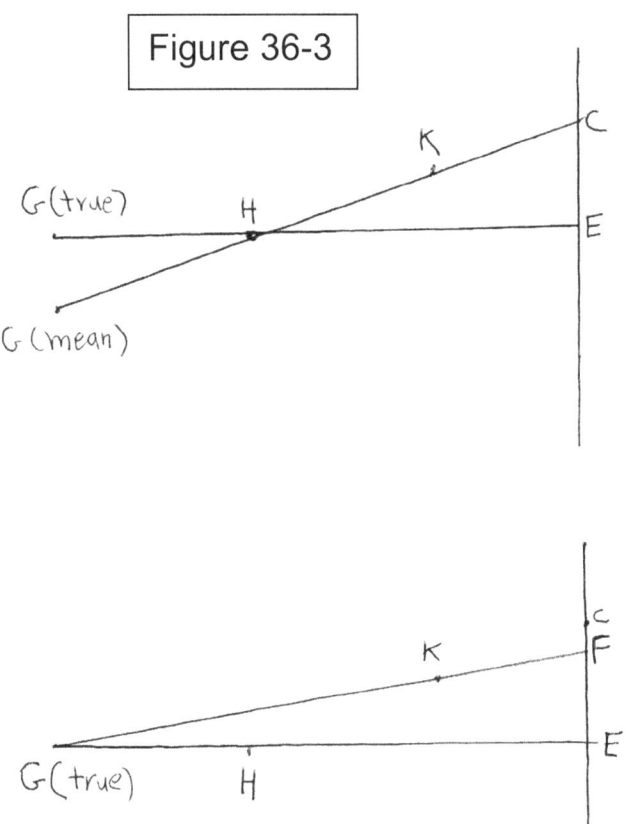

Figure 36-3

We start with Figure 36-2 and Figure 36-3.   In this figure we are viewing from the south ecliptic pole with the true sun at G.   Kepler assumes the sun is stationary. The intersection of the plane of the ecliptic and the orbit of Mars is along line GNMA, the line of nodes. The orbit of the earth follows arc NH, and because we are viewing from the south ecliptic pole the earth moves clockwise along arc NH.   Likewise the orbit of Mars is clockwise along arc MK which we see is north of the plane of the ecliptic which is arc AE connected to G. The projection of the plane of the orbit of Mars which contains the true sun onto the celestial sphere is arc AF connected to G.

Now if the earth were at N and Mars at M we would have an opposition along the line of nodes and Mars would be seen from earth to have celestial latitude of 0°, and its celestial longitude will be the position of point A on the ecliptic.   As seen from the true sun, G, Mars would lie along line GNMA and would have the same celestial longitude as seen from the earth.

Next consider the situation of an opposition where the earth is at H and Mars at K. This is shown in both Figure 36-2 and 36-3.   Now the observer on the earth sees Mars in opposition to the true sun, G, along line GKF, so that GHE and GFF are in the same plane. In the upper part of Figure 36-3 line HK is extended to the mean sun marked G(mean). However, Kepler believed this line would not be in the same plane as GHEFK, but would make an angle at H, then HG(mean) would go to the left or right of G, the true sun, in Figure 36-2.   In Figure 36-3 G(mean) would be in front or back of the plane of the drawing.

The observer records Mars' celestial latitude as angle CHE in Figure 36-2.   Kepler had to find the time of such an opposition with the true sun, G, and the position of Mars, point C, as seen from earth on the celestial sphere at this time.   He used Tycho's registry of data on the observations of Mars to find the position of Mars at the time of the opposition to the true sun when Mars is seen at point C on the celestial sphere and GKF is a straight line. What Kepler then wished to determine for his table of oppositions was the length of arc AF projected onto the ecliptic, which would be the true celestial longitude of Mars at the time of opposition with the true sun, G. To understand what is meant by the projection of Mars onto the ecliptic see Figure 36-4.

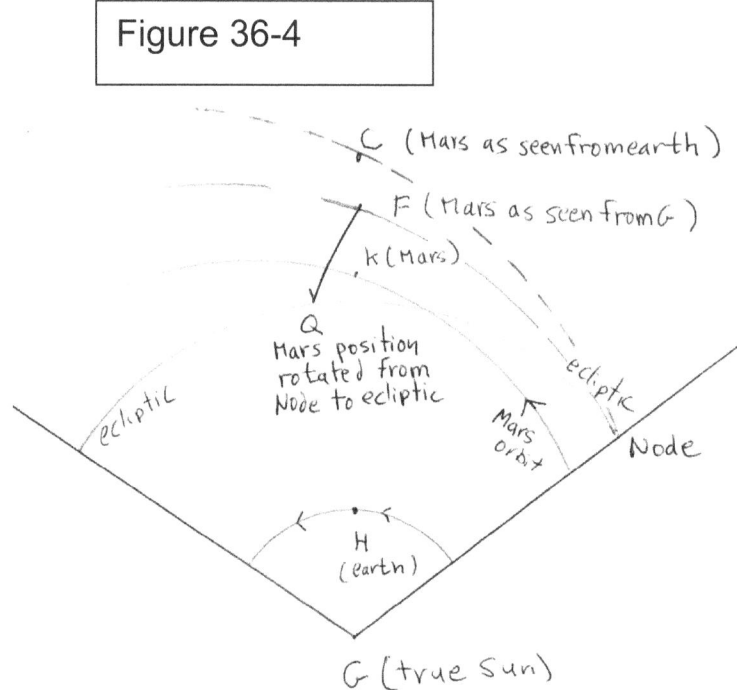

Figure 36-4

In this figure we show the earth's orbit with earth at H. We show the ecliptic and the reader should visualize G, H and the ecliptic as on a plane. Above the line of the ecliptic is latitude north of the ecliptic. We see Mars' orbit staring at the node on the right and Mars is at K on its orbit as in Figure 36-2. The observer on earth sees mars at C, and an observer on the true sun would see Mars at F, lower celestial latitude than position C. If we took the arc length from the node to F, and moved the arc with F to the ecliptic it would be at point Q which would have a greater celestial longitude than what the observer at earth records at C. By moving F to Q then the line GQ is on the plane containing the sun and earth correcting the celestial longitude error caused by Mars having celestial latitude above the ecliptic.

In Kepler's theory in Figure 36-2, the celestial latitude of Mars as seen from the true sun is angle FGE which is equal to arc EF. This is because the line of sight from the sun through Mars, K, onto the celestial sphere is line GKF. Kepler had found angle EAF equal to 1° 50', the angle between the plane of the ecliptic and the plane of Mars' orbit. Kepler had to determine the position of point F on the celestial sphere for each of Tycho's oppositions. He could then project F onto the ecliptic to get the corrected celestial longitude for the opposition. If he could solve spherical triangle AEF he could determine arc EF and arc AF. He knew angle AEF = 90°, angle EAF = 1° 50', and the value of arc AE from the celestial coordinates of Mars at the time of the opposition of true sun.   He could use the law of cosines for the angle of spherical triangle we determined in chapter 15:

cos C = - cos A cos B + sin A sin B cos c. For triangle AEF, angle AEB would be B, angle EAF would be A, and c would be arc AE.   Using these know values Kepler would determine C, angle EFA.

Then using:

$$\cos A = -\cos B \cos C + \sin B \sin C \cos a,$$

he uses the value he determined for C and the other known values to find a, which is arc EF.   He then would have the value of the celestial latitude of Mars for the opposition.

He then can determine arc b, (arc AF) by using the law of sines for a spherical triangle: sin A/sin a = sin B/sin b.

In this way Kepler could find points along the orbit of Mars with reference to the true sun and develop his table of oppositions of Mars with the true sun.

We should note that in Tycho's earth centered system for Mars, Tycho regarded the orbit of Mars to lie along arc AC whereas Kepler regarded the orbit of Mars to lie along arc AF.   Tycho knew the values of angles EAC and E(90°), the value of arc AE from the celestial coordinates of Mars at the time when Mars is in opposition to the mean sun.   He then could solve the spherical triangle for arc AC for each of the 10 oppositions in his table and define his orbit for Mars along arc AC. As we know he took the opposition to occur when G is the mean sun and not the true sun unlike Kepler.

We will not attempt to show the details of Kepler's calculations to set up his own table of oppositions of Mars with the true sun.   We will present Kepler's table of oppositions so the reader can compare the results with Tycho's table in chapter 35.   Kepler also added two more oppositions to the 10 of Tycho which he had observed himself in 1602 and 1604. Furthermore, Kepler states that he used the observations of another astronomer, David Fabricius, who made his observations of some of the oppositions in East Friesland to confirm his data.

Table 2

Date	Celestial Longitude	Celestial Latitude
1. 1580 Nov 18 1h 31 m	66° 28' 35" (Gemini)	1° 40' 0" N
2. 1582 Dec 28 3h 58 m	106° 55' 30" (Cancer)	4° 6' 0" N
3. 1585 Jan 30 19 h 14 m	141° 36' 10" (Leo)	4° 32' 10" N
4. 1587 Mar 6 7h 23 m	175° 43' 0" (Virgo)	3° 41' 0" N
5. 1589 Apr 14 6h 23m	214° 23' 0" (Scorpio)	1° 12' 45" N
6. 1591 Jun 8 7h 4 3m	266° 43' 0" (Sagittarius)	4° 0" 0" S
7. 1593 Aug 25 17h 27m	342° 16' 0" (Pisces)	6° 2' 0" S
8. 1595 Oct 31 0h 39m	47° 31'40" (Taurus)	0° 8" 0" N
9. 1597 Dec 13 15h 44m	92° 28' 0" (Cancer)	3° 33' 0" N
10. 1600 Jan 18 14h 2m	128° 38' 0" (Leo)	4° 30' 10" N

11. 1602 Feb 20   162° 27' 0" (Virgo)   4° 10' 0" N
    14h 13m

12. 1604 Mar 28   198° 37' 10" (Libra)   2° 26' 0" N
    16h 23m

Table 1

Date	Celestial Longitude	Celestial Latitude
1.1580 Nov 17	66° 46' 10" (Gemini)	1° 40' 0" N
9 h 40 m		
2.1582 Dec 28	106° 46' 10" (Cancer)	4° 6' 0" N
12 h 16 m		
3.1585 Jan 31	141° 10' 26" (Leo)	4° 32' 10" N
19h 35m		
4.1587 Mar 1	175° 10' 20" (Virgo)	3° 38' 12" N
17 h 22 m		
5.1589 Apr 15	213° 58' 10" (Scorpio)	1° 6' 45" N
13h 34 m		
6.1591 Jun 8	266° 32' 0"(Sagittarius)	3° 59" 0" S
16h 25		
7.1593 Aug 24	342⁰ 43' 45" (Pisces)	6° 3' 0" S
2 h 30 m		
8.1595 Oct 29	47° 56' 15" (Taurus)	0° 5' 15" N
21 h 22 m		
9.1597 Dec 13	92⁰ 28' 0" (Cancer)	3° 33' 0" N
13 h 35m		
10.1600 Jan 19	128° 18' 0" (Leo)	4° 30' 50" N

We see in comparing Table 2 with Table 1, from
chapter 35, that the times of opposition with the true sun
314

are very close to the times of oppositions with the mean sun, and that the celestial longitudes are usually well within 1°. Kepler believed these were significant differences because of his theory and his belief that Tycho's observation were accurate to about 2' of arc.

Once Kepler had his table he could construct what was to be his first theory for the orbit of Mars so as to explain the first and second inequalities of the planet's motion. He, like his predecessors Brahe, Copernicus, and Ptolemy, believed he could find a theory which involved circular motion for the planet. Because of the inequality with respect to the ecliptic, he knew the sun could not be at the center of Mars' motion, and he adopted the plan indicated in Figure 36-5 (Note this is the same plan shown in Figure 36-1, but we have omitted the mean sun from consideration and do not show the orbit of the earth).

Figure 36-5

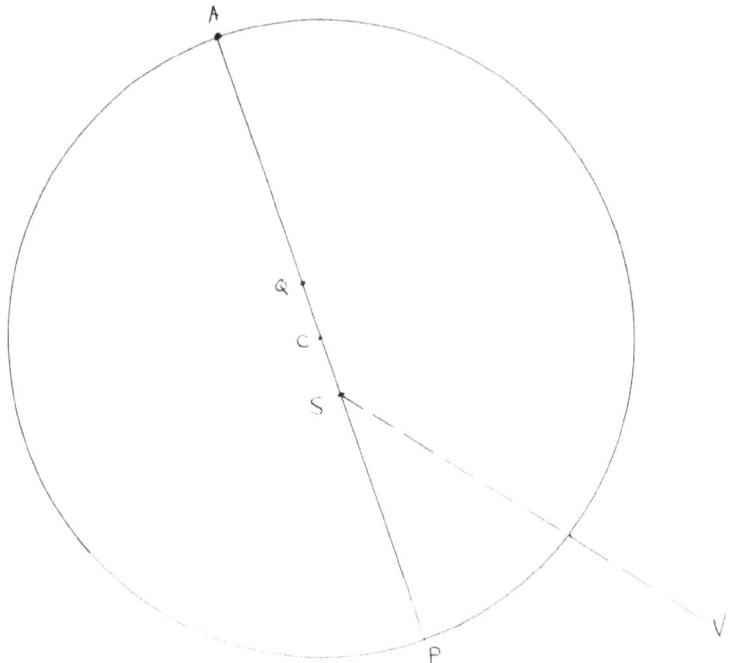

We see the center of Mar's circular orbit at C as viewed from the north ecliptic pole. The true sun, S, is stationary and eccentric to the center of the orbit as is the equant point Q. The line of apsides is PSCQA where A is the aphelion, P the perihelion, and V the first point of Aries. In this model aphelion refers to the farthest point in Mars' orbit from the sun, and the perihelion is the closest point in Mars' orbit to the sun. Kepler then tried to calculate the parameters for his model using data from the table of 12 oppositions of Mars with the true sun. He wanted to find the radius of the orbit, CA, the angle of the line of apsides from the first point of Aries, angle VSA, and the values of the ratios CS/SQ, CQ/SQ, and SQ/CA. In his work Kepler did not assume point C was half way between S and Q.

We recall from our work with Ptolemy's theory that for the superior planets, and Mars in particular, the bisection of the eccentricity, that is the assignment of the center of the deferent to a position half way between the earth and the equant, was an arbitrary choice of Ptolemy's.   Kepler had come to believe that this feature of Ptolemy's theory was a cause of imperfection in the theory, and he did not wish to bisect the eccentricity for his model by placing the center of Mars' orbit, C, half way between the sun, S, and the equant, Q.   When Kepler first arrived at Benatek he found Tycho was also convinced the eccentricity for Mars should not be bisected.

However, the problem of finding the parameters became more complicated so that Kepler had to use four oppositions from his revised table to determine the parameters.   He had to perform a long series of iterative calculations using ingenious arguments from geometry and trigonometry based on a geometrical construction involving four of the oppositions from the revised table.

At this point we refer the reader to chapter 24 and the work of Ptolemy when he found the eccentricity for the superior planets using Figures 24-2, 24-3, and 24-4. Ptolemy required data from three oppositions.   He solved a series of triangles and chords in a circle to obtain the eccentricity.   In the following discussion of

Kepler's work we will see why Kepler needed four oppositions.

We construct Figure 36-6 which we have modified very little from the figure used by Kepler in chapter 16 of the second part of the "Astronomia Nova".

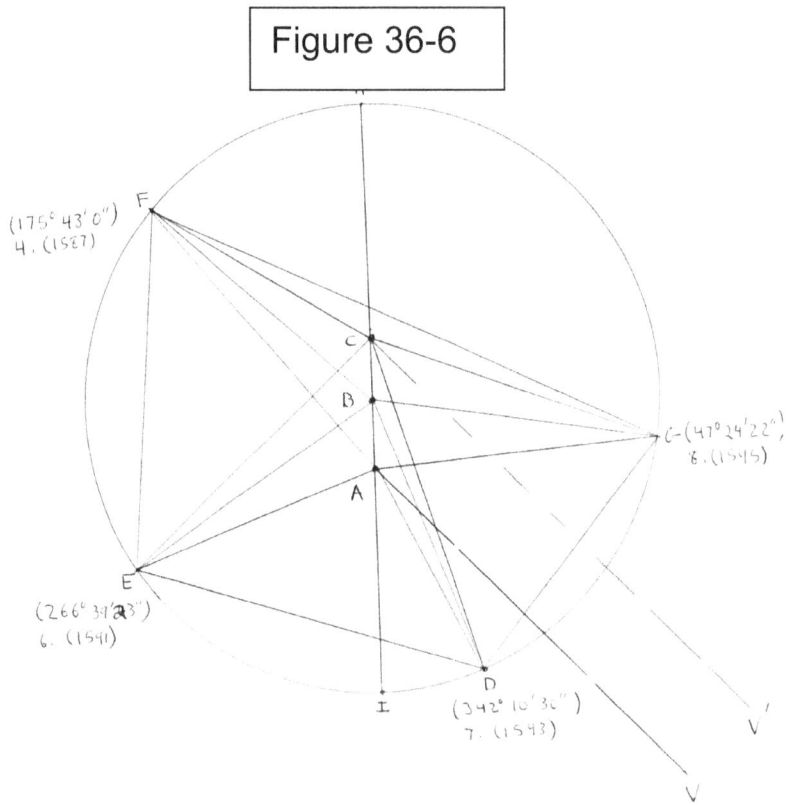

Figure 36-6

The orbit of Mars is seen as a circle viewed from the north ecliptic pole with center at B eccentric to the sun at A. The orbit of the earth is not shown. The equant point is at C on the line of apsides HI, where I is the perihelion and H is the aphelion. Broken line AV is the line from the sun to the first point of Aries (0° celestial longitude). Points F, E, D, and G are the points on

318

Mars' orbit as seen from the sun where the four oppositions chosen by Kepler occur.   Kepler corrects their values to positions they would have in 1587 by correcting for the precession of the later oppositions. Thus their celestial longitudes as seen from the sun are slightly different from the values in the revised table of Kepler except for F which occurred in 1587.   The number of the opposition from the revised table and the date are given next to the points.

Angle HCF is the mean anomaly for the fourth opposition since C is the equant around which uniform or mean motion is to occur in the theory.   The true anomaly is angle HAF, since the reference point for Kepler's theory is the position of the true sun.   Kepler knew the mean angular velocity of Mars was 1.881 years or 686.98 days to move 360° around its orbit as seen from point C. If he knows the celestial longitude of point H, angle HCF, angle HAF, and the ratios AB/AC, BC/AC, and ABBH, (AC is assigned the value of 10,000 units), he can then calculate the position of Mars for any time after position F. But first he must determine all these parameters.

Kepler begins by making several assumptions based on Tycho's work.   He calculates from these assumptions and then sees if the assumptions fit his theory or not.   If they don't fit he starts over with new assumptions and recalculates the parameters until his results agree with his starting assumptions.

He assumed the celestial longitude of H as seen from the sun A (angle VAH) is 148° 44' 0". This value came from Tycho's work. He can then calculate the values of the true anomalies, angles HAF, HAE, HAD, and HAG. For example, angle HAF = angle VAF (175° 43' 0") - angle VAH (148° 44' 0") = 26° 59'.

Now Tycho had also made a table of mean longitudes for the oppositions which Kepler had corrected so that Kepler could make assumptions for the mean longitudes for each of the four oppositions. These values are the celestial longitudes of Mars as seen from the equant point C. Thus Kepler assumed angle V'CF = 180° 50' 56", angle V'CE = 275° 40' 14", angle V'CD = 339° 49' 34", and angle V'CG = 37° 6' 51". (Note in Figure 36-6 to form angles V'CF, VC'E, VC'D, and VC'G, we draw line V'C parallel to line VA, since the first point of Aries on the ecliptic is so far away from the orbit of Mars. lines AV, CV' can be considered parallel).

With these assumptions Kepler can now calculate the mean anomalies for the oppositions, namely angles HCF, HCE, HCD, and HCG. For example, angle HCF = angle VCF (180° 50' 56") - angle VCH (148° 44' 0") = 32° 6' 56". Angle HCE = VCE (275° 40' 14") - angle VCH (148° 44' 0") = 126° 56' 18". (In the Latin first edition of the "*Astronomia Nova*" a mistake occurs and angle HCE is labeled angle HCD. Other mistakes and inconsistencies also occur from time to time in the details of Kepler's work).

Kepler then proceeds to solve a series of triangles in Figure 36-6 to determine the values for angles GFE, FED, EDG, and DGF.   We note these are the angles of the inscribed quadrilateral FEDG.   From our knowledge of geometry we can show that for a quadrilateral inscribed in a circle the sum of the opposite angles equal 180°.   That means angle GFE + angle EDG equal 180°, and angle FED + angle DGF equal 180°.

   1. Angle EFG intercepts arc EDG and we have learned an angle inscribed in a circle is equal to half the intercepted arc so angle EFG = ½ arc EDG.

   2. Likewise angle EDG = ½ arc EFG.

   3. We see arc EFG + arc EDG = 360°.   We then add the results of steps 1 and 2.

   4. angle EFG + angle EDG
      = ½ arc EDG + ½ arc EFG,
      angle EFG + angle EDG
      = ½(arc EDG + arc EFG),
      angle EFG + angle EDG = ½(360°) = 180°.
      By the same argument,
      angle FEG + angle DGF = 180°.

With this proposition in mind Kepler, using the assumptions above, solves a series of triangles.   We will not follow all the details of Kepler's approach, but instead will indicate how the problem can be solved by the laws of sines and cosines for plane triangles.

   We can calculate the values of angles FAE, EAD, DAG, and GAF by the differences in the celestial longitudes between points F, E, D, and G as seen from the true sun. For example, angle FAE = 266° 39' 23" (celestial longitude of E) - 175° 43' 0" (celestial longitude of F) = 90° 56' 23".

From the values obtained for angles HCF, HCE, HCD, and HCG we can calculate angles FCE, ECD, DCG, and GCF. For example, angle FCE = angle HCE (126° 56' 18") - angle HCF (32° 6' 56") = 94° 49' 22".

We consider triangle CAF shown separately in Figure 36-7.

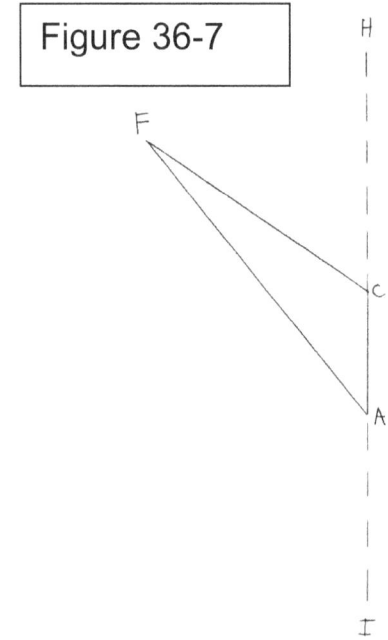

Kepler assigns length AC the value of 10,000 units. We wish to find the value of length AF.

5. We find angle C, angle C = 180° - angle HCF(32° 6' 56") =147° 53' 4".

6. Angle F = 180° - angle C - angle A, angle A = angle HAF (26° 59'),

322

angle F = 180° - 147° 53' 4" - 26° 59"

= 5° 7' 56".

7. The law of sines will give us AF:

AF/sin C = AC/sin F,

$$AF = \frac{(10,000)(.53163)}{.08945} = 59,433\,units.$$

In the same manner we can consider triangles ACE, ACD, and ACG in Figure 36-6.   We can solve triangle ACE for length AE, triangle ACD for length AD, and triangle ACG for length AG.   Kepler found AE = 52,302 units, AD = 48,052 units, and AG = 50,703 units.

We then consider triangle AFE as part of quadrilateral FEDG.   This is shown in Figure 36-8.

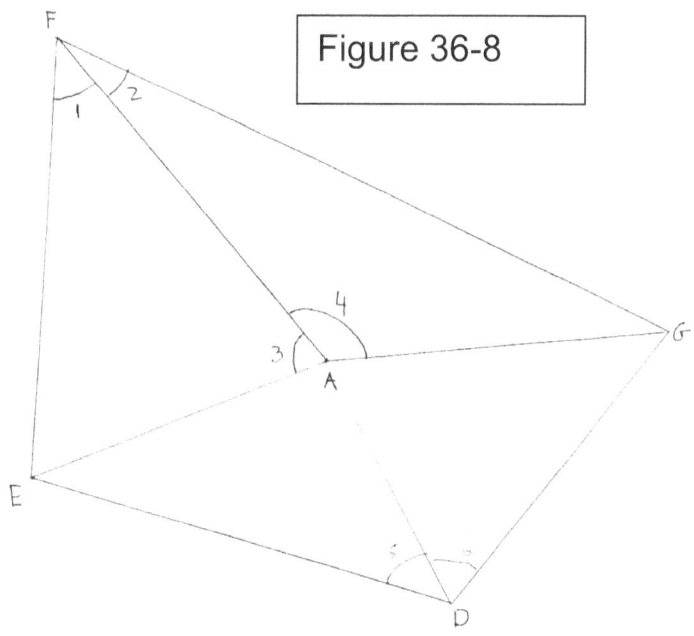

Figure 36-8

Angles 1, 2, 3, and 4, are as indicated.   We solve this triangle for angle 1.   From the above work we have found lengths AE and AF, and angle 3 (angle FAE). We use the law of cosines to find length FE.

8. $(FE)^2 = (AF)^2 + (AE)^2 - 2(AF)(AE)$ cos angle 3.

Using the values, AF = 59,433, AE = 52,302, and angle 3 = 90° 56' 23", we find FE = 79,810 units. We then use the law of sines to find angle 1.

9. AE/sin 1 = FE/sin 3, or, sin 1 = (AE)(sin 3)/FE. Using AE = 52,302, FE = 79,810, and angle 3 = 90° 56' 23", we find angle 1 = 40° 56' 17".

We then turn to triangle AFG in Figure 36-8. We have found the values for lengths AF, AG, and angle 4 (angle GAF). We solve the triangle for angle 2. We first use the law of cosines to find length FG.

10. $(FG)^2 = (AF)^2 + (AG)^2 - 2(AF)(AG)$ cos 4. Using AF = 59,433, AG = 50,703, and angle 4 = angle GAF = angle VAF (175° 43') - angle VAG (47° 24' 22") = 128° 18' 36", we find FG = 78,123 units. We then use the law of sines to find angle 2.

11. AG/sin 2 = FG/sin 4, 0r, sin 2 = (AG)(sin 4)/FG

Using AG = 50,703, angle 4 = 128° 18' 36", and FG = 78,123, we find angle 2 = 30° 36' 53".

We see from Figure 36-8 that angle 1 + angle 2 = angle EFG, one of the angles of the quadrilateral. We can continue with the same mathematical approach to solve triangles AED and ADG and thus find the three remaining angles of the quadrilateral, FED, EDG, and DGF. We can then test to see if the sums of the opposite angles equal 180°.

324

We find angle 1 + angle 2 = 30° 36' 53" + 40° 56' 17" = 71° 33' 10".   When we solve triangles EAD and DAG for angles 5 and 6 we find, angle 5 = 52° 22' 14", and angle 6 = 59° 47' 10".   The sum of 5 and 6 equals 112° 9' 24", which is angle EDG. The sum of EDG + EFG should equal 180°, if Kepler's theory and assumptions are correct.   However, 112° 9' 24" + 71° 33' 10" = 183° 42' 34".

Although Kepler used a different mathematical method than the laws of sines and cosines, he had found that the opposite angles of the quadrilateral did not equal 180°. This meant he had to revise his assumptions from Tycho's data for the celestial longitude of the line of apsides.   It took many repetitions of the above calculations to find that the correct value for angle VAH was 148° 48' 55".

Of course changing this angle changes the mean and true anomalies for each of the oppositions so the values of the lengths of AF, AE, AD, and AG all will have new values as well as all the angles. This in turn affects the results in solving the series of triangles in Figures 36-7 and 36-8 which all have to be solved anew.

Kepler went through this process again and again until he finally found values for angle VAH, the celestial longitude of the line of apsides, sides FE, ED, DG, and GF which gave him results within 2' arc for the angles of the quadrilateral. Once he had the correct values he used angle FEG (in the figure line EG is not connected) in Figure 36-6 to state:

12. angle FEG = ½arc FHG, and since arc FHG equals central angle FBG, angle FBG = 2(angle FEG).

We now examine triangle FBG.   We see it is an isosceles triangle, since BF = BG.   Thus angle BFG = angle BGF.

13.

angle FBG + angle BFG + angle BGF = 180°,

Or,

2(angle BFG) = 180° - angle FBG (which we know). For triangle BFG we now have the values for angle BFG = angle BGF, and length FG. We solve the triangle for length BF, the radius of Mars' orbit by the law of sines.

14. BF = (FG)(sin BGF)/sin FBG .

We now have the values of angles BFG and BGF, and length BF.   We see from Figure 36-6, angle AFG = angle AFB + angle BFG.   We would have determined the value of angle AFG in solving the triangles in Figure 36-8 (angle 2), so we can now find angle AFB.

15. angle AFB = angle AFG - angle BFG.

We now turn to triangle AFB drawn separately in Figure 36-9.

Figure 36-9

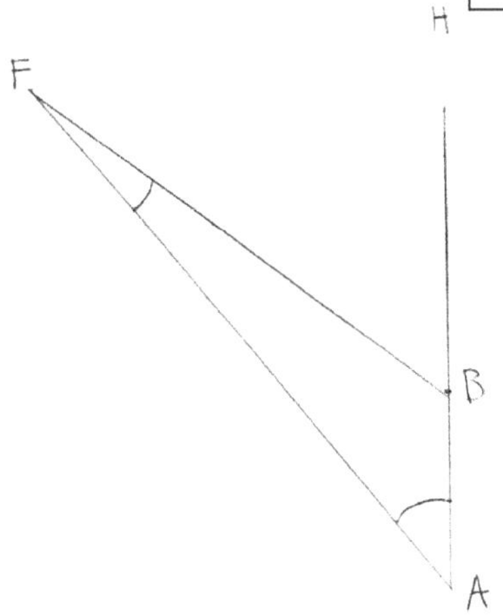

We know angle F and sides AF and BF. We can use the law of cosines to find AB.

16. $(AB)^2 = (AF)^2 + (BF)^2 - 2(AF)(BF)(\cos F)$.

We can then find BC in Figure 36-6 by,

17. $BC = AC - AB$.

In this manner we can determine the parameters for Kepler's longitudinal theory of Mars. He found the radius of the orbit BF = BE, etc. to be 53,868 units, AB to be 6,104 unit and BC to be 3,896 units, and the celestial longitude of the line of apsides for opposistion 4 in Mar of 1587 to be 148° 48' 55".

If instead of the units used by Kepler (AC = 10,000 units), we assign the radius, BF, the value of 1 unit then by proportion,

18. AC/BF = 10,000/53,868 = .18564 . This is called the eccentricity.

AB/BF = 6,104/53,868 = .11332.

BC/BF = 3,896/ 53,868 = .07232.

Kepler in commenting on his method which took so many calculations to find these results wrote,

"If you are wearied by this tedious procedure take pity on me who carried out at least seventy trials."

After determining the parameters of his theory from four oppositions, Kepler wanted to test the theory for all 12 oppositions in his revised table. He used the opposition of 1587 as his epoch date.   We will show how the theory can be tested for opposition 11 of Feb 20, 1602 which was observed by Kepler himself.   Here Kepler predicts the celestial longitude of Mars for Feb 20, 1602 and compares it to the observed value.   From the theory of the sun and his theory for Mars he hopes to predict that the earth will be 180° from the celestial longitude of Mars on that date.

From the theory of Mars on Mar 6 1587, the epoch date, angle VAH in Figure 36-6 is 148° 48' 55". Kepler knew that this angle is increasing after the epoch date from two factors.   First is the precession of the equinoxes which Tycho had determined to be 51" per year.   This means the line of apsides moves eastward or increases in celestial longitude 51" each year. Second is the finding from the work of Ptolemy and Tycho that the line of apsides is moving eastward slowly in relationship to the stars on the ecliptic.

We recall in our work with Ptolemy in chapter 20 that the star Regulus, also known as Cor Leonis   (heart of the lion, since it is located at the position of the heart of the constellation Leo), was found to have a celestial

328

longitude of 122° 30' (Leo 2° 30') in 139 A.D.   Kepler determined from the work of Ptolemy on Mars and his work that the line of apsides must include the true sun. He found that in 140 A.D the celestial longitude of the line of apsides was 122° 3', or 27' east of Regulus.

Tycho, however, had determined the celestial longitude of Regulus in 1587 to be 144° 5', and since angle VAH that year was 148° 49' (rounded off from 148° 48' 55"), then the line of apsides lies, 148° 49' - 144° 5' = 4°   44' west of Regulus.   Thus in 1,447 years the line of apsides had moved from 27' east of Regulus to 4° 44' west of Regulus, or a total of, 4° 44' + 27'= 5° 11' eastward.   This amounts to an eastward movement of the line of apsides around the ecliptic of 13" per year in addition to the 51" per year due to the precession, or a total of 1' 4" eastward movement per year.

The time from the epoch date of Mar 1587 to Feb 1602 is about 15 years so that in this time the angle VAH has increased by (15)(1 ' 4") = 16'.   Because the time is slightly less than 15 years, Kepler used the value of 15' 56", which is to be added to the epoch value of 148° 48' 55" to give 149° 4' 56" as the celestial longitude for the line of apsides for opposition 11.

We now construct Figure 36-10 in which we show the orbit of Mars with line of apsides HI with the true sun at A, center at B, and the equant at C.

329

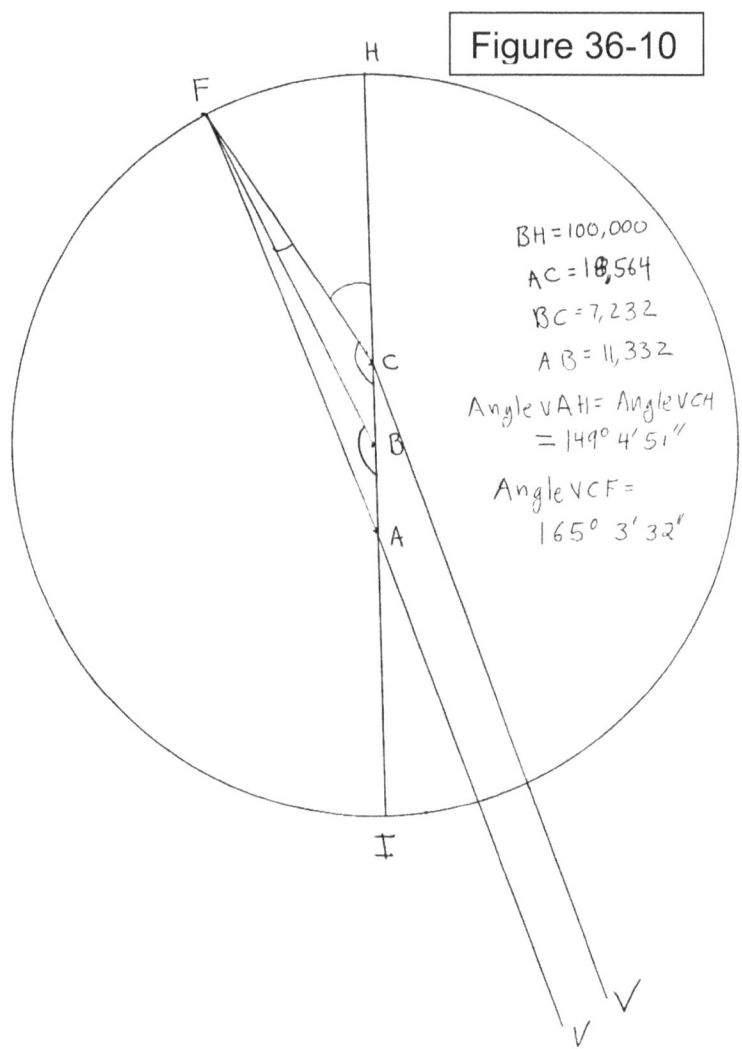

Figure 36-10

BH = 100,000
AC = 18,564
BC = 7,232
AB = 11,332
Angle VAH = Angle VCH
 = 149° 4′ 51″
Angle VCF =
 165° 3′ 32″

The opposition of 1602 is at F.

We know the celestial longitude at H = 149° 4' 56", BF is assigned the value by Kepler of 100,000 units so that AC = 18,564 units, AB = 11,332 units, and BC = 7,232 units.

Kepler needs angle HCF, the mean anomaly for Mars on the date of the opposition. Kepler uses Tycho's data to determine angle VCF equals 165° 3' 32".

Thus, angle HCF = angle VCF (165° 3' 32") - angle VCH (149° 4' 51") = 15° 58' 41".

Kepler then uses the law of sines to solve triangle BCF.

19. BF/sin C = BC/sin F, or,

sin F = (BC)(sin C)/BF.

We recall sin (180° - A) = sin A, so,

sin C in triangle BCF = sin angle HCF

= sin 15° 58' 41".

$$\sin F = \frac{(7,232)(.27527)}{100,000} = .01991.$$

F = 1° 8' 26".

Since an external angle of a triangle equals the sum of the two opposite angles, then angle HCF = angle B + angle F, for triangle BCF.

20. angle B = angle HCF (15° 58' 41")

- angle F (1° 8' 26"),

Or, angle B = 14° 50' 15".

Next Kepler turns to triangle AFB in Figure 36-10. Angle FBA in this triangle equals 180° - angle HBF, since ABF is a straight angle. Angle HBF is angle B in step 16 so,

21. angle FBA = 180° - 14° 50' 15" = 165° 9' 45". Kepler's goal is to solve triangle AFB for angle BAF which we note is the true anomaly, angle HAF, for the position of Mars at F. We see angle VAH + angle HAF = the celestial longitude of the opposition calculated by Kepler's theory. Once calculated he will compare the

result to the observed value he himself determined in 1602 to see how accurate his theory is.

Thus far in our work in solving triangles we have used the law of sines and the law of cosines for plane triangles, and we could use them in this case also. The law of cosines would give us the value of AF since,

22. $(AF)^2 = (BF)^2 + (BA)^2 - 2(BF)(BA) \cos FBA$.

We could then use the law of sines to find angle HAF since

23. BF/sin HAF = AF/sin FBA , or,

sin HAF = (BF)(sin FBA)/AF

However, in solving this triangle, as well as in solving the triangles in Figure 36-6, Kepler uses another trigonometric law called the law of tangents for plane triangles. This law can be derived from some trigonometric identities and from the law of sines.

Throughout our study of mathematical astronomy we have sought to learn the mathematics used by the great astronomers, and we will take time now to derive the law of tangents, and then use it to find the value of angle HAF as did Kepler.

In chapter 13 we learned the two following trigonometric identities,

24. $\sin (A + B) = (\sin A)(\cos B) + (\cos A)(\sin B)$

25. $\sin (A - B) = (\sin A)(\cos B) - (\cos A)(\sin B)$.

We add the two equations 19 and 20,

26. $\sin (A + B) + \sin (A - B) = 2(\sin A)(\cos B)$.

We then subtract equation 25 from equation 24,

27. $\sin (A + B) - \sin (A - B) = 2(\cos A)(\sin B)$.

We now let, C = A + B, and D = A - B. Then,

28.

$$\frac{C+D}{2} = \frac{(A+B)+(A-B)}{2} = \frac{2A}{2} = A.$$

29.

$$\frac{C-D}{2} = \frac{(A+B)-(A-B)}{2} = \frac{2B}{2} = B.$$

We substitute the results of steps 28 and 29 into equations 26 and 27:

30. sin C + sin D = 2[sin ½(C + D)][cos ½(C - D)],

31. sin C - sin D = 2[cos ½(C + D)][sin ½(C - D)].

These formulas hold for any value of C or D so we could use A in place of C and B in place of D in equations 30 and 31.   Thus these formulas can be applied to a triangle, ABC, with angles A, B, and C with sides a, b, and c where a is opposite A, b opposite B, and c opposite C.   See Figure 36-11.

Figure 36-11

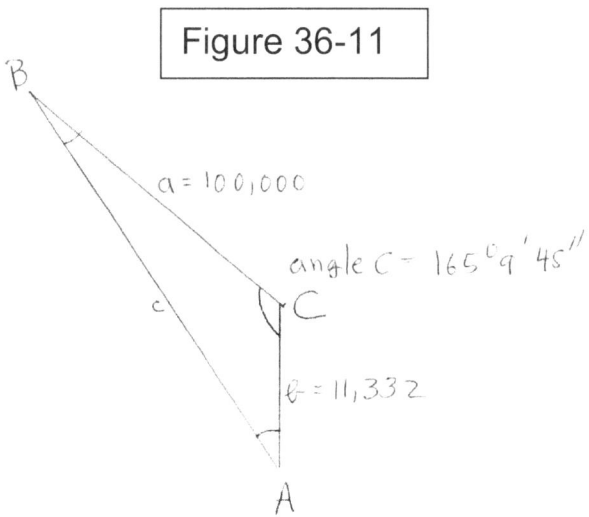

For triangle ABC , the law of sines states,

333

32.   a/sin A = b/sin B, or, a/b = sin A/sin B.

We recall our work with proportions in chapter 1 taught that if, a/b = c/d, then we can change the proportion by composition and division to get,

33. (a + b) / (a − b) = (c + d) / (c − d)   .

We let c = sin A, and d = sin B, in the proportion and apply it to triangle ABC:

34.

$$\frac{a+b}{a-b} = \frac{\sin A + \sin B}{\sin A - \sin B}.$$

We now use equations 30 and 31 in the right side of equation 34, and use A in place of C, and B in place of D.

35.

$$\frac{a+b}{a-b} = \frac{2\,[\sin \tfrac{1}{2}(A+B)][\cos \tfrac{1}{2}(A+B)]}{2\,[\cos \tfrac{1}{2}(A+B)][\sin \tfrac{1}{2}(A-B)]}.$$

Since, sin/cos = tangent(tan),

36.

$$\frac{a+b}{a-b} = \frac{\tan \tfrac{1}{2}\,(A+B)}{\tan \tfrac{1}{2}\,(A-B)}.$$

This is the law of tangents for solving triangles. To use it to solve for angle A of triangle ABC the law is written as follows by solving equation 31 for tan ½(A - B):

37.

$$\tan\frac{A-B}{2} = \frac{a-b}{a+b} \cdot \tan\frac{A+B}{2}.$$

In triangle ABC we know, A + B + C = 180°, or A + B = 180° - C. Thus, ½(A + B) = 90° - ½C.   Equation 32 becomes

38. tan (A – B)/2   = (a – b)/(a + b)   • tan (90° - ½C)

If we start knowing the values of two sides of the triangle, a and b, and the value of angle C, we first solve equation 33 for tan ½(A - B).

Since (90° - ½C) = ½(A + B) , we then know the value of both ½(A - B), and ½(A + B).   If we add these two results we have angle A since,

39.

$$\frac{A+B}{2} + \frac{A-B}{2} = \frac{2A}{2} = A.$$

Kepler uses the law of tangents in this way to solve triangle ABF in Figure 36-10.   We recall, BF = 100,000 units, BA = 11,322 units, and angle FBA = 165° 9' 45".   We then label triangle ABF as in Figure 36-11. BF = a, BA = b, and angle FBA = C.   We use the law of tangents as given in equation 37.

40.

$$\tan\frac{A-B}{2}$$

$$=\frac{100,000-11,332}{100,000+11,332}\cdot\tan(90^0-\frac{165^09'45''}{2}).$$

$$\tan\frac{A-B}{2}=\frac{88,668}{111,332}\cdot\tan7^025'8''=$$

(.79643)(.13021) = .10370.

Using the arctangent, ½(A – B) = 5^0 55' 14".

We find A by equation 39. First we need

(90^0 - ½C), C = 165^0 9' 45' = 165.1625^0.

½C = 82.58125^0,

90^0 - 82.58125^0 = 7.41875^0 = 7^0 25' 8".

41. ½(A + B) + ½(A - B) = (90° - ½C) + ½(A - B) = A,

(7° 25' 8")   + (5° 55' 14") = 13° 20' 22" = A.

Going back to Figure 36-10, we see angle HAF is the same as angle A in Figure 36-11. The celestial longitude for F is,

42. angle VAF = angle VAH + angle HAF,

angle VAF = (149° 4' 51") + (13° 20' 22"),

angle VAF = 162° 25' 13" (Virgo 12° 25' 13").

Kepler's observed value for opposition 11 in Table 2 was 162° 27' 0" (Virgo 12° 27' 0").   The difference between the observed and calculated value is 1' 47" within the error Kepler will accept for his theory.

Kepler went through the process of checking all 12 oppositions with his theory using the law of tangents.

The largest difference between his theory and the observations was 2' 12" for opposition 5 in 1589.   Most of the differences were less than 1' of arc.

At this point we note that in calculating the predicted value for the celestial longitude of the opposition, angle VAF in Figure 36-10, Kepler starts his calculation with the value of angle VCF, the mean celestial longitude or the celestial longitude as seen from the equant C.   This value is determined from Tycho's theory of the mean motion of Mars where Mars moves around the equant at uniform angular velocity which we designate as ω_c.   Today the value of ω_c is 360° in 686.98 days.   Kepler does not tell us the value Tycho used but a close approximation is 360^0 in 686.75 days.

We recall that when Ptolemy developed his theory he had to have very exact values for the uniform angular velocities around the deferent and the epicycle to make prediction of future planetary positions and to check the other parameters of his theory.   For the sake of a mathematical exercise let us start with all the parameters of Kepler's theory for the fourth opposition on Mar 6, 7 hours 23 minutes, 1587.   We will then calculate the value of angle VCF for the time of the eleventh opposition on Feb 20, 14 hours, 13 minutes, 1602.   This means we will establish the fourth opposition as the epoch date for Kepler's theory.   See Figure 36-12.

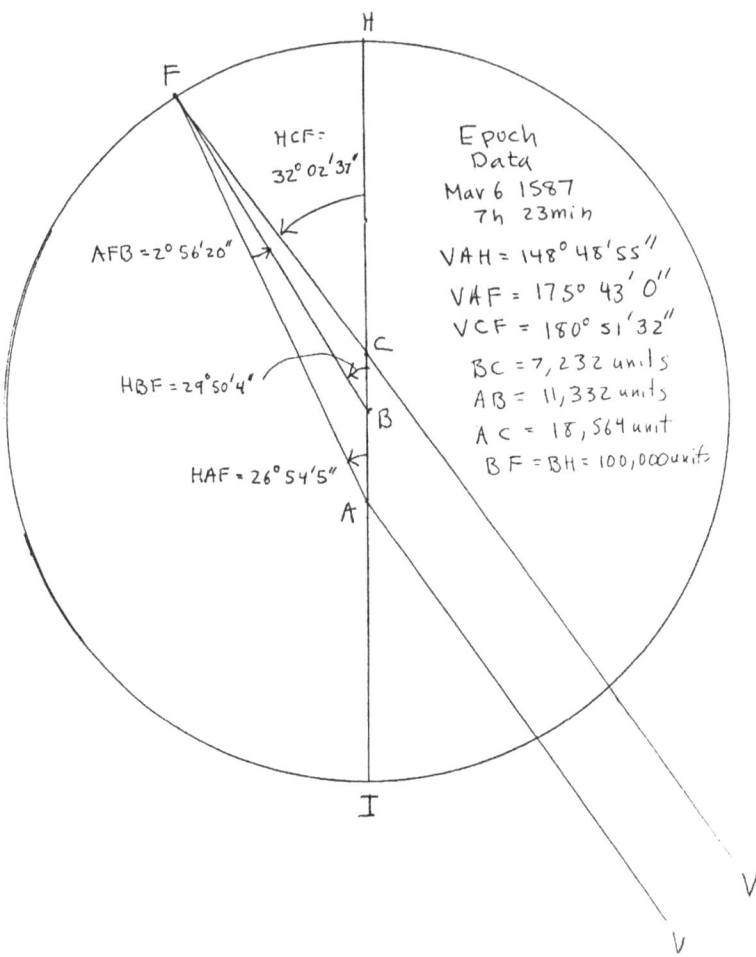

Figure 36-12

HCF = 32° 02' 37'

Epoch Data
Mav 6 1587
7h 23min

AFB = 2° 56' 20"

VAH = 148° 48' 55"
VAF = 175° 43' 0"
VCF = 180° 51' 32"
BC = 7,232 units
AB = 11,332 units
AC = 18,564 unit
BF = BH = 100,000 units

HBF = 29° 50' 4'

HAF = 26° 54' 5"

Kepler's model states that on the epoch date in 1587 angle VAH = 148° 48' 55", while the position of Mars as seen from the true sun, angle VAF is 175° 43' 0" (taken from Table 2). Remember that at the time of opposition with the true sun the earth lies along line AF between the sun at   A and Mars at F. We first find angle HAF.

43. angle HAF = angle VAF - angle VAH

angle HAF = 175° 43' 0" - 148° 48' 55",

338

angle HAF = 26° 54' 5".

We recall Kepler had determined BF = 100,000 units, AB = 11,332 units, and BC = 7,232 units. We seek to solve Figure 36-12 for the value of angle VCF. To do so we begin by solving triangle ABF for angle AFB. We use the law of sines.

44. BF/ sin HAF = AB/sin AFB, or, sin AFB

= AB sin HAH/BF.

sin AFB = (11,332)(sin 26^0 54' 5")/100,000

= .051272.

Angle AFB = 2° 56' 20", which is marked in the figure. Since an external angle of a triangle equals the sum of the two opposite interior angles,

45. angle HBF = angle HAF + angle AFB,

angle HBF = 26° 54' 5" + 2° 56' 20",

angle HBF = 29° 50' 41 ".

We will use this value in solving triangle CBF for angle FCB. As was Kepler's practice in solving such triangles we will use the law of tangents. We label triangle CBF in Figure 36-13.

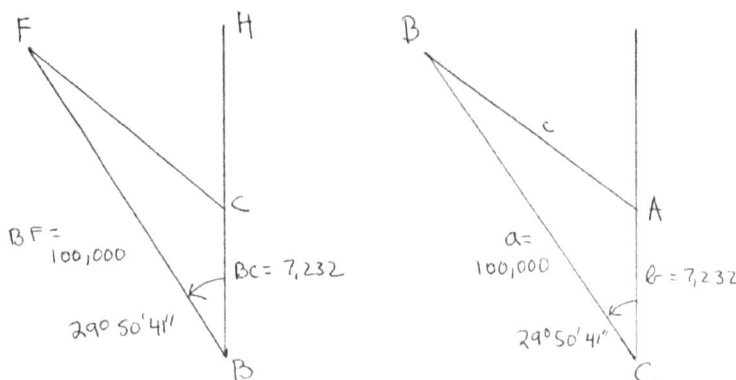

Figure 36-13

Then we can use the law of tangents in the form,

46. tan (A – B)/2 = (a – b)/(a + B)· tan   (90° - ½C),

where, (90° - ½C) = ½(A + B).

We see from Figure 36-13, a = 100,000,

b = 7,232, and

C = 29° 50' 41".

47. tan ½(A - B) = (100,000 – 7,232)/100,000 +

7,232) • tan [90° - ½(29° 50' 41")]

tan ½(A - B) = (.86511)(tan 75° 4' 40"),

tan ½(A - B) = 3.24624,

½(A - B) = 72° 52' 43". Since,

(90° - ½C) = 75° 4' 40", then

½(A + B)   = 75° 4' 40".

48. ½(A - B) + ½(A + B) = A = 72° 52' 43" + 75° 4' 40",

A = 47° 57' 22".

From Figure 36-13, A = angle FCB.

49. In figure 36-12,

340

angle FCH = 180° - angle FCB

angle FCH = 180⁰ - 147° 57' 22"

angle FCH = 32° 2' 37".

50. We see,

angle VCF = angle FCH + angle VCH,

angle VCF = 32° 2' 37" + 148° 48' 55",

angle VCF = 180° 51' 32".

We now have all the parameters for the epoch date.   We will use the value of 360° for 686.75 days for ω_c, the uniform angular velocity of Mars, point F, around its orbit as seen from C.

The opposition of 1602 (Feb 20, 14h 13m) occurs 5,464.285 days after the epoch date, Mar 6, 1587 7h 23 m.   This value is calculated from the calendar. Between 1587 and 1602 there are 11 years of 365 days, 3 years of 366 days (remember from our work on the calendar that the year 1600 is not a leap year), and 1 year of 351.285 days.

5,464.285 d x 1 revolution/686.75 d = 7.9567 revolutions.

.9567 revolutions x 360°/ revolution = 344.4232° beyond

7 revoultions.   We add this to the starting value of angle

VCF of 180° 51'32" (180.8589°).

344.4232° + 180.8589° = 525.2820°.

525.2820° - 360° = 165.2820°, is the value of angle VCF in 1602 without correction for the precession of the equinoxes and the movement of the line of

apsides. We recall Kepler's correction is 15' 56" =
.26556°. Then,   165.2820° - .26556° = 165.0165° =
165° 1'.

This then is the predicted value of angle VCF for
the time of the eleventh opposition in 1602.   We recall
Kepler used the value of 165° 3' 52" based on Tycho's
calculation of the mean motion of Mars around an
equant point when he began his calculation for the
predicted value of angle VAH in Figure 36-10.

Kepler now had what appears to be a highly
successful theory for the celestial longitude of Mars.
However, when he tested his theory further he ran into
trouble which was to lead him on a further search for the
correct theory for the motion of Mars.   Nonetheless, the
theory for the celestial latitude given in chapter 35, and
theory for the celestial longitude in this chapter were
major steps in the history of astronomy.   Kepler's
theories were far superior to Ptolemy's and Copernicus'
for predicting the celestial longitudes and latitudes of
Mars.   His idea of making the true sun the basis for the
theory is one of the great advances in his diligent work.

Most books which discuss Kepler's work in the
"*Astronomia Nova*" do not give any detailed calculations,
but instead summarize his results.   Readers then know
little of how Kepler actually proceeded and how
important the observations of Tycho were in his theory.
Although, we have only shown a few calculations from
the hundreds performed by Kepler, we have shown the
geometric and trigonometric considerations needed to

understand how Kepler did his work as well as some of the numerical data from the "*Astronomia Nova*".

In the next chapter we will see where problems occurred with the theory for Mars. Kepler, however, was so persistent in his endeavors that eventually he found what today we call his law of areas and his great discovery that Mars moves in an elliptical orbit.

Bibliography for Chapter 36

1. Kepler, Johannes,   "Astronomia Nova", Pars Secunda, Chapters CVI and XVIII, Latin edition printed by Culture et Civilisation, 115, Avenue, Gabriel Lebon, Bruxelles, 1968.

2. Small, Robert, "An Account of the Astronomical Discoveries of Kepler", A reprinting of the 1804 text by The University of Wisconsin Press, Madison Wisconsin, 1963.

3. Klaf, A.A., "Trigonometry Refresher for Technical Men", Dover Publications, New York, 1946.

4. Kepler, Johannes, "New astronomy", translated by William H. Donahue, Cambridge University Press, 1992.

Chapter 37

The *Hypothesis Vicaria* and the Need for Revision

Kepler's theory for the celestial longitude of Mars based upon a circular orbit with the center eccentric to the true sun with an equant point as described in the last chapter was satisfactory to an unprecedented degree of accuracy for the twelve oppositions from 1580 to 1604. However, Kepler found the theory was unsatisfactory for prediction of the celestial latitudes at the times of the oppositions when considering his theory for the celestial latitudes.   Kepler came to call his theory for celestial longitudes the *'hypothesis vicaria'*, which is Latin for the vicarious theory, since it stood in place of the physical theory he was seeking based on the true sun.

In this chapter we shall explore the problems that arose with the *'hypothesis vicaria'*, and the path Kepler pursued to resolve them.   Kepler studied the *'hypothesis vicaria'* for two oppositions, numbers 3 and 7, in Table 2 of chapter 36. Here he was interested in calculations based on the celestial latitudes of the oppositions, and the relative distances of the earth and Mars from the sun.   Let us start by giving the data for these two oppositions.

	Date	Celestial longitude	Celestial latitude
3	1585	141° 36' 10"	4° 32' 10" North
7	1593	342° 16' 0"	6° 2' 0" South

We recall to the nearest degree the aphelion for the *hypothesis vicaria'* is 148°, and the perihelion is 148° + 180° = 328°. We see that that opposition 3 is near aphelion, and opposition 7 is near perihelion. We recall also from chapter 35 that the line of nodes runs from celestial longitude 47° to 227° in Kepler's theory for the celestial latitude of Mars. We show these relationships in Figure 37-1.

Figure 37-1

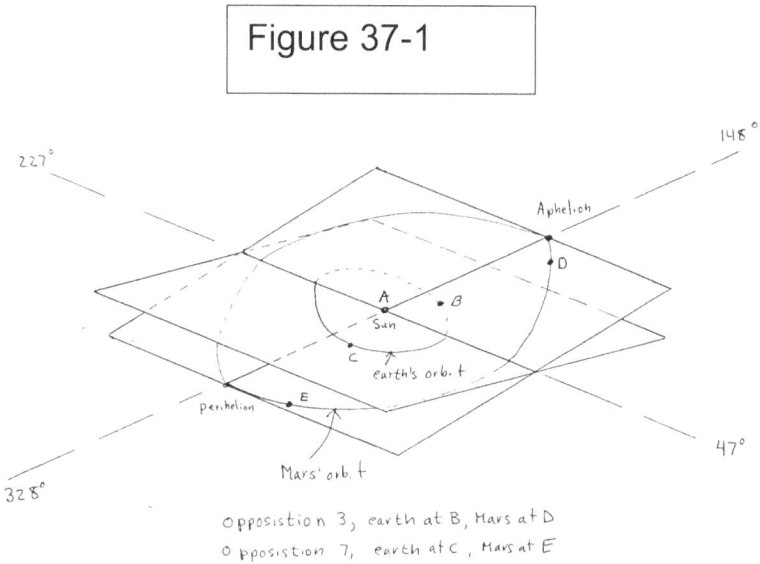

Opposistion 3, earth at B, Mars at D
Opposistion 7, earth at C, Mars at E

345

Figure 37-1 Enlarged

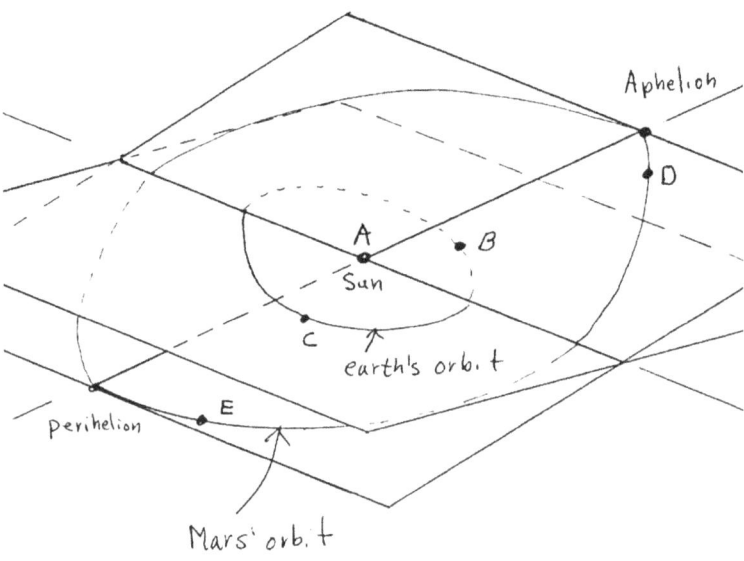

Opposistion 3, earth at B, Mars at D

Opposistion 7, earth at C, Mars at E

In this figure we have indicated the plane of the ecliptic and its extension to the celestial sphere represented by the outermost circle. The lines connecting the aphelion to the perihelion and the line of nodes 47^0 to 227^0, are indicated. The positions of Mars in its orbits for oppositions 3 and 7 are points D and E respectively. The corresponding positions of the earth for these oppositions with the true sun at A are points B and C.

In Figure 37-2 we have constructed a cross section through opposition 3 where the relative positions of the sun, A, earth, B, and Mars, D are marked.

346

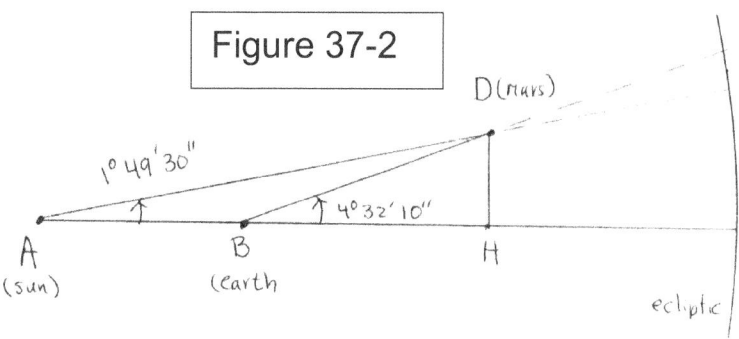

Figure 37-2

1° 49' 30"

D (mars)

4° 32' 10"

A
(sun)

B
(earth

H

ecliptic

Kepler determined that if Mars were observed from the sun at aphelion (celestial longitude 148°), angle DAH in Figure 37 -2 would be 1° 50'.   Because at opposition 3 the celestial longitude of Mars is 141°,   7° east of the aphelion,   angle DAH should be 1° 49' 30" at the time of opposition 3.   We see in the figure the celestial latitude at opposition 3 is angle DBH = 4° 32' 10".   We can calculate angle DBA as,

    1. angle DBA = 180° - 4° 32' 10" = 175° 27' 50".

    2. If angle DAH = 1° 49' 50" then,

    angle BDA = 180° - angle DAH (1° 49' 50") - angle DBA (175° 27' 50") = 2° 42' 40".

Kepler can now solve triangle ABD for side AD using the law of sines provided he has a value for the length of side AB, the distance from the earth to the sun at the time of the opposition.   Kepler determines this value from considerations shown in Figure 37-3.

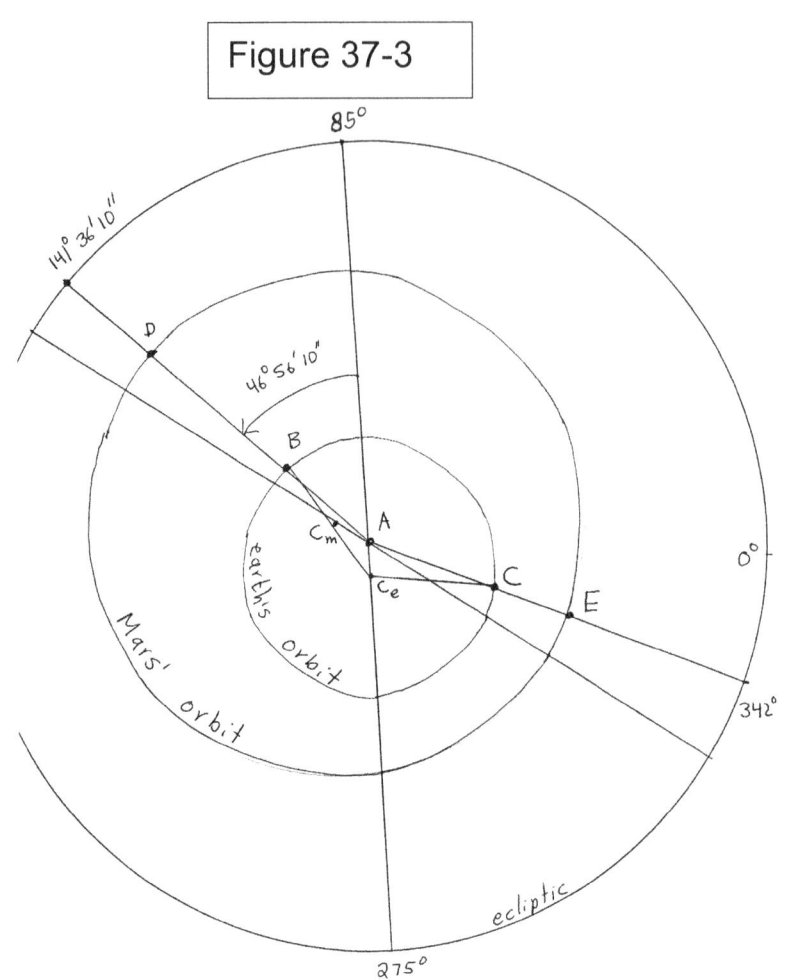

Figure 37-3

348

Figure 37-3 Enlarged

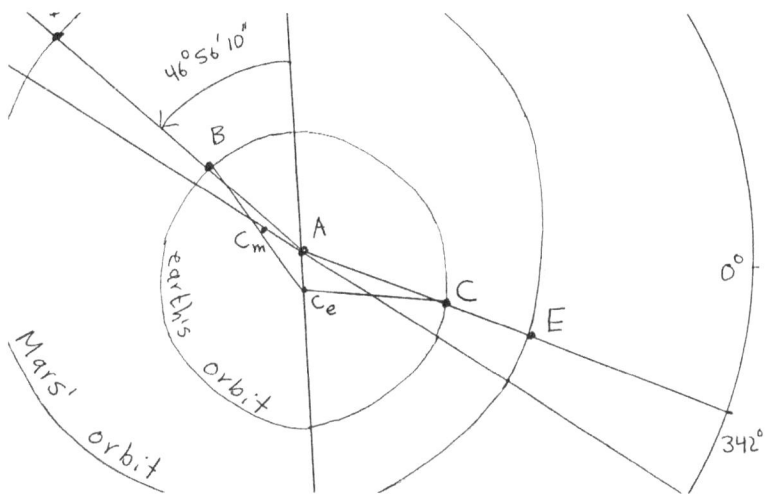

In this figure we view from the north ecliptic pole. The center of the earth's orbit is C_e with the true sun eccentric at A. The line of apsides for the earth's orbit comes from Tycho's solar theory which in 1585 was 275°. This means-line C_eA has a celestial longitude of 275°. The earth's orbit is the smaller circle with the earth's position at the third opposition at B, and the seventh opposition at C.

The orbit of Mars for the *'hypothesis vicaria'* has its center at C_m, eccentric to the sun at A and with its line of apsides having celestial longitude of 148°. At the third opposition Mars is at position D, and at the seventh opposition at position E. Kepler assigns the value of 100,000 units for the radius of the earth's orbit, C_eB. Tycho had found the eccentricity, C_eA, for the earth's orbit was 3,584 units. Kepler uses these values to solve triangle AC_eB for side AB, since he can find the value of angle BAC_e from the celestial longitude of D.   AB was found to be 97,500 units.   He also found the value of AC to be 101,400 units by solving triangle AC_eC

We now return to Figure 37-2 and use the value of 97,500 units for AB.

3. AD/sin ABD = AB/sin BDA, or, AD = (AB)(sin ABD)/sin BDA.

$$AD = \frac{(95,700)(\sin175^0 27'50'')}{\sin2^0 42'40''}$$

$$= 163,023\ units$$

where the radius of the earth's orbit equals 100,000 units. By the same mathematical argument length AE in Figure 37-3 can be found.   Kepler found AE = 139,000 units.   Kepler then finds the length of AD if AD in Figure 37-3 coincided with the line of apsides.   We see from the figure that if AD coincided with the line of apsides it would lie about 7° further to the east and hence would be slightly longer.   Kepler added 150 units to its value so that if AD lay along the line of apsides its value would be 163,173 units.   If AE in Figure 37-3 coincided with the line of apsides it would be less in length.   Kepler subtracted 300 units from his calculated value so that AE equals 139,000 units if it coincided with the line of apsides.   Furthermore, if D and E were on the line of apsides then,

4. DE = AD + AE = 302,173 units.

The radius for Mars' orbit is ½DE = C_mD = 151,087 units so that   the eccentricity of the center of the orbit of Mars would be,

5. AC_m = AD – C_mD = 163,173 - 151,087,

AC_m = 12,086 units, where C_eB, the radius of the earth's orbit is 100,000 units.

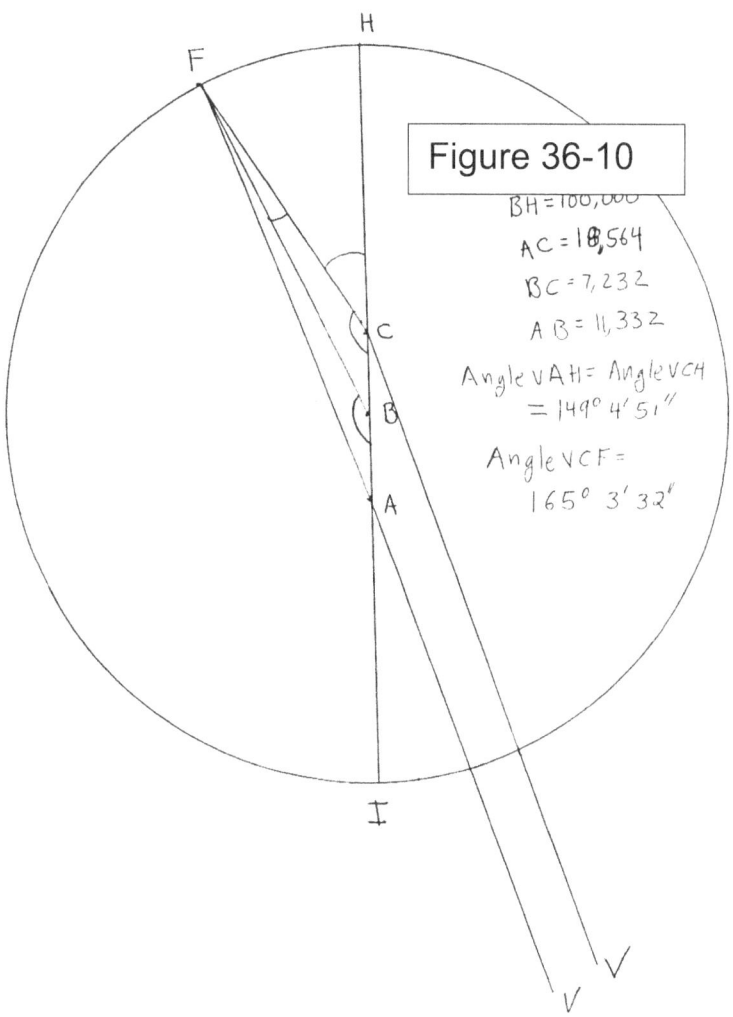

Figure 36-10

BH = 100,000
AC = 18,564
BC = 7,232
AB = 11,332

Angle VAH = Angle VCH
= 149° 4' 51"

Angle VCF =
165° 3' 32"

To find the value of AC_m in Figure 37-3 when the radius of Mars is assigned the value of 100,000 units as in Figure 36-10, we solve the proportion,

351

$$\frac{AC_m}{100,000} = \frac{12,086}{151,087}$$, since when AC_m is 12,086

units the radius of Mars orbit,, ½DE, is 151,087 units. AC_m = 7,999 units with the radius of Mars' orbit is 100,000 units.

In chapter 36 we found Kepler's value for AC_m in the 'hypothesis vicaria' was 11,332 units when the radius of Mars orbit is assigned the value of 100,000 units (see Figure 36 -10 for length AB). This disagreement between the value determined by the observation of the third opposition and Kepler's theory was too great, and indicated to Kepler some feature of the 'hypothesis vicaria' was not correct.

Kepler considered that one principle of his theory that could be incorrect was that the true physical orbit of Mars may not be a perfect circle as in the 'hypothesis vicaria". Another principle of the theory that could be incorrect was that uniform motion about an equant point with the eccentricities as determined could be incorrect.

Kepler went on to test the 'hypothesis vicaria' from observations Tycho had recorded when Mars was not in opposition with the sun and when the planet had a celestial longitude close to the line of apsides. He again calculated the eccentricity from the data of the observations and found that it did not agree with the eccentricity, 11,332, from his theory.

Kepler had approached his new astronomy starting with a physical hypothesis that the body of the sun caused the planets to move in their orbits according to a mathematical plan instituted by God. He felt with diligence and perseverance man could find the mathematics of the true orbit if he followed the physical principle that the sun was the source of the force that moved the planets. This concept was a marked departure from his predecessors Ptolemy, Copernicus, and Tycho Brahe, who had worked with geometric models based on the position of the mean sun, and on the concept of circular motion. They were satisfied to be able to explain the planet's position to within 10' of arc and did not hypothesize a physical cause for the motion in the body of the sun. Kepler's insistence in a physical theory caused him to continue in his work despite the setback he found in working with the relative distances of the earth and Mars from the sun which produced the error in the *'hypothesis vicaria'*.

Kepler turned to "The Almagest" and noted that Ptolemy had bisected the eccentricity in his theory for Mars. We recall that Ptolemy did not clearly explain why he did so, and Kepler came to believe that Ptolemy did this because of the same type of problem that he, Kepler, had with the celestial latitudes during the oppositions. Kepler wondered if he bisected the eccentricity in his theory by placing the center of Mars' orbit half way between the equant and the true sun, would the problem with the latitudes be solved while still maintaining an accuracy within 10' of arc for the celestial longitudes of the oppositions. Kepler made the calculations with this revised model of the 'hypothesis vicaria'. In Figure 36-10 this means AB = BC = ½(11,332 + 7,232) = 9,282, when BF = 100,000.

When he tested this revised theory with the oppositions he found that if the opposition occurred around 90° or 270° from the line of apsides, the celestial longitudes agreed within 2' of arc, but for oppositions near the line of apsides such as the ones in 1585 and 1593 errors of about 8' of arc were evident. Kepler wrote of this discrepancy,

"Divine providence granted us such a diligent observer in Tycho Brahe that his observations convicted the Ptolemaic calculation [the calculation made by bisecting the eccentricity] of an error of 8'; it is only right that we should accept God's gift with a grateful mind, because these 8' have led to a total reform of astronomy".

He was referring to his eventual discovery of the elliptical orbit for Mars with the sun at one focus and the law of areas. An error of 8' would not have distressed Ptolemy or Copernicus, but it was the turning point for Kepler.

Kepler's next step was to reconsider the solar theory of Tycho, that is the motion of the earth as seen from the mean sun. Tycho, like Ptolemy, had used an eccentric circle theory for the apparent motion of the sun as seen from earth like the model we described in chapter 18. This model has no equant point since the motion of the sun is considered to have uniform angular velocity as seen from the center of the eccentric circle, or we could say the equant is located at the center of the circular orbit.   It occured to Kepler that if the planet Mars has an equant point as in the *'hypothesis vicaria'*, then the earth, since it is a planet in the Copernican system, should also have an equant point eccentric to the center of its orbit just as the true sun is eccentric to the center of the orbit.   Kepler still considered the basic orbits of the planets at this stage of his thinking to have circular orbits. He set about the task to determine if the earth indeed had an eccentric equant point using observations of the sun's position from the data of Tycho.   Historians consider this phase of Kepler's work as his attack on the second inequality of planetary motion, since it is the motion of the earth around the sun relative to Mars which causes the retrograde arcs of Mars during oppositions as seen from earth, and this phenomenon is the second inequality of planetary motion.   Kepler felt if he could establish that the earth like Mars had an eccentric equant point then he might be able to correct the errors in the *'hypothesis vicaria'*.

In order to understand how Kepler studied the motion of the earth in its orbit around the sun, we must take up the subject of the parallax of Mars caused by the earth's motion relative to a stationary sun.   We construct Figure 37-4.

Figure 37-4

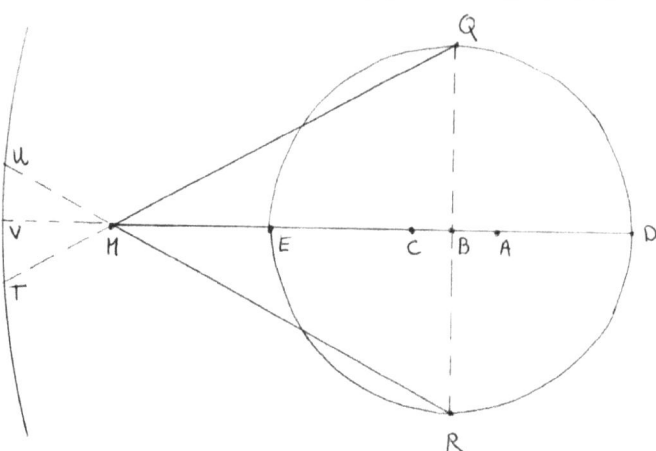

Here we view the orbit of the earth from the north ecliptic pole. Point A represents the true sun which is eccentric to the center of the earth's circular orbit located at B.   The line of apsides for the earth's orbit is DE which we extend to the ecliptic on the celestial sphere. We assume Mars is observed when the earth is at position Q,   90° celestial   longitude from the line of apsides, and Mars at position M such that the heliocentric celestial longitude of Mars is the same as the heliocentric longitude as the line of apsides DE. The observer on earth would see Mars along line of sight QM, and would record its celestial longitude at point T on the ecliptic.   If Mars is subsequently observed from point E on the earth's orbit, the celestial longitude would be point V on the ecliptic.   Angle QME is defined as the parallax angle for Mars for the observation from point Q.   Since the celestial sphere is so far from the solar system, angle VMT, which equals angle QME, can be taken as the central angle for arc TV on the ecliptic so the difference between celestial longitudes T and V is the value of the parallax angle QME.   Likewise if Mars at position M is observed on earth from position R on the earth's orbit, the observer on the earth at R would record the celestial longitude of Mars as point U on the ecliptic.   The difference in celestial longitudes between V and U is the parallax angle, angle RME.   Because M is on the line of apsides of the earth's orbit the two parallax angles, QME and RME will be equal, since triangles QME and RME are congruent because side RB = BQ, the radius of the earth's orbit, both triangles share side MB, and the included angles between the corresponding equal sides, right angles MBR and MBQ are equal.

Kepler theorized that if there were an equant point along the line of apsides eccentric from the center, B. This is represented by point C. Then Kepler could calculate different parallax angles for Mars with the proper observations.   To illustrate this concept we construct Figure 37-5.

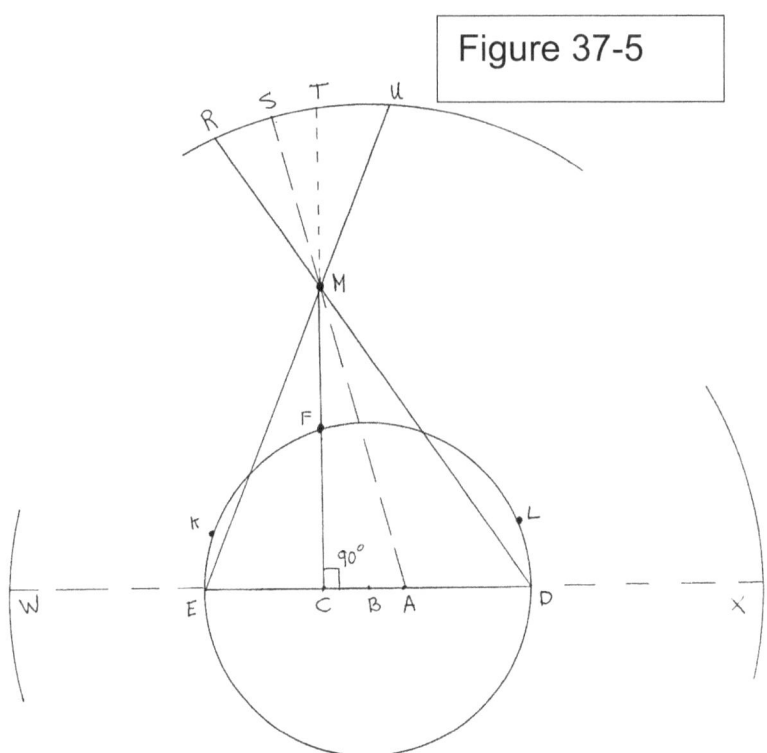

Figure 37-5

If C is an equant point, then the earth moves along its orbit with uniform angular velocity, ω_e , as seen from C,

$$\omega_e = 360^0/365.25d = .9856°/d.$$

We now envision the earth moving from D, perihelion (the sun at A, the center of the earth's orbit at B), to F, a 90° rotation as seen from C. This will take,

$$90° \times 1d/.9856^0 = 91.3125 \text{ d}.$$

Consider an observation of Mars when the earth is at F and Mars is at M, such that line of sight FM, when extended back to the line of apsides is perpendicular to DE at C. This means Mars will have celestial longitude at T on the ecliptic as seen from earth at F, and celestial longitude at S as seen from the true sun.

358

We then need two more observations of Mars when it is at M with heliocentric celestial longitude at S. The first one is when the earth is at D. Mars will have celestial longitude at R on the ecliptic. We then calculate the parallax for Mars for this opposition as angle DMC, which equals the difference between celestial longitudes R and T.

The second observation we need is when the earth is at E and Mars at M. The calculated parallax for this observation is angle EMC, which equals the difference between celestial longitudes U and T.

If the center of uniform motion for the earth, C, is eccentric to the center of the orbit, B, then we see from the figure that EC does not equal DC, so triangles ECM and DCM are not congruent and the parallax angles, EMC and DMC, cannot be equal. Conversely if the parallax angles are found to be significantly different from each other by calculations made from the observations, then the center of uniform motion, C, is not coincident with B, the center of the earth's orbit.

Kepler could not find observations by Tycho exactly corresponding to the above, but found two observations when the earth was at K and L and Mars at M. The calculated parallax angles for these two observations were markedly different, which convinced him that the earth's orbit could not be described as an eccentric circular orbit with uniform motion about the center. This discovery represented another of the findings for the new astronomy of Kepler.

His next task was to determine the numerical values for the eccentricity. He found that if BD, the radius of the earth's orbit, is assigned the value of 100,000 units that BC = 1,837 units. Using Tycho's work, AB was 1,747 units so that AC = 3,584 units. This means the center of the earth's orbit nearly bisected AC, the total eccentricity. Kepler had then discovered that the earth's orbit was very simiiar to the situation he had found for the 'hypothesis vicaria' for Mars, which further convinced him that the earth behaved as a planet in a manner similar to Mars.

This discovery was so important for planetary theory that Kepler decided he needed to demonstrate its validity with more observations and calculations. The demonstration that both earth and Mars had the same basic theory argued strongly that the physical cause of planetary motion lay within the body of the sun, and the same theory for one planet would serve for all.

To further confirm the theory of the eccentric equant for the motion of the earth, Kepler recognized that Mars would occupy the same heliocentric longitude every 687 days, the period for Mars. He then studied four observations of Mars by Tycho 687 days apart, and corrected the data for the precession of the equinoxes. We will show his calculations for the first of these which occurred on Mar 5, 1590 at 7 h 10 m. See Figure 37-6.

Figure 37-6

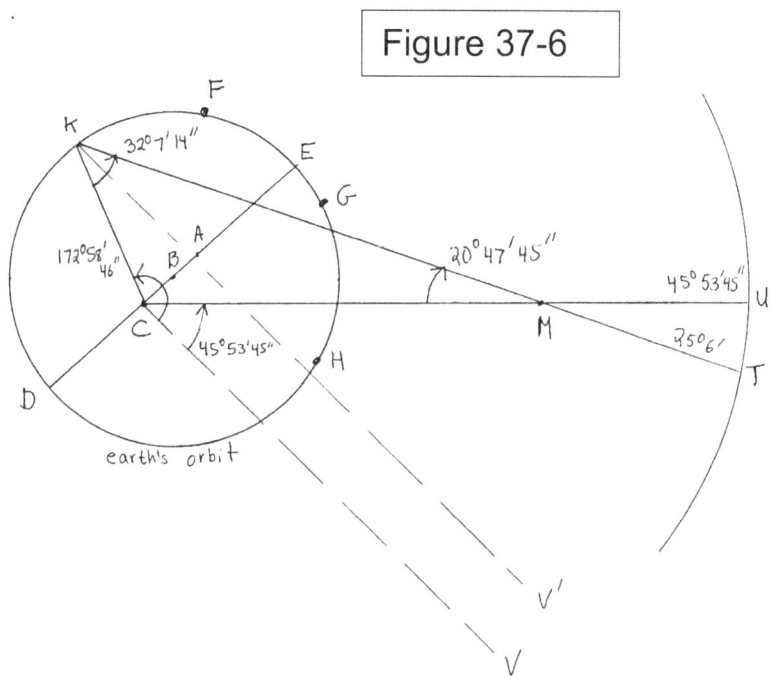

The earth's orbit is viewed from the north ecliptic pole with DE the line of apsides, B the center, A the sun, and C the presumed equant point.   For the observation Mars is at position M and the earth at K.   From the *'hypothesis vicaria'* for this date, Kepler knew the heliocentric longitude of Mars, angle VAM equivalent to angle VCM, since the celestial sphere is so far from the solar system, was equal to 45° 53' 45".   Kepler knew the longitude of the earth from Tycho's theory for the observation to be 172° 58' 46" which is angle VCK. This means angle KCM in triangle KCM is equal to 172° 58' 46" - 45° 53' 45" = 127° 5' 1".

As seen from the earth at K, the observation indicated Mars had a celestial longitude of 25° 6'.   This observation runs along line KM which extended to the ecliptic is point T. We see from the figure CM extended to the ecliptic is point U, and that arc TU then equals 45° 53' 45" - 25° 6' = 20° 47' 45". We can think of this as parallax angle UMT = angle KMC. Kepler now has two angles for the triangle KCM so he can calculate angle CKM as equal to:

180° - angle KMC - angle KCM, or,

angle CKM = 180° - 20° 47' 45" - 127° 5' 1 "

= 32° 7' 14".

Kepler's goal is to find length CK for the observation in terms of length CM, and then compare this result with calculations for the other three observations when the earth is at points F, G, and H on its orbit. He assigns length CM the value of 100,000 units which is constant for the four observations. Length CK can be found by the law of sines:

CK/sin KMC = CM/sin CKM , or, CK = (CM)(sin KMC)/sin CKM.

$$CK = \frac{(100,000)(\sin 20^0 47'45'')}{\sin 32^0 7'14''}$$

$= 66,774\, units.$

He then considered the three other observations when Mars is at M. See Figure 37-7.

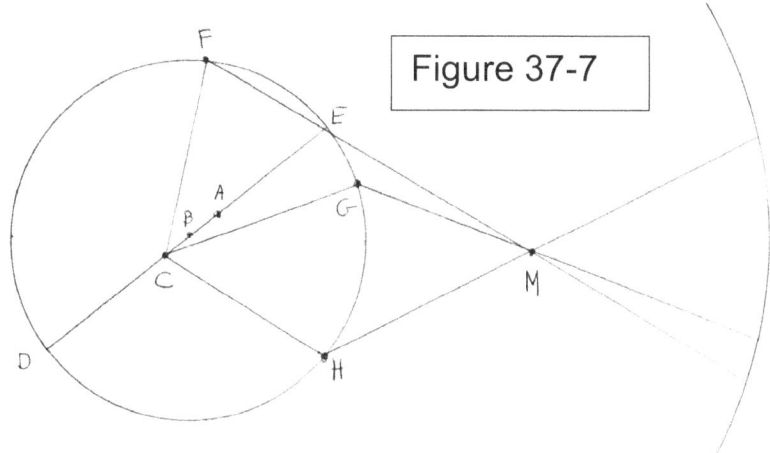

Figure 37-7

Mars is again at point M and has the same heliocentric celestial longitude as in Figure 37-6. The three observations were made when the earth was at F, G and H, each 687 days apart. Kepler then uses the same procedure for the first observation to solve triangles FCM, GCM, and HCM for sides CF, CG, and CH, respectively. If the center of uniform motion, C, is coincident with B, the center of the earth's circular orbit, then CF = CG = CH, and equals CK from the first observation in Figure 37-6. Kepler found CF = 67,467, CG = 67,794, and CH = 67,478 when CM has the value of 100,000. We recall CK = 66,774.

The finding that all these distances varied one from another indicated to Kepler that the theory for the earth's orbit includes an equant point eccentric to the center of the orbit just as in the case of Mars and confirmed his earlier work on the earth's orbit. However, Kepler still was not satisfied that he had a full demonstration for the eccentric equant for the earth and so he produced several more arguments for it in the "*Astronomia Nova*".

363

After he was convinced that both earth and Mars had eccentric equants, he was sure that if the orbits of the other planets were studied in detail then they too would show this phenomenon. He still had the problem of the 'hypothesis vicaria' to deal with. His values for the eccentricities did not produce the correct distances for the relative distance of the earth and Mars from the sun as we have discussed. Kepler then returned to his starting point for planetary theory, that being the principle that the body of the sun itself is responsible for the motion of the planets including the earth. He reasoned that although an equant point for the planets serves the mathematician as a basis for calculation, it is separate from the sun at an empty point in space and has no physical significance. He wanted to show a mathematical relationship for the planets' motion directly connected with the body of the sun. He reasoned that as seen from the true sun a planet's velocity in its orbit is variable, and the variations may be related to the distance of the planet from the sun at any point in the planet's orbit. The planet moves with its greatest speed as seen from the sun at perihelion, and at its slowest speed as seen from the sun at aphelion. To put it another way a planet's angular velocity increases in its orbit as it approaches perihelion from aphelion, and then after perihelion its angular velocity along its orbit decreases to a minimum value at aphelion, after which the cycle is again repeated. The equant point then becomes a mathematical device that 'evens out' the changes in planetary angular velocity so that from the equant point the planet appears to be moving uniformly.

Showing that the earth also requires an equant eccentric to the sun was a major step because the earth's behavior in regard to variable speed was the same as Mars and presumably all the planets. Prior to Kepler's demonstration mathematicians and astronomers had argued the earth was different than the other planets, since the solar theory of Ptolemy, Copernicus, and Tycho Brahe did not require an equant for the earth. The new theory is a result of Kepler's mathematical analysis and the extraordinary careful observations of Tycho which were used by Kepler in his analysis.

Kepler then made a bold hypothesis. See Figure 37-8.

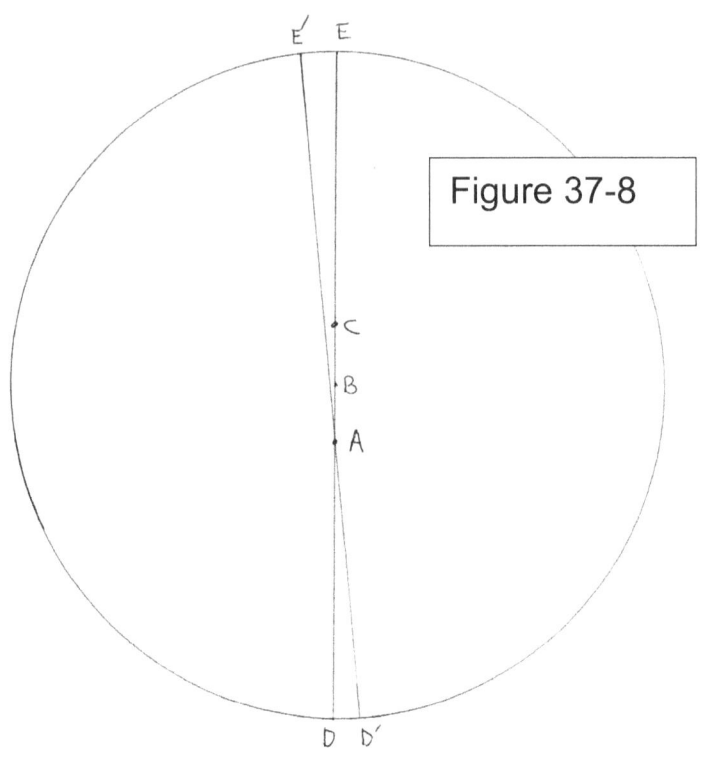

Figure 37-8

In the figure B is the center of the planet's circular orbit, A the body of the sun, and C the equant.   Kepler argued that if the sun, A, is the physical cause for the planet's motion, then the time for the planet to move through a small arc such as EE', or DD', when the small arc is near aphelion or perihelion is directly proportional to the distance of the small arc from the sun.   The problem with this hypothesis is to decide what length should be taken as the distance from the sun.   Because A, the sun, is eccentric to the radius of the orbit, lengths AE and AE' are not equal.   They would be equal only if arc EE' had A for its center instead of B.   Likewise AD and AD' are not equal for the same reason.

Kepler tried to argue around this problem by introducing a new mathematical argument, the argument

366

of infinitesimals. Greek mathematicians and mathematicians until the 17th century avoided infinitesimal quantities (quantities smaller than any measurable amount or smaller than any numerical value that is chosen), because of certain logical problems which occur when reasoning with infinitesimally small quantities of distance or time. The most famous of these logical problems were the paradoxes of Zeno of Elea(490-430 BC), a Greek philosopher before the time of Socrates and Plato. Zeno had argued that if there were in reality times or distances which were infinitesimally small, the following paradox of Achilles and the tortoise would result.

Assume Achilles and the tortoise are having a race, and that because Achilles moves much faster, the tortoise is allowed to start a given distance ahead of Achilles. By the time Achilles has traversed the distance to the starting point of the tortoise, the tortoise will have moved some distance beyond his starting point. Then by the time Achilles traverses this distance once again the tortoise has moved a smaller distance farther. Achilles then moves this small distance in a shorter time than used in covering the previous distance, but the tortoise in that small time interval has moved a small distance farther ahead. The process keeps repeating as the times and distances become smaller and smaller until they reach the infinitesimally small, yet Achilles cannot catch up to the tortoise and pass him in the race. Of course, in reality Achilles passes the tortoise, and faster objects pass by slower objects, so something is wrong with the argument of Zeno, and mathematicians felt it lay in the concept of the infinitesimally small.   Zeno also had other paradoxes of space which also made the notion of infinitesimals suspect.

The ancient Greek mathematicians felt the problem with the argument of Zeno was that times and distances could not be made infinitesimally small, and that arguments using infinitesimals must therefore be avoided in mathematical reasoning. In modern mathematics it is shown the times between the distances decrease to zero so that Achilles catches up to and passes the tortoise.

By the 17th century Kepler and other mathematicians introduced into mathematical arguments the concept of the infinitesimally small, and Kepler did just that in his idea that Mars and the other planets move through infinitesimally small arcs such that the distances from the arcs to the eccentric sun remain the same.   This idea of Kepler's that the times for a planet to move through infinitesimal arcs is directly proportional to the distance of the arc from the sun is known today as Kepler's distance law of planetary motion.   It turns out the law is false when actual measurements are applied to it, and eventually Kepler recognized this, and revised the distance law into what today we call the law of areas of planetary motion which he confirmed after he discovered the elliptical orbit.   We recall from the preface of Volume 1 that Kepler's law of areas for planetary motion states that the areas swept out by a line from the sun, at the focus of the elliptical orbit to the planet, means equal intervals of time sweep out equal areas.

At first Kepler confined his distance law to just that part of the planet's orbit near the line of apsides as in Figure 37-8. Later he extended it to the entire orbit of a planet, although he had to admit at one point it is approximate at points away from the apsides.

Let us see how Kepler could argue for the distance law near the line of apsides. We construct Figure 37-9.

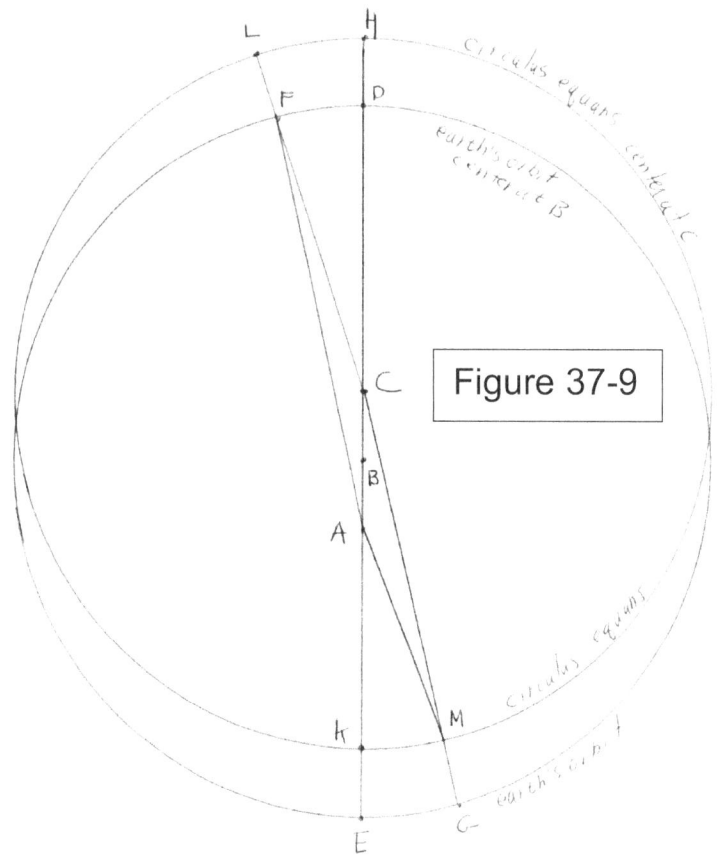

Figure 37-9

As in in previous figures B is the center of the earth's orbit, A is the sun, and C is the equant. In this figure we have drawn in the circulus equans with center at C, and radius CH equal to the radius of the earth's orbit, BD. Kepler in his theory for the distance law bisected the eccentricity so that AB = BC.   The earth is presumed to move through an infinitesimally small arc DF (we show DF as a significant arc to make the figure understandable) in an infinitesimally small time.   We take CDF to be a right triangle with the right angle at CDF.   Because arc DF is considered infinitesimal it is considered as a close enough approximation to a side of a right triangle.   Likewise CHL is taken as a right triangle with right angle at CHL.   These right triangles have equal angles so their sides are in proportion,

   6. CD/CH = DF/HL, or, (CD)(HL) =   (CH)(DF).

     Since AB = BC, and BD = CH, then DH = KE, so

   7. CD = AE.   We substitute into step 1.

   8. (AE)(HL) = (BD)(DF).

CKM and CEG are also taken to be right triangles, since arcs EG and KM are considered infinitesimal, so by the same argument used in steps 1   through 3,

   9. CE/CK = EG/KM , or, (CE)(KM) = (CK) (EG) .

     Substitute AD for CE and BD for CK,

   10. (AD)(KM) = (BD)(EG).

     Next we take infinitesimal arc DF equal to arc EG. Then,

11. (BD)(DF) = (BD)(EG),   since we are multiplying the same

quantity, BD,   by equal quantities, DF and EG.

Using steps 3 and 6 we can write,

12. (AE)(HL) = (AD)(KM), and then using this with AD and KM as the means in a proportion and AE and HL as the extremes,

HL/ KM = AD/AE.

Since HL and KM are arcs with their center at C, the center of uniform motion, and AD and AE are taken as the distances from the sun to the earth for the infinitesimal arcs DF and EG, then since the arcs HL and KM are directly proportional to the times to move through the arcs, these times are directly proportional to the distances of sun from the earth.

Since,

speed = distance/time, then time = distance/speed.

Kepler can then argue the speeds of the planets near the apsides are inversely proportional to the distance from the sun as another way of expressing his distance law.

This is the type of argument Kepler used to justify his distance law.   The distance law appealed to Kepler because he could relate it to his physical theory of planetary motion caused by the body of the sun.   In chapter XXXIV in part III of the "Astronomia Nova", Kepler describes the type of force the sun has which acts in accordance to his distance law.   He considers that the sun gives off emanations which flow from the body of the sun to the planet like a spoke on a wheel which causes the planet to move in its orbit. The longer the 'spoke' the slower the planet moved. Also Kepler considered the body of the sun to be rotating around an axis so as to send out its emanations.

He pointed out that the farther a given planet is from the sun the slower its speed in its orbit and thus the longer its period, i.e. Saturn takes 30 years for one orbit of the sun, Jupiter, 12 years, Mars 23 months, earth 12 months, Venus 8½ months, and Mercury 3 months. Kepler then considers what kind of body the sun to be, from which the motive emanation proceeds, and finds the idea of a magnet most likely. This idea came from his study of the recently published book, "*De Magnete*", by William Gilbert who had done work on the magnetic properties of loadstone, and found that the earth itself is a large magnet such that compass needles point to the north magnetic pole. It was not hard for Kepler to argue that the sun is most likely a magnet also, which issues magnetic emanations to the planets which diminish the farther the planet lies from the sun accounting for the distance law. Kepler wrote,

"the example of the magnet I have hit upon is a very pretty one, and entirely suited to the subject; indeed it is a little short of being the very truth...For by the demonstration of the Englishman William Gilbert, the earth itself is a big magnet, and is said by the same

author, a defender of Copernicus, to rotate once a day, just as I conjecture of the sun."

We have pointed out that the distance law is incorrect because for any measurable arc such as DF in Figure 37-9 the distance AD will not equal AF, since A is not the center for arc DF, B is. If B is very close to A, the distance law is a close approximation, and when Kepler tested the distance law for the earth it appeared to work well even for arcs chosen away from the apsides. The success of the distance law for the earth occurs because the eccentricity of the earth's orbit is so small relative to the radius. When tested against Mars the distance law produces significant errors, and as noted previously. Kepler did eventually replace the distance law with his next innovation which today we know as the law of areas, although at first it was applied to circular orbits.

Before we describe the area law we should point out a philosophical difficulty Kepler had with his distance law. Since Kepler had proposed that the sun caused the motion of the planets by a magnetic force extending from the sun to the planet, he asked himself why each planet had its center of rotation eccentric from the sun. The magnetic force should in theory be uniform throughout the orbit and should produce a circular orbit with the sun at the center. Because the sun was eccentric to the center of the planet's circular orbit Kepler speculated that the planets themselves were endowed with an intrinsic force or energy of their own, which accounted for the variable distances from the sun in the course of their orbit, which in turn caused the variation in their speeds as seen from the sun due to the distance law. This speculation resulted in other hypotheses which Kepler tried, even reintroducing an epicycle for the planet just as Ptolemy and Apollonius had done in ancient times. All of this proved unsatisfactory. Kepler was the sort of investigator that left no stone unturned in his search for the true motion of the planets and a good deal of the "*Astronomia Nova*" is taken up by his work along paths which later had to be rejected. Kepler presents us in the "*Astronomia Nova*" with all his work so his we can understand his thinking and the trials and errors which occurred on the path to his discoveries.

We now turn to a discussion of the area law for planetary motion. Kepler proposed this relationship (he never called his discoveries 'laws' in the "*Astronomia Nova*") as a modification of the distance law to aid in his calculations, since it much easier mathematically to check an area law with the observations. Nonetheless, his physical hypothesis for planetary motion at this stage of his work was the distance law.

To understand the area law and how Kepler might calculate from it consider Figure 37-10.

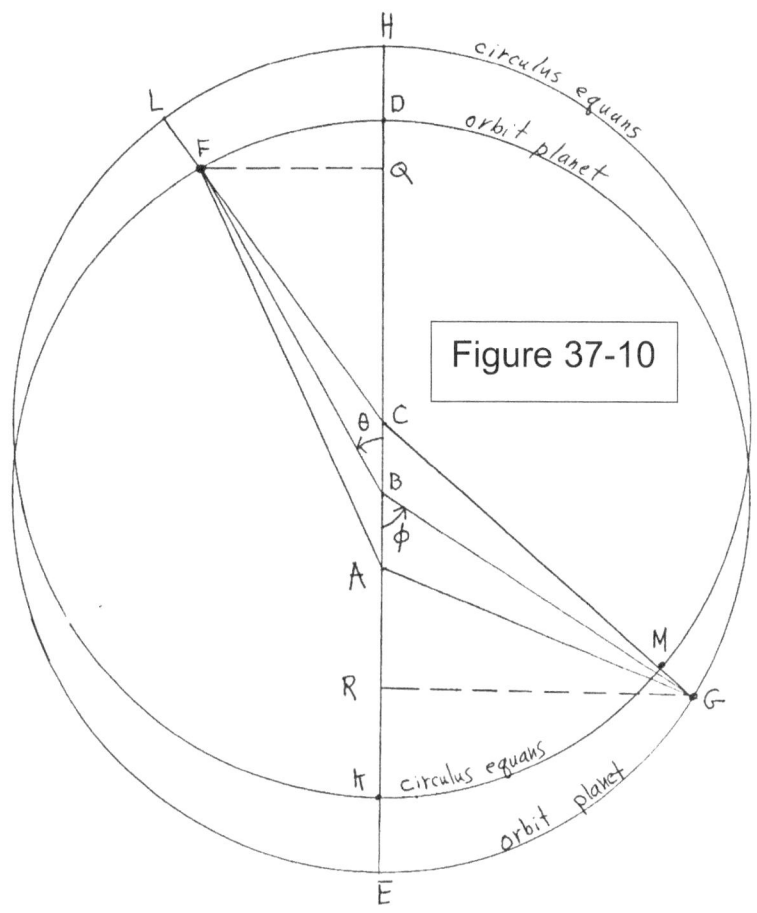

Figure 37-10

We again view from the north ecliptic pole. A is the sun, B the center of the circular planetary orbit with radius BD = BE, DE the line of apsides, and C the equant point. In Kepler's work with the area and distance law, he assumed the eccentricity AC was bisected by B, so AB = BC. We also show the circulus equans with center at C and radius CH equal to CK. Note also the radius of the circulus equans equals the radius of the planet's orbit, so BD = CH.

If ω_p is the uniform angular velocity of the planet around its orbit, then we can call t_1 the time for the planet to move from D to F.   As seen from C this will

375

result in arc HL on the circulus equans. Arc HL = $(t_1)(\omega_p)$. If t_2 is the time for the planet to move from E to G on its orbit, then as seen from C arc KM is traced on the circulus equans. Arc KM = $(t_2)(\omega_p)$. We can then write the ratio of the times for arcs HL to KM as,

13.

$$\frac{HL}{KM} = \frac{(t_1)(\omega_p)}{(t_2)(\omega_p)} = \frac{t_1}{t_2}.$$

Thus we can take HL/KM as equal to the ratio of the times for the planet to move through arcs DF and EG.

We can see from the figure that when the planet moves through arc DF the line AD sweeps out the area from AD to AF bounded by arc FD. Likewise, as the planet moves from E to G line AE moves through the area bounded by line AE, line AG and arc EG. Since these areas are swept out in times t_1 and t_2 respectively, then the ratio of t_1 to t_2 should be equal to the ratio of area ADF to area AEG. We can then write,

14.

$$\frac{t_1}{t_2} = \frac{HL}{KM} = \frac{area\ ADF}{area\ AEG}.$$

In words we have argued that the areas swept out by the line from the sun to the planet are proportional to the times to sweep out these areas. The value of the area law is that we can calculate areas ADF and AEG from observations, take their ratio and see if it is the same as the ratio of the times.

Kepler's area law as it is usually stated says in equal time intervals the areas swept out by the line from the sun to the planet are equal. Thus in Figure 37-10 if area AFD equals area AEG the time for arc HL and KM are equal.

We may ask how do we calculate the areas? We see area ADF is made up of the sum of the circular sector BDF plus the area of triangle AFB.

15. Area ADF = sector BDF + triangle AFB

The area of a circular sector can be given in terms of the angle of the sector, which we have indicated by θ (Greek letter theta) in the case of sector BDF, and the radius of the sector. We designate radius BD as r. If angle θ is measured in the radian system of measurement, and we recall there are 2π radians in 360°, we can write the proportion,

16.

$$\frac{\theta}{2\pi} = \frac{area \, sector \, BDF}{area \, circle \, DFE} = \frac{area \, sector \, BDF}{\pi r^2}.$$

$$area \, sector \, BDF = \frac{\theta \pi r^2}{2\pi} = \frac{1}{2} \cdot \theta r^2.$$

To find the area of triangle AFB we drop perpendicular FQ from F perpendicular to DE. We see,

17. sin θ = FQ/BF , and since BF = BD = r,

FQ = r(sin 0).

We recall the area of a triangle equals ½ base times height. Let AB equal the base and FQ equal the height, so,

18. area AFB = ½(AB)(r)(sin θ).

Returning to step 3,

area AFD = ½ θr^2 + ½ (AB)(r)(sin θ).

Kepler can calculate this area for Mars or the earth since he has determined AB by his assumption that the eccentricity is bisected, the radius has the value of 100,000 units, and θ can be determined from the observational data and the solution of triangles BFC and ABF.

We can calculate the area of AEG as well.   Area AEG equals the area of circular sector BEG minus the area of triangle BAG. The central angle for the sector is φ (Greek letter phi), the base of triangle BAG is AB, the height is RG = r(sin (φ)), so

19. **Area AEG = ½φr^2 - ½(AB)(r)(sin φ) .**

Kepler can determine the values of φ, r, and AB from the observation of G and his previous work.

To convince ourselves that the area law is correct for equal times let us do a calculation when arcs HL and KM are equal as shown in figure 37-11.

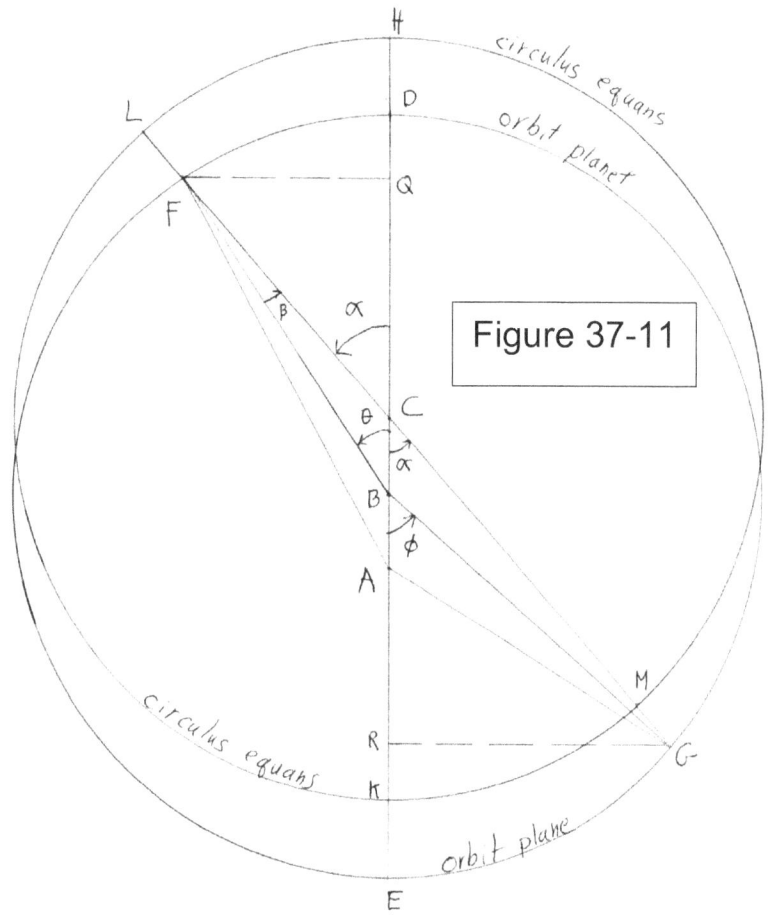

Figure 37-11

The figure is the same as Figure 37-10 except in Figure 37-11 angle HCL, the mean anomaly, equals angle KCM.   We designate this angle by α (Greek letter alpha). For ease in our calculation we choose the radius BD equal to 1 unit, and length AC = .2 units so AB = BC = .1 unit. Let us do the calculation for angle α = 40°.   Then angle FCB = 180° - 40° = 140°.   We first solve triangle BFC for angle BFC designated by β (Greek letter beta).   We use the law of sines.

20. BF/sin FCB = BC/sin β, or,

sin β = (.1)(1)sin 140°,

sin β = .06428, or, β = 3.685^0.

21. We calculate θ as 180^0 – 140^0 – 3.685^0

= 36.315^0.

Converted to radians,

(36.315°)x 1 radian/ 57.2958^0 = .6338 radians.

We use step 7 to find the area of FAD:

22. Area AFD =

½(. 6338)(1^2) + ½(.1) (1) (sin 36.315°),

area AFD =.3169 + .0296 = .3465 square units,

where the radius BF equals 1 unit.

To find area AEG we first note triangle FBG is an isosceles triangle since BF = BG so angle BGC = β = 3.685°.   We see angle GBC equals 180° - φ, and thus,

23. 180° - φ = 180* - β – α,

180° - φ = 180° - 3.685° - 40°

180° - φ = 136.315°, or, φ = 43.685°

= .7624 radians.

Using step 8 we find area AEG.

24. Area AEG =

½(1^2) (.7624) - ½(.1)(sin 43.685°),

area AEG = .3812 - .0346 = .3466 square units.

Thus the areas, AFD and AEG when calculated in this hypothetical example are equal to a high degree of accuracy as they should be when arc HL = arc KM = 40°.

However, Kepler did not calculate areas by the method we just used.   He calculated the areas by a method of analogy which we will look into.   Before he began his area calculations he developed a table of earth-sun distances for each degree of true anomaly from the line of apsides in chapter XXX of the "*Astronomia Nova*". See Figure 37-12.

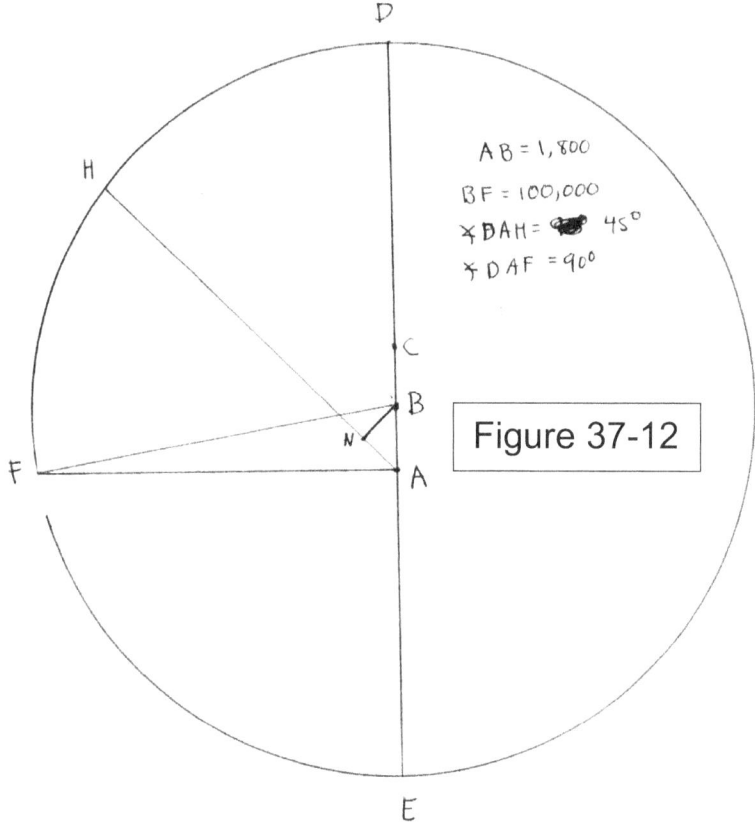

AB = 1,800
BF = 100,000
∡DAH = 45°
∡DAF = 90°

Figure 37-12

Once again A is the sun, B the center of the earth's circular orbit, C the center of the circulus equans, and DE the line of apsides. The total eccentricity for the earth's orbit was taken as 3,600 units when the radius BD, BE is 100,000 units. The value of 3,600 units came from Tycho's theory. We recall in chapter 18 Ptolemy had calculated the eccentricity for the solar theory as .04167 when the orbit's radius was 1 unit, which is equivalent to 4,167 units when the radius is 100,000 units.   Kepler bisected the total eccentricity so AB = BC = 1,800 units.   This means AD (aphelion distance) = BD + AB, or, AD = 100,000 + 1,800 = 101,800 units, AE (perihelion distance) = BE - AB = 100,000 - 1,800 = 98,200 units.

Kepler then calculates the earth-sun distance at 90° of true anomaly when the earth is at F and angle DAF = 90°.

25. Consider triangle BAF.   sin BFA = BA/BF = 1,800/100,000 = .01800, and, arcsin .01800 = 1° 1' 53"
 = angle BFA.

26. The distance of the sun to earth, AF, is found by,

cos BFA = AF/BF , or, AF = (BF)(cos BFA).

AF = (100,000)(cos 1°1' 53")

= (100,000)(.99984)

AF = 99,984 units.

Let us solve for the earth sun distance AH when the true anomaly, angle DAH, equals 45°.   The perpendicular BN is dropped from B to A.

27. cos BAN = cos 45° = BN/AB , or, BN

= (AB)(cos 45°),

BN = (1,800)(.7071) = 1,273.

28. Imagine line BH, then,

sin BHA = BN/BH = 1,273/100,000 = .01273.

arcsin .01273 = 0° 43' 45" = angle BHA.

29. HN in right triangle BNH is found by,

cos BHA = HN/BH , or, HN = (BH)(cos BHA),

HN = (100,000)(cos 0° 43' 45")

  = (100,000)(.999992),

HN = 99,999.2

30. We see the earth-sun distance is,

AH = HN + AN.

AN = BN since triangle BNA is an isosceles triangle.   AH = 99,999.2 + 1,273 = 101,264 units. (Kepler's value was, 101,265).

Kepler proceeds to solve a series of triangles after assigning the values for the true anomaly, angle DAH, values from 1° to 179° in 1° increments.   This gives him the earth-sun distance for each degree of true anomaly through 360°, since values from 180° to 360° are symmetric to the values from 0° to 180°.   He then constructs his table which, however, is given in a somewhat different form, because later in his work he realized the earth's orbit like Mars is not circular but elliptic.

It is in chapter XL of the "*Astronomia Nova*" that Kepler gives his method of analogy for calculating the areas swept out by the line from the sun to the planet. He constructs Figure 37-13 which we have copied from Kepler.

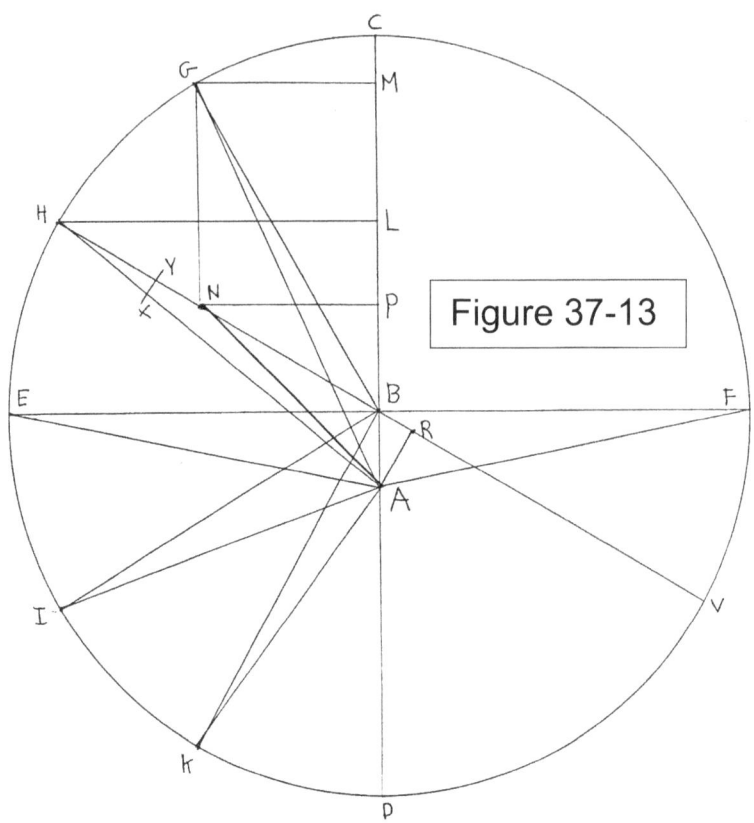

Figure 37-13

The circle represents the earth's orbit with center at B and the sun at A along the line of apsides DC. Kepler considers the semicircle CED to be divided into an infinite number of equal arcs of which Kepler shows arcs CG = GH = HE = EI = IK = KD. We note point E is chosen so angle CBE = 90°.

To each point on the semicircle two lines are drawn, one from A and one from B. This creates a series of triangles: BGA, BHA, BEA (a right trianlge), BIA, and BKA.

We also have a series of equal circular sectors: CBG, GBH, HBE, EBI,IBK, and KBD. Kepler argues that the area of any one circular sector, say CBG, is

384

made up of the sum of the infinite number of lines which run from B to all the points on arc CG.

Today we might say that the area CBG is swept out by line BC as it moves from BC through angle CBG, and thus encompasses all the possible lines from B to arc CG. Likewise, the area of the semicircle CED is equal to the sum of all the circular sectors.

We have mentioned the problems of arguing using the concept of the infinitely many, and in this case we have the problem that Euclid had defined a line as a breadthless length so the sum of any number of lines will have zero area. Lines for Euclid can enclose an area but cannot be part of an area.

Kepler would be correct if he meant that he can divide the semicircle with more and more circular sectors which increases the number of lines necessary to mark off the sectors. The sum of the areas of all the circular sectors will be equal to the area of the semicircle.

If n is the number of circular sectors between BC and BD, then (n - 1) is the number of lines which must be drawn from B to the circumference to create the circular sectors. We let n approach toward the infinitely large, or increase without bound. Then the area of the semicircle will be equal to the sum of the n circular sectors no matter how large n is chosen. Even though the area of the semicircle will be marked off with larger and larger numbers of lines it is the sum of the areas of the circular sectors and not the lines which equal the area of the semicircle.

385

Kepler calls the area of sector CBG the eccentric anomaly for position G on the earth's orbit. Eccentric refers to the theory for the orbit of the earth where B, the center, and A, the sun, are not at the same point. For position H the eccentric anomaly is the area of circular sector CBH, for position E the eccentric anomaly is area CBE, etc. In our work up to this point the term anomaly represented an angle, or the phenomenon of non uniform motion of the planets as seen from the earth (the first and second anomalies of planetary motion). Kepler now calls these areas anomalies because angles CBG, CBH, etc. are directly proportional to the respective areas CBG, CBH, etc., so the areas can be used as a measure of the eccentric anomaly for any position of the earth on its orbit.

Kepler next argues that the series of areas CAG, GAH, HAE, etc. also add up to the area of semicircle CED. This argument is correct for any finite number of sectors that are made by drawing lines from A to the circumference of the orbit. Kepler then says the area of a sector, for example, CAH, is itself made up of an infinity of lines from A to the infinity of points along arc CH, and that the sum of these infinite lines is the area of the sector CAH. Furthermore, each line from A to any point on the arc CH is the distance from the sun to the earth and hence the sum of these distances is the area of the sector CAH.

Kepler's actual words in translation from the Latin state,

Now seeing that these two assemblies of straight lines, those starting from B, and those starting from A, fill one and the same semicircle CED, and as those which start from A are the very distances whose sum is required, it seemed to me that we could conclude that we should have the sum of the infinite distances from A up to CH or CE, if we calculated the surface CAH or CAE; not because the infinite could be reached, but because I considered that this surface contained the measure of the effect of the whole collection of these distances over the accumulated times, and consequently that a knowledge of this surface gave this measure , without the necessity of reckoning up smaller parts".

Note that Kepler recognizes the problem of adding up an infinity of parts. He argues that he can substitute the area of sector CAH for the series of lines from A to points on the arc CH. Since his distance law states the times to move through equal small arcs on the orbit are proportional to the distances of the arcs from the sun, then if Kepler has an infinite number of small sectors CAG, CAH, etc., the areas of these sectors will be proportional to the times of travel through the small arcs CG, GH, etc., and the areas of the sectors can be used as a measure of the mean anomaly.

Thus for position G area CAG is called the mean anomaly, and likewise the mean anomaly for position H is area CAH. Kepler's use of these areas take the place of the arcs on the circulus equans as in Figure 37-9. In Kepler's words, *"whence the ratio of the area CAE is to one half the time of restitution [one half of the time for one revolution], which we designate by 180°, as the areas CAG, CAH are to the times of travel over CG and CH.   Thus, the area CAG becomes a measure of the time or mean anomaly, which corresponds to arc CG of the eccentric, seeing that the mean anomaly measures time."*

The last sentence in the original Latin of Kepler reads,

"Itaque CGA area siet mensura temporis seu anomaliae media, quae arcui eccentrici CG respondel, cum anomalia media tempus metiatur."

Although this is not the final form Kepler's area law would take, this statement is the first time in the history of mathematical astronomy that the areas within the orbits are related mathematically to the times of planetary motion.

Kepler wants the area of sector CAG to represent the mean anomaly, and circular sector CBG to represent the eccentric anomaly for position G. The difference between these two areas is triangle ABG, and Kepler calls the area of triangle ABG the physical equation of position G.   Furthermore, he calls angle BGA in triangle ABG the optical equation of position G.   Likewise, for position H of the earth's orbit, triangle ABH is the physical equation and angle BHA, the optical equation. Kepler then devises a mathematical method to express the areas in measurements of degrees, minutes, and seconds instead of square units.   We will show Kepler's method.

Consider triangle BEA where angle CBE = angle EBA = 90°. The area of triangle BEA = ½(base)(height). For the earth we recall the radius of the eccentric orbit, BE, is 100,000 units and AB = 1,800 units.   AB = base, and BE = height for triangle BEA.

31. Area BEA = ½(1,800)(100,000) = 90,000,000 square units.

32. Area of circle CEDF = π (BE)^2 = π(100,000)^2,

Area of circle CEDF = 31,415,926,536 square

units.

Kepler then set up the following proportion,

33. area circle/area triangle

= degrees in circle/ degrees for triangle

31,415,926,536/ 90,000,000

= 360°/degrees for triangle,

= (360°)(90,000,000)/ 31,415,926,536

= 1° 1' 53" .

Thus, the area of triangle BEA, the physical equation, is expressed in degrees by Kepler.   He adds the value of the physical equation expressed in degrees to the area of circular sector CBE also expressed in degrees to get the mean anomaly for position E. Circular sector CBE expressed in degrees is 90° since it is ¼ of the circle (¼ times 360° = 90°).   The mean anomaly by Kepler's method is then,

 90° + 1° 1' 53" = 91° 1' 53".

Next Kepler calculates angle AEB in right triangle ABE.

34. tan AEB = AB/BE = 1800/100,000 = .01800.

angle AEB = arctan .01800 = 1°1' 52".

This is the optical equation for position E.

35. angle BAE in right triangle ABE =

180° - 90° - 1° 1' 52",

angle BAE = 88° 58' 8".

This is the true anomaly of position E. When the true anomaly is added to the celestial longitude of the line of apsides, AC, we then have the celestial longitude of position E.

Kepler uses his solution of right triangle ABE as a means to solve for the mean anomaly, the physical equation, the optical equation, and the true anomaly for any other position on the earth's orbit given the value of the eccentric anomaly. He uses a geometrical argument involving proportions. We now let G represent any position of the earth between C and H. H can be any position on the orbit between C and E where E is 90° from C.

From G draw line GN parallel to AC which meets BH at N. Then draw GM, HL, and NP perpendicular to the line of apsides AC. Consider triangles GAB and HAB. They have the same base AB. The height of triangle GAB is GM. If we consider right triangle GMB, we see sin GBC = GM/GB, or, GM = (GB)sin GBC. The height of triangle HAB is HL. If we consider right

390

trianlge HLB, we see sin HBC = HL/HB, or, HL = (HB)sin HBC. Since both GB and HB are radii of the circle they are both equal to 100,000 units.

36. Area triangle GAB = ½(AB)(GB)(sin GBC)

Area triangle GAB =

½(AB)(100,000) (sin GBC),

Area triangle HAB = ½(AB)(100, 000) (sin HBC)

We take the ratio of the areas to set up the proportion,

37. Area HAB/Area GAB = sin HBC/sin GBC .

As an example of how to use this proportion let the earth be at the position where angle GBC =45°. We then use for position H the values we have already found for position E.   That is assume H to be at E where angle HBC is angle EBC = 90°. Then,

38. Area EAB/Area GAB = sin EBC/sin GBC

= sin 90° /sin 45°

We found the area of EAB expressed in degrees (the physical equation) in Kepler's method to be 1° 1' 53". Thus,

39. Area GAB = (1° 1' 53") x sin 45^0/sin 90^0

= (1°1' 53")(.7071).

Since 1° = 3600", and 1' = 60", then,

1° 1" 53" = 3,713".

Area GAB = (3,713")(.7071) = 2,625"

= 0° 43' 45".

This is the physical equation for position G when angle GBC = 45°.   The mean anomaly is area sector CAG which equals area GAB + area GBC.   Area GBC

391

expressed in degrees is the value of central angle GBC or 45°. Thus the mean anomaly is 45° + 0° 43' 45" = 45° 43' 45".

To find the true anomaly and optical equation Kepler solves triangle GAB using the law of tangents. We recall from chapter 36,

40.

$$\tan \frac{A-B}{2} = \frac{a-b}{a+b} \left[\tan \left(90^0 - \frac{1}{2}C \right) \right],$$

and,

$$90^0 - C = \frac{A+B}{2}.$$

Also,
$$\frac{A+B}{2} + \frac{A-B}{2} = A.$$

A, B, and C are the angles of the triangle, and a, b, c are the sides opposite the respective angles.

We let BG = a = 100,000, angle GAB = A, AB = b = 1,800, angle BGA = B, and angle GBA = C = 180° - 45° = 135°, so ½C = 67° 301.

41.

$$\tan\frac{A-B}{2}$$

$$= \frac{100,000 - 1,800}{100,000 + 1,800}\left[\tan\left(90^0 - 67^0 30'\right)\right]$$

$$= (.96464)\left(\tan 22^0 30'\right) = .39957.$$

arctan .39957 = 21° 46' 48".

A = 22° 30' + 21° 46' 48" = 44° 16" 48"

= angle GAB, the true anomaly.

B = 180° - C - A = 180° - 135° - 44° 16' 48" ,

B = 0° 43' 12" = angle BGA, the optical

equation.

Thus we have shown Kepler's method of determining areas by his area law expressing the areas in degrees, minutes, and seconds, starting from an assigned value of the eccentric anomaly, in this case 45°.   Kepler then tested his area law for his theory of the earth's orbit to see if the predicted values for the true anomaly agree with the values in Tycho's data. Kepler found excellent agreement.   He could predict the sun's position as seen from earth with his theory with errors of about   9" of arc which were well within the observational error.

Yet Kepler knew in Chapter XL of the "*Astronomia Nova*" that his area law and his distance law are not mathematically equivalent.   He acknowledge this in the title of the chapter when he described the area law as, *"an imperfect method of calculating the equation on the basis of the physical hypothesis [his distance law] sufficing, however, for the path of the sun or the earth".*

In Figure 37-13 he notes line EF is a diameter of the circle and considers the area of the circle to be equal to the sum of all the lines, infinite in number, which could be drawn as diameters.   In this line of reasoning each diameter is considered as an infinitesimal area and when all such infinitesimal areas are summed the area of the circle is formed.   Although we have mathematical trouble in summing lines which have no breadth, we can perhaps think of the area of the circle as being swept out by rotating diameter EF through 360°.   By sweeping out the area of the circle in this manner we would include all possible diameters, and in that way understand what Kepler was thinking when he summed up all the diameters.

Kepler then considered lines AE and AF which represent the distance lines from the sun to the earth at positions E and F. Kepler argues that if we summed up all such lines as EF is rotated through 360°   an area larger than that of the circle will result, since AE + AF is greater than BE + BF for all position except along the line of apsides,   i.e. when E is at D and F is at C.   At this stage of his thinking the distance law was the correct physical law of planetary motion, and he decided to use the area law as his approximation for the distance law for the theory of Mars to see if the results agreed with the predictions from the 'hypothesis vicaria', which

Kepler already knew gave values for the celestial longitudes of Mars to within 2' of arc of Tycho's observations. In the next chapter we will take up his results and see that discrepancies led to the abandonment of the circular orbit for Mars, and the discovery that an elliptical orbit best fit the data with the area law being vindicated when the sun is at one focus of the ellipse.

Before we close this chapter it may be well to list the steps Kepler had taken thus far in our study of the "*Astronomia Nova*".

1. Kepler started with the hypothesis that the sun is the physical cause of the motion of Mars and the planets, and is stationary as is the celestial sphere.   Planetary theory is to be based on measurements from the true sun and not the mean sun as in Ptolemy, Copernicus, and Tycho Brahe.

2. The angle between the plane of the ecliptic, which includes the sun and earth, and the plane of the orbit of Mars, which includes Mars and the sun, is determined to have an inclination of 1° 50'.

3. The *'hypothesis vicaria'* for the celestial longitudes of Mars is worked out from Kepler's revised table of oppositions based upon the heliocentric celestial longitudes.   The theory includes an eccentric circular orbit for Mars with an equant point.   The total eccentricity is 18,564 units from the sun to the equant point when the radius of the circle is 100,000 units. The

eccentricity is not bisected as is shown in Figure 36-8. However, Kepler finds the *'hypothesis vicaria'* gives incorrect values for the celestial latitudes for the oppositions, indicating the relative distances between the sun and earth and the sun and Mars are incorrect. When Kepler tries to correct for these errors by bisecting the total eccentricity as Ptolemy had done in his theory, he finds the predictions for the celestial longitudes are incorrect by 8' of arc. Kepler knows Tycho's data are accurate to 2' of arc, and he knows he must do more work.

4. Kepler restudies the solar theory of Tycho which did not include an equant point.   He demonstrates the theories of Tycho and Ptolemy are incorrect and that the earth like Mars should have an equant point.

5. Kepler is now convinced the earth, Mars, and all the planets are similar in their theory because they must obey the same physical law of motion caused by magnetic forces from the sun.   He postulates his distance law as the fundamental physical law of planetary motion wherein the distances of the planets from the sun are inversely proportional to the speeds of the planets through infinitesimal arcs.

6. To aid in calculation Kepler introduces the area law which states areas swept out by the line from the sun to the planet are proportional to the times required to sweep out the areas.   In so doing he develops a method of calculating areas where the areas are

converted into measurements of degrees, minutes, and seconds.

7. Kepler tests the area law on the earth's orbit where the distance from the sun to the center of a circular orbit of the earth is 1,800 units when the radius of the orbit is assigned the value of 100,000 units.   He finds his area law works well and plans to apply to the orbit for Mars, even though he believes the area law is an approximation for the true physical law: the distance law.

Bibliography for Chapter 37

1. Kepler, Johannes,   "Astronomia Nova",   Pars Tertia, Chapters XXII through XL,   Latin edition printed by Culture et Civilisation,   115, Avenue, Gabriel Lebon, Bruxelles, 1968.
2. Small, Robert, "An Account of the Astronomical Discoveries of Kepler",   A reprinting of the 1804 text by The University of Wisconsin Press, Madison Wisconsin, 1963.
3. Aiton, E.J., "Kepler's Second law of Planetary Motion", "Isis 60:   75-88, 1969.
4. Wilson, Curtis, "How Did Kepler Discover His First Two Laws?" "Scientific American", 226, March 1972, no.3,   93-106, 1972.

Chapter 38

The Oval Orbit and the Auxiliary Ellipse

In our work with Kepler's "*Astronomia Nova*" we have seen how Kepler develops and then rejects models for the motion of Mars, because the models do not fit the observations from Tycho[1]s data. In the fourth part of the book Kepler applies his area law to a new theory for Mars. To develop this new theory for Mars he makes several assumptions. He again assumes a circular orbit with the sun stationary, but eccentric to the center of the orbit. He then proceeds to determine the values for the distances of the earth and Mars from the sun using his theory for the earth's orbit which we discussed in chapter 37. He used a series of observations as shown in Figure 38-1, which is taken from chapter XLII of the "*Astronomia Nova*".

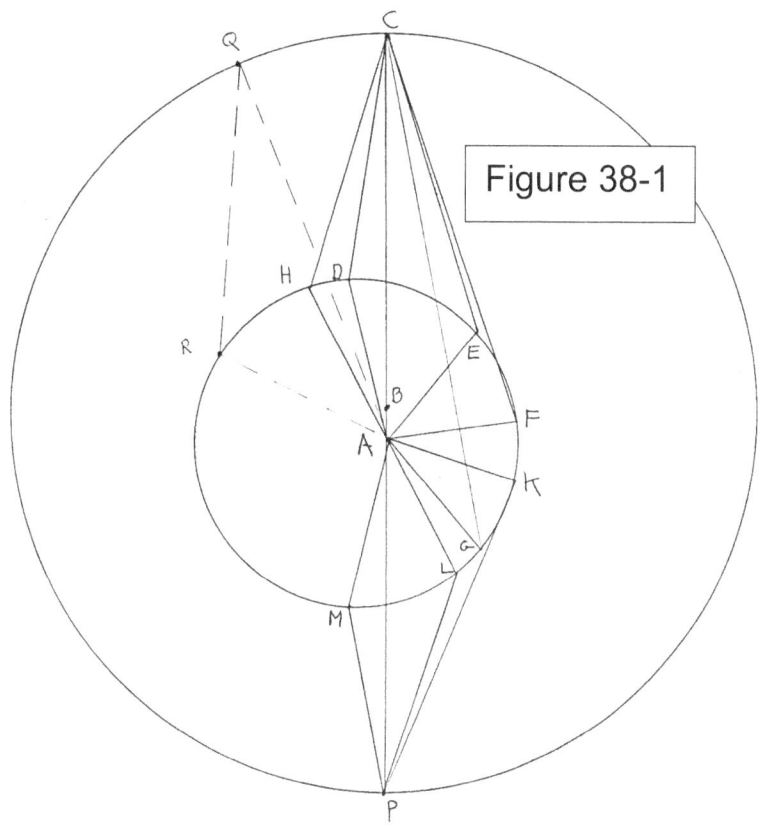

Figure 38-1

In this figure Kepler shows the orbits of both the earth and Mars as seen from the north ecliptic pole. The sun is at A and the line of apsides for Mars is CAP with B at the center of Mars' circular orbit. The center for the earth's orbit is not indicated. Kepler was able to find in Tycho's data a series of observations where the earth was at positions H, D, E, F, and G, while Mars was essentially at aphelion (about 148° 53' heliocentric longitude) point C. He then assumed a value for the distance AC and then tested it using data from the observations.

For example, in triangle CAH he knew length AH from his theory of the earth's orbit. He knew the angle

399

AHC from the observation, and the value of angle CAH. He then solves the triangle for length AC. In doing these calculations he assigned the value of 100,000 units for the radius of the earth's orbit. He then solves triangles CAD, CAE, CAF, and CAG to confirm the value of AC. He determined AC to be 166,780 units.

He again turned to Tycho's data and found observations when the earth was at positions K, L, and M and Mars was at perihelion P. He then calculated AP by solving the required triangles and found AP to be 138,500 units. He could now calculate the radius of Mars' orbit in the new theory to be 166,780 + 138,500/2, or, 152,640 units. The eccentricity AB for the new theory is equal to BC - AP, or 152,640 - 138,500 = 14,140.

Kepler then assigned the value of 100,000 units for the radius of Mars' orbit, BC, and then calculated the value of AB by solving the proportion,

1. AB/ BC = 14,140/152,640 = .09264, and since BC = 100,000

   AB = (100,000)(.09264) = 9,264 units.

AB represents one half of the total eccentricity if Kepler had included the equant point in this new theory for Mars. He did not use the equant point, since he planned to use the area law to determine the positions of Mars. We note that Kepler can discard the equant point when he uses the area law to calculate the position of Mars. We contrast the value of 9,264 for AB with ½

400

the total eccentricity of the *'hypothesis vicaria'*, which we found in chapter 36, Figure 36-8, is ½(18,564) = 9,282.

He now is ready to apply the area law to his new theory. The model has the same geometry as the earth's orbit so we can use Figure 37-13.

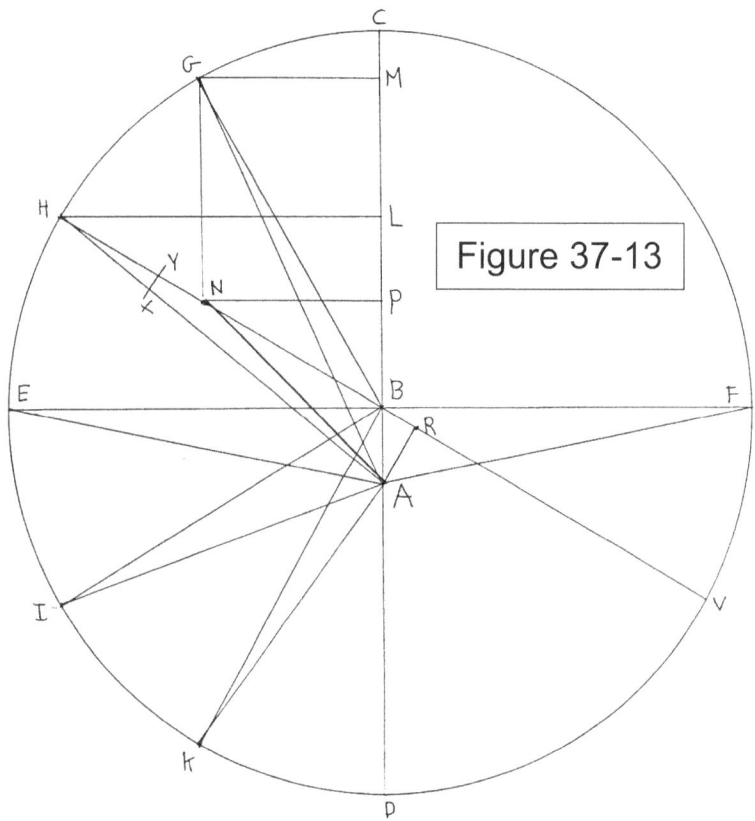

Figure 37-13

This time circle CEDF represents the orbit of Mars with radius BC = 100,000 units, and AB = 9,264 units.   We see immediately that Mars has a much greater eccentricity than the earth where AB = 1,800 units.   We consider first the situation when Mars is at quadrature which means angle CBE = 90°.

2. tan AEB = AB/BE = 9,264/100,000   = .09264

angle AEB = arctan .09264 = 5° 17' 34". This is the optical equation for position E of Mars' orbit in the new theory.   The physical equation is the area of triangle BEA expressed by Kepler's method in degrees, minutes, and seconds.

    3. Area BEA = ½ (base) (height) = ½(AB)(BE),

       Area BEA = ½(9,264)(100,000) = 463,200,000

       square units.

Area circle/Area BEA = degrees for circle/degrees for triangle BEA, or degrees triangle BEA = (degrees circle)(Area BEA)/Area circle.

    We recall the procedure from step 22 in chapter 37:

   degrees for triangle = $(360^0)(463,200,000)/ \pi R^2$

   = $166,752,000,000/\pi(100,000)^2$

   = 166,752,000,000/31,415,926,536,

   degrees triangle = 5° 18' 28"

    The mean anomaly for position E equals the area of circular sector CBE (90°) + area triangle BEA (5°18' 28"), or, 95° 18' 28".

4. The true anomaly is angle BAE in right triangle BEA and equals

   180° - 90° - angle AEB.

   angle BAE = 180° - 90° - 5° 17' 34" = 84° 42' 26".

    This value is the same as angle CAE, and is the heliocentric angle beyond the line of apsides of point E when angle CBE = 90°.

By the 'hypothesis vicaria" when Mars is at E the calculated value of angle CAE was found by Kepler to be 84° 42' 2".   Thus the area law calculation for the new theory is only in excess by 24" of arc.   Because of the symmetry of the model the value for the area law at F where angle CBF = 270° is also within 24" of arc.

However, when Kepler applied the area law for positions of Mars such that angle CBG = 45°, or angle CBG = 135°, the area law method of calculation did not agree with the calculation using the "hypothesis vicaria". The method of calculation is the same we used in chapter 37 steps 25 through 30.   The results were: for angle CBG = 45°, the mean anomaly (area CBG) = 48° 45' 12", the true anomaly (angle BAG) = 41° 28' 54"; for angle CBG = 135°, the mean anomaly = 138° 45' 12", the true anomaly = 130° 59' 25".   Comparing the true anomalies to the vicarious theory:

Eccentric anomaly	True anomaly by area law	True anomaly by "hypothesis vicaria"
45^0	41°28'54"	$41^020'33"$
135^0	$130^059'25"$	$131^0 7'26"$

These are significant variations of about 8' of arc. We recall that the values calculated by the "hypothesis vicaria" are within 2' 12" of the observed values in Tycho's data.

The problem Kepler now faced was that his new theory for Mars was accurate for celestial longitudes of

Mars for positions near the line of apsides, C and D, and for the quadrants, E and F, in Figure 37-13. However, the new theory gave positions for Mars too advanced by 8' 21" at 45° beyond the line of apsides as seen from B, and 8' 1" too retarded at 135° beyond the line of apsides.   Because the new theory and the "*hypothesis vicaria*" are symmetric to the line of apsides, we could show that if angle CBG = 180° + 45° = 225° the new theory would advance Mars 8' 1" beyond the position calculated by the "*hypothesis vicaria*", and at angle CBG = 360° - 45° = 315°, the new theory would retard Mars by 8' 21" compared to the value calculated by the "*hypothesis vicaria*".   The situation is presented in Figure 38-2.

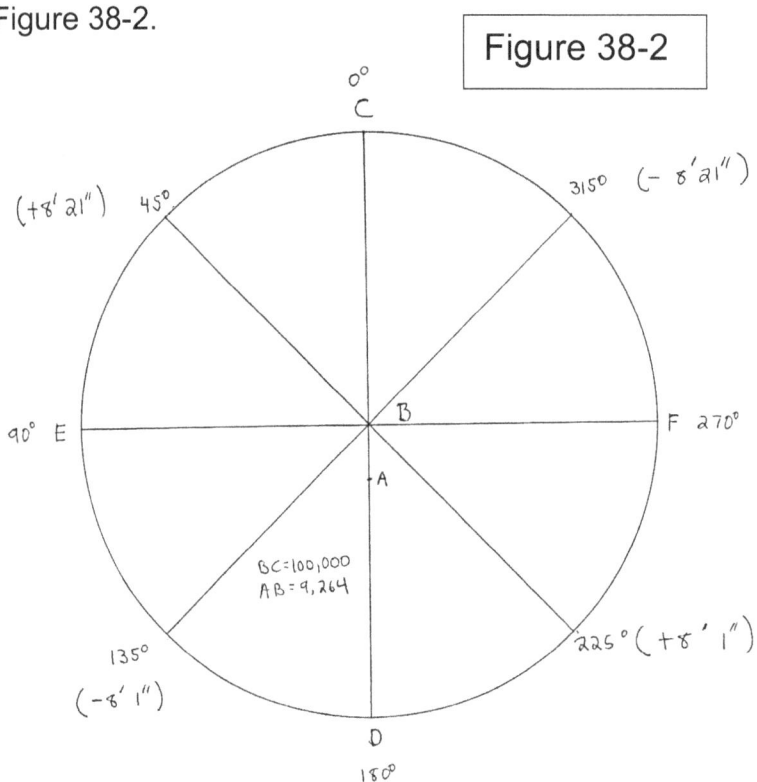

Figure 38-2

Points E and F are the quadrant points, and the 45°, 135°, 225°, and 315° points are called the octant points, since 8 times 45° equals 360°.

As we inspect the results shown in Figure 38-2, we can see the new theory as calculated by the area law has Mars moving too rapidly in its orbit after it passes the line of apsides at C or D such that its position 45° after passing the line of apsides is too advanced.   The new theory slows the motion of Mars down too much after Mars passes the quadrant points so that 45° beyond the quadrant points E and F its motion is retarded.

This discovery made Kepler realize that either the area law is incorrect (he was not prepared to abandon the distance law as the correct physical law), or that Mars does not follow a circular orbit.   If the area law is a valid method for calculation as Kepler had hoped it would be, since it is an approximation to his distance law, then in order to correct for the variations at the octant points Kepler would have to bring the path of Mars inside the circle at the quadrant points E and F.

If the path of Mars is brought inside the circle at the quadrant as shown by line CG'D in Figure 38-3, then angle BAG' is the true anomaly.

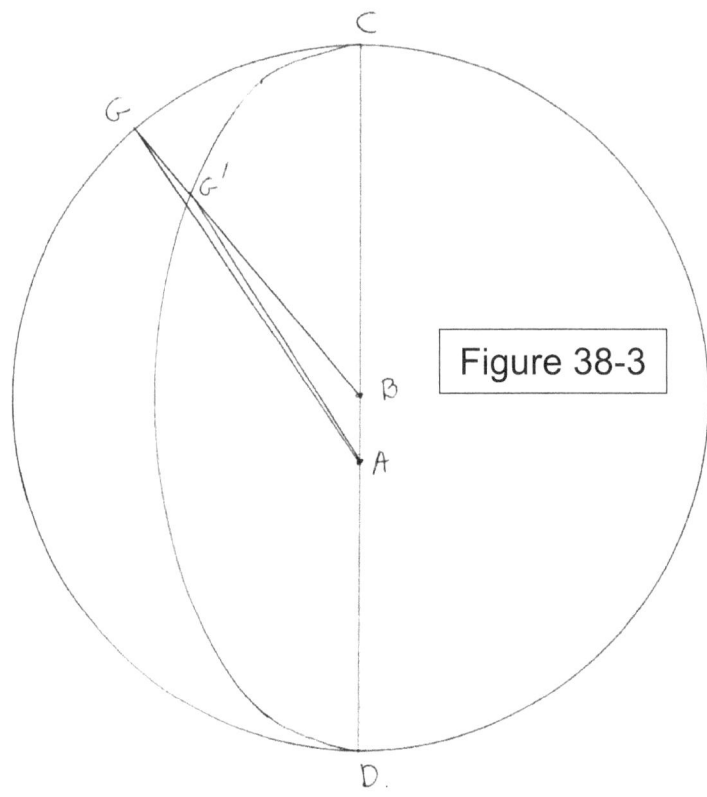

Figure 38-3

We see it is smaller than angle BAG, the true anomaly, if position G were on the circular orbit. This change has the effect of slowing the motion of Mars as it passes aphelion and thus will correct the error of 8' 21" in the first octant. Since the time for 1 orbit is the same for the circular orbit and the orbit within the circle, then because the motion is retarded by bringing Mars' path within the circle, the speed of Mars must increase after passing the quadrant points E and F so that the retardation of 8' 1" in the new theory will tend to be corrected. The amount the orbit must be brought inside the circle is very small to correct for errors as small as 8' of arc.

Kepler used several methods to demonstrate that the path of Mars must move inside the eccentric circle. We will describe one of them using Figure 38-4.

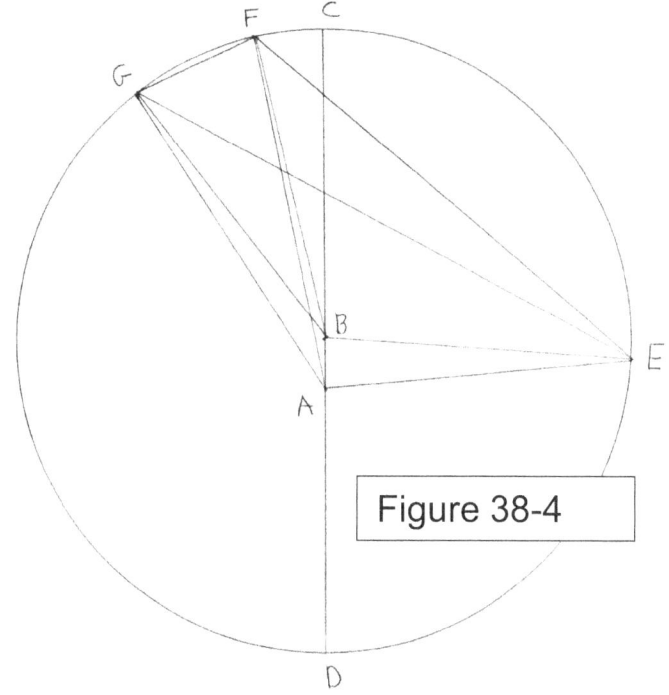

Figure 38-4

This figure shows the assumed eccentric circular orbit with the sun at A, the center of the circle at B, and the line of apsides CD. Kepler chose three observations of Mars made by Tycho in 1590 which had the heliocentric longitudes at C, G, and E. For his work with this figure he assigned the value of 100,000 units for the earth's radius. We recall from earlier in this chapter when the earth's radius is 100,000 units the radius of the eccentric circle, BC, is 152,640 units, and AB = 14,140 units. Kepler calculates the values of distances AG, AF, and AE by solving the three triangles AGB, AFB, and AEB.

407

For example, consider triangle AGB. AB = 14,140, BG = 152,640 if the eccentric circle theory is correct, and angle BAG is the true anomaly. Kepler knew the value of ABG from the data of the observation. We have already learned Kepler's method of solving triangles using the law of sines and the law of tangents. In this manner Kepler finds the distances AG, AF, and AE based on the eccentric circle theory.

Next he calculates the values for AG, AF, and AE from the observations based on Figure 38-1 which involves considerations of the earth-sun distances to solve for the sun-Mars distances. For example, in Figure 38-1 point G of Figure 38-4 is represented as lying at Q in figure 38-1.

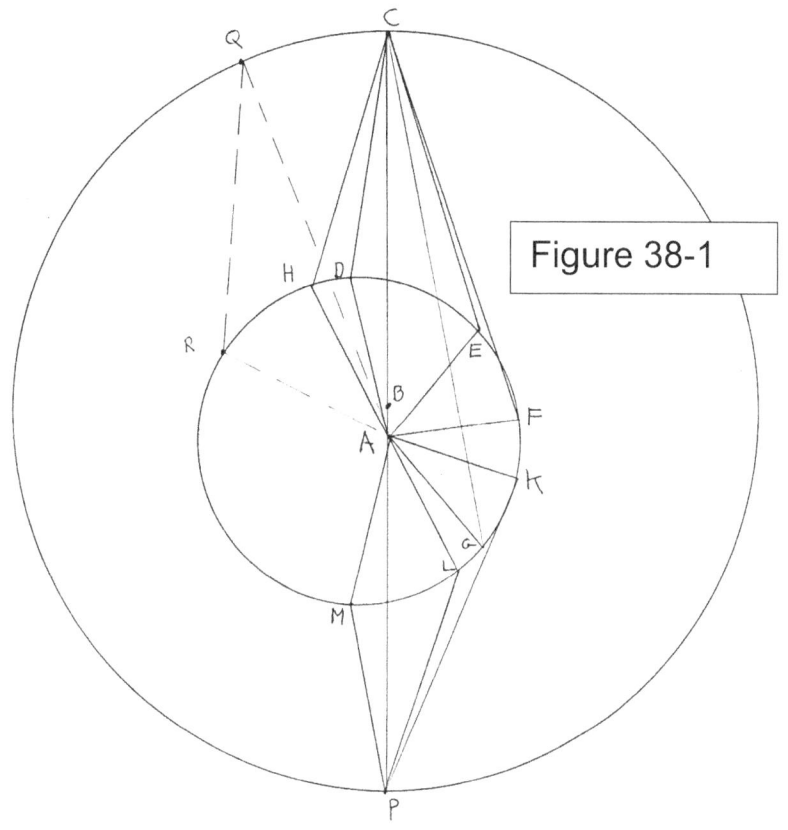

Figure 38-1

For this observation the earth is at R in Figure 38-1. Kepler wants to solve triangle QRA for the sun-Mars distance AQ. Kepler determined AR from the parameters of his solar theory. He determined angles QRA from the observation. Angle QAR can be determined from the "*hypothesis vicaria*" by finding the heliocentric longitude of Q, finding the heliocentric longitude of the earth and taking the difference between them. Kepler then had two angles of the triangle QAR and the length of one side, AR, and thus solved the triangle for AQ.

409

This method should give values for the sun-Mars distances that agree with the values of AG, AF, and AE determined from Figure 38-4, if the orbit of Mars follows the eccentric circle.   If, however, Mars moves inside the circle of Figure 38-4 between aphelion and perihelion, then the values for AG, AF, and AE would be less when determined using Figure 38-1 compared with the method of Figure 38-4.   We will give Kepler's results in the following table.

Distances calculated from the circle theory.	Distances calculated. from observation.	Variance
Figure 38-4	Figure 38-1	
AG = 163,883	AG = 163,100	783
AF = 166,225	AF = 165,845	380
AE = 148,539	AE = 147,750	789

Since the values are less using the method of Figure 38-1, the circular orbit, Kepler found that Mars moves within the eccentric circular orbit of Figure 38-4.   Mars moves farther in at positions about 90° from the line of apsides as at E in Figure 38-4 (E was 104° from aphelion).   We note G was 37° from aphelion, and it is almost in from the circle as much as E.   Kepler had to conclude there must have been some error in that observation since his idea was that it should not have been that far away from the circle 37° from aphelion.

Kepler used this method and others to give him compelling evidence that Mars path as seen from the north ecliptic pole moved inside the eccentric circle and had to be some kind of oval shape.   But what oval could it be?

He thought perhaps the oval could be generated by the combination of two circular motions, the epicycle deferent theory used by Ptolemy.   He placed Mars on an epicycle whose center moved around the deferent. It is possible to show that the resultant path of an epicycle deferent theory will move inside the circle at the quadrants and touch the circle along the line of apsides at aphelion and perihelion.

It may be worth a digression from Kepler's work to show mathematically that an epicycle deferent system can be constructed such that the path of the planet traces out an ellipse, even though this is *not* how Kepler came upon the idea of the ellipse.   This demonstration will show that the combination of two circular motions, the motion of a planet around the center of an epicycle, and the motion of the center of the epicycle around the circumference of a deferent can result in the path of an ellipse with a major axis equal to the sum of the radius of the deferent and the radius of the epicycle, and the minor axis equal to the radius of the deferent minus the radius of the epicycle.   See Figure 38-5.

411

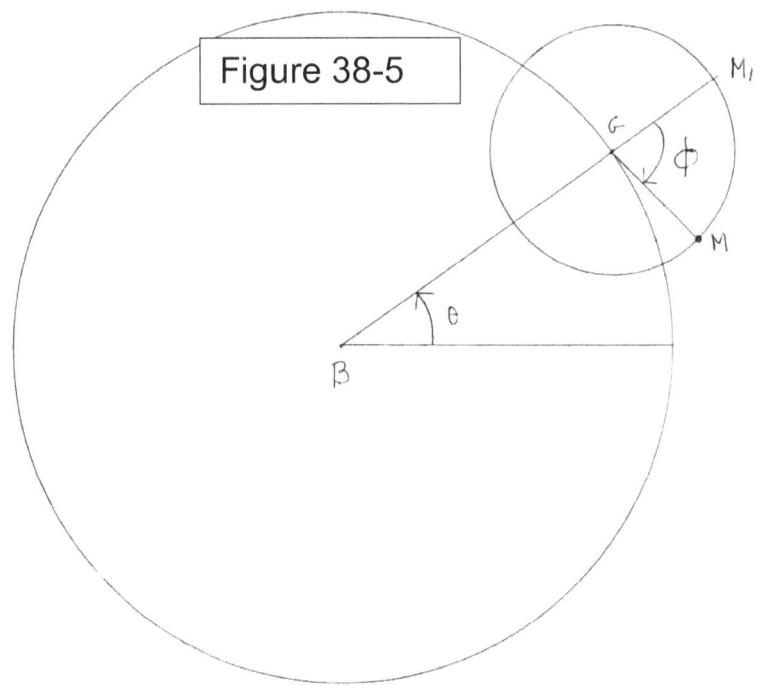

Figure 38-5

M is the planet moving uniformly in the clockwise direction around the circumference of the epicycle. We designate the uniform clockwise angular velocity by ω_e. We designate φ (Greek letter phi) as the angle through which M has moved from its initial position when it was on line BG where B is the center of the deferent and G is the center of the epicycle. If t represents the time it takes to move through φ, then,

$\varphi = \omega_e t.$

Point G, the empty center of the epicycle, in time t rotates counterclockwise around the circumference of the deferent through angle θ. If we designate the uniform angular velocity in the counterclockwise direction of G around the deferent by ω_d, then,

$\theta = \omega_d t.$

412

To produce an elliptical path for the motion of M we choose the following parameters for the epicycle deferent system.

5. The initial values for θ and φ are both 0°.

6. $\varphi = 2\theta$ at all times, and ω_e is clockwise while ω_d is counterclockwise. This means since $\omega_e = \varphi t$, or,

$\varphi = \omega_e/t$, and since,

$\omega_d = \theta t$, or, $\theta = \omega_d/t$, and, $\varphi = 2\theta$,

then, $\omega_e/t = 2(\omega_d/t)$, or,

$\omega_e = 2\omega_d$.

This means M is rotating twice as rapidly in the clockwise direction on the epicycle as G is rotating in the counterclockwise direction on the deferent.

7. The radius BG of the deferent is designated by R, and the radius of the epicycle GM is designated by r which is less than R.

With these parameters we can demonstrate that M follows the path of an ellipse with major axis equal to (R + r), and minor axis equal to (R - r). We recall in chapter 33 and in Figure 33 3 the formula for an ellipse with its center at the origin of a Cartesian coordinate system is $x^2/a^2 + y^2/b^2 = 1$, where a is the major axis and b is the minor axis.

Figure 33-3

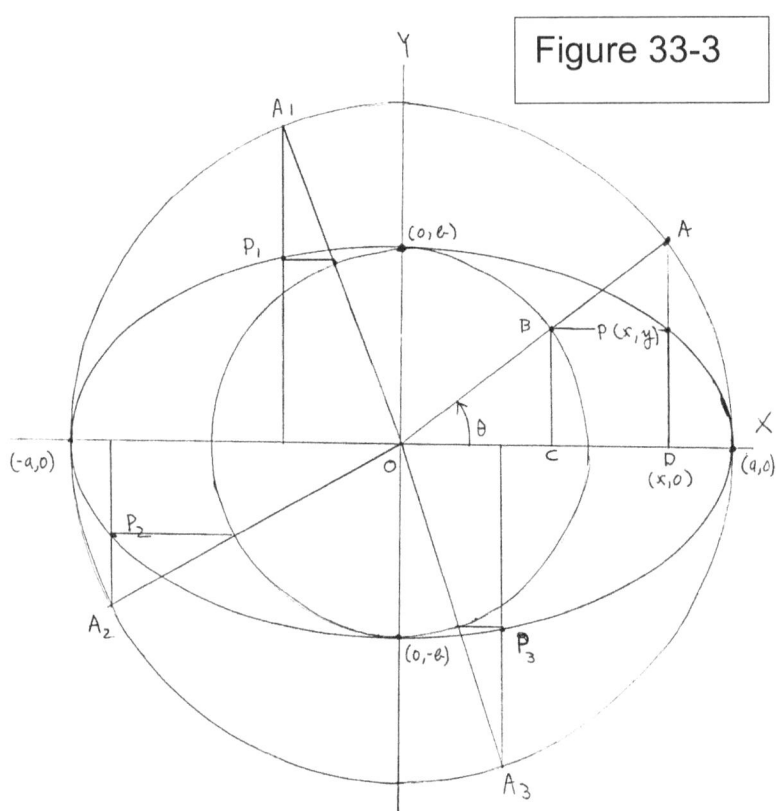

We place our epicycle deferent system on a Cartesian coordinate system so that B in Figure 38-5 is at the origin, and the initial values for θ=0°, and φ = 0° lie along the X axis as shown in Figure 38-6.

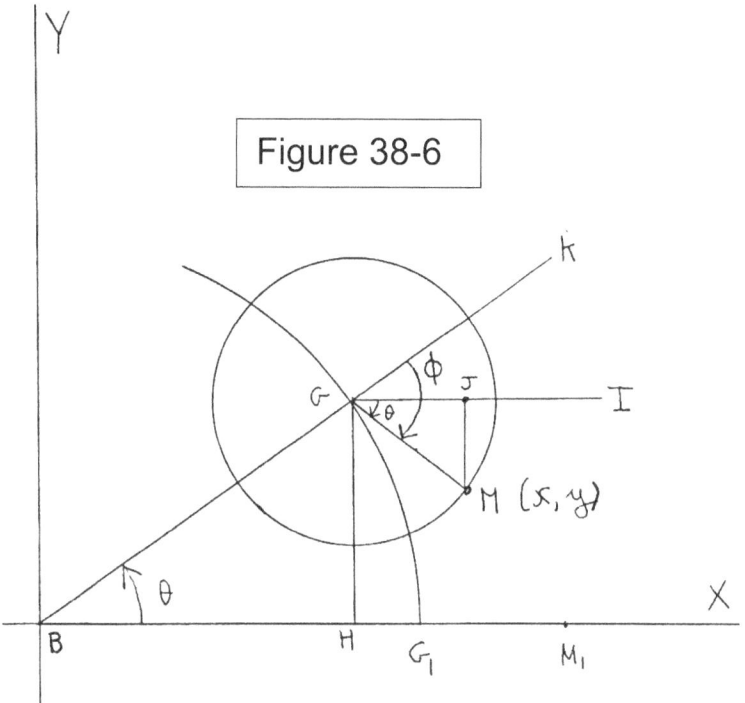

Figure 38-6

We let M be the position of the planet on the epicycle at time t after the initial values. We note G_1 is the initial value of G, and M_1 is the initial value of M. $BG_1 = R$, and $G_1M_1 = r$. After time t, M has the coordinates (x, y). Recall $\varphi = 2\theta$.

We drop perpendicular GH from G to the X axis at H. We draw GI parallel to the X axis from G, and draw MJ perpendicular to GI. We extend BG to K. We see angle KGM = φ = 2θ. From our work on parallel lines we argue angle KGI = θ. Thus, angle IGM = angle KGM - angle KGI, or angle IGM = $2\theta - \theta = \theta$.

8. The x coordinate of M equals BH + GJ.

9. In right triangle BHG, cos θ = BH/BG , or,

   BH = R cos θ.

415

10. In right triangle MJG, cos θ = GJ/GM , or, GJ = r cos θ.

11. Thus, x = R cos θ + r cos θ,

    x = (R + r)cos θ.

12. The y coordinate of M equals GH - MJ.

13. In right triangle MJG,

    sin θ = MJ/GM, or, MJ = r sin θ.

14. In right triangle GHB,

    sin θ = GH/BG, or, GH = R sin θ,

15. Thus, y = R sin θ - r sin θ, y = (R - r)sin θ.

16. We let (R + r) = a, and, (R - r) = b. Thus,

    x = a cos θ, and y = b sin θ.   Squaring both sides of

    the two equations,

    $x^2 = a^2 \cos^2 \theta$, or, $x^2/a^2 = \cos^2 \theta$.

    $y^2 = b^2 \sin^2 \theta$, or, $y^2/b^2 = \sin^2 \theta$.

    We add these two equations:

    $x^2/a^2 + y^2/b^2 = \cos^2 \theta + \sin^2 \theta$.

    We recall from chapter 13 the trigonometric identity,

    $\sin^2\theta + \cos^2\theta = 1$, for an angle, thus,

    $x^2/a^2 + y^2/b^2 = 1$.

    We know the loci of points with this equation is
an ellipse with major axis, a = R + r, and minor axis, b =
R - r. We have shown Kepler's idea of an epicycle
deferent system to bring in the orbit of Mars within an
eccentric circle is mathematically valid, and further that
an ellipse is one possible path.

    In order to determine more about the shape of his
oval Kepler had to perform many calculations with the

416

epicycle deferent system, but in the end he abandoned the epicycle deferent scheme because of physical considerations.   The center of the epicycle is an empty point and Kepler would have had to have the solar magnetic force pull an empty point which was unacceptable to him.   Furthermore, Mars was rotating in the opposite direction around the epicycle compared to rotation of the epicycle's center, and Kepler had to postulate that Mars had a force or virtue of its own to account for such motion.

He then tried to work out the shape of the oval by a complicated system of reasoning.   We shall only take up one of the arguments he presented to obtain the shape of the oval path.   In this method he makes use of the "*hypothesis vicaria*" of which we are familiar.   Kepler uses this theory in conjunction with an eccentric circle with a bisected eccentricity.   He constructs Figure 38-7.

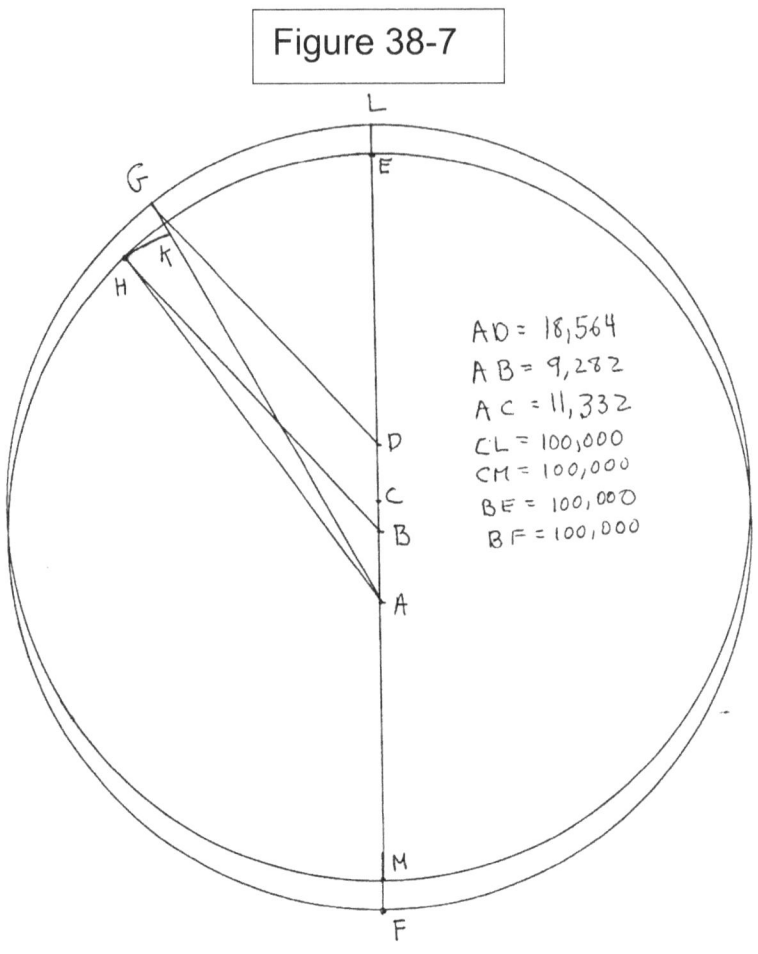

Figure 38-7

$$AD = 18,564$$
$$AB = 9,282$$
$$AC = 11,332$$
$$CL = 100,000$$
$$CM = 100,000$$
$$BE = 100,000$$
$$BF = 100,000$$

The sun is at A and the center of the circle for the 'hypothesis vicaria' is at C with the equant point at D. The circle has points L, G, and M on its circumference. Using the data from Figure 36-10 we have CL = CM = 100,000 units, and AC = 11,332 units.   Kepler knows the "hypothesis vicaria" will give the correct celestial longitudes of Mars for any time, but also knows it gives the wrong distances of Mars from the sun.

Let us consider a position of Mars such as G on the circle of the "hypothesis vicaria".   The true anomaly is angle LAG, and the mean anomaly is angle LDG since D is the equant point.

Kepler knew the longitude given by the theory is correct, but distance AG from the sun to Mars is incorrect.   Mars is closer to the sun when it lies along AG.   Kepler bisected the total eccentricity AD = 18,564 units, to find point B so that AB = 9,282 units.   He then constructed the circle with radius of 100,000 units and center at B.   This circle has points E and F on its circumference.   This circle is essentially the eccentric circle theory Kepler had been working with when he used the area law as described earlier in this chapter. He constructed angle EBH equal to angle LDG.   H is located on the circle with center at B.

Since Mars can be considered as moving with uniform angular motion around D, the equant point of the "*hypothesis vicaria*", the time for Mars to move through angle LDG is the same to move through EBH. Kepler argues the correct distance of Mars from the sun along line AG is distance AH.   He then makes circular arc HK with radius AH and center at A intersect line AG at K.   AK is taken as the correct distance of the Mars from the sun when it lies along AG, that is, AK is the distance of Mars from the sun when the true anomaly is angle LAG.   Kepler then plotted the shape of the oval by assigning multiple values for angle LDG, the mean anomaly, based on the "*hypothesis vicaria*".   The oval will lie inside the circle EHM, and was found to be egg

419

shaped, and not an ellipse by this method.   The oval
was wider toward aphelion, and slimmer toward
perihelion.   We must remember that the oval is very
close to being a circle.   Kepler found, as we shall show,
that when Mars was 90° from the line of apsides it only
came in from the circle about 858 units in a circle with a
radius of 100,000 units.   We show in Figure 38-8 in a
greatly exaggerated way the shape of an egg shaped
oval within a circle with center at B eccentric to A.

Figure 38-8

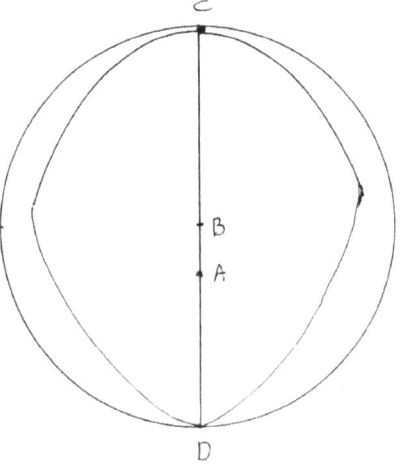

Kepler was now faced with the task of calculation
with such an orbit.   How could he calculate the mean
anomaly, the physical equation, the optical equation,
and the true anomaly with such a model?   After a
number of trials of summing 180 sun-Mars distances, he
could not get a satisfactory answer.   He then recalled
the work of Archimedes which allowed him to calculate
areas if he could substitute for his egg shaped oval an
420

ellipse with equivalent area inscribed within the circle with radius BC. He then could calculate with his area law. At this stage he did not believe the actual path of Mars was such an ellipse, so he called the ellipse he used for calculations the auxiliary ellipse. We will show shortly how Kepler determined the dimensions of the auxiliary ellipse, but at this point we need to understand Archimedes' work on calculating the area of an ellipse with reference to its circumscribed circle.

Archimedes had demonstrated in his book, "On Conoids and Spheroids", that the ratio of the area of a circle with a radius of length 'a' taken to the area of an ellipse with semimajor axis of length 'a', and semiminor axis ' b', is equal to the ratio of 'a' to 'b'. We will give a demonstration of this proposition similar to that given by Archimedes, but taking advantage of our previous work on the ellipse from chapter 33. We construct Figure 38-9 which is similar to Figure 33-3.

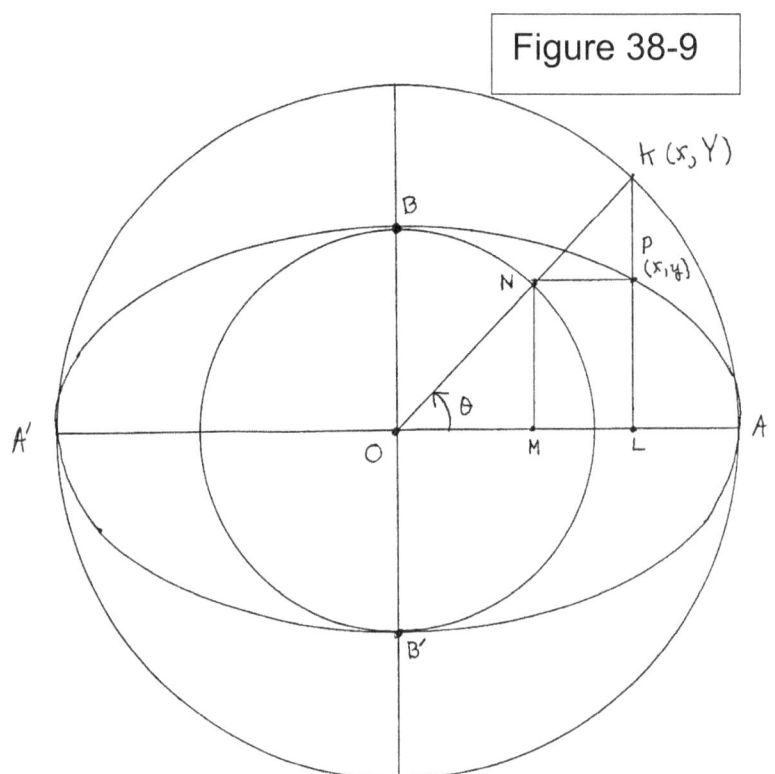

Figure 38-9

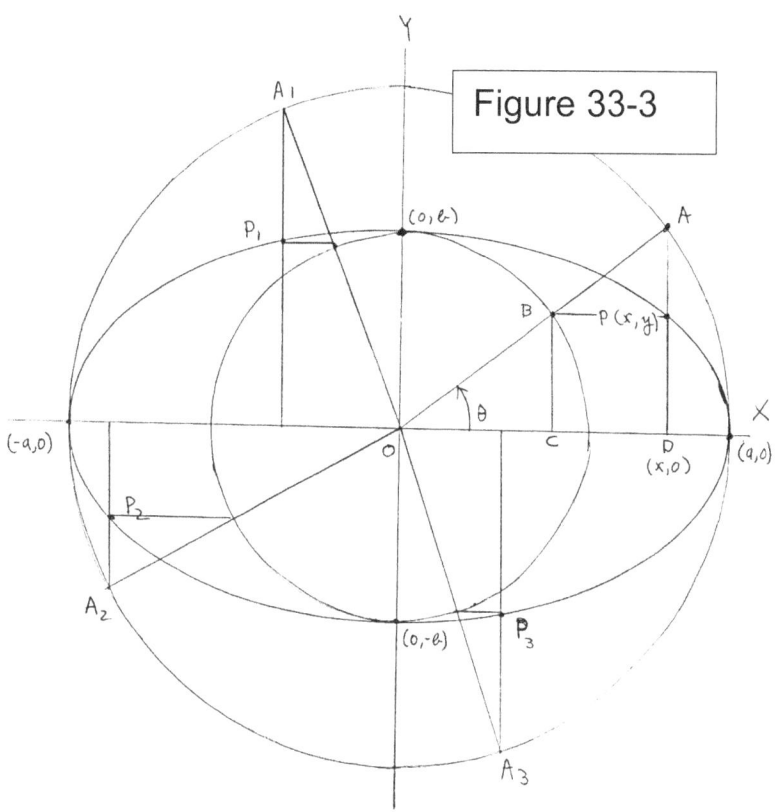

Figure 33-3

In Figure 38-9 the ellipse A'BAB' is circumscribed by the circle with radius OA = a, the semimajor axis of the ellipse. We also draw the circle with radius OB = b, the semiminor axis of the ellipse, inscribed within the ellipse. We draw KL perpendicular to OA through any point P on the ellipse, and NP parallel to OA, where N is on the inscribed circle.   θ is the angle from OA to OK. From our work in chapter 33 we know N lies along OK as shown.

Archimedes claims that for any point P on the ellipse: KL/ PL = a/ b .

We can demonstrate this by using trigonometry. In right triangle OLK,

17. sin θ = KL/ OK , or, since OK = a,   sin θ = KL/a .
We drop perpendicular NM to OA. In right triangle OKN,

18. sin θ = NM/ON, or since NM = PL, and ON = b,

sin θ = PL/b . Then,

19. KL/ a = PL/b . Reversing the means of the proportion

KL/PB = a/b. If we assign K the coordinates (x, Y), and P the coordinates (x, y), then, Y/y = a/b.

Next Archimedes inscribes a polygon in the circle with radius 'a'.   See Figure 38-10.

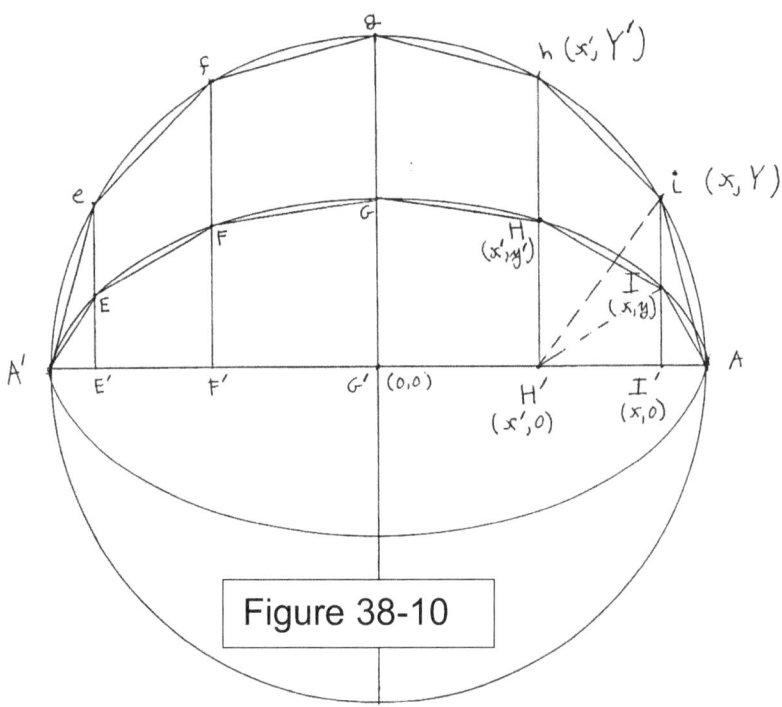

Figure 38-10

We have indicated the polygon in the upper half of the circle.   Perpendiculars are dropped from the vertices of the inscribed polygon to the diameter A'A. The vertices are labeled as noted, and the

424

perpendiculars intersect the ellipse at E, F, G, H, and I. These points are connected with chords.   We then place G' at the origin of a Cartesian coordinate system so G'A is the positive X axis, and G'g is the positive Y axis.   We then assign coordinates to H', (x', 0), for H, (x', y'), for h, (x', Y'), for I', (x, 0), for I, (x, y), and for i, (x, Y).   These coordinates are marked on the figure.

We now consider the area of the four sided figure H'hiI' inscribed in the circle.   A four sided figure with two of its sides parallel is called a trapezoid.   In this case H'h is parallel to I'i.   The other two sides, hi and H'I', are clearly not parallel.   The area of a trapezoid can be found by considering its area as the sum of the areas of two triangles. In this case the two triangles are formed by the broken line H'i.

20. Area trapezoid (H'hiI')

= area triangle (hH'i) + area triangle (H'iI')

Area trapezoid (H'hiI') = ½(hH')(H'I') + ½(iI')(H'I').

Since (H'I') is the height of both triangles while (hH') is the base of triangle (hH'i), and (iI') is the base of triangle (H'iI'):

Area trapezoid (H'hiI') = ½(H'I')(hH' + iI').

We now use the coordinates to represent these lengths.

Area trapezoid (H'hiI') = ½(x - x')(Y + Y').

We next consider the area of trapezoid (HH'I'I) which is inscribed within the ellipse.   It's area is the sum of the two triangles (HH'I) and (H'II').

425

21. Area trapezoid (HH'I'I) = ½ (HH')(H' I') + ½ (I I')(H' I')

Area trapezoid (HH'I'I) = ½(H' I')(HH' + I I').

Using the coordinate values.

Area trapezoid (HH'I'I) = ½(x - x')(y' + y).

We now take the ratio of area trapezoid (H'hiI') to the area of trapezoid (HH'I'I).   We call the former 'trapezoid (circle)', and the latter 'trapezoid (ellipse)'.

22.

$$\frac{trapezoid(circle)}{trapezoid(ellipse)} = \frac{\frac{1}{2}(x - x')(Y + Y')}{\frac{1}{2}(x - x')(y + y')}$$

$$= \frac{Y + Y'}{y + y'}.$$

We next evaluate, Y + Y'/ y + y'.   We know from our work on Figure 38-8 that Y/y = a/b, and, Y'/y' = a/b, so that Y/y = Y'/y'.   Thus, Y' = Yy'/y, and, y' = yY'/Y. Then,

23.

$$\frac{Y + Y'}{y + y'} = \frac{Y + \dfrac{Yy'}{y}}{y + \dfrac{yY'}{Y}} = \frac{\dfrac{Yy + Yy'}{y}}{\dfrac{Yy + Y'y}{Y}}$$

$$= \frac{Yy + Yy'}{y} \bullet \frac{Y}{Yy + Y'y}.$$

426

Consider the term Y'y.   Since ,Y/y = Y'/y', then, Y'y = Yy'. Then in above equation we substitute Yy' for Y'y, so we have,

$$\frac{Y + Y'}{y + y'} = \frac{Yy + Yy'}{y} \cdot \frac{Y}{Yy + Yy'} = \frac{Y}{y} = \frac{a}{b} = \frac{trapezoid(circle)}{trapezoid(ellipse)}.$$

The preceding argument can be used for all the other pairs of trapezoids made from the sides of the inscribed polygons in Figure 38-9, so that the ratios of all the trapezoids inscribed within the circle taken to the corresponding trapezoids within the ellipse are equal to a/b.   We can then imagine increasing the number of sides of the inscribed polygons larger and larger so that as the number of sides increases toward the infinitely large, the sum of their lengths approaches the circumference of the circle in the case of the inscribed polygons in the circle, and approaches the circumference of the ellipse in the case of the inscribed polygons within the ellipse.   Then the sum of the areas of all the inscribed trapezoids in the circle approaches the area of one half of the circle, and is equal to the area of one half of the circle in the limit as the number of trapezoids increase towards the infinitely large. Likewise, the area of all the inscribed trapezoids in the ellipse approaches the area of one half of the ellipse, and is equal to the area of one half the ellipse in the limit as their number approaches the infinitely large.

427

Since the ratio of areas of each pair of trapezoids (circle to ellipse) is equal to a/b, then

24. area of ½circle/area of ½ellipse = a/b, or,

   area of circle/area ellipse = a/b.

25. The area of a circle with radius 'a' is:

   Area circle = πa^2. Thus,

   πa^2/area of ellipse = a/b,

   area of ellipse = $\pi a^2 b/a$ = πab,

   where a is the semimajor axis, and b,

   the semiminor axis of the ellipse.

Kepler was aware of the formula for the area of an ellipse because of his study of Archimedes. Because he could not readily calculate the area of sectors of his egg shaped oval, but could do so for an equivalent ellipse, he replaced the egg shaped oval with the auxiliary ellipse. But how did Kepler determine the length of the semimajor axis for the auxiliary ellipse?

Kepler gives his answer in chapter XLVII of the "*Astronomia Nova*". Consider Figure 38-11.

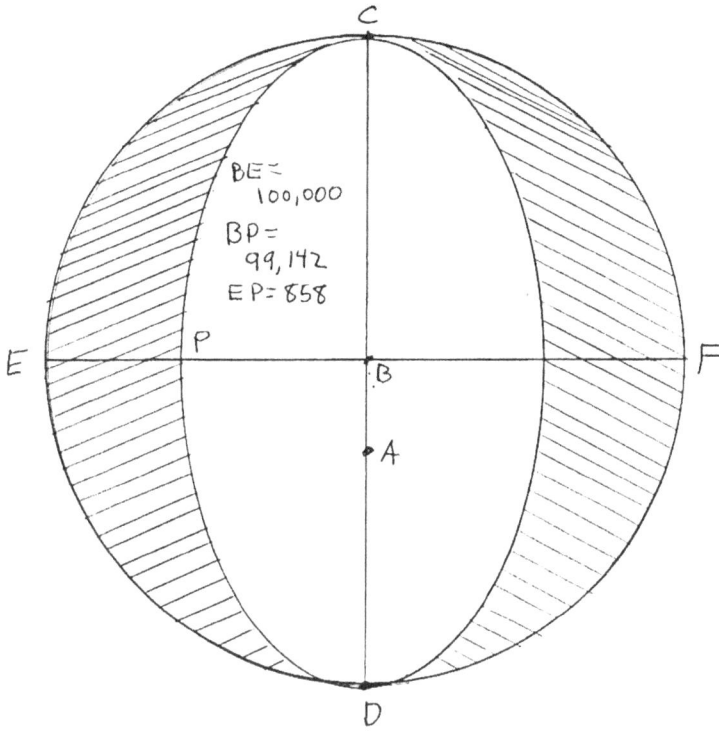

Figure 38-11

The circle CEDF represents the circular orbit which was determined from the work with Figure 38-1, where B is the center of the circle, A the sun, BC = BE = BD = 100,000 units. The eccentricity of the sun is AB = 9,264 units.

The ellipse is inscribed within the circular orbit so that its major axis lies along the line of apsides CD, and is equal to the diameter of the circle. The area between the ellipse and the circle is shaded on both sides of the line of apsides. On each side this area is shaped like a horned moon, and is called a lunula.

Kepler determined for his auxiliary ellipse that the area of the two lunulae should be approximately equal in area to a circle with a radius of AB = 9,264 units.

429

26. Area of 2 lunulae = $\pi(AB)^2$,

Area of 2 lunulae $\cong (3.14159)(9,264)^2$

= 269,617,440 square units.

Kepler, however, determined the area of the 2 lunulae should equal 269,500,000 square units for his approximation.

27. Area of circle = $\pi(BC)^2$ = $(3.14159)(100,000)^2$

Area of circle = 31,415,900,000 square units.

28. Area of ellipse = area of circle - area of 2 lunulae,

Area of ellipse

= 31,415,900,000 - 269,500,000(Kepler's

approximation).

Area of ellipse = 31,146,400,000 square

units(approximation).

Since the area of an ellipse by Archimedes' formula is πab, and in this case a, the semimajor axis equals 100,000 units.

29. πab = 31,146,400,000 square units,

b = 31,146,400,000/(3.14159)(100,000)

= 99,142.07 units.

In Figure 38-11, b = BP, and EP = BE - BP, so

30. EP = 100,000 - 99,142.07 = 857.93 units.

Kepler rounded this to 858 units. Thus, the auxiliary ellipse had semimajor axis equal to 100,000 units, semiminor axis equal to 99,142 units, and the maximum distance of the ellipse from the circumference of the circumscribed circle is 858 units when angle CBE = 90°.

430

We note in the above calculation of the auxiliary ellipse that Kepler had not used the exact value for the area of the ellipse, but rounded off the value of the 2 lunulae to 269,500,000 square units to calculate the area of the ellipse. In chapter XLVII of the "*Astronomia Nova*" Kepler gives a geometric argument as to why he chose this particular ellipse, and why $\pi(AB)^2$ was chosen for the value for the area of the two lunulae.

He constructed Figure 38-12, and gave the following argument.

Figure 38-12

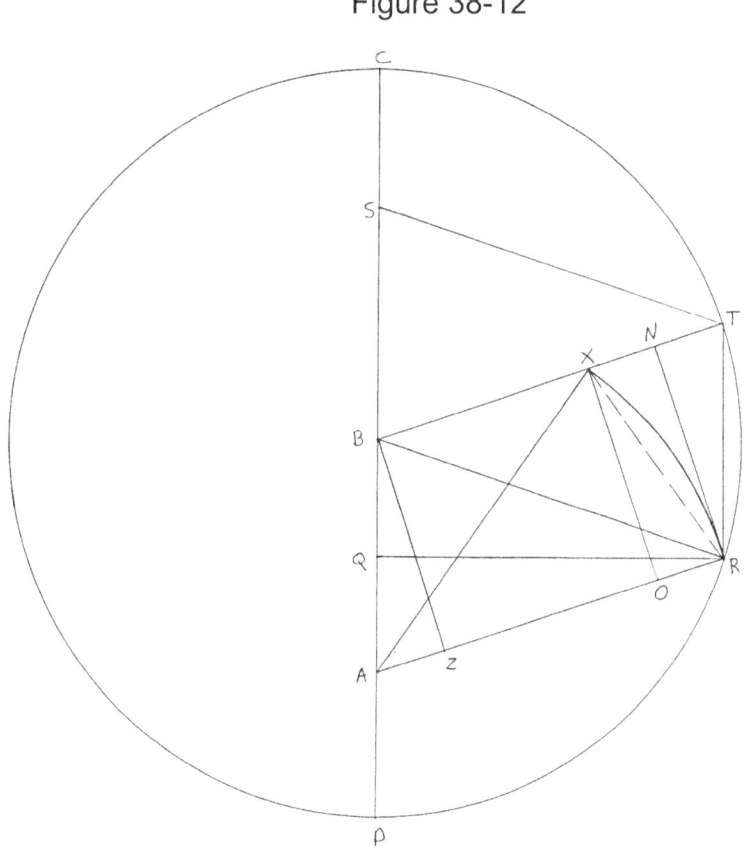

The larger circle is the circle represented in Figure 38-11 which circumscribes the ellipse. As before

A is the sun, B the center of the circle, with AB = 9,264 units, and BC = 100,000 units. We have greatly exaggerated length AB relative to BC so the figure can more readily be interpreted.

AB is bisected at Q so AQ = QB, and QR is drawn perpendicular to AB. AR and BR are connected. Since QR is the perpendicular bisector from the vertex R of the triangle ARB, then we know AR = BR. This means triangles BQR and AQR are congruent. Next the equant point S is located such that AB = BS. ST is constructed parallel to BR, and BT is connected.

31. Angle BST = angle ABR (from parallels ST, BR),

angle BAR = angle BST (angle BAR = angle ABR),

then BT is parallel to AR

Thus,

angle BAR = angle SBT, so,

angle BST = angle SBT.

This means triangle BST is an isosceles triangle, so, BT = BS. Triangles BST and ABR are congruent since, AB = BS, and the angles opposite the equal sides are equal. Thus, BT = AR, so that when TR is joined it is parallel to AB, and equal to AB.

With A as a center and AR as radius arc RX is drawn to BT. Thus, AX = AR. RN is drawn perpendicular to BT at N. Kepler then claimed,

2(TN) = TX.

32. RX is connected as a broken line. XO is drawn from X perpendicular to AR at 0. Then OX = RN.

432

Consider triangles AOX and BNR. They are both right triangles, they have corresponding sides equal, OX = RN, and AX = BR (BR is a radius and we know, BR = AR = AX). This means the remaining sides BX and AO are equal and the triangles are congruent. Then, angles RBT and RAX are equal.

33. Consider triangles ARX and BRT. They are congruent since two corresponding sides, BR, AR, and AX, BT, and the angles between them, angles RBT and RAX are equal.

Thus, TR = RX.

34. Then triangle XRT is an isosceles triangle, and we know the line from the vertex between the equal sides, when perpendicular to the third side, bisects the third side. Thus, NX = TN. Since, NX + TN = TX, 2(TN) = TX.

Kepler considered AX the distance of Mars from the sun when it lies along AX. The argument is the same as he used for the sun-Mars distance in Figure 38-7. That is, since AR equals AX when Mars is along line AX, and X is on BT parallel to AR, its distance from the sun is AX. For Kepler that meant X is on the auxiliary ellipse. Kepler then demonstrated, $(AB)^2 = (BT)(TX)$.

For this demonstration Kepler uses the proposition from Euclid we already established in our work with Ptolemy regarding the intersection of two chords within a circle. To make this clear we construct

Figure 38-13, where radius BN is extended to W, and RN is extended to U.

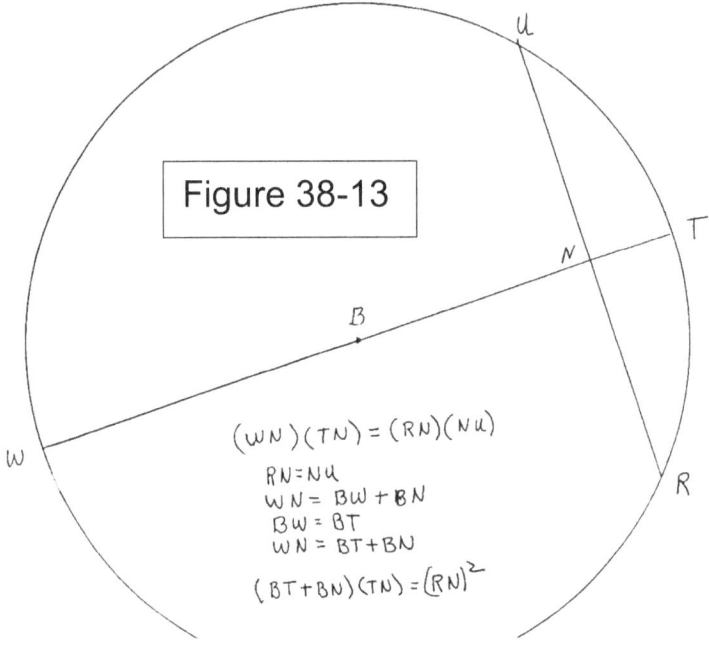

Figure 38-13

$$(WN)(TN) = (RN)(NU)$$
$$RN = NU$$
$$WN = BW + BN$$
$$BW = BT$$
$$WN = BT + BN$$
$$(BT + BN)(TN) = (RN)^2$$

The proposition we established states,

35. (WN)(TN) = (RN)(NU).

Since RN is perpendicular to WT, RN = NU.

WN = BT + BN, since, BT = BW.

Then, $(BT + BN)(TN) = (RN)^2$.

Going back to Figure 38-12 and right triangle RNT, Kepler used the Pythagorean Theorem to state,

36. $(RN)^2 + (TN)^2 = (TR)^2$, or since, TR = AB,

$(AB)^2 = (RN)^2 + (TN)^2$.

Then the value for $(RN)^2$ we determined can be substituted:

37. $(AB)^2 = (BT + BN)(TN) + (TN)^2$,

$(AB)^2 = TN(BT + BN + TN)$.

We see from Figure 38-12, BN + TN = BT.

434

Then,

$(AB)^2 = TN(BT + BT)$

$(AB)^2 = (TN)(2)(BT) = 2(TN)(BT).$

Since Kepler had showed, $2(TN) = TX$, then

38. $(AB)^2 = (BT)(TX).$

We recall Archimedes result:

Area circle/Area ellipse = $\pi a^2/\pi ab$, where a is the radius BT, and b the semiminor axis of the ellipse. In Figure 38-12 if line BXT were perpendicular to diameter CD, then since X would be on the auxiliary ellipse, and BX would be equal to b. Because Kepler realized BXT is not perpendicular to the line of apsides he used BX as an approximation to b. Thus,

39. Area ellipse = $\pi ab \cong \pi(BT) (BX)$. (BT is a radius equal to a).

From Figure 38-11,

40. Area of 2 lunulae = area circle - area ellipse

Area of 2 lunulae $\cong \pi(BT)^2 - \pi(BT) (BX)$.

Recall, $(AB)^2 = (BT)(TX)$, and from Figure 38-12 TX = BT - BX, we have.

$(AB)^2 = (BT)(BT - BX) = (BT)^2 - (BT)(BX).$

We solve this equation for (BT)(BX).

$(BT)(BX) = (BT)^2 - (AB)^2.$ Then,

41. Area of 2 lunulae = $\pi (BT)^2 - \pi[(BT)^2 - (AB)^2]$,

Area of 2 lunulae = $\pi(BT)^2 - \pi(BT)^2 + \pi(AB)^2$,

Area of 2 lunulae = $\pi(AB)^2$.

This is the result Kepler wanted for his auxiliary ellipse. We see then why Kepler made the

435

approximation of 269,500,000 for the area of the 2 lunulae when he calculated the area for the ellipse. He now was in a position to test the auxiliary ellipse with his area law to see if it agreed with the *"hypothesis vicaria"* for the true anomaly at various positions on the orbit.

To understand Kepler's method of calculating with his auxiliary ellipse we construct Figure 38-14.

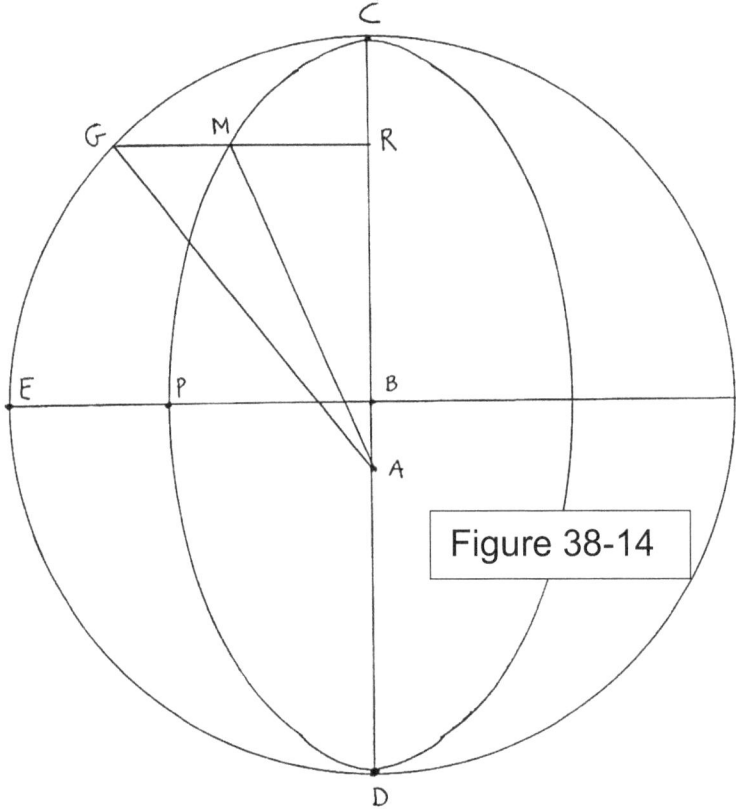

Figure 38-14

In this figure the auxiliary ellipse is inscribed within the circle with diameter CD, the line of apsides. B is the center, and A is the sun. The auxiliary ellipse is inscribed in a circle with radius BC = 100,000 units, and AB = 9,264 units. Recall BP, the semiminor axis of the ellipse is 99,142 units.   We need to demonstrate the proportion,

  42. area sector CAG /area sector CAM = a/b, where
      a = BE,and b = BP.

  43. area CAG = area triangle GRA + area CRG,
      where GMR is the perpendicular from G to CD.
      area CAG = ½(AR)(GR) + area CRG,
      where AR is the base, and GR the height of
       triangle GRA.

  44. area CAM = area triangle MRA + area CRM
      area CAM = ½(AR)(RM) + area CRM.

We can think of areas CRG and CRM as being made up of multiple segments made by inscribed polygons within the circle and the ellipse as in Figure 38-10.   From the argument pertaining to that figure then,

  45. area CRG/area CRM = a/b, and from our work
  on Figure 38-9, GR/RM = a/b. Thus,

      area CRG = a/b(area CRM), and

      GR = a/b(RM). Then we have rom steps 43, 44:
  46.

$$\frac{area\ sec\,tor\ CAG}{area\ sec\,tor\ CAM} =$$

$$\frac{\frac{1}{2}(AR)(\frac{a}{b})(RM) + \frac{a}{b}(area\,CRM)}{\frac{1}{2}(AR)(RM) + area\,CRM}.$$

$$\frac{area\ sec tor\ CAG}{area\ sec tor\ CAM}$$

$$= \frac{a/b[\frac{1}{2}(AR)(RM) + area\ CMR]}{[\frac{1}{2}(AR)(RM) + area\ CMR]}.$$

The terms in the brackets cancel so,

$$\frac{area\ sec tor\ CAG}{area\ sec tor\ CAM} = \frac{a}{b}.$$

We recall when Kepler tested his eccentric circle theory in Figure 38-2 he was interested in comparing the results with the "*hypothesis vicaria*", since the "*hypothesis vicaria*" does give the celestial longitudes of Mars to within 2' of accuracy by Tycho's observations. In his comparison Kepler chose points at the octants and quadrants. He sought to test the theory of the auxiliary ellipse with the area law in the same way. When he tested the eccentric circle theory for the quadrants he found the mean anomaly was 95° 18' 28" when the eccentric anomaly is 90°.

To see how to calculate with the auxiliary ellipse we construct Figure 38-15.

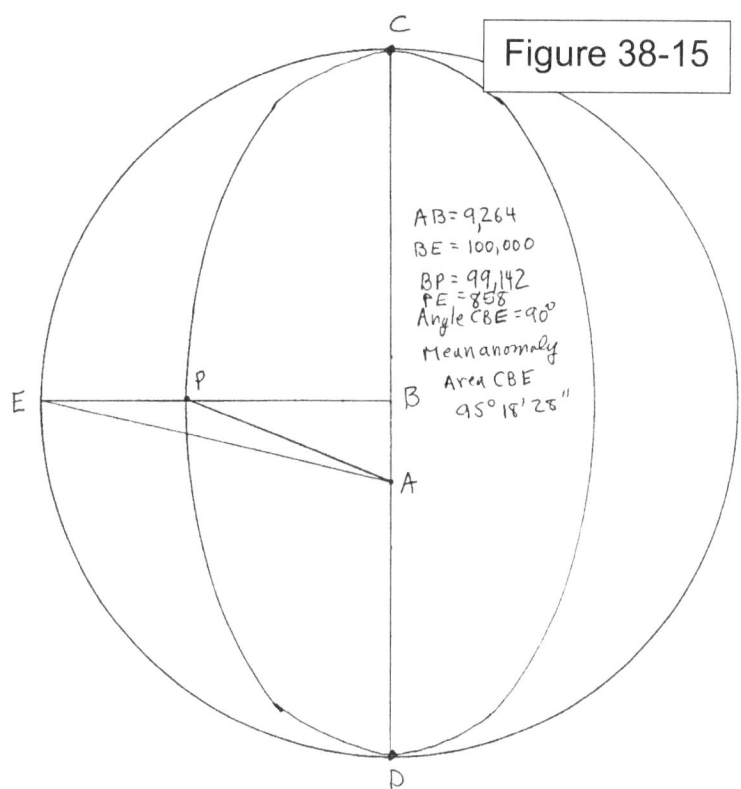

Figure 38-15

AB = 9,264
BE = 100,000
BP = 99,142
PE = 858
Angle CBE = 90°
Mean anomaly
Area CBE
95° 18' 28"

The mean anomaly for the eccentric circle is area CAE = 95° 18' 28" when angle CBE = 90°. We can convert the value of 95° 18' 28" to square units by the following proportion,

47. area CAE/area circle = 95° 18' 28"/360°,

area CAE = $[\pi(BE)^2(95.30778°)/360°$,

area CAE = $\pi(100,000)^2$ (.264744) =

(31,415,926,536)(.264744) = 8,317,178,056.

(Kepler got 8,317,172,671 square units).

We can also determine area CAE by a second method which should agree very nearly.

48. Area CAE = area triangle ABE + area circular

439

sector CBE,

area ABE = ½(AB)(BE) = ½(9,264)(100,000),

area ABE = 463,200,000 square units.

area sector CBE = [90°/(360°)] x(area of circle),

area CBE = ¼(31,415,926,536)

= 7,853,981,634 square units.

Kepler's value was 7,853,981,670 square units.

area CAE = 463,200,000 + 7,853,981,634 =

8,317,181,635 square units. Kepler's value:

8,317,181,670 square units.

When we compare this with the first method we see the results are very nearly equal.

The true anomaly for the eccentric circle theory was found by solving triangle ABE for angle BAE, and Kepler found the value of 84° 42' 26".

The inscribed auxiliary ellipse has semimajor axis a = BC = 100,000 units, and semiminor axis b = BP = 99,142 units.

The mean anomaly for the auxiliary ellipse is area CAP bounded by lines CA, AP, and the arc PC of the ellipse.   We know

49. area CAE/area CAP = a/b,

area CAP = (b/a)(area CAE) =

(99,142/100,000)(8,317,178,056).   We used

the value for area CAE from step 47. Then,

area CAP = 8,245,816,668 square units.

To express this in degrees, minutes, and seconds

50. area CAP/area ellipse = degrees of mean

440

anomaly/360^0.

degrees mean anomaly

= (area CAP)(360^0)/(area of ellipse).

The area of ellipse = b/a(area circle) =

(.99142)(31,415,926,536) = 31,146,377,886

square units.

degrees mean anomaly =

(8,245,811,285)(360^0)/31,146,377,886

= 95.30778^0,

degrees mean anomaly = 95°18'28".

This shows the mean anomaly is the same for the eccentric circle and for the ellipse.

The physical equation for the auxiliary ellipse is the area of triangle ABP.

51. area ABP = ½(AB)(BP) =

½(9,264)(99,142),

area ABP = 459,225,744 square units.

degrees for physical equation =

(area ABP)(360^0)/(area ellipse).

physical equation = (459,225,744)(360^0)/

31,146,377,886, physical equation

= 5°18' 28".

The optical equation is angle BPA.

52. tan BPA = AB/BP = 9,264/99,142 = .093442.

arctan .093442 = angle BPA = 5° 20' 18".

The true anomaly for the auxiliary ellipse is angle BAP.

53. tan BAP = BP/AB = 99,142/9,264 = 10.70186

arctan 10.70186 = angle BAP = 84° 39' 42".

When Kepler calculated the true anomaly using the *"hypothesis vicaria"* starting with mean anomaly 95° 18' 28", he found the value 84° 42' 2" for the true anomaly. The difference between the value calculated with the auxiliary ellipse and the value from the *"hypothesis vicaria"* is 2' 20", which meant the auxiliary ellipse is satisfactory at the quadrants.

To be sure we understand how to calculate with an elliptical orbit we will do the calculation for the first octant. Kepler tested the auxiliary ellipse for a mean anomaly of 48° 45' 12". We recall this is the mean anomaly for the eccentric circle as shown in Figure 38-16 when angle CBG = 45°.

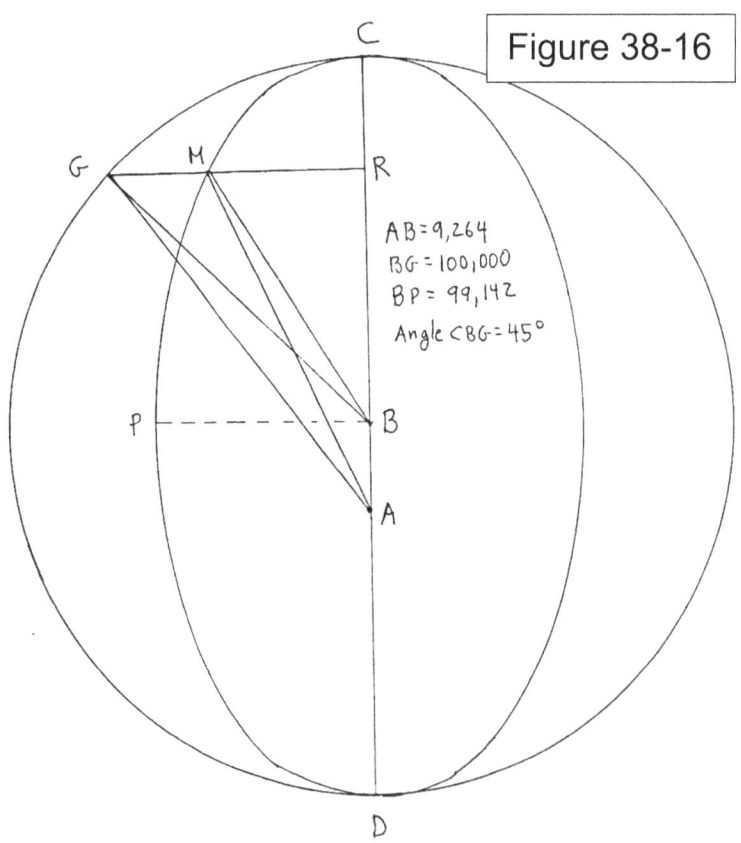

Figure 38-16

AB = 9,264
BG = 100,000
BP = 99,142
Angle CBG = 45°

We first find the area of sector CAG in square units.

54. area CAG/area circle = $48^0 45'12''/360^0$,

area CAG = (31,415,926,536)(.135426)

= 4,254,530,940 square units.

55. The second method of finding area CAG:

area CAG

= area triangle ABG + area circular sector CBG.

area ABG = ½(AB)(GR),

since AB is the base of the triangle,

and GR is the height.

sin CBG = sin 45° = GR/GB, or,

GR = (100,000)(sin 45°),

GR = 70,711.

area ABG = ½(9,264)(70,711)

= 327,533,352 square units.

area CBG = (45^0)(area circle)/360^0,

area CBG = 1/8 x (31,415,926,536)

area CBG = 3,926,990,817 square units.

area CAG = 327,533,352 + 3,926,990,817

= 4,254,524,169 square units.

This result agrees closely with the first method.

The mean anomaly for the ellipse is area CAM.

56. area CAM = (b/a)(area CAG).

area CAM = (99,142)

(4,254,530,940)/100,000,

area CAM = 4,218,027,065 square units.

degrees of mean anomaly

= (area CAM)(360^0)/area ellipse.

degrees mean anomaly

=(4,218,027,065)(360^0)/31,146,377,886),

degrees of mean anomaly = 48° 45' 12",

the same as the eccentric circle, as it should

be.

57. The physical equation for the ellipse is the area of triangle ABM,

area ABM = ½(AB)(MR),

MR = b/a (GR)

= (99,142)(70,711)/100,000 = 70,104.

area ABM = ½(9,264)(70,104)

= 324,721,728   square units.

degrees for physical equation

= (area ABM)(360^0) /area ellipse.

degrees for physical equation =

(324,721,728)(360^0)/31,146,377,886,

degrees for physical equation = 3° 45' 12".

The true anomaly is angle BAM.   We find this angle by first finding the length of RB.

58. tan RBG = tan 45° = GR/RB, since,

tan 45 ^{0} = 1,

RB = GR = 70,711.

59. tan BAM = tan RAM = MR/AR

= 70,104/(AB + RB) =

70,104/(9,264 + 70,711) = .876574.

arctan   .876574 = angle BAM = 41° 14' 13",

(Kepler got 41° 14' 9").

444

We have shown the method Kepler used to solve for the true anomaly for the theory of the auxiliary ellipse. Kepler then gave a table to compare the values in the first two octants, and the first quadrant for the true anomaly as calculated by the theory for the auxiliary ellipse and the "*hypothesis vicaria*".

mean anomaly	true anomaly from auxiliary ellipse	true anomaly from *hypothesis* vicaria
48° 45' 12"	41° 14' 9"	41° 20' 33"
		variance - 6' 24"
95° 18' 28"	84° 39' 42"	84° 42' 2"
		variance - 2' 20"
138° 45' 12"	131° 14" 5"	131° 7" 26"
		variance + 6" 39"

Because of the symmetry of the auxiliary ellipse we can construct Figure 38-17 to show the errors produced by the auxiliary ellipse in predicting the true anomalies relative to the "*hypothesis vicaria*".

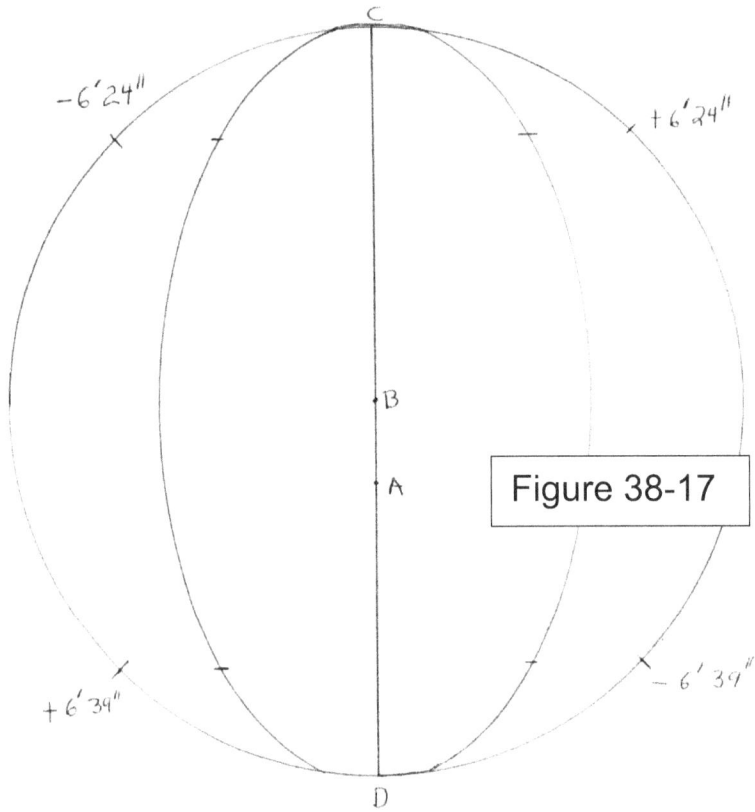

−6′24″ +6′24″

·B

·A   Figure 38-17

+6′34″   − 6′34″

C

D

These errors are of nearly the same magnitude as the eccentric circle of Figure 38-2, but in the opposite direction. That is in the auxiliary ellipse Mars moves too slowly after passing the apsides and speeds up too much after passing the quadrants.

Kepler was undoubtedly discouraged. He had spent hours working out the distances for his oval, he had developed an ingenious method to calculate the equations using the auxiliary ellipse, but the longitudes for the true anomalies would not agree with the *"hypothesis vicaria"* (which gave the true longitudes of Mars to 2' arc). Kepler then had to start again, and, as we shall see in the next chapter, he would discover, as if by chance, the correct orbit.

Before we close this chapter we might consider why we are presenting the theories of Kepler which had to be abandoned. We could have given the final theory of Kepler, i.e. an elliptical orbit with the sun at one focus with the parameters for the semimajor axis, semiminor axis, and distance of the center to the focus. Instead we have taken the reader through some of the laborious work of Kepler in his five year attempt to find the correct orbit. We did this because for one to understand how science and mathematics proceed together, we need to understand that a scientist must often try many false hypotheses before he finds one that make sense with the data of his observations. Scientists are problem solvers as much as they are philosophers on the beauty and symmetry of nature, and how the relationships in natural phenomena have mathematical expression.

In the *'Astronomia Nova'* Kepler has given us a masterpiece of the problem solving aspects of the working scientist. At times his calculations overwhelm us. We only selected a small part of them for our work, but we have seen how Kepler developed his theories starting with his deeply held belief that there is a mathematical solution for the orbit of Mars that will make physical sense, that is be based on an attractive force from the true sun that varies is some way to the distance of the planet in its orbit. Going through the trials of

Kepler's work and his mathematical methods we become better mathematicians ourselves, and appreciate more deeply the hidden secrets of nature and how the scientist probes their mystery.

Bibliography for Chapter 38

1. Kepler, Johannes, "Astronomia Nova", Latin Edition printed by Culture et Civilisation, 115, Avenue, Gabriel Lebon, Bruxelles, 1968.

2. Small, Robert, "An account of the Astronomical Discoveries of Kepler", A reprinting of the 1804 text by The University of Wisconsin Press, Madison Wisconsin, 1963.

3. Aiton, E.J., "Kepler's Second Law of Planetary Motion", "Isis" 60: 75-88, 1969.

4. Wilson, Curtis, "How Did Kepler Discover His First Two Laws?", "Scientific American", 226, March 1972, no.3, 93-106, 1972.

5. Koyre, Alexandre, "The Astronomical Revolution", translated by Dr R. E. W. Madison, Cornell University Press, Ithaca New York, 1973.

Chapter 39

The Discovery of the Elliptical Orbit

Discouraged by the results of the auxiliary ellipse calculations presented in chapter 38, Kepler considered that he may have calculated the sun-Mars distances incorrectly in setting up his egg shaped oval. We recall also he made approximations when determining the parameters of the auxiliary ellipse. This led him to the laborious work of calculating many distances from observations for different parts of Mars' orbit. He did this for positions of Mars away from the celestial longitude of the line of apsides as he had previously done in Figure 38-1.

When he did these calculations he assigned the radius of the eccentric circle for the earth's orbit the value of 100,000 units as in the work with Figure 38-1.

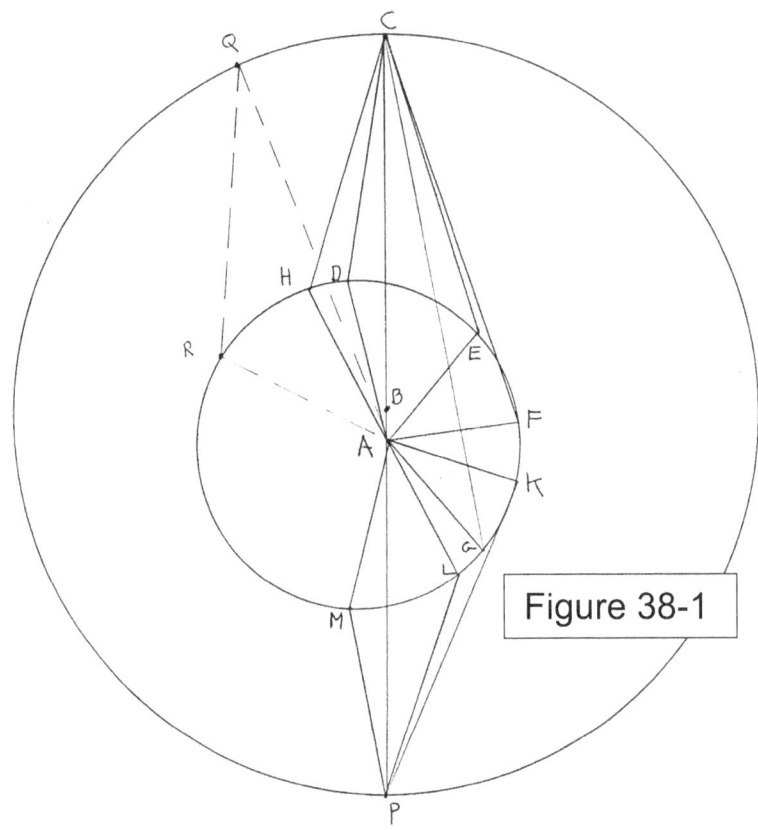

Figure 38-1

For Figure 38-1 Kepler had found the mean distance for the line of apsides to be 152,640 units. With the new calculations the distance of the sun from the aphelion position for Mars was 166,465 units, and the distance from the perihelion position was 138,234 units. The mean distance was then, (166,465 + 138,234)/2 = 152,349.5, or, 152,350 units. The eccentricity from the sun to the midpoint of the line of apsides was, 152,350 - 138,234 = 14,116 units.

Kepler then converted these figures after assigning 100,000 units for the mean distance for the orbit of Mars along the line of apsides. See Figure 39-1.

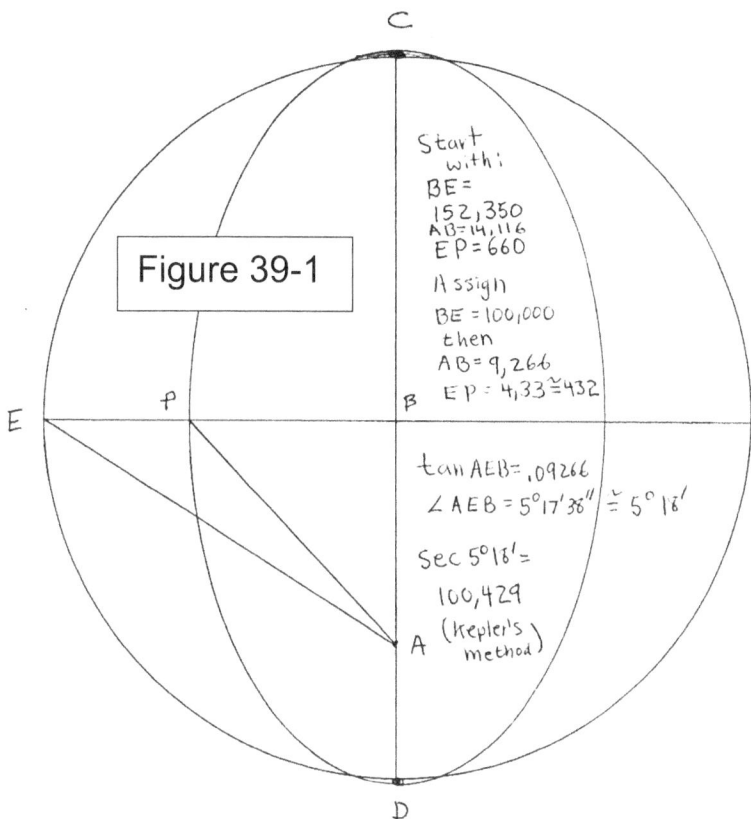

Figure 39-1

Start with:
BE = 152,350
AB = 14,116
EP = 660

Assign
BE = 100,000
then
AB = 9,266
EP = 4,33 ≈ 432

tan AEB = .09266
∠ AEB = 5°17′36″ ≈ 5° 18′

Sec 5°18′ =
100,429
(Kepler's method)

Thus, 100,000 /152,350 = AB/14,116 , or, AB = 9,266 units.

His calculation for the distance of Mars at quadrature, when angle CBE = 90° was 660 units in from the circle.   This is distance PE in Figure 39-1. Converting this value:

100,000/152,350 = PE/660 , or, PE = 433 units. Kepler got 432 units. BP then equals, 100,000 - 432 = 99,568 units.

Considering triangle AEB we see,

1.	tan AEB = 9266/100,000 = .09266, and,

arctan .09266 = angle AEB = 5° 17' 38" (the optical equation for the eccentric circle). To Kepler this meant the optical equation for Mars was very close to 5° 18'.

2.   If we consider triangle APB we see,

tan APB = 9,266/99,568 = .093062, and thus,

angle APB = 5° 19'.

When he considered these results, and the problems he had found with the eccentric circle theory, and the theory of the oval calculated with the auxiliary ellipse, something strange happened to Kepler.   In a dramatic section of chapter LVI of the 'Astronomia Nova' Kepler writes in Latin (with translation added),

"Qua in cogitatione dum versor anxie,...[thinking on these matters]... forte fortuito incido in secantem anguli 5° 18'...[and by chance on the secant of 5° 18"]...Quem cum viderem esse 100429, hie guasi e somno expergefactus, & novum lucem intuitas,   sic coepiratiocinari...[the secant here is AE, equal to 100,429,(sec 5° 18'[angle AEB in figure 39-1] = AE/BE. AE = (BE)(1.00429) = 100,429, since BE =100,000); it was like being awakened from a dream and comprehending a new light]."

Recall secantθ equals 1/cosθ. Cos 5°18' =.99573, so secant 5°18' = 1.00429.

Because of this value of AE = 100,429, it occurred to Kepler that the correct distance for PE in Figure 39-1 should be 429 units instead of 432 units, and that BP should be 100,000 - 429 = 99,571 units.   What he had

452

noticed was that in the theory of the auxiliary ellipse which had overcorrected the errors for the motion of Mars from the eccentric circle, the value of PE was 858 units (step 30, chapter 38), and 429 units is exactly ½ of 858 units.   The true orbit for Mars had to be in between the egg shaped oval (calculated by the auxiliary ellipse), and the eccentric circle of Figure 38-2.

After some prolonged considerations Kepler then considered a figure similar to Figure 39-2.

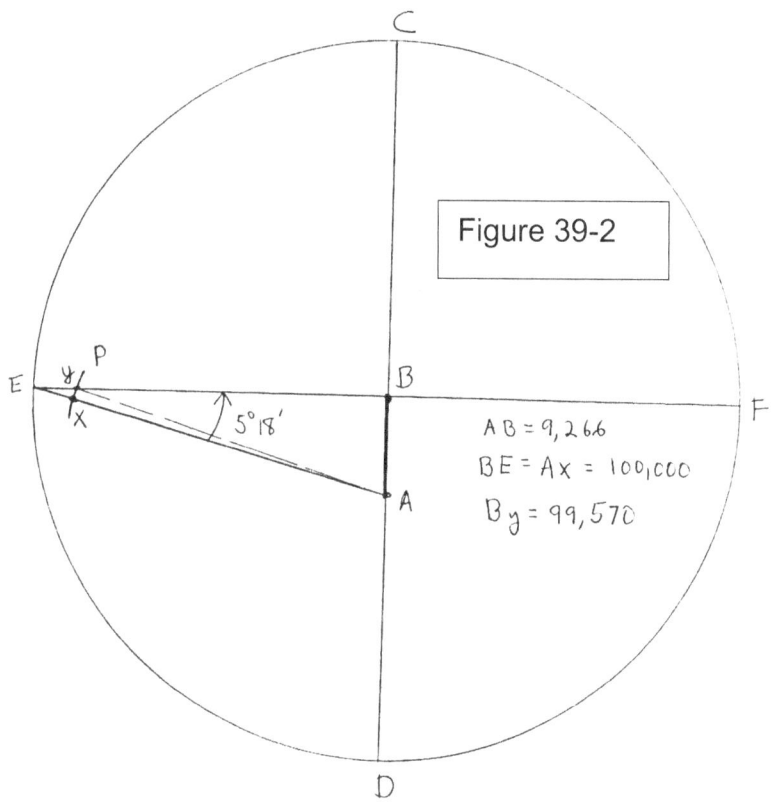

Figure 39-2

$AB = 9,266$
$BE = Ax = 100,000$
$By = 99,570$

Here angle CBE = 90°, AB = 9,266, BE =100,000, angle AEB = 5°17' 38" $\cong$ 5° 18', and PE should be 429.

3. cos AEB = BE/AE, or, AE = BE/cos AEB , or.

   AE = (BE) (sec AEB),

AE = (100,000)(1.00429) = 100,429 units.

Kepler with his *'novum lucem intuitas'* (new intuitive light) as he called it, decided to define a length he called the diametrical distance of the sun when the eccentric anomaly, angle CBE = 90°. He does this by drawing the perpendicular from the sun A to the diameter EF. In this case, line AB. When angle CBE = 90⁰, the perpendicular from A to diameter AF crosses EF at B. BE = 100,000 units. He then takes A as a center and 100,000 units as a radius and makes arc xy from AE to BE. x is on line AE and y is on BE. Ax = Ay = BE = 100,000 units. y is the position of Mars along BE by Kepler's new theory of the diametrical distance. However, this method changes the distance PE in Figure 39-1 slightly, for,

4. sin AyB = AB/Ax = 9,266/100,000,

   which results in,

   angle AyB = 5.31665°

   cos AyB = cos 5.31665° = By/Ay, or,

   By = (Ay)(cos 5.31665°)

   = (100,000)(.995698) $\cong$ 99,570 units, or,

   PE = 100,000 – 99,570 = 430 units, instead of

   429 units.

How do we calculate the diametrical distance when the eccentric anomaly is not 90⁰? See Figure 39-3.

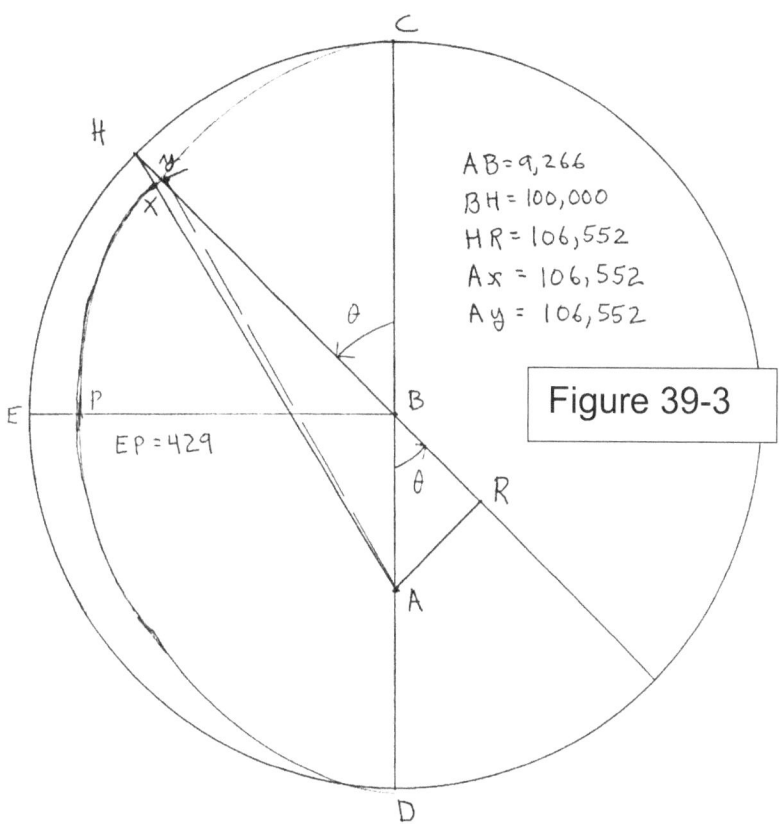

Figure 39-3

AB = 9,266
BH = 100,000
HR = 106,552
Ax = 106,552
Ay = 106,552

EP = 429

In this figure the eccentric anomaly is angle CBH less than 90°. We draw perpendicular AR from the sun A to diameter HBR. We designate angle CBH, the eccentric anomaly, as θ. We see angle ABR is also equal to θ since they are vertical angles. In right triangle ARB,

5. cos θ = BR/AB, or, BR = (AB)cos θ = (9,266)cos θ.

In Figure 39-3 the projection of AH onto diameter HBR is HR. Kepler draws AR perpendicular to diameter HBR so HR is the diametrical distance. We see,

6. HR = BH + BR = 100,000 + (9,266)cos θ.

455

(Some sources call the radius of the eccentric circle R and designate eccentricity of the sun AB by e. Then the diametrical distance in the general case is equal to, R + e cos θ).

As an example let us calculate the position and distance of Mars when θ is 45°.   Then HR = 100,000 + (9,266)(cos 45°).

7. HR = 100,000 + (9,266)(.70711) = 106,552 units.

We then measure 106,552 units along AH to point x. With radius Ax and center A, arc xy is drawn so y is on diameter HV.   Point y according to Kepler's diametrical distance rule is on the orbit of Mars.   Note the distance of Mars from the sun is 106,552 units, and the true anomaly is angle BAy.

The true anomaly is found by solving triangle BAy. We know AB = 9,266, Ay = 106,552, and angle ABy = 180° - CBy = 180° - 45° = 135°.

8. $\sin θ = \sin 45^0 = AR/AB = .70711$, so,

AR = $(\sin 45^0)(AB)$,

AR = (.70711)(9,266) = 6,552.

sin Ayb = AR/Ay (right triangle ARy),

sin AyB = 6,552/106,552 = .06149.

angle AyB = 3° 31' 32".

angle BAy = 180° - 135° - 3° 31' 32" = 41° 28' 28", the true anomaly when the eccentric anomaly is 45°.

Kepler thought the diametrical distance rule was caused by the sun's magnetic force acting on Mars making Mars move closer to the sun in its path from C to

D.   Then as Mars moved from D back to C the diametrical distance rule resulted in Mars receding farther from the sun.   In the general the rule states the diametrical distance equals (R + e cos θ), where R is the radius BC, and e is the eccentricity of the sun AB.   We see the diametrical distance depends on the cosine function which we know is periodic, that is repeats itself as θ goes through each 360°. That is the cosine goes from + 1 to 0 then to - 1 then back to 0 and then to + 1 again in the course of 360°.   This accounts for the symmetric motion of Mars first moving closer to the sun after aphelion to its closest distance at perihelion when 8 = 180° and cos θ has its minimum value of - 1.   Mars then recedes from the sun back to aphelion when 6 = 360° and cos 0 = + 1, its maximum value.

Another way of expressing this phenomenon is to say the distance of Mars from the sun oscillates back and forth from a maximum distance to a minimum distance back to the maximum distance as determined by the diametrical distance rule.   Kepler calls this oscillating movement of Mars along the diameter HV the *libration* of Mars.   Kepler considered this the correct physical law of planetary distance. The discovery of this law, as if by chance, awakening him from his sleep, as he described it, was occasioned by considering the secant of 5°18' in Figure 39-1.

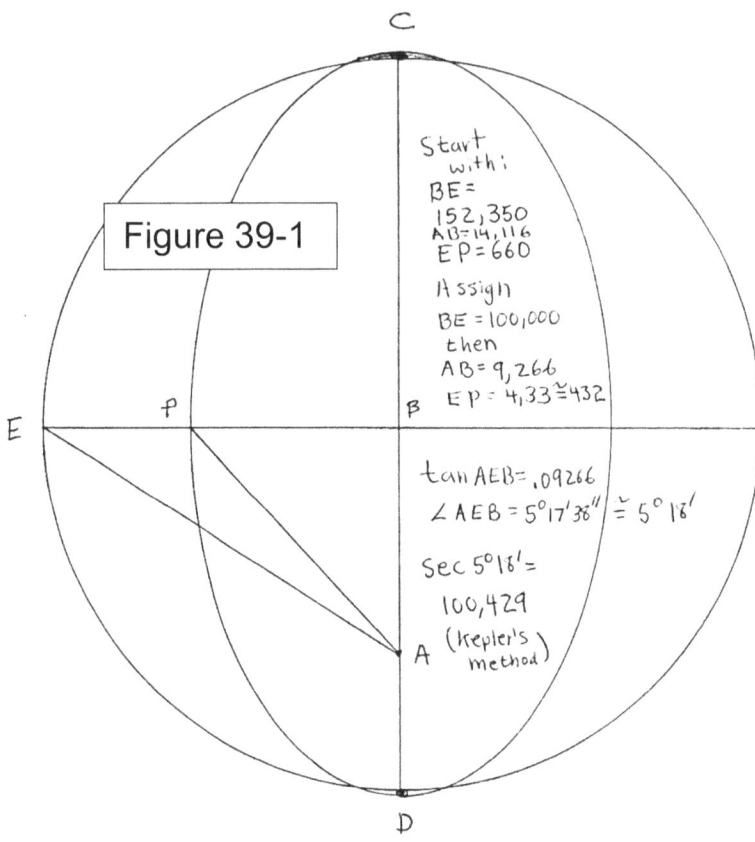

Figure 39-1

Start with:
BE = 152,350
AB = 14,116
EP = 660

Assign
BE = 100,000
then
AB = 9,266
EP = 4,332432

tan AEB = .09266
∠ AEB = 5°17'38" ≐ 5° 18'

Sec 5°18' = 100,429

(Kepler's method)

E   P   B   C   D   A

We must remark that this idea of determining the orbit of Mars by the diametrical distance rule is both subtle and ingenious. It is one of those flashes of insight that cannot be derived in a logical deductive manner. Kepler discovered the diametrical distance rule in 1605, after more than 4 years of work on the orbit of Mars. We might call the diametrical distance rule an intuition of a prepared mathematical mind.

Of course, the true test of this rule would be its ability to give Kepler the celestial longitudes or true anomalies of Mars as based on comparison with the *hypothesis vicaria*, and the correct sun-Mars distances when compared with the trigonometric calculations

458

discussed in chapter 38.   His calculations with the diametrical distance rule at 45° and 225°, the octants after the line of apsides, showed Mars to be 5' 30" ahead of the values from the *"hypothesis vicaria"*.   At 135° and 315°, the octants after the quadrature points, the diametrical distance rule gave values about 4' retarded from values of the *"hypotheisi vicaria"*.   Despite his chance insight in finding the diametrical distance rule, the orbit calculated from the rule was not the correct one.

Kepler in October 1605 wrote a letter to the astronomer David Fabricus explaining that he had made a mistake when calculating with the diametrical distance rule.   By then he had realized that the eccentric circle theory had given him an orbit where the sun-Mars distances were too great, and the auxiliary ellipse had given sun-Mars distances too small.   The true orbit had to lie between them.   In chapter LVIII of the "*Astronomia Nova*" Kepler wrote in Latin, with translation added:

"Circulus cap. XLIII peccat excessu   (the eccentric circle theory of chapter XLIII   (Figure 38-2) was in excess of the true orbit] ellipsis capitis XLV peccat defecta (the auxiliary ellipse of Figures 38-15 and 38-16 had errors such that the orbit was inside the true orbit]. Et sunt exessus ille & hie defectus aequales. (the excess of the former (eccentric circle theory) equals the defect of the latter (auxiliary ellipse)].   Inter circulum vero & ellipsin nihil mediat ellipsis alia.   Ergo ellipsis est Planetar iter; & lunula a semiciculo refecta habet

459

dimidiam prioris latitudinem scilicet 429. (The only figure in the middle between the eccentric circle and the auxiliary ellipse is another ellipse. Therefore, the ellipse is the path of the planet, and the lunula cut off from the semicircle has half the breath of the auxiliary ellipse, namely 429 parts]."

With these words Kepler announced for the first time that the path of Mars was a true ellipse. Kepler still had to explain how the diametrical distance rule works with an ellipse so that no significant differences in true anomalies occur with the ellipse and the "*hypothesis vicaria*". He did explain this in chapter LIX of the "*Astronomia Nova*". He also had to show that the sun-Mars distances calculated by the true ellipse agreed with the calculations from Tycho's observations. Kepler stated that at some points on the orbit he was uncertain of the sun-Mars distances by about 200 units.

Before we take up the correct diametrical distance rule and how it is applied to the ellipse, let us define the ellipse Kepler chose for Mars and show that the sun is located at one focus of the ellipse. If ½ semimajor axis for the ellipse is assigned the value of 100,000 units then ½ semiminor axis has the value of 99,571 units. When we circumscribe the ellipse with a circle with its diameter coincident with the major axis, which lies along the line of apsides, then at quadrature the ellipse lies, 100,000 - 99,571 = 429 units from the circle as shown in Figure 39-4. In the figure EP is greatly exaggerated.

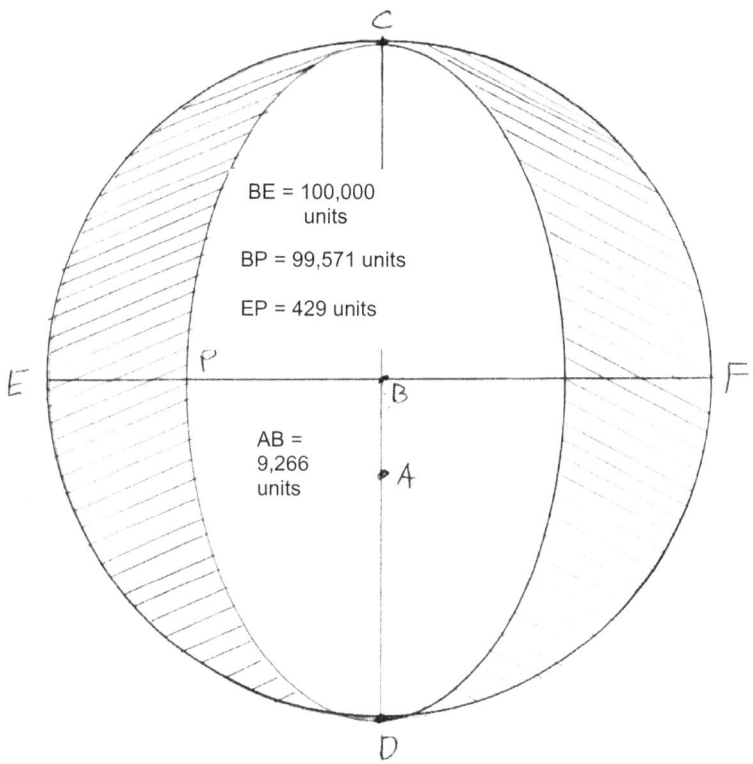

Figure 39-4

AB, the distance of the sun from the center of the ellipse, equals 9,266 units.

We studied the focal properties of the ellipse in chapter 32 (Figure 32-8 and steps 49-55).

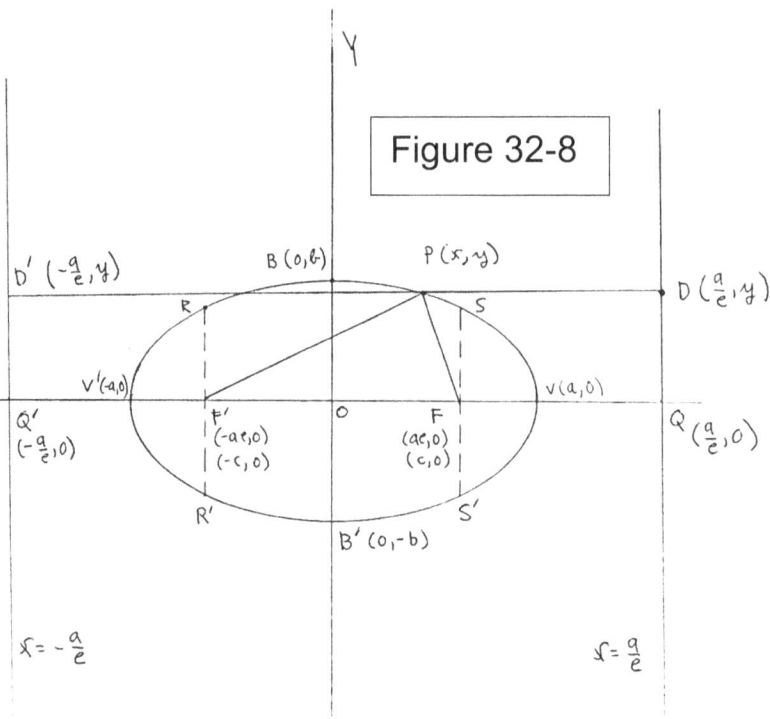

Figure 32-8

We learned, $b^2 = a^2 - c^2$, or $c^2 = a^2 - b^2$, where a is ½ the semimajor axis, b is ½ semiminor axis, and c is the distance from the focus to the center of the ellipse as shown in Figure 32-8. We can determine if the sun lies at a focus in Kepler's elliptical orbit for Mars.

9. $c^2 = (100,000)^2 - (99,571)^2$

$c^2 = 10,000,000,000 - 9,914,384,041,$

$c^2 = 85,615,957$, or c = 9,252 units.

This means AB should be 9,252 units instead of the 9,266 units Kepler found by using Tycho's observations. This value, however, is close enough to demonstrate that the sun is at one focus.

Actually Kepler did not introduce the idea of the focus of the ellipse in the "*Astronomia Nova*", which was

462

published in 1609. However, in the fourth and fifth books of his work the "*Epitome Astronomiae Copernicanae*", published in 1621 eleven years after the "*Astonomia Nova*", Kepler stated the sun was at the focus of the ellipse closest to perihelion. In Book V Kepler states that an ellipse has two 'hearths' or foci such that the center of the major axis is midway between them just as we learned in our work with Figure 32-8.

In chapter LIX of the "*Astronomia Nova*" Kepler explains how the diametrical distance should be applied, and corrects the mistake he made when he first calculated with it. Kepler introduces his most famous figure from the "*Astronomia Nova*" to demonstrate the properties of the ellipse. We show in Figure 39-5 a copy of the first page of chapter LIX from the "*Astronomia Nova*".

PROTHEOREMATA.

I.

SI intra circulum deſcribatur ellipſis, tangens verticibus circulum, in punctis oppoſitis; & per centrum & puncta contactuum ducatur diameter; deinde a punctis aliis circumferentiæ circuli ducantur perpendiculares in hanc diametrum: ex omnes a circumferentia ellipſeos ſecabuntur in eandem proportionem.

Ex l. 1. Apollonii Conicorum pag. XXI. demonſtrat COMMANDINVS *in commentario ſuper* V. *Sphæroideon* ARCHIMEDIS.

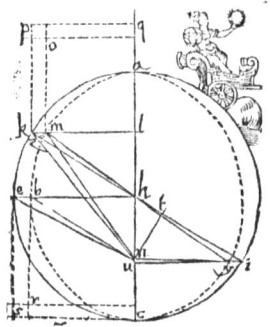

Sit enim circulus AEC. *in eo ellipſis* ABC *tangens circulum in* AC. *& ducatur diameter per* A. C. *puncta contactuum, & per* H *centrum. Deinde ex punctis circumferentiæ* K. E. *deſcendant perpendiculares* KL, EH, *ſectæ in* M.B. *a circumferentia ellipſeos. Erit ut* BH *ad* HE, *ſic* ML *ad* LK. *& ſic omnes aliæ perpendiculares.*

II.

Area ellipſis ſic inſcriptæ circulo, ad aream circuli, habet proportionem eandem, quam dictæ lineæ.

Vt enim BH *ad* HE, *ſic area ellipſeos* ABC *ad aream circuli* AEC. *Eſt quinta Sphæroideon* ARCHIMEDIS.

Figure 39-5

Kepler has drawn the ellipse within the circumscribed circle, and next to the figure the Greek goddess of victory is drawn symbolizing that Kepler had finally found the true orbit of Mars. We note in passing he has changed the labeling of the figure using *h* for the center, *n* for the sun, and *ac* for the line of apsides.

To demonstrate the revised diametrical distance rule we redraw Figure 39-5 as Figure 39-5a with the ellipse more elongated so we can see better the geometrical relationships.

Figure 39-5a

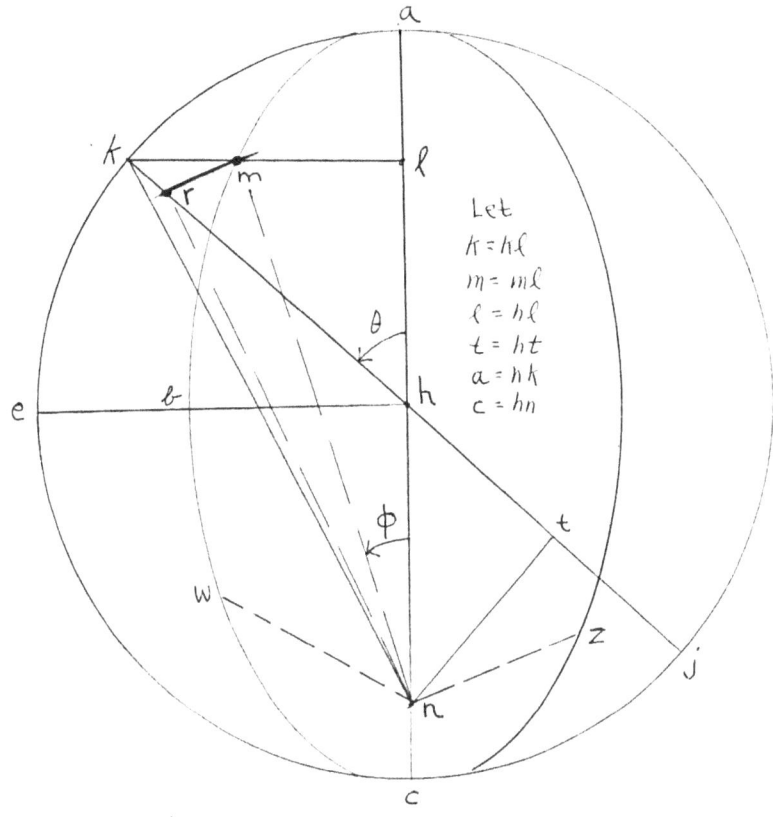

Let
$k = h\ell$
$m = m\ell$
$\ell = h\ell$
$t = ht$
$a = hk$
$c = hn$

$he = ha = a$ (semi major axis)
$h\ell = \ell$ (semi minor axis)
$hn = c$ (distance from center to focus)
$tk = nr = nm$
θ = eccentric anomaly
ϕ = true anomaly

We will follow the labeling of Kepler's figure.   The center of the ellipse is h, the line of apsides is ac with aphelion at a, perihelion at c, and the sun at n. Although in the "*Astronomia Nova*" Kepler does not use the concept of the focus, we will use this concept and our work from chapter 32 on the ellipse. Thus, the sun is

465

at one focus, and we designate length *hn* as egual to *c*. We draw the perpendicular he to ac. Thus, he is the length of the radius of the circumscribed circle and ½ the semimajor axis ac of the ellipse which we designate by a. ½ the semiminor axis of the ellipse is length hb which we designate by b. From our work on the ellipse we know, $a^2 = b^2 + c^2$.

Kepler changed the diametrical distance rule as follows. He marks off arc *ak* with eccentric anomaly angle θ, angle *ahk*. Then from *k*, a perpendicular is dropped to ac at *l*. Then he draws diameter *khj*, and *nt* is drawn perpendicular from the sun to the diameter, *kj*. He then measures length *tk*, and with *n* as center and length *tk* as radius, arc *rm* is drawn with r on diameter *kj* and m on perpendicular *kl*. This means, *tk* = *nr* = *nm*. Kepler makes the following claims: *m* is the position of Mars in its orbit for any given value of θ. nm is the correct sun-Mars distance. *m* is on the ellipse which is the true orbit of Mars with ac the semimajor axis, *bh* is ½ the semiminor axis, and c is the distance of the sun from the center of the ellipse, *h*, to one focus, point *n*, the sun.

We will present a demonstration to show that *m* is on the ellipse for any value of angle θ. But note the difference between the first diametrical distance rule and the rule in Figure 39-5a. With the first rule Kepler would have placed Mars at point r on the axis *khj* with *nr* = *tk*. For the revised diametrical distance rule he places Mars at *m* on the perpendicular *kl*, giving a true anomaly of

466

angle *anm* which we designate by φ.   This is how Kepler corrected the mistake of the first diametrical distance rule.   In the first rule the orbit of Mars would be an oval but not an ellipse, whereas we will demonstrate that the revised rule means Mars must always be on an ellipse. It was this discovery that convinced Kepler that the ellipse was the true orbit of Mars around the sun n.

In order to show *m* is on the ellipse we must show the proportion we proved for the ellipse in chapter 38 holds, that is,

10. *kl/ml = he/hb = a/b* .

To simplify the algebraic expressions we will use the following symbols: let *k = kl, m = ml, l = hl, t = ht,* a *= hk = ah, hb = b,* and *c = hn.*   Then we wish to show,

11. *k/m = a/ b*

12. The Pythagorean Theorem for right triangle *hlk* states,

$l^2 + k^2 = a^2$,

$$\frac{l^2}{a^2} + \frac{k^2}{a^2} = 1, \; or, \; \frac{k^2}{a^2} = 1 - \frac{l^2}{a^2}.$$

13.   In right triangle *hlk* we see sin θ = *k/a* , or $sin^2 \; \theta = k^2/a^2$.   In right triangle *htn,* cos θ = *t/c* (angle *nht* = θ, since θ is the vertical angle for angle *nht.*

$cos^2 \; \theta = t^2/c^2$.

Adding these results, and using the trigonometric identity:

$$\sin^2 \theta + \cos^2 \theta = \frac{k^2}{a^2} + \frac{t^2}{c^2} = 1.$$

$$\frac{k^2}{a^2} = 1 - \frac{t^2}{c^2}.$$

Using the result of step 12:

14.

$$1 - \frac{l^2}{a^2} = 1 - \frac{t^2}{c^2}, \ or, \ \frac{l^2}{a^2} = \frac{t^2}{c^2}, \ so,$$

$$t^2 = \frac{l^2 c^2}{a^2}, \quad t = \frac{lc}{a}.$$

15. Consider right triangle mln.

We know $mn = tk$, and

$tk = hk + ht$, or,

$tk = a + t$, from the figure.

Thus, $mn = a + t$.   We see $nl = c + l$.   The Pythagorean Theorem for triangle mln gives:

$m^2 + (nl)^2 = (mn)^2$,

$m^2 = (mn)^2 - (nl)^2 = (a + t)^2 - (c + l)^2$,

$m^2 = a^2 + 2at + t^2 - [c^2 + 2cl + l^2]$,

$m^2 = a^2 + 2at + t^2 - c^2 - 2cl - l^2$.

Using results from step 14:

$m^2 = a^2 + 2a[(lc/a)] + (l^2c^2)/a^2 - c^2 - 2cl - l^2$,

$m^2 = a^2 + 2cl + (l^2c^2)/a^2 - c^2 - 2cl - l^2$,

$m^2 = a^2 - c^2 + (l^2c^2)/a^2 - l^2$.

From our work on the ellipse, $a^2 = b^2 + c^2$, so, $a^2 - c^2 = b^2$.

$m^2 = b^2 + (l^2c^2)/a^2 - l^2,$

$$m^2 = \frac{b^2a^2 + l^2c^2 - l^2a^2}{a^2} =$$

$$\frac{b^2a^2 + l^2(c^2 - a^2)}{a^2}.$$

Since, $a^2 - c^2 = b^2$, then, $-a^2 + c^2 = -b^2$.

$$m^2 = \frac{b^2a^2 - b^2 l^2}{a^2} = \frac{b^2(a^2 - l^2)}{a^2}.$$

16. From step 12,

$$l^2 + k^2 = a^2, \text{ or, } k^2 = a^2 - l^2. \text{ Then,}$$

$$m^2 = \frac{b^2k^2}{a^2}, \text{ or, } \frac{k^2}{m^2} = \frac{a^2}{b^2}, \text{ thus },$$

$$\frac{k}{m} = \frac{a}{b}.$$

This establishes that m must be on the ellipse for all values of θ. The revised diametrical distance rule is mathematically equivalent to an elliptical orbit with the sun at n and ½semiminor axis hb, and ½semimajor axis he. Another way of saying this is that Kepler's diametrical distance rule produces the very ellipse through which the planet moves.

Before Kepler developed the revised diametrical distance rule he had trouble accepting an ellipse for the orbit of Mars, since he could not relate an ellipse to a physical theory of motion originating in the sun. With the demonstration that the diametrical distance rule produces an ellipse, he could now give his physical theory of Mars' motion. The physical theory resulted for Kepler in the diametrical distance rule, and the diametrical distance rule resulted in an ellipse. Kepler was then convinced that an ellipse was the true path of Mars' motion.

Today we know Kepler's physical theory of planetary motion is not correct. As we shall learn in later chapters, it is the gravitational interaction of the planets and the sun that causes the motion of the planets to be in elliptical orbits. Nonetheless, he was the first of the great astronomers who insisted that the true body of the sun is the source of planetary motion, and that this resulted in the planets moving in elliptical orbits.

In the "*Epitome Astronomiae Copernicanae*" he developed his physical theory. We will try to present a very simplified account of it. The theory has two parts which fit with the diametrical distance rule. The first part envisions the sun sending out rays which lead the planet along the path of the circle which circumscribes the ellipse. In Figure 39-5 the rays act along line nm. However, the planet has magnetic fibers which are aligned either to be attracted toward the sun or repelled from it. In a sense the planet has a kind of inertia that alters the circular motion which would otherwise occur. The magnetic fibers act to cause the planet to oscillate along the perpendicular kl.   From a to b the planet moves outward along kl toward k.   At point b the planet's magnetic fibers point so as to be attracted toward the sun thus oscillating toward l on kl.   All this in accord with the diametrical distance rule, which we know produces the elliptical path.   Although Kepler's physical ideas were to be replaced by the systematic study of motion by Galileo and Newton, such that the correct derivation of the area law and the elliptical orbit would occur.   Kepler's physical hypothesis encouraged future astronomers and mathematicians to develop a physical hypothesis for planetary motion based on the body of the sun.   Furthermore, the development of a physical hypothesis compatible with the diametrical distance rule meant Kepler could accept the elliptical orbit as the true path of Mars.

Kepler could then check the accuracy of the ellipse by the area law and compare the results with the "*hypothesis vicaria*".   Kepler found that the distances calculated by the revised diametrical distance rule agreed with those calculated by Tycho's observations within 100 to 200 units at points away from b in Figure 39-5, which was within the accuracy of Kepler's method.

The true anomalies calculated by the ellipse agreed with the *'hypothesis vicaria'* within 3/4' of arc, an astounding success. In chapter LX of the "*Astronomia Nova*" he found that the celestial latitudes predicted by the ellipse agreed with the latitudes of the observations. The new ellipse worked.   Never before had a theory of

471

planetary motion given such accurate predictions. Kepler, moreover, had given the true path of the planet's motion and linked it to a physical rule, the diametrical distance rule.

Kepler had also given a means of calculating the mean motion or Mars: the area law.   Calculating with the area law means using the area of an elliptical sector, such as amn in Figure 39-5, by the methods we discussed in chapter 38.   No equant point is required to find the mean motion.   A clear description of the area law as we give it today is not found in the "*Astronomia Nova*".   In the "*Epitome Astronomiae Copernicanae*" did describe the area law with clarity.   He stated the areas of sectors of the ellipse are a means of measuring the times of arcs through which the planet moves.   If there are two elliptic sectors of equal area with unequal arcs, then the time to move through the corresponding arcs will be equal.   Thus, in Figure 39-5 if elliptical sector *amn* equals sector *wczn* then the time for the planet to move through arc am equals the time to move through arc *wcz*.

To simplify calculation Kepler did bring back the idea of the equant.   He placed it at the empty focus of the ellipse which is at a distance c from the center toward aphelion.   This fits in with the idea that the elliptical orbit has a total eccentricity bisected by the center.   Actually Mars and the other planets do not move uniformly as seen from the empty focus.   To do so would mean that a planet as seen from the empty focus would move through equal angles in equal times, and the observations show slight variations.   Still, astronomers in the 17th century did use the empty focus to calculate planetary positions adding a correction factor.   In subsequent chapters we will show how both the area law and the elliptical path (as well as Kepler's third law) are mathematical consequences of the fundamental laws of motion, and the law of universal gravitation first elaborated by Newton.

Kepler's laws were discovered, as we have seen, by the lengthy process of trying multiple hypotheses and comparing their predictions with the "*hypothesis vicaria*", which Kepler knew gave accurate values for the true anomalies.   Kepler's perseverance kept him working for nearly five years until he finally found the elliptical orbit and diametrical distance rule which convinced him his laws were correct.   We recall that Kepler did not regard his discoveries as laws (this designation has been given by history), but rather as explanations for the observations based on a physical cause of motion in the body of the sun.

Kepler could now apply his theory to the other planets and begin work on tables of planetary positions. By 1627 he did produce the tables based on his laws in the "Rudolphine Tables".   His tables improved the accuracy of planetary positions 30 fold over previous tables, thus fulfilling Tycho's dream of accurate planetary tables.   For example, in the previous tables for Mars the planets celestial longitude during the course of several years would often be incorrect by 3° to 4° and occasionally by 10°.   In Kepler's tables Mars would be found where the tables predicted almost always within 2' of arc, an astounding accomplishment.

As another example Kepler worked out the orbit of Mercury from Tycho's observations, and found it had an even larger eccentricity than Mars. The "Rudolphine Tables" predicted that in 1631 Mars would pass in front of the sun.   By then telescopic astronomy made it possible to actually observe Mars in front of the sun just as Kepler had predicted.

Kepler's final major discovery was his third law which was first published in another book the "*Harmonice Mundi*" (Harmony of the World) in 1619. He wrote in Latin,

"quod proportio quae est inter binorum quorumcunque Planetarum tempora periodica sit praecise sesquialtera proportionis mediarum distantiarum, id est Orbium ipsorum". (the periodic times of any two planets are in the sesquialteral ratio to their mean distances, i.e. of their orbits).

The term 'sesquialteral' refers to the ratio of 3/2 to 1.

The third law means that if T_1 is the period or time for one orbit of planet A, and R_1 is the mean distance of planet A from the sun, and if T_2 is the period of planet B, and R_2 its mean distance, then

$$\frac{(R_1)^3}{(T_1)^2} = \frac{(R_2)^3}{(T_2)^2}.$$

Kepler did not provide a table giving the respective values of the periods, the mean distances, and the ratios of R^3 to T^2 for the third law to see if the ratios are essentially equal. We can give such a table based on Kepler's data.

Planet	Period	Distance	(Period)^2	(Distance)^3	$(R)^3/(T)^2$
Mercury	88.4d	.388	7,815	.0584	7.47×10^{-6}
Venus	225d	.724	50,625	.3795	7.496×10^{-6}
Earth	365.25d	1.000	133,408	1.000	7.496×10^{-6}
Mars	687d	1.524	471,969	3.5396	7.500×10^{-6}
Jupiter	4,332d	5.200	18,766,224	140.608	7.493×10^{-6}
Saturn	10,713d	9.510	114,768,369	860.090	7.494×10^{-6}

To put the law another way: for any planet, $R^3/T^2 = K$, where K is a constant numerical value. In our table we used 1.000 units for the mean distance of the earth from the sun. The mean distance is equal to ½ of the length of the line of apsides, or ½ the major axis of the planet's ellipse, the semimajor axis a. For the time of the period we used measurement in days. With these choices of numerical units, the value of K $\cong$ 7.49 x 10^{-6}. K would have a different numerical value if R were measured in miles, and T were measured in years. However R and T are measured the value of K is the same for each planet and does not depend on the eccentricity or other parameters of the planets elliptical orbit.

Kepler in the "*Epitome Astronomiae Copernicanae*" linked the third law to supposed magnetic emanations from the sun. The shorter the planet's period the greater the magnetic emanation would be. The matter within the planet has an effect as well. The larger and more massive a planet is, the greater is its resistance to the magnetic effect of the sun. All of this acted in accord with the third law.

Kepler was greatly pleased with his hypotheses of planetary motions, his ability to demonstrate their conformity with Tycho's careful observations, and the vindication that the fundamental premise of Copernicus' theory was correct: the earth moves as a planet around the sun according to the same laws as the other planets.

It was crucial to Kepler's belief that mathematical demonstrations would be necessary in the study of planetary motion, and that there would be found mathematical demonstrations that explain the observed phenomena. For Kepler, God created the universe according to mathematical principles that man can understand with diligence and perseverance, and that he, Kepler, had the mission of showing the rest of mankind the fundamental celestial harmony.   Near the end of the "*Harmonice Mundi*" Kepler gave thanks to God for his mathematical harmony,

"I give thanks to Thee, 0 Lord Creator...I have completed the work of my profession, having employed as much power of mind as Thou didst give to me; to the men who are going to read those demonstrations I have made manifest the glory of Thy Works."

Kepler believed the universe and the solar system operated in accordance with divinely imposed laws which were beautiful and harmonious for those who would contemplate them.   A physical system connected these laws, and Kepler had proposed magnetism as the important physical force, combined with inertia in the planets themselves, which resulted in the oscillations related to the diametrical distance rule resulting in the elliptical orbit.

The correct physical laws were not to be those of Kepler.   It would take the work of Galileo, who studied the behavior of accelerated motion on earth, and Newton who extended the principles of Galileo to the heavens, and introduced universal gravitation to show that all three of Kepler's laws are consequences of fundamental laws of motion and the law of universal gravitation.   Kepler along with Galileo and Newton were the pioneers of the new astronomy, the new mathematics, and the new physics of the 17th century.

Footnote

We will give a table of modern data for the planets known in Kepler's time as well as those discovered since then.

Name	Distance	Periodic Times		Inclination	Eccentricity
		days	years	of orbit	of ellipse
Mercury	.3871	87.97	.241	7°0 '	.206
Venus	.7233	224.70	.615	3°24 '	.007
Earth	1.0000	365.25	1.000	0° 0'	.017
Mars	1.5237	686.98	1.881	1°51 '	.093
Jupiter	5.2082	4,322.59	11.862	1 °18'	.048
Saturn	9.5388	10,759.2	29.457	0°29 '	.056
Uranus	19.1910	30,685.9	84.015	0°46'	.047
Neptune	30.0707	60,187.6	164.788	1°47'	.009
Pluto	39.4574	90,741.3	247.697	17° 9 '	.249

Bibliography for Chapter 39

1. Kepler, Johannes, "Astronomia Nova", Latin Edition printed by Culture et Civilisation, 115, Avenue, Gabriel Lebon, Bruxelles, 1968.

2. Small, Robert, "An Account of the Astronimical Discoveries of Kepler", A reprinting of the 1804 text by The University of Wisconsin, Madison Wisconsin, 1963.

3. Koyre, Alexandre, "The Astonomical Revolution", translated by Dr R. E. W. Madison, Cornell University Press, Ithaca New York, 1973.

4. Barlow, G. W. C, Bryan, G. H., "Elementary Mathematical Astronomy", revised by Sir Harold Spencer Jones, University tutorial Press, Ltd., London.

5. Kepler, Johannes, "Epitome of Copernican Astronomy: IV and V"., "The Harmonies of the World: V.", in "Great Books of the Western World, Vol. 16", Encyclopaedia Britannica, Inc., Chicago, 1952.

Chapter 40

Galileo, the Study of Free Fall, and the Differential
Calculus

We now move from the work of Kepler to his
great contemporary in Italy, Galileo Galilei, known to the
world as Galileo.   He lived from 1564 until Jan 8, 1642
(Gregorian calendar).   He was born at Pisa in Italy as
the son of a musician.   In 1581 Galileo studied at the
University of Pisa to pursue a career in medicine.
However, he had an intense curiosity about objects in
motion.   Even before his formal mathematical training
was completed, he discovered that the period of a
pendulum remained the same no matter how large the
amplitude of the swing.   He noticed this while in the
cathedral at Pisa watching a swinging lamp in set in
motion by the wind.

While at the university Galileo was so attracted to
mathematics that he gave up medicine to study
mathematics and science against the wishes of his
father.   He was so brilliant that by 1589 he was
awarded prizes for his mathematical papers. It was
about this time he began a long series of independent
studies on motion which would occupy him for twenty
years.

In 1592 Galileo was appointed chairman of
mathematics at the University of Padua.   He stayed in
Padua 18 years.   It was in this period that he made his

great discoveries on the motion of freely falling objects, the principle of inertia, and the motion of projectiles. His great work on these subjects was not published, however, until 1638 when he was in his seventies.

In 1609 his work on motion was interrupted for many years when word reached him that the telescope had been invented in Holland.   He began to build his own telescopes which were three times as powerful in magnification and resolution as previous telescopes from Europe.   With his improved telescopes Galileo could obtain magnification of about twenty times. Galileo turned his telescopes on the heavens and made many new discoveries which he published to achieve wide renown.

By 1613 Galileo became personally convinced that the Copernican system was the correct explanation for the motion we see in the planets and stars.   This resulted in great controversy for him after he taught and argued for the reality of the earth's motion.   From 1624 to 1630 he worked on a long treatise comparing the Ptolemaic system to the Copernican system clearly favoring the latter which we know as, "The Dialogue Concerning the Two Chief World Systems".   This book was published in 1632.   He used arguments from his previous work on motion to support the Copernican view, and was effective in refuting arguments used by the Aristotelian philosophers who maintained the earth was stationary.   The book was very popular, but soon after its release church authorities in Rome banned its

481

publication as arguing against the doctrines of the church. Galileo himself was brought before the Inquisition for suspicion of heresy, because he went far beyond calling the Copernican system just a hypothesis. Galileo underwent a trial in Rome, was forced to recant his teachings, and was forced to a lifetime confinement at his estate.   He was forbidden to publish books as well, although he wrote his masterpiece on the study of motion and had it published in Holland. This work is known as the, "Discourses and Mathematical Demonstrations Concerning Two New Sciences", and was published in Leyden, Holland in 1638 thus avoiding the church censors.   Not long after this Galileo went blind and died at his estate.

To make Galileo's work more comprehensible to the modern reader we will begin by discussing some of the fundamental physical quantities used in the study of motion.   First is the measurement of length or distance. Moving objects can follow paths either of straight lines or curves.   We will often use symbol s to represent lengths or distance through which objects move. In the United States we are familiar with the system of measurement using inches, feet, yards, and miles to represent lengths. However, the scientific community today uses the metric system which we will follow as our main system for physical measurements in the rest of this book.   The metric system has advantages particularly when dealing with the study of motion.   The basic unit of length is the meter.   So that we will become familiar with the meter we include the following table showing conversion from the English system of lengths measurement to and from the meter.

Table for Length Conversion

Length		meter	inch	foot	mile
1 meter	=	1	39.37	3.281	.0006214
1 inch	=	.0254	1	.0833	.00001578
1 foot	=	.3048	12	1	.0001894
1 mile	=	1,609	63,360	5,280	1

When an object has motion, it has velocity. Knowing the velocity tells us the direction of motion, either along a curve or along a straight line, and the magnitude of motion in terms of the distance covered per unit of time. Using the metric system,

$v = s/t$, where v is the magnitude of the velocity measured in meters per second, s the distance measured in meters traveled, and t is the time of travel measured in seconds. In this equation we mean instantaneous velocity at a given instant of time t. The other quantity we will consider in our study of motion is acceleration. We all have an intuitive understanding of acceleration when, for example, we recall the feeling of moving off from a stop sign in a car, or the feeling of taking off in an airplane. Acceleration is the rate of change of velocity when velocity is changing either in magnitude of direction or both.

The magnitude of the acceleration is symbolized by a, and it is measured in units of velocity over time:

$a = v/t$, where v is measured in meters per second, and t is in seconds. The units of measurement of acceleration are: meters/(second x second), or meters per (second)^2.

One of Galileo's great discoveries was the principle that as objects released near the earth's surface to fall to the earth, a motion called free fall, the objects experience uniform acceleration, acceleration with constant magnitude and constant direction. Newton found that the force of gravity is responsible for the uniform acceleration in free fall near the earth's surface.

Galileo did not introduce the idea of a gravitational force causing free fall, but he demonstrated the quantitative aspects of uniform acceleration. Galileo's plan was to give a description of what happens to objects in free fall near the earth's surface based on experimental researches which gave him the mathematical rule or law for objects undergoing free fall. He developed his mathematical expressions assuming no resistance to free fall motion. He ignored the effect or air resistance in his formulations. However, he realized the medium through which an object falls affects the results; for example, water greatly slows the motion compared to air. He designed experiments so that impediment to free motion was minimal and could in effect be ignored. His law of free fall thus applies to an ideal state. Such a law is the basic physical law. If impediments to free fall are present their effect could be added to the basic law.

Consider a heavy object such as a spherical brass ball released to fall freely through the air from a height of 300 meters. Such an experiment shows that the air offers virtually no impediment to free fall. The object starts its motion with 0 velocity (0 meters per second). After 1 second of free fall experiments done in the modern era show the object has a velocity of 9.807 meters per second in the direction toward the earth perpendicular to the earth's surface assuming a level surface on the ground. After 2 seconds of free fall the brass ball has a velocity of (9.807 + 9.807 = 19.614 meters per second) toward the earth.

Thus, we find if there is no detectable air resistance for every second of free fall before the brass ball hits the ground, the brass ball gains velocity at the rate of 9.807 meters per second. This is the situation of uniform acceleration due to the force of gravity acting on the brass ball near the earth's surface. This law of free fall was first worked out and given by Galileo. We shall see how Galileo discovered this fundamental law of motion in his years at Padua.

Of course, there is always some air resistance in real experiments, and there is great difficulty in actually measuring the velocity at an instant of time at 1 or 2 seconds after the motion has begun.

It is customary to use the following abbreviations for the metric system of measurement: m for meters, and s for seconds. Thus, velocity is abbreviated as m/s, and acceleration as m/s^2.

Since acceleration is uniform in free fall the formula for the velocity at any instant of time after the beginning of free fall is,

1. $v = (9.807)t$, where t is measured in seconds, and the term (9.807) is in m/s^2. If we look at two separate times we can write,

2. $v_1/v_2 = (9.807)t_1/(9.807t_2, \quad v_1/v_2 = t_1/t_2.$

We note that velocity at any instant is directly proportional to the time of free fall.   Galileo investigated free fall with the mathematics of proportion which he learned from the fifth and sixth books of Euclid's "Elements", which we recall were based on the work of Eudoxus.   Galileo did not, of course, know the modern value of 9.807 m/s^2 for the magnitude of the uniform acceleration since he did not have instruments to measure instantaneous velocity, and how it changes second to second.

We see in the formula, $v = 9.8t$, that we have a functional relationship where v is the dependent variable, and t is the independent variable, while 9.8 m/s^2 is a constant of proportionality for all values of v and t for which the function applies.   Note that the function is independent of the weight or heaviness of the object in free fall.   This means that if there is no impediment to free fall a light object, like a feather, will fall as fast as a heavy object, like a stone.

Even before Galileo's work Simon Stevins of Bruges had noted that when two objects of different weight, one 10 times heavier than another, were released simultaneously from the same height of about 9 meters, both objects hit a plank on the ground at the same time as judged by the sound of impact. This result was reported in 1586.

There is the famous story that Galileo did research on free fall by dropping objects from the Leaning Tower of Pisa. Galileo was familiar with Stevin's results and confirmed them for his students and several Aristotelian philosophers. He wished to show them that Aristotle's notion that heavier objects always reach the ground in significantly less time than lighter objects was incorrect, provided there was no significant impediment due to air resistance. Aristotle had argued in his works that objects made of the same material but of different weights would acquire velocity based on their weights, the heavier object acquiring velocity faster than the lighter object in free fall. Galileo knew this was incorrect, and according to a contemporary of Galileo's named Vivani, Galileo did drop objects of the same material from the Leaning Tower of Pisa, and he found that the heavier object reached the ground just a few inches ahead of the lighter object. Galileo probably used lead or brass spheres for this demonstration. The very slight difference found can be accounted for as follows.

There is a discussion about the Leaning Tower of Pisa experiment reported by I. Bernard Cohen and others who tried to repeat the demonstration from the Leaning Tower of Pisa in the modern era. They found when releasing two objects, one heavy and one lighter, at the same time from the tower, the experimenter must extend both arms from the top of the tower at once. The result depends on physically being able to release the objects at the same instant. Galileo had reported in his notes that the heavier sphere seemed to start out a bit slower than the lighter sphere. According to the formula for free fall there should be no perceptible difference, yet Galileo reported a slight difference even without measuring instruments on repeated trials. Furthermore, the modern experimenters reported the same phenomenon.

In 1983 Thomas Settle demonstrated that if two spheres of different weights are held palms downward while leaning out of the tower, the objects cannot be released simultaneously. The lighter sphere is always released first just before the heavier sphere. Settle explains this is due to a difference in muscle coordination between the two arms of the experimenter. Thus Galileo's observation in his notes is indeed correct, but it is not due to a law of motion, but instead the experimenter's muscle incoordination.

Today we have very accurate data from uniform acceleration experiments due to gravity near the earth's surface. Experiments done in a vacuum with no air resistance will give a reproducible value of 9.80665 m/s2 under standard conditions for the uniform acceleration due to gravity near the earth's surface. Later in our work we shall find the value varies at different latitudes north or south of the equator, and varies if experiments are done high above sea level in the mountains.

Today we know the best mathematical approach to the study of variable motion, motion where the velocity is changing, is to use the mathematics of the calculus. Although the calculus was not developed until some years after Galileo's work, we will introduce basic elementary calculus so that we can better understand the work of Galileo.

Numerous mathematicians are responsible for the calculus, but foremost recognition has been given to the work of Isaac Newton and Gottfried Wilhelm von Leibniz who did their work in the latter part of the 17th century.

We begin by taking up the functional relationship between distance and time in free fall motion. Early in Galileo's work before he realized that velocity is directly proportional to time in free fall, Galileo wondered which relationship, velocity directly proportional to time or velocity proportional to distance, was the correct physical relationship. In modern terms is, $v = kt$, or is $v = ks$, correct, where k is the constant of proportionality. Aristotle had proposed that v was proportional to s in free fall, and in his early years Galileo favored this relationship as his first hypothesis. We shall see in due course that actual experiments performed by Galileo taught him the correct relationship that v is proportional to t and not to s. Before he found this, his experiments showed that the distance traveled in motion with uniform

489

acceleration was directly proportional to the square of time, or as we write today, $s = kt^2$, where k is the proportionality constant.   In the metric system the equation is, $s = 4.9t^2$, where s is measured in meters, and t is measured in seconds.   The units of measurement for the proportionality constant are m/s^2.

Let us make a table of values for the first 5 seconds of an object released in free fall near the earth's surface.

Distance (s)	Time (t)
0	0
4.9 m	1 s
19.6 m	2 s
44.1 m	3 s
78.4 m	4 s
122.5 m	5 s

Of course, this table represents the idealized situation of no impediment to free fall from air resistance or measurement error.   We note that in the formula, $s = 4.9t^2$, the value of the constant is ½ the value of the uniform acceleration of gravity, $9.8\ m/s^2$.   But let us suppose we start with the relationship, $s = 4.9t^2$, and do not know the relationship between velocity and time, nor the value of the uniform acceleration due to gravity.   We ask, can we derive the velocity-time relationship starting with the distance-time relationship, $s = 4.9t^2$?

The use of the calculus allows us to accomplish this derivation mathematically.   We start by trying to find the velocity of an object in free fall after 3 seconds of fall.

490

We seek what is called the instantaneous velocity at the instant of 3 seconds after the start of free fall. We see from the table that the object has fallen a distance of 44.1 meters 3 seconds after free fall has begun, according to the formula, $s = 4.9t^2$.

But what do we mean by 'instantaneous velocity', and how do we calculate its numerical value? Velocity means distance traveled divided by the time interval to travel the distance. But such a calculation really gives the average velocity for the time interval chosen. For example, between 3 and 4 seconds in the table we see the distance traveled has gone from 44.1 meters to 78.4 meters. In the 1 second time interval between 3 and 4 seconds of free fall the average velocity equals:

3.

$$v(average) = \frac{s \text{ at } 4\sec - s \text{ at } 3\sec}{1\sec}$$

$$= \frac{78.4 - 44.1}{1} = 34.3 \ m / s.$$

We can also calculate the average velocity for the time interval between the 2nd and 3rd second:

4.

$$v(average) = \frac{44.1 - 19.6}{3 - 2} = 24.5 \ m / s.$$

We conclude that the instantaneous velocity at exactly 3 seconds must lie between the values of *24.5 m/s* and *34.3 m/s*. We see also that the magnitude of the velocity is continually changing during the motion. The direction does remain constant towards the center of the earth, perpendicular to the earth's surface taken as a horizontal plane.

We can calculate values closer to the instantaneous velocity at 3 seconds by taking the smaller time intervals before and after the 3 second value.   If we calculate the average velocity between 3 and 3.1 seconds, we find the following: s at 3.1 seconds is,

5. $s = 4.9(3.1)^2 = 47.09$ meters. Then,

$$v(average) = \frac{47.09 - 44.1}{3.1 - 3} = 29.9 \; m \, / \, s.$$

We also calculate the average velocity for the interval of time between 2.9 and 3 seconds by first finding s at 2.9 seconds.

6. $s = 4.9(2.9)^2 = 41.2$ meters.

$$v(average) = \frac{44.1 - 41.2}{3 - 2.9} = 29.0 \; m \, / \, s.$$

From this we conclude the instantaneous velocity at 3 seconds of free fall lies between 29.0 and 29.9 m/s.   It would seem we can get closer and closer to the numerical value for the instantaneous velocity at 3 seconds by continuing to make the time intervals about 3 seconds smaller and smaller.   In one sense we might object that the instantaneous velocity at exactly 3 seconds cannot be found by this process since our method of calculating always involves division of differences in distance by a small but finite interval of

492

time. At exactly 3 seconds after the start of free fall, or any other instant after the start of free fall, there is no interval of time, and calculating by the process,

$$v = \frac{s(dis\,tan\,ce\ 1 - dis\,tan\,ce\ 2)}{t_1 - t_2}$$

becomes meaningless when the time interval is zero.

The creators of the calculus recognized this difficulty, and to overcome it they developed the method of increments, followed by the process of finding the limit as the increments approach zero. Although the formal process of finding limits came after the era of Newton and Leibniz, the germ of the idea was in their work and is now standard in the calculus. We shall see how the method of increments and the process of finding limits can give us the numerical value for the instantaneous velocity.

The symbol used by mathematicians today for a small increment of change in a variable is the Greek letter delta in the large case: Δ. For an incremental (meaning extremely small) change in distance we write Δs (read as delta s). For an incremental change in time we write Δt. Because the function, $s = 4.9t^2$, involves two variables, we say an incremental change in the time. Δt, results in a corresponding incremental change in distance, Δs. Leibniz called these extremely small changes in the variables infinitesimal changes, but today it is preferred to speak of 'incremental changes'. In

493

this way we emphasize the change in the variable is a real, although very small, change in the numerical value of the variable. For example, we might think of 1/1,000,000 of second as an incremental change of the time at the value of 3 seconds in the function for free fall. It is a change so small that it cannot be readily measured, yet it is a real change, and will produce a correspondingly small incremental change in the distance fallen after 3 seconds of free fall.

We now proceed with the method of increments to determine the instantaneous velocity at 3 seconds of free fall.

7.   We know, $s = 4.9t^2$, and for $t = 3$, $s = 44.1$. (We understand we are working in the metric system). We calculate the value of s when t increases from 3 to $3 + \Delta t$.   This means the corresponding increase in s is to increase from 44.1 to $(44.1 + \Delta s)$.   Thus, the relation, $s = 4.9t^2$, becomes,

8. $(44.1 + \Delta s) = 4.9(3 + \Delta t)^2$,

$44.1 + \Delta s = 4.9(9 + 6\Delta t + \Delta t^2)$

Note we always consider Δs or Δt as a single quantity so that Δt^2 means, $\Delta t \times \Delta t$, and $6\Delta t$ means, $6 \times \Delta t$. Then,

$44.1 + \Delta s = 44.1 + 29.4\Delta t + 4.9\Delta t2$, or

$\Delta s = 29.4\Delta t + 4.9\Delta t^2$.

This last equation gives the fundamental relationship between Δs and Δt at 3 seconds after the start of free fall. Here we say Δs is the dependent variable and Δt is the independent variable.

Let us choose Δt = .000001 sec, and then calculate the value of Δs at the point, t = 3 sec. We then use the final equation in step 8.

9. Δs = 29.4(.000001) + 4.9(.000001)$^{2+}$

Δs = .0000294 + 4.9 x 10^{-12} =

.000 029 400 00490

Δs = .000029 400 004

The ratio, $\Delta s / \Delta t$ = .0000294000049/.000001 = 29.4000049.

We see that when we square the value of an extremely small increment and then use it in the formula the numerical result is very much less, i.e. 4.9 x 10^{-12} versus .000001.   The term with the squared value of the increment is so small it does not significantly affect the final value of Δs.   When we examine the equation, Δs = 29.4Δt + 4.9Δt^2 derived when t = 3 sec, we see that we can divide each term by Δt, since Δt is a finite real quantity.

10.

$$\frac{\Delta s}{\Delta t} = 29.4 + 4.9 \, \Delta t.$$

This equation gives us the average velocity, $\Delta s / \Delta t$, for any small increment in time after 3 sec of free fall. The mathematician now says let us make the increment, Δt, smaller and smaller such that its value approaches zero.   As this occurs the term, 4.9Δt, must become smaller and smaller in comparison to 29.4, which remains constant as Δt becomes progressively smaller.

495

We know already when Δt is .000001, then, $\Delta s/\Delta t$ is 29.4000049. This suggests that as Δt becomes smaller and smaller, the limiting value for $(\Delta s/\Delta t)$ is 29.4. If this is true, we should define the instantaneous velocity as the limiting value of $\Delta s/\Delta t$ as Δt approaches zero, which has the numerical value of 29.4 m/s. We can now see the idea of a limit in assessing incremental changes. The instantaneous rate of change of one variable with respect to another is defined as the limit of the ratio of the incremental change of the first variable to the incremental change in the second as the incremental change in the second variable approaches zero. In free fall the instantaneous rate of change of Δs with respect to Δt is the limit of $\Delta s/\Delta t$ as Δt approaches 0. The value for this limit at $t = 3$ is 29.4, because Δs we found $\Delta s/\Delta t =$ 29.4 + 4.9Δt, for $t = 3$.

The instantaneous velocity is then defined as the limit of $\Delta s/\Delta t$ as Δt approaches zero, where Δs and Δt are incremental values of the two variables s and t. Mathematicians in the modern era have created symbolism for the limit process as follows:

$$v_i = \lim_{\Delta t \to 0} \frac{\Delta s}{\Delta t}.$$

This equation means that the instantaneous velocity, v_i, is the limit of the average velocity when the time interval, Δt approaches zero at the particular time for which the instantaneous velocity is desired.

496

To find the numerical value of the limit we do not divide a value of Δs by a value of Δt. We have found that procedure approximates the limit and may even intuitively tell us what the limit should be(in our case we realized the limit must be 29.4 m/s). Instead to find the limit we allow Δt to approach 0 in the equation, $(\Delta s/\Delta t) = 29.4 + 4.9\Delta t$, without actually becoming 0. By doing this we see that the term $4.9\Delta t$ must approach 0 when Δt approaches 0. Since this is true, then $(\Delta s/\Delta t)$ must approach 29.4 as Δt approaches 0. Thus 29.4 is the limit we seek and is the instantaneous velocity during free fall 3 seconds after free fall has begun.

To find the instantaneous velocity we had to go through the method of increments to find the equation for $(\Delta s/\Delta t)$, and then apply the reasoning behind the limiting process. The result of all this work thus far has given us the value for the instantaneous velocity at only one instant in time, i.e. 3 seconds after free fall has begun. It is desirable to develop a general method so we can readily determine the instantaneous velocity at any time after free fall begins.

We note the distance formula for free fall, $s = 4.9t^2$, is a function where the dependent variable, s, is directly proportional to the square of the independent variable, t. In general, we can state the relation to be of the form,

11. $y = ax^2$, where a is the proportionality constant.

We will use the method of increments on this general function. We let any value of x say x_1 increase by the increment Δx, which causes the corresponding increase, Δy for y_1 (the value of y when $x = x_1$). x_1 becomes $(x_1 + \Delta x)$, and y_1 becomes $(y_1 + \Delta y)$. Then,

12. $y_1 + \Delta y = a(x_1 + \Delta x)^2$

$y_1 + \Delta y = a(x_1)^2 + 2ax_1 \Delta x + a(\Delta x)^2.$

We subtract the equation, $y_1 = a(x_1)^2$, from the above equation, the left side from the left side, and the right side from the right side.

13. $(y_1 + \Delta y) - y_1 =$

$[a(x_1)^2 + 2ax_1 \Delta x + a(\Delta x)^2] - a(x_1)^2,$

$\Delta y = 2 ax_1 \Delta x + a(\Delta x)^2.$

We divide both sides by Δx.

14. $\Delta y / \Delta x = 2ax_1 + a\Delta x.$

This equation gives the average rate of change of y with respect to x over the interval of change Δx. To find the instantaneous rate of change we find the limit of $(\Delta y / \Delta x)$ as Δx approaches 0. It can be demonstrated in the study of limits that the limit of a sum of terms is equal to the sum of the limits of each term so we can write.

15.

$$\lim_{\Delta x \to 0} \frac{\Delta y}{\Delta x} = \lim_{\Delta x \to 0} 2ax_1 + \lim_{\Delta x \to 0} a\Delta x.$$

The limit on the left side is the instantaneous rate of change of y with respect to x at the value x_1 of x. The term, $2ax_1$, is constant and does not vary as Δx approaches 0. The limit of a constant term then is unchanged by taking the limit. The limit of the term,

$a\Delta x$, as Δx approaches 0 must be 0, since $a\Delta x$ gets closer and closer to 0 as Δx approaches 0. This in turn means that the instantaneous rate of change of y with respect to x is equal to $2ax_1$. Since x_1 can represent any value of x the function may take on, then in general the instantaneous rate of change of y with respect to x is equal to $2ax$ for all values of x.

There is a special name for the instantaneous rate of change of one variable with respect to another. It is called the derivative of y with respect to x. The process of finding the derivative is called differentiation. The branch of the calculus of learning how to find derivatives for various functions is called the differential calculus. There are a number of symbols for the derivative used in calculus books. We will use primarily the symbol introduced by Leibniz:

$$\frac{dy}{dx}.$$

Thus, the instantaneous rate of change of y with respect to x, when y is the dependent variable and x the independent variable in a function relating y and x, is symbolized by dy/dx , which is the limit of an incremental change, Δy, of y, divided by an incremental change. Δx, in x as Δx approaches 0.

For the function, $y = ax^2$, the derivative is,

$dy/dx = 2ax$.

We can now give the derivative of the distance function, $s = 4.9t^2$, for free fall. Here s takes the place of y in the general function, $y = ax^2$, $4.9 = a$, and t takes the place of x. Thus, the instantaneous rate of change of distance with respect to time, i.e. the velocity, is equal to the derivative of s with respect to t.

16.

$$v_i = \frac{ds}{dt} = 2(4.9)\, t = 9.8\, t.$$

We have found the instantaneous velocity function with respect to time using the calculus starting with the distance function.   We know from our previous work that at 3 seconds of free fall the instantaneous velocity should be 29.4 m/s.   Using our derived function in step 16, we see when $t = 3$ then,

17. $v_i = (9.8)(3) = 29.4\ m/s.$

We also see that the derivative of the distance function in step 16 is also a function relating instantaneous velocity with time since the beginning of free fall.   Note again the derivative of distance with respect to time given by the symbol, ds/dt, is the same as the instantaneous velocity which we have been symbolizing by v_i.   It is customary to symbolize the instantaneous velocity by just v, and if we are dealing with average velocity to indicate that by a separate notation.

We can also find the value of the uniform acceleration due to gravity near the earth's surface by using the differential calculus.   The uniform acceleration is the rate of change at any instant of the instantaneous velocity.   The calculus tells us then that the instantaneous rate of change of the velocity with respect to time, symbolized by, dv/dt, is equal to the acceleration at that instant of time.   In free fall we already know that the acceleration at each instant is the same, since Galileo found the acceleration to be uniform or constant during free fall. The function for instantaneous velocity is,

18. $v = 9.8t$.   A general function of this form is, $y = ax$, where a is a constant.

We will use the method of increments to find the derivative, dy/dx, for this function.   Let x_1 be the value of x where we wish to find the derivative.   We let x_1 increase by the incremental value Δx,   which results in an incremental change in y_1, the value of y when x is x_1. Then when x becomes $x_1 + \Delta x$, y_1 becomes $y_1 + \Delta y$. Then,

19. $y_1 + \Delta y = a(x_1 + \Delta x) = ax_1 + a\Delta x$.   We subtract $y_1 = ax_1$ from this is equation; left side from left side and right side from right side to get,

20. $\Delta y = a\Delta x$.

The average rate of change of y with respect to x in the interval Δx is found by dividing both sides of step 20 by Δx.

21. $(\Delta y/\Delta x)$ $= a$.   We take the limit of the left side of this equation as Δx approaches 0 to get the instantaneous rate of change of y with respect to x at the value for x of x_1, and the value of y of y_1.

$$\lim_{\Delta x \to 0} \frac{\Delta y}{\Delta x} = \frac{dy}{dx} = \lim_{\Delta x \to 0} a, \ or, \ \frac{dy}{dx} = a,$$

since the limit of a constant does not change as Δx approaches 0.

Since the result is valid for any value of x, then for the function, $y = ax$, dy/dx, always equals a. In the case of the instantaneous velocity function, $v = 9.8t$, we have, $dv/dt = 9.8$. Because the instantaneous rate of change of velocity with time is the acceleration, the uniform acceleration in free fall is 9.8 m/s^2.

The differential calculus has allowed us to start with the distance function, $s = 4.9t^2$, and derive the velocity function, $ds/dt = v = 9.8t$, and the uniform acceleration, $dv/dt = 9.8$.

Mathematicians also call, dv/dt, the second derivative of the distance function, since it is obtained by taking two successive derivatives of the distance function. The symbol for the second derivative is:

$$\frac{d^2 s}{dt^2}.$$

Here the 'square' notation does not mean multiplication, but that two successive derivatives are taken of the function relating s with t.

We might wonder if there is any point to go further and find the derivative of the acceleration. Since the derivative of the acceleration is the instantaneous rate of change of the acceleration with respect to time, and the acceleration is constant, it is intuitively clear that the acceleration cannot change its value as time passes, and so its derivative must be 0. In general, the derivative of any constant function is 0. We can use the method of increments to establish the general case.

22. $y = a$, is the general function. We can think of this as a function of x, if we write, $y = ax°$.

In mathematics any number to the zero power such as $x°$, is defined as being equal to 1. Thus,

$y = ax° = (a)(1) = a.$

If we choose a value of x, say x_1, for which we wish to find the derivative then we have, $y_1 = a(x_1)°$. As per usual in the method of increments x_1 is increased by the incremental value Δx. However, no matter what the change in x may be, $(x_1 + \Delta x)^0$ always equals 1 and y_1 cannot change in a constant function, $y = ax$. That is like saying the incremental change in y, $\Delta y = 0$. Thus, $y_1 + \Delta y = a(1) = a.$ If we subtract, $y_1 = a$, from this equation, we get, $\Delta y = 0$, as expected. We then divide by Δx:

23. $\Delta y/\Delta x = 0$, the limit of the left hand side as Δx approaches 0 is dy/dx, and the limit of the right hand side as Δx approaches 0 is 0. This holds for all values of x, since the choice x_1 could be any value of x.

Thus, the derivative of any constant function such as, $a = 9.8$ (a being the symbol for the uniform acceleration in free fall), is 0. If we were to use the symbols of the calculus, $da/dt = d^3s/dt^3$, where d^3s/dt^3, means the third derivative of distance with respect to time.

There is no physical utility for the derivative of the uniform acceleration in free fall, so our pursuit of this quantity is a mathematical exercise to introduce us to more procedures in the differential calculus.

To further our understanding of the differential calculus we consider a function of the form, $y = ax^2 + bx + c$. Do we know enough to find the derivative of this function? We know the derivative of ax^2 is $2ax$, the derivative of bx is x, and the derivative of c is 0, where a, b, and c are constants. We mean by the derivative the instantaneous rate of change of y with respect to x, dy/dx.

In formal calculus it can be shown by the method of increments that the derivative of a sum of terms is equal to the sum of the derivatives of each term taken one at a time. Thus, the derivative of,

$y = ax2 + bx + c$, is, $dy/dx = 2ax + b + 0$, or,

$dy/dx = 2\ ax + b$.

It is very useful to know that the derivative has a geometric meaning as well. From our previous work in coordinate geometry we learned that equations of the form, $y = ax^2$, represent parabolas with vertices at the origin. The formula for the distance in free fall, $s = 4.9t^2$, is of this form. We have made a graph of this function in Figure 40-1.

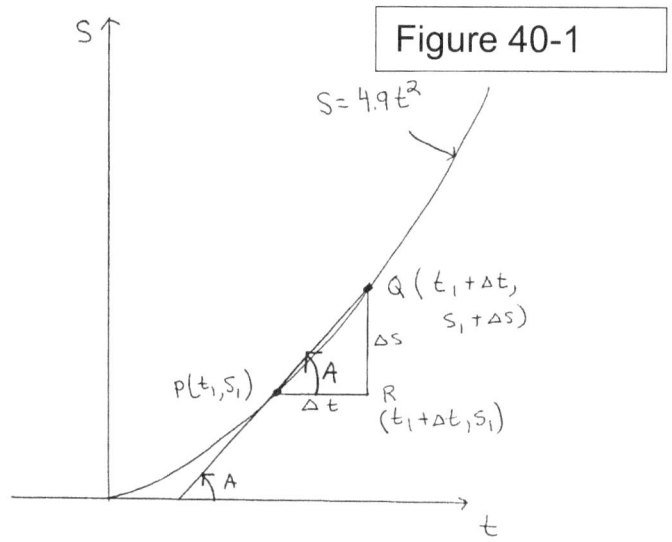

Figure 40-1

The Y axis here is the s axis, and the X axis is the t axis. We have sketched only half of the curve, that half where t is greater than 0, because in physical reality free fall begins when t =0, and continues until the object hits the earth's surface.

Note two points on the parabola, $P(t_1,s_1)$, and $Q(t_1 + \Delta t, s_1 + \Delta s)$. Point Q is the point where t_1 has been increased by the incremental value Δt, resulting in the corresponding increase in s_1 by Δs. We recall from our work with the method of increments, we found in step 10,

$\Delta s/\Delta t = 29.4 + 4.9\Delta t$, when t = 3 seconds.

If we use t_1 instead of 3 seconds this formula would be,

$\Delta s/\Delta t = 9.8t_1$. $\Delta s/\Delta t$ is the average velocity of the object in free fall between times t_1 and $t_1 + \Delta t$.

In Figure 40-1 we have indicated the values for Δt and Δs as perpendicular lines parallel to their respective

505

axes. The two perpendicular lines meet at point R(t_1 +
Δt, s_1). We then connect chord PQ, which gives us right
triangle PRQ. We can consider line PQ a secant line (a
secant cuts the curve at two points). Clearly the length
of PQ depends on the length of the increment Δt. The
smaller Δt is, the shorter PQ is. We can determine the
slope of line PQ from our knowledge of coordinate
geometry in chapter 30.   The slope of PQ is,

    24.

$$m = \frac{(s_1 + \Delta s) - s_1}{(t_1 + \Delta t) - t_1} = \frac{\Delta s}{\Delta t} = \tan A$$
, where

angle A is the angle PQ extended makes with the t axis
as indicated in Figure 40-1.

    We now have a geometric interpretation for the
ratio of the two increments. Δs and Δt.   Now in the
limiting process to find the derivative at $t = t_1$, we let Δt
approach 0.   What this means geometrically is shown in
Figure 40-2.

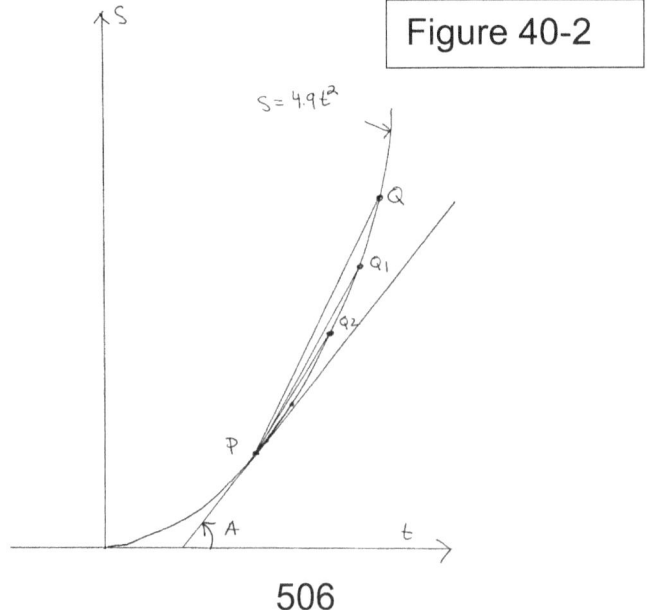

Figure 40-2

For each incremental shortening Δt as Δt approaches 0, the respective secant lines are PQ$_1$, PQ$_2$ etc. as shown in the figure.   We see that the secant lines are getting shorter and shorter and approach as their limit, when Δt approaches 0, the tangent line at P, which makes angle A with the t axis in Figure 40-2.   In effect this means in the limit, the secant line becomes the tangent line at P.   We can then write,

25.

$$\lim_{\Delta t \to 0} \frac{\Delta s}{\Delta t} = \frac{ds}{dt}$$

= slope of the tangent line at P.

Note, (ds/dt) is the derivative of $s = 4.9t^2$,   for all values of t on the half parabola.   We now have learned the numerical value of the derivative at any value of t is equal to the value of the slope of the tangent to the parabola representing the distance function at the value of t we choose.

The discovery by the 17th century mathematicians working on the differential calculus that the derivative is equal to the slope at any point on the curve was a major step in the new mathematics.   Several mathematicians had made this discovery for various curves before the work of Newton and Leibniz.   Now there are functions and curves which do not have derivatives or tangent lines at all values or points on the curve.   Fortunately, in our work with Galileo and Newton this rarely comes up as a problem.

In our earlier work with the distance function we found the value of the derivative at $t = 3$ seconds was, $(ds/dt) = v = 29.4 \ m/s$. In Figure 40-2 this means the tangent of angle A is equal to 29.4, and arctan 29.4 = 88.05°, the angle of inclination of the tangent line to the parabola at $t = 3s$.

We can also speak of the derivative as a derived function. In the case of the distance function, $s = 4.9^2$, the derived function is, $(ds/dt) = 9.8t$. We know that the instantaneous rate of change of distance with respect to time is the instantaneous velocity, so the derived function can also be written, $v = 9.8t$. The derived function is the relation between v and t, and gives us the instantaneous velocity for any value of t. In the case of free fall the values t may take on are limited because objects released in free fall reach the ground in relatively few seconds.

In our introduction to the differential calculus and the concept of the limit and the use of the limiting process to give the definition of the derivative, we have relied upon the intuitive approach to mathematics. Mathematicians in the eighteenth and nineteenth centuries developed a rigorous logical approach to define both the limit and the derivative, which can be found in college level textbooks on the calculus. We list some of these in the bibliography to this chapter. We should realize that the mathematicians of the seventeenth century who first developed the concepts of the calculus did not have rigorous definitions of their concepts and did not even use the term limit as we have done. The concept of what the derivative is and how it can be calculated can be grasped without the highly technical definitions used in many modern textbooks. After one has mastered the elementary calculus and how it solves many of the problems of motion and other problems involving changing quantities, one can then learn the rigorous definitions of the limit and the derivative, and appreciate better why there was a need for a firm logical basis for the calculus. At our stage of development, we rely on the intuitive approach combined with geometrical interpretation to understand the elements of the calculus.

Much of the early work on finding tangents to curves in the 17th century was done by Pierre de Fermat (1601-1665). We have already seen some of his contributions to coordinate geometry. Isaac Barrow (1630-1677) who was the first Lucasian Professor of mathematics at Cambridge university, also used an essentially incremental method to determine the tangent to a curve. Since finding the tangent is finding the instantaneous rate of change of the curve at the point where the tangent is found, the process is that of finding the derivative at that point. We will look into Barrow's method which is similar to Fermat's, as it serves as an introduction the subject of differentials which we will use frequently in our work with the calculus. See Figure 40-3.

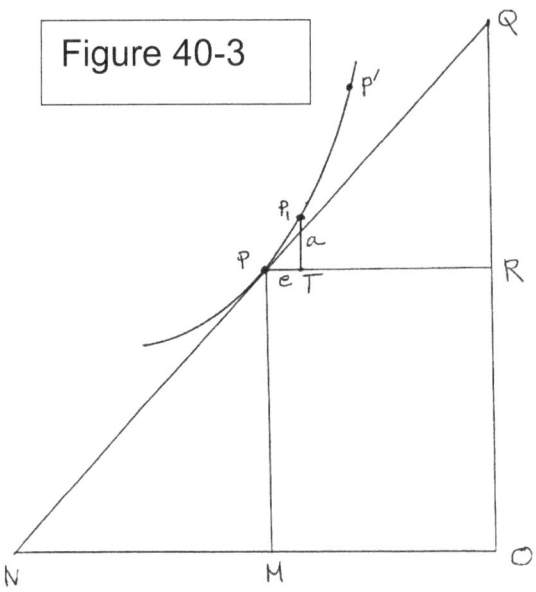

Figure 40-3

In the figure we have drawn a curve, PP'P¹, for which Barrow wished to find the tangent at P. The tangent is the line represented by NPQ which touches the curve at one point only, namely P, on the curve. We know the tangent line if we know its slope, which is the value of tan angle PNM in the figure. Barrow published a book, "Lectiones Geometria", in 1669 (the same year he resigned his Lucasian Professorship to Isaac Newton) in which he describes his method of finding the tangent.

Barrow makes a right triangle QON with hypotenuse NQ containing P. He then draws perpendiculars PM from P to line NO, and PR from P to QO. Barrow did not use the terminology of coordinate geometry so we should not infer he took NO to be the X axis. However, we can see the analogy of his figure with the figures we have used in coordinate geometry.

Barrow notes triangles NMP and PRQ are similar so, PM/MN = QR/PR = tan PNM.   We next take a point on the curve P_1, which is found by dropping perpendicular P_1 T from P_1 to PR. P_1 is so close to P on the curve that length PT is an incremental length. Actually Barrow considers PT an 'infinitesimal' change compared to length PR.   Likewise P_1T is considered by Barrow to be an 'infinitesimal' length.   Today we would say PT is analogous to our Δx, if P had coordinates (x,y), and P_1T to be the corresponding incremental change Δy. Barrow used the letters e and a respectively for PT and PiT.

Barrow then considers the 'infinitesimal' triangle PTP_1.   He then argues that because the triangle is 'infintesimally' small, arc PP_1 lies along the tangent line, NQ, to the curve at P.   Using the 'infinitesimal' triangle, Barrow states tan PNM = a/e, since triangles PMN and P_1TP are similar.

To show how Barrow gets a quantitative value for the slope we will use his method for the parabola, $y = ax^2$.   We know from the previous work that the derivative is $2ax$ = tangent of the angle the tangent line to the curve at any point makes with the X axis.   Let $a = 1$.   The derivative is then $2x$.   When $x = 3$ the derivative is $2(3) = 6$, which is the slope of the tangent line to the curve at $x = 3$.

Barrow follows the method of increments we have already learned:

26. $y + \Delta y = (x + \Delta x)^2$,   for the function $y = x^2$.

511

$y + \Delta y = x^2 + 2x\Delta x + (\Delta x)2$.   We subtract $y = x^2$ from this equation:

$\Delta y = 2x\Delta x + \Delta x2$.

Here Barrow discards all higher powers of the 'infinitesimal' quantity when they appear.   Note he does not discard Δx when it is not raised to a higher power. Thus,

27. $\Delta y = 2x\Delta x$,

$(\Delta y/\Delta x) = 2x$

Since the 'infinitesimal' triangle is similar to triangle PMN, then, PM/NM = $(\Delta y/\Delta x) = 2x$, which is the slope of the tangent line at P for the parabola $y = x^2$. For $x = 3$, the slope is 6.

The method of Barrow and Fermat came before the limit concept was created, but we can see some similarity to the modern method of finding the derivative. Many other workers on the calculus in the 17th century also used the idea of 'infinitesimal' quantities, and the calculus became known as the 'infinitesimal calculus'. Later on in the 17th century Newton and Leibniz also spoke of 'infinitesimal quantities' and discarded terms which they felt were 'infinitesimally small'.   It was natural that the calculus of this period came under criticism for considering actual quantities so small that they could be discarded if they were squared or raised to higher powers, yet kept them if they were multiplied by real numbers as in the above example.   It seemed arbitrary that mathematicians were discarding terms in their equations.   Today we do not discard terms, but use the limit concept and find that terms do approach 0 in their limits.   In the early part of the 18th century the philosopher Bishop Berkeley attacked Newton and the other mathematicians for using 'infinitesimals' in their arguments.   Berkeley pointed out that the new mathematicians were not using the deductive methods of Euclid and the Greeks.   They proceeded by intuition and induction in discarding certain of their 'infinitesimal' terms and not others.   In particular, the practice of taking the ratio of the two 'infinitesimal' quantities and calling that the derivative was suspect.   After all the 'infinitesimal' quantities were, "neither finite quantities, nor quantities infinitely small, nor yet nothing."   The ratio of the 'infinitesimals' were nothing but, "the ghosts of departed quantities" according to Berkeley.   It was criticism such as this that led to the idea of the limit where an incremental quantity approaches to zero.   In the next chapter we will pursue the ideas of the differential calculus and introduce the concept of differentials.

Bibliography for Chapter 40

1. Galilei, Galileo, "Dialogue Concerning the Two Chief World Systems-Ptolemaic & Copernican", translated by Stillman Drake, University of California Press, Berkeley and Los Angeles, 1962.

2. Galilei, Galileo, "Discourses and Mathematical Demonstrations Concerning Two New Sciences", translated with a new introduction and notes by Stillman

Drake, Second Edition, Wall & Thompson, Toronto, 1989.

3. Kline, Morris, "Calculus: An Intuitive and Physical Approach", John Wiley and Sons, Inc.,1967.

4. Anton, Howard, "Calculus with Analytic Geometry", John Wiley and Sons, Inc., 1980.

5. Courant, Richard and John, Fritz, "Introduction to Calculus and Analysis", Volume One, Interscience Publishers, A Division of John Wiley and Sons, Inc., 1965.

6. Weisbecker, H., Chief Editor, "The Calculus Problem Solver", Reasearch and Education Association, New York, 1974.

7. Kline, Morris, "Mathematical Thought From Acient to Modern Times", Oxford University Press, 1972.

8. Barrow,Isaac, "The Geometrical Lectures of Isaac Barrow", translated with notes by J. M. Child, Open Court Publishing Company, Chicago, 1916.

8. Berkeley, Bishop George, "The Analyst", in "The Works of Bishop Berkeley", G. Bell and Sons, 1898.

Chapter 41
Differentials and Antiderivatives

We will now see how Barrow's 'infinitesimal triangle' can be used to develop the modern idea of differentials. We will use as our example the familiar function for the parabola, $y = ax^2$. We have learned that the ratio of the increments Δy to Δx gives the average rate of change of the function for the incremental interval of the independent variable x. Thus,

1. Average rate of change = $\Delta y/\Delta x = 2ax + a\Delta x$ (Steps 11 through 14 in chapter 40). Then, by taking the limit as Δx approaches 0,

2. $dy/dx = 2ax$. Note dy/dx is the derivative as we defined in chapter 40, and not a ratio. If we substitute dy/dx for $2ax$ in step 1, we have,

3. $\dfrac{\Delta y}{\Delta x} = \dfrac{dy}{dx} + a\Delta x,\ or,\ \dfrac{\Delta y}{\Delta x} - \dfrac{dy}{dx} = a\Delta x.$

We now designate the term $a\Delta x$ by the Greek letter epsilon: ε. We then can write.

4.

$$\dfrac{\Delta y}{\Delta x} - \dfrac{dy}{dx} = \varepsilon.$$

From this expression and the value of ε, we see that as Δx approaches zero ε also approaches zero. But no matter how close to zero ε becomes, there is always a small difference between $\Delta y/\Delta x$, and the derivative, dy/dx.

This line of reasoning can be applied to functions other than $y = ax^2$, in which case we let ε equal the difference between $\Delta y/\Delta x$

and the derivative dy/dx. Here ε may have several terms or one term, but all the terms will contain Δx so that ε approaches zero as Δx approaches zero. Thus for functions in general which have derivatives we can write the equation in step 4. We now multiply both sides of this equation by Δx.

5.

$$\Delta y - \frac{dy}{dx} \cdot \Delta x = \varepsilon \cdot \Delta x.$$

We now introduce a new term called the differential of y. We define the differential of y as equal to $(dy/dx)\Delta x$(the derivative times Δx). The symbol given for the differential of y is, interestingly, dy. This means,

6.

$$dy = \frac{dy}{dx} \cdot \Delta x.$$

In words, the differential of y is equal to the derivative multiplied by the incremental change in x. Note that the differential of y is not the same as the increment in y, namely Δy, because from step 4 we know there is always a finite difference between the derivative, dy/dx, and, $\Delta y/\Delta x$. Also consider what would happen if $dy = \Delta y$.   We would have, $\Delta y = (dy/dx)\Delta x$, or, $\Delta y/\Delta x = dy/dx$, which we know is not the case from step 4.

We now rewrite step 5 using dy for the differential of y.

7. Δy - dy = $\varepsilon \cdot \Delta x$.

This equation gives the magnitude of the difference between the increment of y, Δy, and the differential of y, dy. Note that the closer Δx is to zero the closer the differential of y is to the incremental change in y. For example, in the function, $y = ax2$, if $a = 4.9$, and the incremental change in x is $\Delta x = .0001$, then,

8. $\varepsilon = a\Delta x$ (from step 3 above).

$\varepsilon = (4.9)(.0001) = .00049$.

The value of $\varepsilon \Delta x$ is

$(.00049)(.0001) = .000000049$.

We use this in step 7:

9. Δy - dy = $\varepsilon \Delta x$ = .00000049.

This indicates for small values of Δx, dy is a very close approximation of Δy.

Let us calculate the value of both Δy and dy, as an example, for the function, $y = 4.9x^2$,

517

at $x = 2$, with the incremental change in x, $\Delta x =$ .0001.

10. $x = 2$, $y = 4.9(2)^2 = 19.6$,

$x + \Delta x = 2.0001$,

$y + \Delta y = (4.9)(2.0001)^2 = 19.60196005$.

Thus, $\Delta y = 19.60196005 - 19.6$

$= .00196005$.

11. To find dy we use the definition,

$dy = (dy/dx)\Delta x$.

Now for $y = ax^2$, the derivative,

$dy/dx = 2ax$,   so when

$a = 4.9$, $dy/dx = 2(4.9) x = 9.8x$.

When,

$x = 2$, $dy/dx = (9.8)(2) = 19.6$.

$dy = (dy/dx)\Delta x = (19.6)(.001) = .00196$.

Thus, $\Delta y - dy = .00196005 - .00196$

$= .00000005$.

We see that indeed dy, the differential of y, is very close to Δy, when small values for Δx are used.   If we were to use even smaller values for Δx, then the approximation of Δy by dy would be even better. The reason dy is used is that in many complicated functions dy is much easier to find then Δy.

Now that we have learned something of the meaning of the differential of *y*, we may ask is there a differential for *x*? The answer is that the differential of *x* is defined as being exactly equal to the increment of *x*, Δx. The symbol for the differential of *x* is *dx*. This in turn means we can rewrite step 6 when *(dy)* is the differential of *y*, and *(dx)* is the differential of *x*.

12.

$$(dy) = \frac{dy}{dx} \cdot (dx). \textit{ Divide by } (dx): \frac{(dy)}{(dx)} = \frac{dy}{dx}.$$

This last result means the ratio of the differential of *y* to the differential of *x* is equal to the derivative. Note the derivative is defined as a limit as we learned in the last chapter, while the differentials are defined as in step 6 and step 12. Because of the way the differentials are defined, it turns out the ratio of the differential of *x* to the differential of *y* is equal to the derivative defined as a limit. Note that the derivative is not defined as the ratio of the differentials. Rather the derivative is defined first, then the differentials are defined using the derivative in the definition for *dy* (step 6), and then defining the differential of *x* as the incremental change in *x*, Δx.

To further clarify the concept of differentials we use a geometric interpretation similar to Barrow's 'infinitesimal' triangle of Figure 40-3. See Figure 41-1.

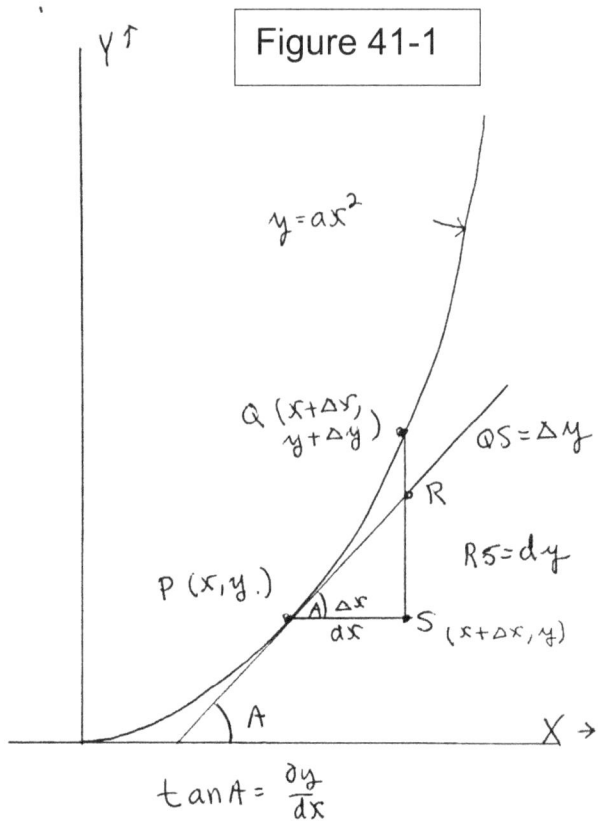

Figure 41-1

$y = ax^2$

$Q (x+\Delta x, y+\Delta y)$

$QS = \Delta y$

R

$RS = dy$

$P (x, y.)$

$A) \Delta x$

dx

$S (x+\Delta x, y)$

A

$\tan A = \dfrac{dy}{dx}$

In the figure the parabola for $y = ax^2$ is drawn with the X and Y axes as indicated. Two points on the parabola are noted, P and Q. P can be any point, although we have shown it as a point with positive x and y coordinates. To move from point P to Q the x coordinate is increased by Δx, and the y coordinate has the resulting increase of Δy. Thus the coordinates of Q are, $(x + \Delta x, y + \Delta y)$. We then construct

520

right triangle PSQ, which is similar to the procedure used by Barrow.   PS is perpendicular to the Y axis and SQ is perpendicular to PS so QS = Δy. Then the tangent to the parabola at P is drawn to intersect the X axis and SQ at R.   In Barrow's case Q and R are the same point.   In Figure 41-1 consider right triangle PSR: side PS is equal to dx = Δx(differential of x), and side RS is equal to dy(differential of y).   In order to find the length of dy, and the location of point R, we must know the derivative, or slope of the tangent at P.   We know from our previous work when $y = ax^2$, then, dy/dx, the derivative, = $2ax$ = tan A, where A is the angle the tangent line at P makes with the X axis as indicated in the figure.

We now see clearly that QS is equal to Δy(the increment of y), which is greater than RS = dy(the differential of y).   Note that PR, the hypotenuse, is part of the tangent line touching the parabola at P.   In right triangle PSR we see tan angle RPS = RS/PS = (dy)/(dx)[the differential of y divided by the differential of x] = tan A = slope of line PR, which we know is the derivative dy/dx.

Barrow's concept was that as dx becomes smaller and smaller, PR is becomes closer and closer to PQ, and essentially becomes the same line when dx is 'infinitesimally' small. Modern mathematicians differ from Barrow because differential dx is always a finite quantity, however small, and PR never coincides with PQ (a straight line never coincides with a curved line). Only in the limit when Δx approaches zero, then the ratio, $\Delta y / \Delta x$, becomes, $(dy)/(dx) = dy/dx$, [meaning the ratio of the differentials equals the derivative]. Thus it is the limiting process which makes $\Delta y / \Delta x$, become the derivative or slope of the tangent line at P.

In step 7 we found. $\Delta y - dy = \varepsilon \Delta x$. We can now write this equation as, $\Delta y - dy = \varepsilon dx$. We see from Figure 41-1 this means,

RQ $= \Delta y - dy = \varepsilon dx$. The geometric interpretation of differentials now shows us why mathematicians chose the symbols dy and dx for the differentials. Geometrically it is clear that for curves such as the parabola the ratio of the differentials will be the slope of the tangent line, and we now know the slope of the tangent line is the derivative of the function at that point. From now on when we see the symbol, dy/dx, we can think of it as meaning either the derivative, or as meaning the ratio of the differentials.

Leibniz introduced the symbolism dy, and dx in his work. He did not consider them as differentials as we have defined differentials. Instead he considered them to be 'infinitesimals' in the same way as Barrow. Both Newton and Leibniz were working on the calculus in the last half of the 17th century, and they used different symbols in their work. Leibniz did not use the limit concept and used "infinitesimals". He did consider the ratio of $(dy)/(dx)$[ratio of infinitesimals] to be the value of the tangent.   He too thought terms could be dropped if they were the product of two 'infinitesimal' quantities.

We now return to our previous work on the velocity of free fall. We recall the velocity function is, $v = 9.8t$(values in meters per second), and the acceleration function is, $a = 9.8$(values in meters per second squared).   We used the derivative to find these two functions starting with the distance function, $s = 4.9t^2$. With the help of differentials and some other concepts of the calculus, we will see how we can start with the acceleration function and then find the velocity function and the distance function. In other words we will learn the reverse process of differentiation known as antidifferentiation.

We start with the velocity function given in terms of ds/dt and t.

13. $ds/dt = 9.8t$.   We seek the function relating s and t.

We recall that we obtained the derivative $dy/dx = 2ax$, from the function, $y = ax^2$, by using the method of increments and taking the limit of the average rate of change.   As we inspect these two formulas we see the derivative is equal to the exponent or power of x in the original function, namely 2, multiplied by the constant a, multiplied by x with its exponent reduced by 1. This suggests a rule for finding the derivative of an algebraic function as follows.

If $y = ax^n$, and n is any whole number then,

$$\frac{dy}{dx} = nax^{n-1}.$$

If this rule is true, we could write, for example, if $y = 4x^3$, then $dy/dx = (3)(4)x^{3-1} = 12x^2$.   It can be proven that the rule is true, and it is an essential rule for finding the derivatives of algebraic functions.   We will refer the reader to standard calculus textbooks for the proof of the rule.   Knowing this rule also suggests that it is possible to reverse the process of finding the derivative so that if we are given the derivative we can find the function from which it was derived.

The rule indicates that to find the function known as the antiderivative, we add 1 to the exponent of x found in the derivative, and then divide x by the new exponent. No change is made in the constant. Thus if,

$$\frac{dy}{dx} = ax^n, \textit{ then, } y = a\frac{x^{n+1}}{n+1}.$$

In the case of the velocity function for free fall, $ds/dt = 9.8t$, so $n = 1$. Using the rule we have.

14.

$$s = 9.8\frac{t^{1+1}}{1+1} = 4.9t^2.$$
This is the distance function which starts when s and t are 0 when the object is dropped in free fall.

If we start with the acceleration function, $d^2s/dt^2 = 9.8 = 9.8t^0$, using the rule we have,

15.

$$\frac{ds}{dt} = v = 9.8\frac{t^{0+1}}{0+1} = 9.8t.$$

We see the antiderivative for the second derivative of the distance function is the first derivative of the distance function, the velocity function.

The rule for finding antiderivatives works except in the case where n = - 1. In that case the denominator of the coefficient would be 0, since, n+1 = -1 + 1 =0. (We cannot have division by zero as it is meaningless).

The rule as we have presented it so far is incomplete. We see that we did get the correct answer for the acceleration and velocity functions, which we can easily verify by taking the derivative. But consider the function, $s = 4.9\ t^{2+} + 2$. The derivative of this function is,

16. $ds/dt = (2)(4.9)t^{2-1} + 0 = 9.8t$, the same result as taking the derivative of $s = 4.9t^2$.

This means that for any constant term added or subtracted to the distance function, or any other algebraic function, the derivative is the same as the derivative of the function without the constant term. This means when we reverse the process of differentiation in finding the antiderivative, we should add a constant term to the result. The rule for finding antiderivatives of algebraic functions such as the functions we have been studying for free fall should be,

17. For a whole number n, not equal to 1,

526

if

$$\frac{dy}{dx} = ax^n, \text{ then, } y = \frac{a}{n+1}x^{n+1} + C,$$

where C represents a constant term.

In problems involving the calculations of the antidervative the value of the constant term can be determined by the initial conditions for the situation being studied. This can be seen in the following example. Let us start with the velocity function, $v = ds/dt = 9.8t$.   In using the rule of step 17, $a = 9.8$, $n = 1$, $y = s$, and $x = t$. The antiderivative is,

18.

$$s = \frac{9.8}{1+1}t^{1+1} + C = 4.9t^2 + C.$$

In the case of free fall when $s = 0$, that is when the object is released, then, $t = 0$.   We can use these initial conditions to calculate the constant C.   We use these values in step 18.

19. $0 = (4.9)(0^2) + C$, or $C = 0$. Thus we confirm that, $s = 4.9t^2$, is the correct distance function.

The operation of finding the antidervative in the calculus is also known as integration, and the antiderivative is also called the indefinite integral, or simply the integral. Leibniz also developed the symbol for the integral which is used today. When we see the symbol in an equation it means that we should find the antidervative of the expression after the integral sign. The integral sign is an elongated 'S' and was the symbol used in 17th century printing for the letter 'S'.

20. The symbol is,

$$\int .$$

If we wish to find the integral of, $dy/dx = 2ax$, we use differentials and multiply both sides by dx to get, $dy = 2ax\,dx$. Then to indicate that we wish to find the integral we place the integral sign before each side of the equation.

21.

$$\int dy = \int 2ax\,dx$$

This equation means we are to find the antiderivative of dy on the left side of the equation, and the antiderivative of $2ax\,dx$ on the right side of the equation. Actually the presence of the differentials dy and following the integral sign do not contribute to the actual process of finding the antidervative or integral. They do remind us of the variable that is involved. Later in our work we shall learn the origin of this notation for integration and how the integral can be regarded as a sum with the differential *dx* playing a role in the sum.   We will see that the result of,

$$\int dy = y, and, \int dx = x.$$

For now the best approach is to consider, $y° = 1$, before the differential *dy* when it follows the integral sign, and then use the rule for finding the antiderivative.   Let us do a few examples.

22.

$$\int dy = \int 1 dy = \int y^0 dy = \frac{y^{0+1}}{0+1} = y + C.$$

In this example we assume the function before *dy* has the value of $y°$ for the purpose of finding the antiderivative, and use the rule (step 17) for finding antiderivatives. If we have the equation, *dy/dx =2ax*, and we wish to find the integral, we can write this in terms of differentials to get: *dy = 2ax dx,* and note the integral of the left side equals *y,* by step 22 and the right side integral is:

23.
$$\int 2ax\, dx = 2a\int x^1\, dx = 2a\frac{x^{1+1}}{1+1} = 2a\frac{x^2}{2}$$

$$= ax^2 + C.$$

The constants of integration may or may not cancel out depending on the initial conditions. We see the constant terms, *(2)(a)*, do not play a role in applying the rule for finding the antiderivative, so they can be moved in front of the integral sign, and later multiplied into the result of integrating *x*. To apply step 17 we used the fact that, $x = x^1$. The differential *dx* does not play a role in calculating the value of the integral.

We have taken time with this example so the reader can go through all the steps of finding the antiderivative. With a little practice many integrations can be performed very quickly.

Thus, if:

dy/dx $= 2x^3 + 4x^4 + 7x^2$, then, by integration,

$y = (2/4)x^4 + (4/5)x^5 + (7/3)x^3 + C$.

We have now introduced the concept that differentiation and integration are reverse processes with the exception that in integration we must add a constant of integration. This concept allows us to write the equality,

24.

$$\frac{d \int ax^n \, dx}{dx} = ax^n.$$

In words this means the derivative of the integral of a function is equal to the function itself. This holds because,

29.

$$\int ax^n \, dx = a\frac{x^{n+1}}{n+1} + C$$

,

and if we let

$$y = \frac{a}{n+1}x^{n+1} + C, then,$$

$$\frac{dy}{dx} = \frac{a(n+1)}{n+1}x^{n+1-1} + 0 = ax^n.$$

We have now given some basic rules of elementary calculus used to find the derivatives and antiderivatives for the fairly simple functions of motion with uniform acceleration seen in free fall near the earth's surface. We have seen how the calculus allows us to determine the velocity function and the acceleration function if we know the distance function. We have also learned how to work in reverse, i.e. starting with the value for the uniform acceleration we have been able to find the velocity function and the distance function, provided we know the time and velocity are both zero at the start of the motion.

Students of physics today apply the calculus to solve problems of motion early on in their studies. However, Galileo, who first discovered the law of free fall, did not have the calculus at his disposal to derive one function from another, and worked with the mathematics of proportion which he had learned from Euclid's "Elements". Galileo did fundamental experiments to learn the relationships of motion under uniform acceleration. In his book, "Two New Sciences", he gave his final theory of motion, but did not discuss in detail the experiments that led to his laws of motion. Recently modern research into his working papers disclosed the key experiments which gave Galileo his profound insights into motion with uniform acceleration. This research has been done by Stillman Drake who published Galileo's methods over the last decades of the 20th century. In the next chapter we will spend some time in seeing how Galileo actually worked out his laws. This will give us a deeper understanding of motion under accelerating forces, which will prepare us for the work of Newton on the accelerated motion of the planets in their elliptical paths.

As a concluding note for this chapter we have a few remarks on the calculus. The subject today is the first branch of higher mathematics known as analysis. We will need only the elementary aspects of the calculus in our work, nonetheless, many have difficulty with the concepts of the calculus, since they are different from the ideas used to develop geometry, trigonometry, and algebra. Much of the difficulty comes in the demonstrations of the numerous theorems regarding limits. We have omitted demonstrations of these theorems, and appeal to the intuitive ideas involved in instantaneous velocity. The 17th century mathematicians worked with intuition and did not prove many of their propositions. Some of the proofs did not come for many years attesting to their difficulty. The limit concept which we have developed intuitively was refined, and that led to the demonstrations of the validity of the derivative and the integral. We have found that many of the difficulties on first encounters with the calculus can be overcome by rereading the discussions, or by consulting one or more of the references given in the bibliography. Also helpful is for the reader to work problems by finding the derivatives and integrals of various functions of the form, $ax^n + bx^m$, etc., using the rules we developed in this chapter.

We shall see it is the calculus which will make the motion of the planets comprehensible. It will ultimately allow us to answer the question posed to Newton: what path or paths will planets or comets follow if it is attracted to the sun by a force that diminishes by the square of the distance of the planet from the sun?

Bibliography for Chapter 41

1. Kline, Morris, "Calculus: An Intuitive and Physical Approach" John Wiley and Sons, Inc., 1967.

2. Anton, Howard, "Calculus with Analytic Geometry",John Wiley and Sons, Inc., 1980.

3. Weisbecker, H., Chief Editor, "The Calculus Problem Solver", Reasearch and Eduction Association, New York, 1974.

4. Edwards, C.H., "The Historical Development of the Calculus", Springer-Verlag, New York, 1979.

Chapter 42
Galileo's Methods

Among Galileo's chief contributions to science were his ideas of how scientific knowledge should be obtained and analyzed. In his age the methods of Aristotle were taught. The Aristotelians believed nature should be studied in a qualitative manner. What mattered most were the causes of phenomena as Aristotle had considered them. All change was attributed to the loss or gain of one of the four basic elements which made up the world. These elements were earth, air, water, and fire. Fire produced heat and dryness; water produced coldness and wetness; air produced lightness, and earth heaviness.

Causes were of four types. The *material* cause determined of what objects were made. The design or shape of objects was determined by the *formal* cause. The process of change was due to the *effective* cause. The *final* cause was the purpose behind the change. For example, the final cause of a tree is the acorn, because it is the purpose of the acorn to become a tree. Determining these causes for the things in the world in a qualitative manner was Aristotle's approach to the study of nature.

In regard to the study of motion Aristotle taught that the natural motion of heavy objects, that is objects made predominantly of earth with little air or fire, was to move to the center of the earth. It was natural that the heavier the object the faster its natural motion, what we would call its acceleration, toward the center of the earth. Another form of motion is unnatural or violent motion. If a horse pulls a cart, the motion of the cart occurs because the horse exerts a pulling force causing the cart to move in an unnatural way. The cart's motion would cease immediately upon cessation of the horse's pulling force. An arrow shot into the air experienced unnatural motion in so far as it moves horizontally above the earth's surface. If the air itself did not push the arrow along, an unnatural motion, the arrow would fall perpendicular to the earth's surface, its natural motion.

Aristotle developed these ideas as a philosopher and did not advocate experimentation to test his ideas. Up until the 17th century European science accepted Aristotle's ideas with very little challenge, and virtually no experimentation was done when it came to the study of motion. As for the motions in the heavens Aristotle argued, as we have mentioned before, that the planets and stars were attached to crystalline spheres which rotated about the stationary earth and were moved by divine power.

By the 17th century the new ideas of Copernicus and others were debated, and Galileo in his early studies realized that a new method for progress in science was required. Sir Francis Bacon in England was another that challenged the Aristotelian view.

Galileo came to believe there were mathematical laws behind phenomena such as the motion in freely falling objects, or the motion of pendulums, or the motion of balls rolling down inclined planes. If these phenomena could be studied with impediments to their motion removed, or reduced to insignificance, then the scientist could learn the mathematical law which determines the motion. The mathematical relationships Galileo sought were idealized in the sense that in nature there are impediments that obscure the underlying relationships. Thus, it is that feathers float slowly to the ground while a brass ball falls quickly. In order to find the mathematical laws of nature the scientist had to create ingenious experiments where the impediments were removed or minimized.

In the study of free fall Galileo had to invent methods which would allow him to record short time intervals accurately. He could not directly measure the velocity of falling or rolling objects to determine that velocity is directly proportional to time in motion with uniform acceleration. But he could measure distances and time intervals and from his results deduce the relationships involving velocity.

Galileo's approach was quantitative. Measurements of distance and time became an essential part of scientific investigation. This meant an abandonment of Aristotelian precepts and a new start with as few preconceptions as possible. Of course, in a sense this is impossible as all scientists start out with some hypotheses on how their experiments might turn out, and often experiments are designed to test a given hypothesis.

Galileo abandoned the notion that rest is the preferred state for matter and that matter seeks a resting place at the center of the earth. Galileo recognized that reasoning alone without confirmation by experiment could be fallacious, and he even had to revise some of his own hypotheses once experimental data failed to confirm them. But once a correct mathematical relationship was confirmed, such as the law that distance in free fall is directly proportional to the square of time elapsed, then by mathematics alone new relationships could be derived that would allow the scientist to make predictions about other physical quantities. Thus, Galileo could derive mathematically that velocity in free fall is directly proportional to the time elapsed without ever having measured velocity.

Galileo believed the physical world made sense and that mathematics was the language whereby the relationships of nature could be understood. In the previous chapters we presented arguments from the calculus to support Galileo's law of free fall. Mathematically we argued, there is such a thing as instantaneous velocity, and that it can be calculated by the derivative. We argued that time could be divided in smaller and smaller intervals, as small as we wished. Also, we argued distances can be divided by smaller and smaller increments, so that in the limit as the increment of time approaches 0 we find the instantaneous velocity.

But do such arguments really apply to the physical world? Today we have the work of atomic physicists who have shown that extremely small distances lead to the inside of atoms, and the laws of nature at the atomic level given by the quantum theory are far different from the relationships given in the formulas for free fall. Thus our arguments are not valid for the physical reality of the atom. Nonetheless, for velocities, times, and distances encountered in everyday experiences the world does follow the laws of motion and force originally derived by Galileo and Newton. Thus, Galileo's scientific approach has been vindicated as has the work of Newton and the use of the calculus to solve problems of motion.

When Galileo thought about the problem of free fall he recognized the interfering problem of air resistance as we have discussed. If he used heavy objects, he also recognized air resistance played very little role in the experimental results. He wrote, *"Having observed this I came to the conclusion that in a medium totally devoid of resistance all bodies would fall with the same speed."* [He means two objects dropped at the same time would acquire velocity at the same rate.]

This is an example of how Galileo approached the problem in its idealized state. He imagines the problem without interfering factors so he can find the essential behavior of motion under study to find its mathematical law. Experiments performed by the mind are often called 'thought experiments', and brilliant thinkers like

Galileo have often discovered their insights into the behavior of nature through them.

Galileo also divorced himself from the notion of cause so important in Aristotle's theory. It was Galileo's task to find the fundamental quantitative relationships which describe how the motion occurs without giving the cause for the motion. If he were told the cause of free fall is gravity which makes all objects move to the center of the earth, he would reply, *'what do we know of gravity? Only that it gives this or that behavior to the motion of objects.'* For Galileo the scientist had no business speculating into the final causes of phenomena. His work was to be descriptive and quantitative, giving mathematical relationships in a manner so that the results of future experiments could be predicted. His program and his methods are given in his great work, "The Two New Sciences" of 1638. He concluded this work with the following remark,

"so that we may say the door is now opened, for the first time, to a new method fraught with numerous and wonderful results which in future years will command the attention of other minds."

We will now take up our study of how Galileo arrived at the three major rules which apply to objects in free fall, or objects under uniform acceleration with no impediments to their motion. We know these three rules by their equations in the metric system: $s = 4,9t^2$, $v = 9.8t$, $a = 9.8$. Galileo used the mathematics of proportion to express these relationships as we shall see. He would state the relations as, (1) motion with uniform acceleration is motion with a constant value for the acceleration such that equal increments in velocity are acquired in equal intervals of time, (2) velocity in uniform acceleration is directly proportional to time, (3) distances moved by objects with uniform acceleration are directly proportional to the squares of the times for the distances.

To perform his studies of motion with uniform acceleration Galileo used polished inclined planes. He did this to confirm earlier studies he had done using pendulums, which also exhibit motion with uniform acceleration. We will begin our work with inclined planes because the mathematical analysis is a bit simpler than pendulum motion. If the inclined plane is polished and adjusted to the right angle the plane offers very little impediment to motion. See Figure 42-1.

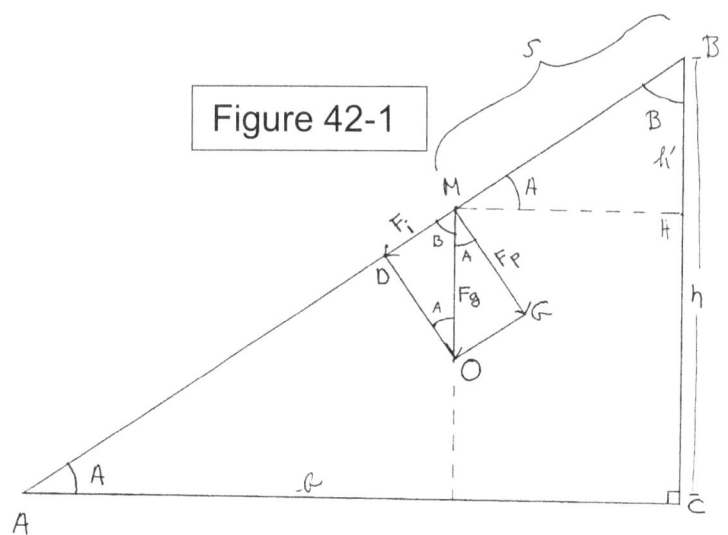

In the figure AB is the inclined plane with AC a horizontal surface on the earth. The height of the plane is CB = h. We see CB is perpendicular to AC. Length AC is designated as b, angle BAC is angle A. We note, sin A = h/AB.

Imagine a spherical ball placed at B on the inclined plane. The force of gravity, which causes uniform acceleration, acts on the ball along the vertical direction BC toward the center of the earth. We will analyze what is happening at a typical point M as the ball rolls down the plane in the experiment. The force of gravity acts along line MO which, if extended, would be perpendicular to AC. Since the ball is rolling down the inclined plane there must be a component or part of the gravitational force acting along the surface of the inclined plane. The scientist says the force can be divided into two components perpendicular to each other. The first component is Fp which is perpendicular to the inclined plane and is represented by line MG in the figure.   The second component is Fi along the inclined plane represented by line segment MD in the figure. The lengths of MG and MD are such that the force of gravity represented by a single line, MO, is the diagonal of a rectangle with MD and MG as its sides.   Although Galileo did not represent the force of gravity in this way, Newton did recognize that forces could be resolved into components, which obey such a rectangular law. In other words, a force of magnitude Fg, represented by line MO can be resolved into two perpendicular forces Fp and Fi such that MO is the diagonal of a rectangle with sides MD and MG, where MD represents Fi and MG represents Fp. Mathematically this means we can use the Pythagorean theorem and write,

$$(Fi)^2 + (Fp)^2 = (Fg)^2.$$

Galileo did find that the uniform acceleration is directly proportional to the force along the direction of motion. This means the component of force Fi along the inclined plane does cause the ball to move with uniform acceleration. We designate angle ABC as angle B. Then, angle A + angle B = 90°. We draw DO parallel to MG so angle MDO is a right angle.   Because MO is parallel to BC, angle DMO equals angle B, and because triangle MDO is a right triangle DOM is equal to angle A. Then,

1. $\sin DOM = \sin A = Fi/Fg$, or, $Fi = Fg(\sin A)$

The forces which cause uniform accelerations are directly proportional to the accelerations. This is one of Newton's laws of motion. If we denote a_i as the acceleration caused by force Fi, and a_g the acceleration caused by Fg, then,

2. $a_i = a_g \sin A$. Since a_g is the acceleration which would act if there were not an inclined plane between M and the earth AC, the value of a_g is 9.8 m/s^2. Thus the value of the acceleration down the plane is less than 9.8 m/s^2, since sin A is less than 1.

We note that this formula is an idealization since there is some friction which impedes the rolling ball, and some part of Fi is required to rotate the ball as it rolls.

Nonetheless, this formula is a good approximation for the acceleration down the inclined plane.

3. Using the calculus, $a_i = dv/dt$ = 9.8 sin A, where we replace a_g by its value 9.8 in the metric system. Using differentials, $dv = 9.8 sin A\, dt.$

    We find the antiderivative,

$$\int dv = \int 9.8\sin A\, dt = 9.8\sin A \int dt.$$

v = 9.8(sin A)t +C. At *t* = 0, *v*= 0, so *C* =0.

v = 9.8t sin A.

We have followed the rule for integration in the last chapter. We know

ds/dt = v=9.8t sin A.

4. Using differentials, *ds = 9.8t sin A dt.*

$$\int ds = 9.8\sin A \int t\, dt,$$

s = 9.8 sin A (t²/2) + C = 4.9t²sin A + C.

When *s*= 0, *t* = 0, so *C* = 0. Then

s = 4.9t²sin A.

This is the distance formula for uniformly accelerated motion down an inclined plane.

We recall sin A = AB/h . As the ball rolls down the plane we can consider the length of the plane traversed after *t* = 0 to be length *s*. We can find the numerical value of *s* by knowing angle A, and the time *t* since the start of the motion using the above distance formula. We also can consider the height of the plane as varying with *s*. In Figure 42-1 for any position of the ball, say M, *s* = BM, and the height of the plane from the perpendicular from M to BC equals BH which we designate by h'. Because MH is parallel to AC angle BMH = A, so sin BMH = sin A = h'/s, or,

s = h'/ Sin A .   Using s = $4.9t^2 sin$ A,

  5. h'/ sin A = $4.9t^2 sin$ A, or, h' = $4.9t^2(sin^2$ A)

We can solve this equation for t, which will tell us how time is a function of the vertical height the ball has traversed. Remember h' is height BH in Figure 42-1, the vertical height when the ball has rolled distance s = BM down the plane after the ball starts from point B.

  6.

$$t^2 = \frac{h'}{4.9 \sin^2 A},$$

$$t = \frac{\sqrt{h'}}{2.214 \sin A}.$$

This formula tells how many seconds it will take the spherical ball to roll down an inclined plane of height h', measured in meters, and which has an inclination of angle A with the horizontal.

  Let us compare two inclined planes with different angles of inclination as shown in Figure 42-2.

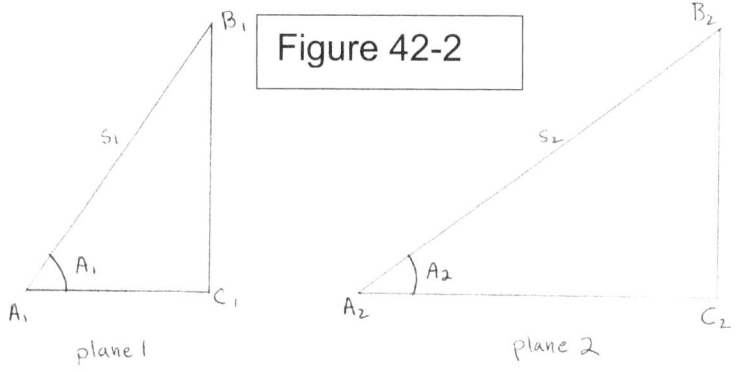

Figure 42-2

plane 1   plane 2

545

Plane 1 has a steeper slope than plane 2 with angle A_1 > angle A_2. We now ask what is the velocity at the bottom of the two planes for spherical balls released from the same height, h? That is h = B_1C_1 = B_2C_2. Using the result of step 3,

7. $v_1 = 9.8t \sin A_1$, and $v_2 = 9.8t \sin A_2$, where v_1 is the velocity on plane 1, and v_2 for plane 2.

8.

$$v_1 = 9.8 \frac{\sqrt{h}}{2.214 \sin A_1} \sin A_1$$

$v_1 = 4.426\sqrt{h}.$

$$v_2 = 9.8 \frac{\sqrt{h}}{2.214 \sin A_2} \sin A_2$$

$v_2 = 4.426\sqrt{h}$

We see $v_1 = v_2$, so the instantaneous velocities at the bottom of the two planes are the same, if the heights of the planes are equal, even though the angles of inclination are different one from another. Note lengths s_1 and s_2 are unequal. Note also that although the magnitudes of the velocities are equal their directions are different. The magnitudes of the velocities depend on the square root of the height multiplied by the constant 4.426, which is based on the metric system. The result is surprising, since in inspecting the two planes in Figure 42-2, our intuition suggests a spherical ball rolling down the steeper plane would have the faster velocity, if both balls start from the same height above the horizontal.

Galileo did reach this result using the mathematics of proportion. We used results we obtained from the calculus and our analysis of Figure 42-1, which introduced Newton's notion of resolving the gravitational force of the earth into two perpendicular components. We will give Galileo's statement of the result (in translation) from the 1655 edition of "The Two New Sciences". He did not include it in his original edition, but had seen to it that it would be added to later editions, which were published after his death. Galileo wrote, *"The degrees of speed acquired by a moveable in descent with natural motion from the same height, along planes inclined in any way whatsoever, are equal upon their arrival at the horizontal, all impediments being removed"*.

Galileo was the first person to discover this phenomenon. The common view before Galileo's work was that the greater velocity would be obtained at the bottom of planes with greater inclination.

We will now consider how Galileo made the discovery of the law of free fall itself, which we know as, $s = 4.9t^2$, in the metric system, where 4.9 is one half the uniform acceleration of gravity near the earth's surface, 9.8 m/s^2. In words the law states that the distance traversed in free fall is directly proportional to the square of the time elapsed.

Recent historical research into Galileo's own notepapers and drawings from the period of 1603-1609 by Stillman Drake in the 1970's revealed new information on Galileo's actual procedures. Drake published his account of these new findings in his books, "Galileo at Work", "Galileo: Pioneer Scientist", and in scientific journals. Before Drake's researches the source for Galileo's methods had been what Galileo wrote in "The Two New Sciences". We have learned this book was written in the 1630's almost 20 years after the original work was completed. The delay in writing his studies of motion for publication is due to Galileo's intense interest in working with the telescope after 1609.

Although all of Galileo's published work, letters, and other manuscripts had been available for many years, there were loose notepapers and drawings which were incomplete, and had remained unpublished. It was this material which Stillman Drake studied and analyzed. The conclusions and final results are seen in the "Two New Sciences", but the complete sequence of Galileo's discoveries was not given in that book.

Galileo had discovered the basic law with pendulum experiments before his work with inclined planes. We will, however, look into the inclined plane work first. We will study an experiment Galileo performed on motion down an inclined plane given in the notes analyzed by Stillman Drake. Galileo was studying the mathematical relationship between the velocity acquired by a ball rolling down an inclined plane and the distance traveled by the ball. In 1603 there is evidence that Galileo's hypothesis was that the velocity acquired is directly proportional to the distance traversed, $v = ks$, where k is a proportionality constant, or as Galileo would have written, $v_1 / v_2 = s_1 / s_2$.   We know this relation is incorrect as velocity is proportional to time not distance in free fall. As noted previously, the velocity acquired by objects in free fall is so fast that Galileo could not measure velocity directly. He developed the idea of rolling spherical balls down a polished wooden inclined plane. The inclined plane would slow the rate at which velocity is acquired, and give Galileo the chance of measuring distances traversed over short but equal time intervals. However, Galileo had to devise a means of measuring short intervals of time so they would be equal.

Drake shows how Galileo did this in one of his manuscript papers from 1604. Galileo used the beat of music for his time intervals. A trained musician can sense the correct beat of a musical piece within 1/64 of a second. That is, if the beat to beat interval of a musical piece is off by more than 1/64 of a second, the trained ear of the musician can sense it. Galileo's father was a musician and maker of musical instruments, and Galileo was undoubtedly trained in music to some extent in his youth.   Galileo could then count the beat of a musical piece, which has about ½ to 1 beats per second, and conclude the beat to beat interval was highly consistent and essentially equal. This would be true even if he did not know the actual interval of time for the beat to beat

549

interval. We must remember accurate clocks for measuring short intervals of time like a second had not been invented in Galileo's time.

Galileo placed frets of catgut string across the groove of his polished inclined plane. The string caused the ball to make clicks as it rolled over each string in its path. The positions of his frets could be adjusted. He then had a musical piece played during the release of the ball while he counted out the beat to see if it would coincide with the clicks the rolling ball made as it moved down the inclined plane. He then adjusted the frets until the sound of the clicks coincided with the beat of the music. Once this is accomplished then the interval of time between clicks is considered to be equal. Does such an experiment produce reproducible results? Stillman Drake repeated just such an experiment, and found he could indeed produce valid results to a high degree of accuracy, if the angle of inclination of the inclined plane is properly adjusted.

With such experiments Galileo verified the basic law of uniform acceleration that the distance traversed is directly proportional to the square of time. Later, using the mathematics of proportion, he could then find the correct relationship between velocity and distance and the relationship between velocity and time.

For his experiment Galileo made a ruler marked in 60 parts, each part was called a punti. In studying this ruler Stillman Drake determined that 1 punti equals .00094 meters in the metric system. The inclined plane used by

Galileo had a height of 60 punti and a rolling length of 2,000 punti. This means if Galileo's plane is the one depicted in Figure 42-1, AB = 2,000 punti, and CB = 60 punti. Then sin A = 60/ 2,000 = .030, or angle A = 1.72°. Obviously the inclined plane is much closer to the horizontal than the one drawn in Figure 42-1. Drake points out that if an inclined plane is made with angle A greater than 2° or less than 1.5°, it is very difficult to adjust the frets of string to get the same time intervals between clicks by Galileo's method. For planes with angle A greater than 2° the rolling ball acquired velocity too rapidly to get reproducible results with multiple trials, and with planes with angle A less than 1.5°, the ball slows down too much after hitting the string frets. Thus, Drake confirmed Galileo's original work that the angle of inclination of the plane was critical for Galileo to get reproducible results so that he could demonstrate the law of motion with uniform acceleration.

In table 1 following we give Galileo's results which Drake found in Galileo's manuscript notes. Note again that Galileo assumes the equality of the time intervals between clicks for his inclined plane. In the distance column s_1 is the distance from 0 punti to the sound of the first click. s_2 is the distance from 0 punti to the sound of the second click, etc. t_1 is the time interval from 0 punti to the sound of the first click, t_2 the second click etc.

	Distance (punti) accumulated	Time
s_1	33	t_1
s_2	130	$2t_1 = t_2$
s_3	298	$3t_1 = t_3$
s_4	526	$4t_1 = t_4$
s_5	824	$5t_1 = t_5$
s_6	1,192	$6t_1 = t_6$
s_7	1,620	$7t_1 = t_7$
s_8	2,104	$8t_1 = t_8$

The first thing Galileo noticed from these results is,

$$\frac{s_1}{s_2} = \frac{33}{130} \cong \frac{1}{4}, while, \frac{t_1}{t_2} = \frac{1}{2}.$$

This told Galileo a simple direct proportion between distance and time is not correct for uniformly accelerated motion. In his work with pendulums, Galileo had found that when,

$$\frac{s_1}{s_2} = \frac{1}{4}, then, \frac{t_1 t_1}{t_2 t_2} = \frac{(t_1)^2}{(t_2)^2} = \frac{1}{4}.$$

This means that if the times for distances traversed in uniformly accelerated motion are squared, then they are in the same ratio as the distances, in this case 1/4. Let us calculate the ratio of distance s_2 to s_3 from Galileo's table.

$s_2/s_3 = 130/298 \cong 4.33/9.33 \cong 4/9$.

We then take the ratio of the square of t_2 to the square of t_3 .

$$\frac{(t_2)^2}{(t_3)^2} = \frac{(2t_1)^2}{(3t_1)^2} = \frac{4}{9}.$$
From the table $t_2 = 2t_1$, $t_3 = 3t_1$.

We find again that the experimental data shows the ratio of any two of the distances is in direct proportion with the ratio of the times squared for the respective distances . The very slight error is within experimental error.

We should point out that Stillman Drake in his book, "Galileo: Pioneer Scientist", published in 1990, argues that Galileo originally treated the distances as a measure of acquired speeds, that is, he thought distances were proportional to speeds, but after his experiments with pendulums, he adopted the correct principle that distances were directly proportional to the square of the times for motion with uniform acceleration.

We will now rewrite the table giving the actual and ideal results and adding a column for the time squared.

	Actual Distance	Ideal Distance	Time elapsed	Time Elapsed squared
s_1	33	$(33)(1)^2 = 33$	t_1	$(t_1)^2$
s_2	130	$(33)(2)^2 = 132$	$2t_1$	$4(t_1)^2$
s_3	298	$(33)(3)^2 = 297$	$3t_1$	$9(t_1)^2$

s_4	526	$(33)(4)^2 = 528$	$4t_1$	$16(t_1)^2$
s_5	824	$(33)(5)^2 = 825$	$5t_1$	$25(t_1)^2$
s_6	1,192	$(33)(6)^2 = 1,188$	$6t_1$	$36(t_1)^2$
s_7	1,620	$(33)(7)^2 = 1,617$	$7t_1$	$49(t_1)^2$
s_8	2,104	$(33)(8)^2 = 2,112$	$8t_1$	$64(t_1)^2$

When the table is written in this form we see in general terms,

$$\frac{s_n}{s_k} = \frac{(t_n)^2}{(t_k)^2}$$

, where s_n and s_k are any two distances, and t_n and t_k are the respective elapsed times. For example, consider the ratio of s_7 to s_2 where $n = 7$, and $k = 2$.

$$\frac{s_7}{s_2} = \frac{(7t_1)^2}{(2t_1)^2} = \frac{49}{4} = \mathbf{12.2500.}$$

Using Galileo's value for s_7, s_2, 1,620/130 = 12.4615. In the ideal column, $s_7 = 1,617$, $s_2 = 132$.

$s_7/s_2 = 1,617/132 = 12.2500$.

By using proportions Galileo did not have to know the numerical value for the time interval between the clicks as the ball rolled down the inclined plane. He just had to know that the intervals were equal. The results of this experiment were generalized into the law for free fall, or the law of motion with uniform acceleration. Galileo gave formal expression to the law in "The Two New Sciences" as proposition II, Theorem II, in the chapter called 'Third Day'. He wrote,

"If a moveable descends from rest in uniformly accelerated motion, the spaces run through in any time whatever are to each other as the duplicate ratio of their times; that is are as the square of the times".

If we turn again to table 2, we see that in the column marked, 'Ideal Distance', the entries are a succession of squares multiplied by the first distance, 33 punti.   Thus we have $1^2, 2^2, 3^2, 4^2, 5^2, 6^2, 7^2$, and 8^2. After squaring, the sequence of numbers multiplied by 33 is: 1, 4, 9, 16, 25, 36, 49, and 64.   This sequence confirms that distance traversed is proportional to the square of time as we saw in the example of the ratio of s_7 to s_2.

Galileo in the "Two New Sciences" notes in a corollary to the proposition on the law of free fall that such a sequence will be seen if the time intervals between the distances are equal. In Galileo's words, *"the spaces will be to one another as are the odd numbers from unity, that is, as 1, 3, 5, 7,...".*

What Galileo meant was the series of square integers, 1, 4, 9, 16, 25, 36, 49, and 64, can be formed by adding the series of successive odd numbers to the preceding square number. Thus,

$$0 + 1 = 1$$
$$1 + 3 = 4$$
$$4 + 5 = 9$$
$$9 + 7 = 16$$

16 + 9 = 25

25 + 11 = 36

36 + 13 = 49

49 + 15 = 64.

We found this sequence in the work of Pythagoras in chapter 1 and Figure 1-1, where the series of odd integers are the gnomens for the square numbers.

Galileo did not mention the experiment with the bronze spherical balls and the catgut frets in "The Two New sciences". Instead he described a method of time measurement using water flowing from a container. In this method he would watch the ball roll down a distance and simultaneously release water from a narrow orifice from his container. The weight of the water released was proportional to the time interval for the distance. He then adjusted the distances that the ball rolled until he had integral multiplies of the weight of water for the first distance. This method of measuring time intervals has also been studied in the modern era to see if reproducible results occur with repeated trials. Thomas Settle in 1960 showed this method does produce reproducible results. Once again this method confirmed the law of uniform acceleration.

Galileo used this method of time measurement for his pendulum experiments, which was his original method of discovering the law of free fall. We can construct a simple pendulum as shown in Figure 42-3.

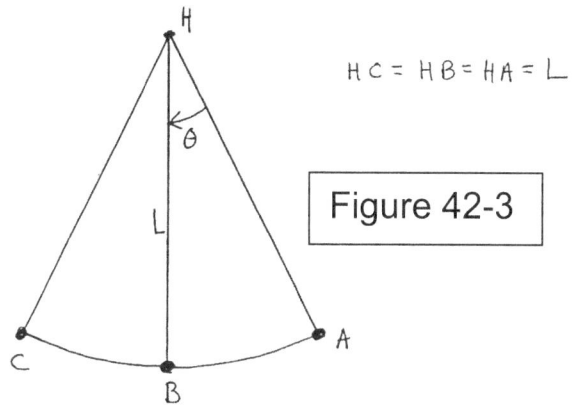

H C = H B = H A = L

Figure 42-3

The bob marked B, is placed on the end of a string of length L, which is attached to a hook and suspended.   Because of gravity the initial position will be toward the center of the earth along line HB.   We can then pull the bob to position A, which means the string will have moved through angle θ (theta).   If we then release the bob, it will commence swinging through arc ABC, and then back through arc CBA.   The time for the bob to move from its starting point at A to C and back to A is called 1 period of the pendulum.   If there were no friction from the air to impede the pendulum's motion, the pendulum would continue to swing back and forth from A to C to A.   The cause for this motion is the force of gravity which we know causes uniform acceleration in the direction perpendicular to the earth's surface taken as a horizontal plane.   Later in this book we will have occasion to use the calculus to derive the equation for the motion of the bob.   We will also see how to mathematically derive the period of the pendulum in terms of the pendulum's length, L, and the value for the acceleration of gravity, designated by g, which we know is approximately 9.8 m/s^2 near the earth's surface.

As we indicated earlier Galileo made observations of pendulum motion and found that the period of a pendulum, designated by T, is independent of the amplitude of the pendulum's swing measured by the value of θ. The period does depend on the length of the pendulum as we shall see. Thus we could choose any relatively small angle θ, find its period, then increase or decrease θ, the angle of the string from the vertical at the time of release, and find the period is the same. Actually, it turns out the period of a simple pendulum does depend on θ, but that for angles less than 17° the dependence is so slight that it cannot be detected except by very accurate clocks. In Galileo's time the dependence could not readily be detected.

Galileo discovered that the periods of simple pendulums did not depend on the value of θ, the amplitude, by observing the motion with variable values of θ, and timing the periods with his pulse rate. He continued his researches into pendulum motion, which resulted in his discovering another law of pendulum motion. This law states the period of a pendulum is directly proportional to the square root of the length of the string, L. To aid us in our understanding of Galileo's work with pendulums, we will give the modern equation for the period of the pendulum using the designation g for the uniform acceleration of gravity. Later in our work with Newton we will show how this equation can be derived from a consideration of the earth's gravitational force.

$T = k\sqrt{L}$, T is the period, L, the length of the string, and k is the proportionality constant. The value of k is $2\pi / \sqrt{g}$.

$$T = \frac{2\pi\sqrt{L}}{\sqrt{g}}.$$

Thus, g is the uniform acceleration of gravity, which we know is approximately 9.8 m/s2.

Stillman Drake points out that Galileo could have derived the law of free fall (the law of motion with uniform acceleration) by doing only inclined plane experiments. However, careful analysis of Galileo's notes revealed to Drake the actual sequence of discovery was that Galileo first discovered the law with pendulum experiments. Drake found Galileo had made a funnel from which water flowed at a steady rate, a device we shall call a water clock. The funnel was filled to a mark, the experiment was started by releasing an object such as a brass ball to fall to the ground in free fall, or in another experiment the bob of the pendulum was released. Simultaneously the funnel was opened and the water allowed to flow into a collection container. When the brass ball hit the floor, or when the bob of the pendulum hit a block placed so that θ= 0 ° (the pendulum had moved through ¼ of its period), then the funnel opening was closed. The water was carefully weighed, and the weight of the water was taken as a measure of the time elapsed for the motion.

Galileo initially wrote down the time as 'grains of water'. Stillman Drake determined from Galileo's notes for Galileo's water clock, 1,472 grains of water flowed in 1 second. Galileo had no method of determining the length of a second.

In one of his first experiments Galileo released a spherical brass ball under free fall to fall a distance of 4,000 punti (4,000 punti x .00094 meters/punti = 3.76 meters). He found this took 1,337 'grains of water'; that is, 1,337 grains of water ran out of the funnel from the time of release until the ball hit the ground. He then released the ball to fall a distance of 2,000 punti (1.88 meters), and it took 903 'grains of water'.

Although he had the water clock as a time keeper, he also wanted to use a pendulum. He placed a block at position B in Figure 42-3 for the pendulum, so the bob after being released from position A, would strike the block after moving through arc AB. As noted this is ¼ of the pendulum's period. The water clock would be used to measure the time for the bob to move through arc AB. Galileo knew the length of the arc, or angle θ, made no difference in the time for the movement. As long as θ is less than 17° the results would be very reproducible.

The time depended on the length, L, of the pendulum. He tried to find the length of the pendulum for which ¼ of its period equals 903 'grains of water', the time it took for a ball in free fall to fall 2,000 punti. It may have taken Galileo many trials and changes of length, L, to finally find the value for L so ¼T = 903 'grains of water'. Galileo's notes indicate the value for L was 1,590 punti (1.495 meters).

Stillman Drake states the modern theory of the pendulum verifies that ¼T for a pendulum of length 1,590 punti (1.495 meters) is 903 'grains of water. (903 'grains of water' x l second/1,472 'grains of water' = .613 seconds). Let us see whether or not we can demonstrate that if L = 1.495 meters, g = 9.8 m/s², then ¼T = .613 seconds.

9.

$$T = \frac{2\pi\sqrt{L}}{\sqrt{g}} = \frac{1}{4}T = \frac{\pi\sqrt{L}}{2\sqrt{g}} =$$

$$\frac{(3.14)(1.22)}{(2)(3.13)} = .612 \; sec \, onds.$$

This result is essentially identical to the 903 'grains of water', which is equivalent to .613 seconds.

We can also use the modern equation for free fall in the metric system, s = 4.9t², to see what the time of fall should be for an object falling 2,000 punti = 1.88 meters.

10. $s = 4.9t^2$, or $t^2 = s/4.9$.

$t^2 = 1.88/4.9 = .384$, or, $t = .619$ seconds.

We see the time of free fall is actually longer than Galileo measured it. Using the value, $t = .619$ seconds for ¼ T, we can find the correct length, L, for the pendulum.

11.

$$\frac{1}{4}T = \frac{\pi\sqrt{L}}{2\sqrt{g}}.$$

We solve for L when $g = 9.8$ m/s², ¼T = .619 seconds.

$$\sqrt{L} = \frac{(\frac{1}{4}T)(2\sqrt{g})}{\pi} = \frac{(.619)(2\sqrt{9.8})}{3.14},$$

$$L = \frac{(.383)(4)(9.8)}{3.14^2} = 1.52 \text{ meters.}$$

L = 1.52 meters x 1 punti/.00094 meters = 1,621 punti.

This would be the correct length of a pendulum such that one fourth of its period is equal to .619 seconds, the time for a brass ball to fall 2,000 punti = 1.88 meters.

Galileo at this point had found the times for two distances of free fall, 4,000 punti and 2,000 punti. He had expressed times in two systems of measurement.

First was the water clock where time is measured in 'grains of water'. He found 1,337 'grains of water' for a fall of 4,000 punti, and 903 'grains of water' for a fall of 2,000 punti. His second clock was the pendulum where the times are expressed by the length L of the pendulum. For the fall of 2,000 punti, $L = 1,590$ punti (we showed that L should be 1,621 punti according to our modern formula for free fall). He next wanted to find the pendulum length for which $\frac{1}{4}T = 1,337$ 'grains of water', the time for a fall of 4,000 punti.

His method was first to find the pendulum length for a time of $\frac{1}{2}$ of the 1,337 'grains of water'; that is to find a pendulum length for a time of 668.5 'grains of water'. He then could double the pendulum length and find the time for that pendulum with the water clock. If it did not equal 1,337 'grains of water', he would double the length again. The results of this process are best seen in a table. We will give in the first column the series of pendulum lengths Galileo tried. Note each successive length is double the preceding one. The corresponding times for $\frac{1}{4}T$ is in the second column. We also convert Galileo's units into the metric system. The series of pendulum lengths are labeled a through f, and the series of times are labeled g through l.

Table 3

L (Pendulum length)	Time (Water clock) for ¼T
(1 punti = .00094 meters)	(1,472 'grains of water = 1s)
a, 870 punti(.818 m)	g, 668.5 'grains' = .454s s
b, 2(870) = 1,740 punti	h, 942 'grains' = .640 s
(1.64 m)	
c, 2(1,740) = 3,480 punti	i, 1,337 'grains' = .908 s
(3.27 m)	
d, 2(3,480) = 6,960 punti	j, 884 'grains' = 1.280 s
(6.54 m)	
e, 2(6,960) = 13,920 punti	k, 2,674 'grains' = 1.82 s
(13.1 m)	
f, 2(13,920) = 27,840 punti	l, 3,768 'grains' = 2.56 s
(26.2 m)	

Galileo probably proceeded in this method of doubling pendulum lengths, because so many trials were needed to find a pendulum length that had ¼T equal to a particular time measured in 'grains of water' from the water clock.

We see from the table that Galileo found the pendulum length for a time of 1,337 'grains of water'. This was the same time for free fall of 4,000 punti(3.76 meters). L was 3,480 punti(3.27meters). Using the data from the table Galileo could take the series of ratios, *a/b, b/c, c/d, d/e,* and, *e/f,* and find they are in continuous proportion with each ratio equal to ½.

a/b = 870/1,740 = 1/2 ; b/c = 1,740/3,480 = ½ , etc.

Galileo could then look at the column of times to see if these values are also in continuous proportion:

g/h = 688.5/942 = .7096 , h/l = 942/1,337 = .7046 , i/j =1,337/1,884 = .7097, j/k = 1,884/2,674 = .7046 , k/l = 2,674/3,678 = .7097 .

These results, although not exactly identical, indicate the ratios are also in continuous proportion. The average value for these ratios is .7076.   We can solve the following proportion:

.7076/1 = 1/x, x = 1/.7076 =1.413.

This maneuver is called finding the inverse ratio. That is the inverse ration of .7046 is 1.413. Galileo would recognize 1.413 as being equal to $\sqrt{2}$, since $\sqrt{2}$ = 1.414 to the third decimal place.

The data from the table then told Galileo that when he doubled the pendulum lengths the times for the periods increase by the square root of 2.   Likewise if the pendulum lengths are quadrupled, for example, from 870 to 3,480 punti, then the time for ¼T is doubled from 668.5 to 1,337   'grains of water'. Again if the pendulum length is quadrupled from 3,480 to 13,920 punti, the time for ¼T is doubled from 1,337 to 2,674 'grains of water'.

This indicated to Galileo that the lengths of the pendulums are proportional to the square of the times, since when we double the length we multiply by 2, and when we quadruple the lengths we multiply by 2x2 = 2^2 = 4. We can write, 3,480/13 920 = ¼ (We took the pendulum length, c, to its quadruple). Then squaring the times taken to doubling the times squared:

$$\frac{(1,337)(1,337)}{(2)(1,337)(2)(1,337)} =$$

$$\frac{1,337^2}{[(2)(1,337)]^2} = \frac{1}{4}$$

In symbols,

$$\frac{L_1}{L_2} = \frac{(\frac{1}{4}T_1)^2}{(\frac{1}{4}T_2)^2}, \text{ or, } L = KT^2,$$

where K is the proportionality constant in this relationship which states that pendulum length is directly proportional to the time of the period squared.

For example, if L_1 =3,489 punti, L_2 = 13,920 punti , then L_1/L_2 = ¼ = .25, and, $(\frac{1}{4}T_1)^2$ = (1,337 grains of water)^2 = 1,787,569, $(\frac{1}{4}T_2)^2$ = (2,674 grains of water)^2 = 7,150,276, and the ratio of 1,787,569/7,150,276 = ¼ =.25.

This is how Galileo found his law of the pendulum.

From our work with the mathematics of proportion we recall,

$$\sqrt{L_1 L_2} \, ,$$

will be the mean proportional between L_1 and L_2, since from chapter 1, if $a/b = b/c$, then $b = \sqrt{ac}$, where b is the mean proportional between a and c. Thus, to find the mean proportional, x, between L_1 and L_2, we write.

$$\frac{L_1}{x} = \frac{x}{L_2}, \, or, \, x^2 = (L_1)(L_2), \, or,$$

$$x = \sqrt{L_1 L_2} \, .$$

By mathematics Galileo then derived the following relationship:

$$\frac{L_1}{\sqrt{L_1 L_2}} = \left[\frac{T_1}{\sqrt{T_1 T_2}} \right]^2 .$$

This proportion states the ratio of a given pendulum length to the mean proportional between the given length and another pendulum length is equal to the square of the ratio of the period for the first length taken to the mean proportional of the period of the first length to the second length.

We can derive this as follows: we know,

$$\frac{L_1}{L_2} = \frac{(T_1)^2}{(T_2)^2} .$$

We multiply the left side by L_1/L_1, and the right side by $(T_1)^2/(T_1)^2$, which is equivalent to multiplying both side by 1.

$$\frac{(L_1)^2}{(L_1)(L_2)} = \frac{(T_1)^2(T_1)^2}{(T_2)^2(T_1)^2}.$$

We take the square root of both sides:

$$\frac{L_1}{\sqrt{L_1 L_2}} = \frac{(T_1)^2}{(T_1)(T_2)} = \left[\frac{T_1}{\sqrt{T_1 T_2}}\right]^2.$$

Galileo then tested his law of the pendulum experimentally by finding the mean proportional between two lengths, and the mean proportional of their times and determined if the predicted value agreed with the measured value.

To begin this process Galileo defined a new unit of time. He called this unit 'tempi'. One 'tempi' of time equaled the time for 16 'grains of water' to flow from his water clock.

1 tempi = 16 'grains of water' x

(1 second)/(1,472 grains of water) = .01087

seconds, or, 1/92 of a second.

He felt the unit of 'tempi' was easier to work with in his calculations than the units of 'grains of water'. He then sought to find the ratio of time j in Table 3, to the time which is the mean proportional between time j and

568

time k.   We know the mean proportional x, between j and k can be given by, $j/x = x/k$, or, $x = \sqrt{jk}$.

j = (1,884 'grains of water) x (1 tempi)/16 'grains of water) = 118 tempi.

k = (2,674 'grains of water')x(1 tempi)/16 'grains of water' = 167 tempi.

Thus,

$$x = \sqrt{(118)(167)} = 140 \ tempi.$$

Galileo then determined the length L of a pendulum with ¼T = 140 tempi.   In order to do this he used his water clock and adjusted the pendulum length until ¼T = 140 tempi. 140 tempi = 140 tempi x 16 'grains of water'/1 tempi = 2,240 'grains of water'.   He found by experiment that L = 9,840 punti (9.25 in).

He next calculated the mean proportional length, x, between d and e in Table 3.

$$x = \sqrt{de} = \sqrt{(6,960)(13,920)} =$$
$$9,843 \ punti \ (9.28 \ m).$$

The experimentally determined value of 9,840 punti, and the value determined by calculating the mean proportional, 9,843 punti, are essentially are equal. Thus by calculating the mean proportional time between two times (140 the mean proportional between 118 and 167), and then calculating the mean proportional length between the corresponding pendulum lengths (9,843 the mean proportional between 6,960 and 13,920), Galileo had found mathematically a value for the pendulum length which equals the value found by experiment. Galileo could conclude that his law of pendulum motion was experimentally valid.

He could also state the law of the pendulum in the following form: the period of a pendulum is directly proportional to the square root of its length.   This follows from the relation, $L = KT^2$, by taking the square root of each side of the equation to obtain, $T = k\sqrt{L}$, where k is a new proportionality constant.

To demonstrate we let L_1 = 6,950 punti, L_2 = 9,840 punti. From the experimental data $\frac{1}{4}T_1$ = 118 tempi, T_2 = 140 tempi. We first find the ratio of $\frac{1}{4}T_1$ to $\frac{1}{4}T_2$, which is equal to the ratio of T_1 to T_2, since,

$$\frac{\frac{1}{4}T_1}{\frac{1}{4}T_2} = \frac{T_1}{T_2} = \frac{118}{140} = .8429.$$

.   Next we find the ratio of $\sqrt{L_1}$ to $\sqrt{L_2}$.

$$\frac{\sqrt{L_1}}{\sqrt{L_2}} = \frac{\sqrt{6.960}}{\sqrt{9,840}} = \frac{83.43}{99.20} = .8140.$$

The result that these values of the two ratios are very nearly equal also confirmed that Galileo had a valid law of the pendulum,

$\sqrt{L_1}/\sqrt{L_2} = T_1/T_2$.

Other values in the table also confirm the relationship. These experimentally determined values of Galileo satisfy the law of the pendulum.

Galileo then related the times of free fall for his spherical balls, released from various distances from the floor, to the times found with various pendulum lengths. To do this mathematically he needed the value for a very important ratio. This ratio is ¼T for a pendulum of length, L, taken to the time of free fall, t, for a distance, s, equal to L, the length of the pendulum. We can write this ratio as (¼T) / L. Once again, ¼T is one fourth the period of a pendulum of length L, and t is the time of free fall over distance, $s = L$.

To find the numerical value of this ratio Galileo chose for s the value of 1,740 punti (1.64 m). He already knew from his previous experiments when a pendulum has L = 1,740 punti, the value of ¼T is 942 'grains of water'. Converted to tempi, this is, 942 'grains of water' x 1 tempi/16 'grains of water' = 58.875 tempi. He then had to determine by another experiment the time, t, for a free fall of 1,740 punti (1.64 m). To do this he used his water clock and found, t = 850 'grains of water'. Converting to tempi, 850 x 1/16

= 53.125 tempi.   With these data Galileo could

calculate the ratio,

$$\frac{\frac{1}{4}T}{t} = \frac{58.875}{53.125} = 1.1082.$$

We can use the modern theory of the pendulum to

calculate what this ratio should be.   The equation for

$\frac{1}{4}T$ is,

$$\frac{1}{4}T = \frac{\pi\sqrt{L}}{2\sqrt{g}},$$

where L is the pendulum length in meters, and g is

the acceleration of gravity in m/s^2.   We know $g = 9.8$

m/s^2 to a good approximation.

The equation for free fall in the metric system is,

$s = 4.9t^2$.   The equation for free fall can be written,

$s = \frac{1}{2}gt^2$.   We solve this equation for t.

$$t^2 = \frac{2s}{g}, \text{ or, } t = \frac{(\sqrt{2})(\sqrt{s})}{\sqrt{g}}.$$

We then can find,

$$\frac{\frac{1}{4}T}{t} = \frac{(\pi/2)(\sqrt{L}/\sqrt{g})}{(\sqrt{2}/\sqrt{g})(\sqrt{s}} =$$

$$(\frac{\pi\sqrt{L}}{2\sqrt{g}})(\frac{\sqrt{g}}{\sqrt{2}\sqrt{s}} = \frac{\pi\sqrt{L}}{2\sqrt{2}\sqrt{s}}.$$

When $L = s$, then

$$\frac{\frac{1}{4}T}{t} = \frac{\pi}{2\sqrt{2}} = \frac{3.1416}{(2)(1,414)} = 1.1107.$$

The value calculated from the modern theory of the pendulum is very nearly equal to the value Galileo determined by his experiments (1.082).   Note that the modern determination of the ratio shows its value is independent of the value of g, and that the ratio is a constant value for all lengths of the pendulum as long as $L = s$.   Galileo believed the value for the ratio which he determined, 1.1082, was constant for the pendulum lengths and distances of free fall with which he was working.   Using this ratio Galileo could calculate the times of free fall for variable distances of free fall. When Stillman Drake studied Galileo's work with the data in Table 2, Drake felt Galileo revised his table of data so the law of the pendulum (square root of L directly proportional to the period, T, holds almost exactly. By revising his table of data to conform to the law of the pendulum, Galileo was correcting for experimental errors. We will rewrite the table of data as Table 4, giving the lengths of the pendulums in punti, and the times in tempi. Each value in Table 4 will be assigned a letter as indicated. Note that the pendulum lengths in any row below the first is double the value of the length in the row above it.

Table 4

Pendulum length L (punti)	¼T (tempi)
a = 869.8125	g = 41.71
b = 1,739.625	h = 58.99
c = 3,479.25	i = 83.42
d = 6,958.5	j = 118.0
e = 13,917	k = 166.8
f = 27,834	l = 236.0

This table could be called Galileo's theoretical table for the law of the pendulum. From the data in the table the law is obeyed almost exactly. For example,

$$\frac{\sqrt{c}}{\sqrt{f}} = \frac{\sqrt{3,479.25}}{\sqrt{27.834}} = \frac{58.9852}{166.835} = .3536.$$

$$\frac{i}{I} = \frac{\frac{1}{4}T \text{ for } c}{\frac{1}{4}T \text{ for } f} = \frac{83.42}{236.0} = .3535.$$

We can now see how Galileo might have calculated the time. t, for a free fall distance of s, when s is chosen to be the length of one of the values for L in Table 4. For example, let distance $s_1 = d = 6,958.5$.   Galileo's ratio states,

$\frac{1}{4}T/t = 1.1082$. We solve for t and call the result t_1.

$t = t_1 = \frac{1}{4}T/1.1082$.

This ratio is valid when $s = L = 6,958.5$.   We use the value j from Table 4 for $\frac{1}{4}T$, since this value corresponds to $d = 6,958.5$.

$t_1 = 118/1.1082 = 106.5 \text{ tempi}$.

This result is the time for a free fall of distance $s_1 = 6,958.5$ punti.

We next calculate the time t_2 for a free fall of distance $s_2 = 27,834$ punti.   When $s_2 = L$, then, $\frac{1}{4}T = 236$ tempi.   We use Galileo's ratio to solve for t_2.

$t_2 = \frac{1}{4}T/1.1082 = 236/1.1082 = 213 \text{ tempi}$.

With the values just calculated for s_1, t_1, s_2, and t_2, Galileo could have looked at the ratios s_1/s_2 , t_1/t_2 ,

$s_1/s_2 = 6,958.5/27.834 = 1/4$, $t_1/t_2 = 106.5/213 = 1/2$.

This suggests that as the distance of free fall is quadrupled the time of free fall is doubled. Just as in the case of the relationship between pendulum length and the period of the pendulum, a direct proportion where the distance of free fall is proportional to the square of time is indicated.

Galileo could test this hypothesis with the data from Table 4. We use the above data:

$s_1/s_2 = 1/4$, $(t_1)^2/(t_2)^2 = (106.5)^2/(213)^2$
$= 11,342.25/45.369 = 1/4$.

Galileo could also verify the law of free fall (distance proportional to the square of time) for other values in Table 4.

Stillman Drake believes Galileo used a different series of calculations than we have shown to find the law of free fall. We have chosen to use the ratio $\frac{1}{4}T/t$, because using this ratio clearly shows how the law of free fall follows as a consequence of the data in Table 4. The pendulum method Galileo used to discover the law of free fall is somewhat complicated, but going through the data and all the mathematical steps gives us an insight into the process of scientific discovery, which would not be clear if only the final formulas are given. After establishing the law of free fall with the pendulum experiments, Galileo then developed his inclined plane experiments to further verify the law. We are now learning the great importance of how the experimental method is combined with mathematical analysis to discover the secrets of nature. As we have noted previously, Galileo sought laws of motion when interfering factors such as the resistance of the medium through which objects move is neglected. Galileo then has the fundamental laws of motion given mathematically. The laws are descriptive. They tell us how objects behave, but make no comment on the cause or the purpose of the motion. Many objected to this approach, but the discoveries made by Galileo, and the extensions of Galileo's work by, were the beginning of the science of physics, or natural philosophy as Newton called it.

Bibliography for Chapter 42

1. Galilei, Galileo, "Discourses and Mathematical Demonstrations Concerning Two New Sciences", translated by Stillman Drake, The University of Wisconsin Press, 1974.
2. Drake, Stillman, "Galileo: Pioneer Scientist", University of Toronto Press, 1990.

3. Drake, Stillman, "Galileo's Discovery of the Law of Free Fall", Scientific American, 228, Number 5, pages 84-92, May 1973.

4. Settle, Thomas, "Galileo and Early Experimentation", in Aris, Rutherford, Davis, and Stueverds, "Spring of Scientific Creativity", University of Minnesota Press, 1983.

5. Drake, Stillman, "Galileo at Work, His Scientific Biography", The Univeristy of Chicago Press, 1978.

6. Drake, Stillman, "The Role of Music in Galileo's Experiments", Scientific American, 232, pages 98-107, 1975.

7. Cohen, I. Bernard, "The Birth of a New Physics", Doubleday and Co., New York, 1985.

Chapter 43

Motion with Uniform Acceleration: The Meaning of Velocity

In the preceding chapters we have made great progress in understanding Galileo's new methods for studying motion. No longer is reasoning alone adequate: experiments must be done with quantitative measurements of time and distance.   Then the situation must be considered in its idealized form where no impediments to motion exist.   Galileo found that this approach led to mathematical relationships which could be expressed in the mathematics of proportion.   We also showed the modern approach to Galileo's work using the calculus to derive the fundamental equations of motion with uniform acceleration.   We found distance traversed to be directly proportional to the time squared, instantaneous velocity directly proportional to time, and acceleration to be constant.   In the case of free fall the value of the constant acceleration is known to be approximately 9.8 meters per second squared in the metric system.

Part of Galileo's approach was to derive from mathematics new information using the basic law that distance is directly proportional to time squared.   As an example of this approach consider the formula for free fall, $s = 4.9t^2$. s in meters, t in seconds, and constant 4.9 in meters/s^2.   Let us solve this equation for t:

$$t^2 = \frac{s(meters)}{4.9m\,/\,s^2}\,,\,or,\,t = \frac{\sqrt{s(meters)}}{\sqrt{4.9m\,/\,s^2}}$$

$$= \frac{\sqrt{s(meters)}}{2.214\sqrt{m\,/\,s}} = .452s$$

We take the positive square root, because time is always positive from the beginning of the motion.   We see that in free fall (or in motion with uniform acceleration) the time is proportional to the square root of the distance.   Note the time does not depend on the nature of the object; in the same time a feather falls as far as a brass ball, as long as there is no impediment to the motion.   Of course, in experiments done by dropping objects through the air, a brass ball clearly moves farther than a feather in the same time, but this is because the air significantly impedes the motion of the feather and not the brass ball.   In a vacuum both would fall the same distance in equal times, as modern experiments have demonstrated.

Galileo did not have the mathematics of the calculus to derive the relation that instantaneous velocity is directly proportional to time.   To show how Galileo determined this relation, as well as the relation between instantaneous velocity and distance, we again turn to the researches on Galileo's manuscript notes by Stillman Drake.   Drake found notepapers in Galileo's handwriting from 1604 which show the probable course of reasoning that Galileo used in understanding the velocity of objects moving with uniform acceleration.

These papers show Galileo searching for a proportional relationship between distance traversed with uniform acceleration and average velocities and acquired velocities. At this stage of his work Galileo did not think that uniform acceleration caused instantaneous changes in velocities. The arguments of philosophers for centuries before Galileo were that as objects move with uniform acceleration small incremental changes in what was called impetus resulted in the changes in velocity. The concept of impetus was not well defined. It seemed to mean the property of a moving body involving both its speed and weight. If only one body is involved, as the body acquires speed, its impetus increases in direct proportion to the speed. If two bodies are involved of different weights and moving at the same speed, the heavier body had the greater impetus because of its greater weight. Objects moving with uniform acceleration meant to the philosophers that the objects are acquiring incremental changes in their impetus. This in turn meant that for short intervals of time the objects are moving with uniform velocity before the changes in impetus occur. The motion could be represented as a summation of average velocities between small increments of distance. Galileo started by reasoning in this manner. To help understand the process we have constructed Table 1 showing units of distance and the associated times for motion with uniform acceleration.

581

Table 1

	Distance (units)	Time elapsed (units)
1st interval	1	2(1x2)
2nd interval	4	4(2x2)
3rd interval	9	6(3x2)
4th interval	16	8(4x2)

In this table the time interval for an object to move the first unit of distance is 2 units of time. Because the velocity is increasing, to move 4 units of distance requires only 2 more units of time, to move from 4 units distance to 9 units distance requires only another 2 units of time.   Galileo did not use the units of punti for distance or 'tempi' for time in his reasoning in this process.

The table is set up to agree with the law of uniform acceleration that distance is proportional to the square of time regardless of the units used for distance and time:

$$\frac{s_n}{s_k} = \frac{(t_n)^2}{(t_k)^2}, \ or, \ \frac{t_n}{t_k} = \frac{\sqrt{s_n}}{\sqrt{s_k}}.$$

For example if we take the distance from the 3rd interval to the distance for the fourth interval, we have, $s_n = 9$, $s_k = 16$, $t_n = 6$, $t_k = 8$.

$s_n/s_k = 9/16$. $(t_n)^2/(t_k)^2 = 6^2/8^2 = 36/64 = 9/16$, equal to s_n/s_k.

Galileo did not write out a table as we have done. The table is our attempt to depict the relationship of distance and time, so that the law of uniform acceleration holds for arbitrary units given in Table 1.

Galileo started by writing 4 units of distance in 4 units of time as in Table 1.   Stillman Drake states the units Galileo wrote for distance were called 'miles', and the units for time were called 'hours'.   They do not represent miles or hours in our modern units, so to avoid confusion, we will call them units of distance and units of time.

Next Galileo refers to units of velocity which he called 'degrees of speed'.   By 'degrees of speed' he meant the average velocity for the distance. He felt if one degree of speed is for the first unit of distance, then two more 'degrees of speed' are the average velocity for the second unit of distance, and three more 'degrees of speed' are the average velocity for the third unit of distance, and finally four more degrees of speed' are the average velocity for the fourth unit of distance.

Galileo meant by this line of reasoning that after 4 units of distance were traversed by an object moving with uniform acceleration, average velocity for the object is, 1 + 2 + 3 + 4 = 10 degrees of speed.   Again, this is the average velocity the object acquires in Table 1 through the second interval which takes 4 units of time.

In the third interval in Table 1 the object moves from 4 units of distance to 9 units of distance, a total of 5 units of distance.   We would expect Galileo to say that in moving from 4 units of distance one more unit to 5 units of distance the object would have 5 more 'degrees of speed' for its average velocity, and continue to acquire more 'degrees of speed' from 6 to 7, 7 to 8, and 8 to 9 units of distance.   However, the note papers Galileo wrote that in moving from 4 units of distance to 9 units of distance, (5 units of distance) the object increases only 5 'degrees of speed' so at 9 units of distance the average velocity for the motion is 10 + 5 = 15 'degrees of speed'. We must consider this an error in Galileo's logic, since, 1 + 2 + 3 + 4 + 5 = 15 degrees of speed, should be the average velocity for 5 units of distance instead of 9 units of distance.

Stillman speculates on Galileo's error in writing 15 'degrees of speed' for 9 units of distance.   Drake felt Galileo became confused by the idea that 1 'degree of speed' occurs for each unit of distance since this had been used by previous investigators.   He then added 1 'degree of speed' for each of the 5 units of distance from 4 units to 9 units of distance.   We see from this analysis that even great scientists can make errors in their work, at least in their preliminary notes.

He did not correct the error and drew a line on his paper such as shown in Figure 43-1.

Figure 43-1

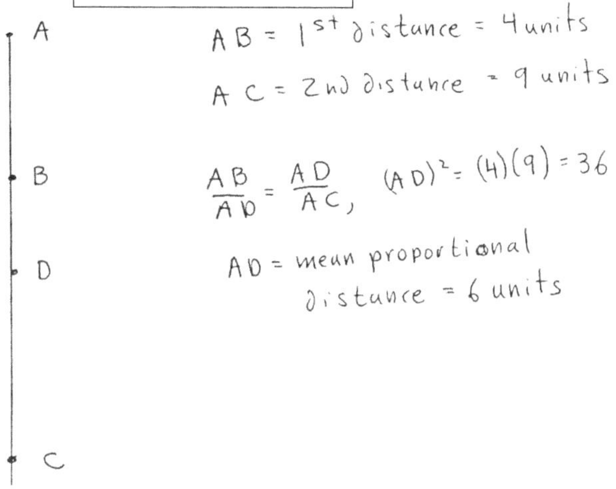

A

$AB = 1^{st}$ distance $= 4$ units

$AC = 2nd$ distance $= 9$ units

B

$\dfrac{AB}{AD} = \dfrac{AD}{AC},$ $(AD)^2 = (4)(9) = 36$

D

$AD =$ mean proportional distance $= 6$ units

C

The segments of the line represent distances moved by an object with uniform acceleration. Distance AB is the distance after the second interval in Table 1, that is, 4 units of distance which is traversed in 4 units of time. Distance AC is the distance after the 3rd interval in Table 1, and is 9 units of distance traversed in 6 units of time.   Thus distance BC equals, AC - AB = 9 – 4 = 5 units of distance. Distance AD is chosen such that AD is the mean proportional between AB and AC, that is,

$$\frac{AB}{AD} = \frac{AD}{AC}, \, or, \, (AD)^2 = (AB)(AC) = (4)(9) = 36,$$

$$AD = 6.$$

If we take the ratio of the square roots of AB and AC, we have for the ratio of the square roots of distances traversed.

$$\frac{\sqrt{AB}}{\sqrt{AC}} = \frac{\sqrt{4}}{\sqrt{9}} = \frac{2}{3}.$$

Galileo took the ratio of his speeds, which he considered to be average velocities instead of instantaneous velocities, which we recall were logically inconsistent. Speed A to B / speed A to C = 10 'degrees of speed' / 15 'degrees of speed' = 2/3.

This is the same ratio as the square roots of the distances traversed. Galileo then made the hypothesis that the ratio of average velocities for motion taken from the start of the motion to two distances equals the ratio of the square roots of the distances.

It turns out that this is the correct relationship between average velocities and distances for uniformly accelerated motion, even though Galileo had found this relationship by logical inconsistency in his reasoning.

To show this is true consider the following example in the differences in distance and the differences in average velocities in free fall using our equations in the metric system for 3 and 4 seconds of free fall:

$s = 4.9t^2$, for $t = 3$, $s = 44$ meters, av velocity = 44/3 = 14.67m/s. For $t = 4$, $s = 78.4$ m, av velocity = 78.4/4 = 19.6m/s. Ratio av velocities 14.67/19.6 = .749, ratio square roots distances $\sqrt{44}/\sqrt{78.4}$ = .749.

If Galileo had gone a step further to the fourth interval in Table 1, he might have written 16 units of distance take 8 units of time. Let us define a distance AF equal to the 16 units of distance. If we then add 1 degree of speed acquired for the 7 added units of distance from the 3rd to the 4th interval as Galileo did in going from 4 to 9 units of distance, then the degrees of

586

speed acquired at 16 units of distance would be 15 + 7 = 22 degrees of speed. The ratio of is average speeds would be:

Average velocity from 0 to 9 units distance / average velocity from 0 to 16 units distance = 15/22 .

The ratio of the square roots of distances would be,

$$\frac{\sqrt{AC}}{\sqrt{AF}} = \frac{\sqrt{9}}{\sqrt{16}} = \frac{3}{4}.$$

3/4 does not equal 15/22, and Galileo would not have been able to maintain his hypothesis on the relationship between average velocities and distances in motion with uniform acceleration by his original method of reasoning.

We can verify that the square root of distance is directly proportional to the average velocity for uniformly accelerated motion by using our equations for free fall in the metric system in general terms.

1. $s = 4.9t^2$
2. $average\ velocity = s/t = 4.9t^2/t = 4.9t$
3. $t = \sqrt{s} / 2.214$, so, $t = .452\sqrt{s}$
4. $average\ velocity = (4.9)(452\sqrt{s}) = 2.214\sqrt{s}$
5. $\sqrt{s} = 1 / 2.214 = .452(average\ velocity)$.

This shows the square root of the distance is directly proportional to the average velocity in free fall with the proportionality constant of .452 using the metric system. The relation holds for other uniformly accelerated motions such as the pendulum and the inclined plane, but with a different proportionality constant.

Galileo did take the relationship between the square root of distance and the average velocity as being correct, and then considered the relationship between average velocity and what we call instantaneous velocity. Before we consider Galileo's reasoning let us show the correct relationship using the equations for free fall in the metric system.

6. From step 2, average velocity = *4.9t,* when, $s = 4.9t^2$. We know instantaneous velocity, $ds/dt = v = 9.8t$. We then see the average velocity equals ½ the instantaneous velocity as 4.9 is ½ of 9.8.

7. In general if *s* is directly proportional to time squared, then, $s = kt^2$, where k is the proportionality constant.

Average velocity = $s/t = kt^2/t = kt.$.

Instantaneous velocity,

$v = ds/dt = 2kt,$ when $s = kt^2.$

Again we see in the general case of uniform acceleration that the average velocity at an interval distance near *s* is ½ the instantaneous velocity. Galileo did know this relationship in 1604, however, Stillman Drake did not find it written out in Galileo's manuscript notes.

Historians point out that the relationship had been derived as far back as the 14th century by William Heytesbury and Richard Swineshead of Merton College Oxford, England, and by Nicole Oresme independently in Paris. Galileo may have heard of their arguments known as the Merton Rule, but the Oxford scholars had

not applied the rule to free fall motion.   In Galileo's book
of 1638, "The Two New Sciences", he demonstrates the
proposition using an approach similar to the early work
on the calculus in the 17th century.

Formally the proposition is given by Galileo as
follows:

*"If the time a certain distance traversed by a body
starting from rest under uniform acceleration is equal to
the time which the same distance would be traversed by
the same body moving at a uniform speed, then the
value of the uniform speed is ½ the maximum speed of
the uniformly accelerated body beginning at rest."*

He constructs Figure 43-2 to demonstrate the
proposition.

Figure 43-2

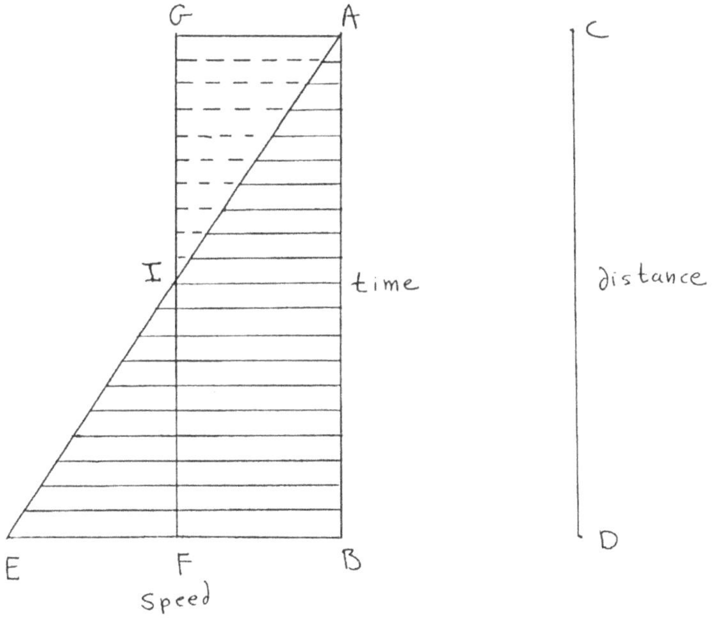

In the figure line CD represents the distance
traveled in time AB by an object moving with uniform
velocity.   Note time is represented on the vertical line

589

AB. The speed at time B is drawn as line BE perpendicular to AB. Line EA is connected, and a series of parallel lines are drawn from from AB to AE, all of them perpendicular to AB. The distances between these parallel lines measured on AB represent small increments of time. Galileo argues that there are an infinite number of instants of time between time A and time B so that the increments of time marked off by 'infinitely many' parallel lines can be considered 'infinitesimally' small.

The length of each parallel line from AB to AE represents the instantaneous velocity at any point along AB.

Galileo does not, however, use such modern language to describe his figure. He calls each parallel line "the moments of speed consumed" by the object in motion. With our knowledge of the calculus we recognize that time is plotted against speed in the figure, and the product of the length of any one of the parallel lines (which we think of as the instantaneous velocity) times the increment of time equals the increment of distance moved during that increment of time, that is, if we think of each parallel line as having a thickness or width which is a small increment of time, and then multiply this small increment of time by the length of the line, which is the instantaneous velocity, as the increment of time approaches 0, then the product of the increment of time by the instantaneous velocity at that instant of time equals the increment of distance moved at that instant of time.   Then by adding together all the parallel lines (increments of distance) up to any point along AB will give us the total distance traversed up to that point of time on AB.

We recall from the calculus we obtained the distance, s, in free fall from the formula, $v = 9.8t$, $v = ds/dt = 9.8t$, $ds = 9.8t\ dt$.

$$\int ds = \int 9.8dt, \; s = (9.8\,/\,2)t^2 + C = 4.9t^2 + C.$$

We recall if, $v = 0$ when $t = 0$, then the constant of integration, C, equals 0, so, $s = 4.9t^2$, as expected.   In other words Galileo's process of adding up an infinite number of parallel lines, each representing what we call the instantaneous velocity, gives him the distance traversed, just as we find the distance traversed by integrating the velocity function with the calculus. Note that the sum of 'infinitely' many parallel lines is an area. Thus the sum of all the parallel lines in Figure 43-2 fills up the area of triangle ABE.   This suggests that the integral, $\int ds = \int 9.8t\ dt$, can be represented geometrically as an area, when velocity is a function of time.   In the figure we see we would have to orient the triangle ABE so that B would be at the origin of a coordinate system with BE lying along the positive Y axis for 'speed' (instantaneous velocity) to become the dependent variable, and time to become the independent variable along BA the positive X axis. Later in our study of the integral we will see that the

591

integral can indeed be studied using the geometric representation of area.

Galileo marked off length FB on EB such that 2(FB) = BE. He wished to show that if the object moved at uniform speed with zero acceleration equal to magnitude FB, ½ the maximum instantaneous velocity BE, when the object moves with uniformly accelerated motion, then after time AB the distance traversed will equal the same distance that the object would move with uniformly accelerated motion. He calls the distance traversed by the object moving with uniform speed for time AB distance CD, represented by line CD drawn to the right of triangle ABE. FG is drawn parallel to AB with FG equal to AB, and AG is connected.

Area rectangle FBGA = (AB)(BF) .

Area triangle ABE = ½(AB)(BE), and since,

> 2(BF) = BE, area triangle
>
> ABE = ½(AB) 2 (BF) = (AB)(BF), or,
>
> area triangle ABE = area rectangle FBGA.

Galileo extends the series of parallel lines from AB to GI as broken lines, and argues that if an object moved at uniform speed FB for time AB "the moments of speed consumed" would equal the area of rectangle FBGA, since there are an infinite number of the parallel lines which fill up the rectangle each one representing an infinitesimal distance moved for each infinitesimal increment of time.

In the uniformly accelerated motion the parallel lines increase in length each instant of the motion so that the distance after time AB equals the area of the triangle ABE. Galileo has shown the area of the triangle equals the area of the rectangle so the distances moved in motion with uniform speed at a value equal to ½ the maximum velocity of motion with uniformly accelerated motion are the same, i.e., BF (the motion with uniform speed) = ½(BE)[½ the maximum instantaneous velocity obtained in time AB with uniformly accelerated motion].

Galileo's method can be criticized by the same arguments that were raised when other mathematicians introduced infinitesimal time and distance intervals. He did, however, introduce several new concepts not seen in the manuscript notes from 1604. In "The Two New Sciences" of 1638 he introduces the concept of an instantaneous speed, or as we would say instantaneous velocity. He also introduced the concept of an instant of time in considering problems of moving objects. And he has introduced the idea that the distance traversed can be represented as a geometric area when speed is plotted against time. We learned that the distance function is the integral of the velocity function, and to represent an integral as an area became very important in the development of the calculus by Newton and Leibniz later in the 17th century.

Galileo had then worked out the correct relationships between distance and time for uniformly accelerated motion, and distance and average velocity in uniformly accelerated motion. We now return to his work first revealed to us by Stillman Drakes analysis of the 1604 manuscript notes. Galileo constructed a figure similar to Figure 43-3.

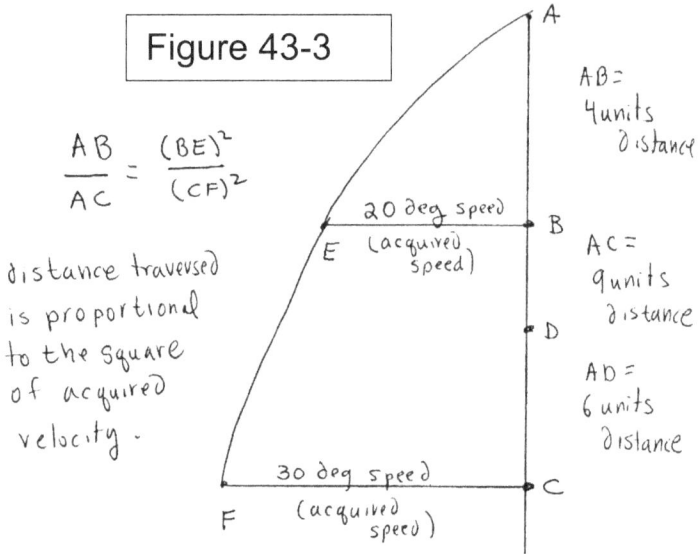

Line ABDC represents distance traversed in uniformly accelerated motion just as in Figure 43-1. As in Table 1, AB = 4 units of distance which is traversed in 4 units of time, AC represents 9 units of distance traversed in 6 units of time.   We recall Galileo had argued the average velocity through distance AB was given the value of 10 'degrees of speed', while the average velocity through distance AC was 15 'degrees of speed'.

Galileo then supposed that if the average speed to move from rest at A to B is 10 'degrees of speed', then the instantaneous velocity (the speed acquired by the object at B) should be 2 times 10 'degrees of speed', or 20 'degrees of speed'.   He applied here the proposition just demonstrated that the average speed for a uniformly accelerated motion is ½ the maximum instantaneous velocity obtained when the object is at B.   Likewise, if the average speed from rest at A to C is 15 'degrees of speed', then the instantaneous velocity at C should be twice 15 degrees of speed or 30 degrees of speed. Galileo indicates the instantaneous velocity at B (acquired speed = 20 'degrees of speed') by the horizontal line BE in Figure 43-3.   BE is perpendicular to AC.   The instantaneous velocity at C (30 'degrees of speed') is indicated by line CF perpendicular to AC.

Galileo then states that the ratio of the two acquired speeds (instantaneous velocity), BE/CF = 20/30 = 2/3, equals the ratio of the two average speeds for the respective distances, 10 'degrees of speed' / 15 'degrees of speed' = 10/15 = 2/3.   Since Galileo had previously argued that the ratio of the average speeds was equal to the square roots of the distances, Galileo's hypothesis was,

$$\frac{\sqrt{AB}}{\sqrt{AC}} = \frac{\sqrt{4}}{\sqrt{9}} = \frac{2}{3}, \text{ and, } \frac{BE}{CF} = \frac{20}{30} = \frac{2}{3}.$$

The relationship holds for the data used in Galileo's analysis.

We recall Galileo had chosen the value of 15 degrees of speed for the average velocity for distance AC by faulty reasoning.   Nonetheless, his hypothesis turns out to be correct; that is, the ratio of the instantaneous speeds is equal to the ratio of the square roots of the respective distances, and equal to the ratio of the average speeds for the respective distances. This in turn means the square root of the distance in uniformly accelerated motion is directly proportional to the instantaneous velocity. We can confirm this hypothesis by using the equations for free fall in the metric system.

8. $s = 4.9t^2$, so $\sqrt{s} = 2.214t$

9. $v = 9.8t$, or, $t = .102v$, where v is the instantaneous velocity.

Using this value of t in step 8,

10. $\sqrt{s} = (2.214)(.102)v$, or, $\sqrt{s} = .226v$, which means the square root of distance is directly proportional to the instantaneous velocity with the proportionality constant of .226 using the metric system in free fall motion.

The discovery that the square root of distance is directly proportional to the instantaneous velocity was a major advance, since before Galileo those philosophers who had speculated on the relationship had always maintained instantaneous velocity increased in direct proportion to distance traversed.

Galileo then proposed that the ratio of the acquired speeds (instantaneous velocities) for two distances equals the ratio of the initial distance to the mean proportional distance between the two distances. In Figure 43-3 distance AD is the mean proportional between AB and AC. The hypothesis is then,

$$\frac{speed\ BE}{speed\ CF} = \frac{dis\,tance\ AB}{dis\,tance\ AD}.$$

596

BE/CF = 2/3, and, AB/AD = 4/6 = 2/3, showing the ratios are equal (recall we found the value of AD to be 6 in our discussion of Figure 43-1). Once again Galileo's hypothesis is verified for the numbers with which he was working. We can also verify the hypothesis with our equations for free fall. We state the hypothesis as.

$$\frac{s_1}{\sqrt{s_1 s_2}} = \frac{v_1}{v_2}.$$

$\sqrt{s_1 s_2}$, is the mean proportional between s_1 and s_2.
To show,

$$\sqrt{s_1 s_2}$$

is the mean proportional let x be the mean proportional so, $s_1/x = x/s_2$, or, $x^2 = s_1 s_2$, so,

$$x = \sqrt{s_1 s_2}$$

We square both sides of the proportion:

$$\frac{(s_1)^2}{s_1 s_2} = \frac{(v_1)^2}{(v_2)^2}, \text{ or, } \frac{s_1}{s_2} = \frac{(v_1)^2}{(v_2)^2}.$$

If we use $s_1 = 4.9(t_1)^2$, $s_2 = 4.9\ (t_2)^2$, $v_1 = 9.8t_1$, and $v_2 = 9.8t_2$ in the above proportion we have,

$$\frac{(4.9)(t_1)^2}{(4.9)(t_2)^2} = \frac{(9.8t_1)^2}{(9.8t_2)^2}.$$

We see both sides are equal as the numerical terms cancel out verifying the hypothesis.

Using the proportion in Figure 43-3 we have,

$$\frac{dis\tan ce\ AB}{dis\tan ce\ AC} = \frac{(speed\ BE)^2}{(speed\ CF)^2}, \frac{4}{9} = \frac{20^2}{30^2}, .444 = .444.$$

To Galileo this meant the relationship between distance and instantaneous velocities could be represented geometrically by a parabola, drawn as curve AEF in Figure 43-3. The formula for a parabola is, $y = ax^2$, in the Cartesian coordinate system where a is a constant. From our equation, $v = 9.8t$, we have, $t = .102v$. We substitute this value of t into the equation, $s = 4.9t^2$, to get, $s = .05v^2$. This is the equation for a parabola where y is replaced by s, a by .05, and x by v. In Figure 43-4 we show a graph of this function indicating the parabolic increase of distance as the instantaneous velocity increases in free fall.

> Figure 43-4

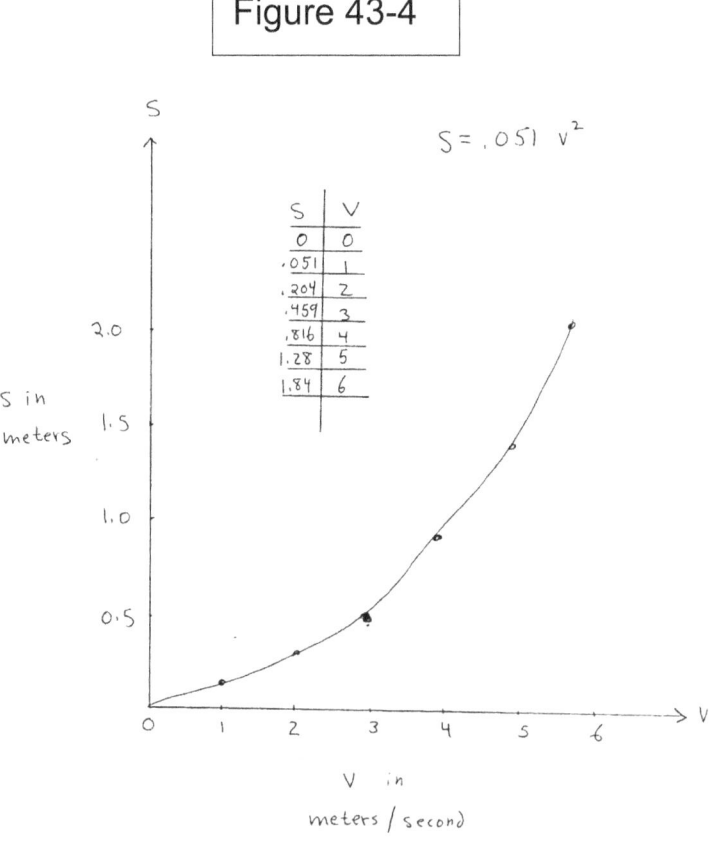

The hypothesis given by:

$$\frac{s_1}{\sqrt{s_1 s_2}} = \frac{v_1}{v_2},$$

can be further evaluated mathematically by relating time in free fall to the instantaneous velocity. From Figure 43-3 we start by evaluating the following proportion:

$$\frac{dis\tan ce\ AB}{dis\tan ce\ AD} = \frac{speed\ BF}{speedCF}, or, \frac{s_1}{\sqrt{s_1 s_2}} = \frac{v_1}{v_2}.$$

In his papers from 1604 Stillman Drake could not find that Galileo actually wrote out the relationship between instantaneous velocity and time, although it appeared as a major principle in "The Two New Sciences" of 1638. We know the relationship from our previous work with the calculus, $ds/dt = v = 9.8t$. This means instantaneous velocity, v, is directly proportional to time with proportionality constant of 9.8 in the metric system.

We can derive the relationship that instantaneous velocity is directly proportional to the time in uniformly accelerated motion by the mathematics of proportion and the basic equation for uniformly accelerated motion relating distance and time, namely, $s = kt^2$, where k is the proportionality constant. We substitute $k(t_1)^2$ for s_1, and $k(t2)^2$ for s_2 in the above proportion.

$$\frac{s_1}{\sqrt{s_1 s_2}} = \frac{v_1}{v_2}, \text{ becomes, } \frac{k(t_1)^2}{\sqrt{k(t_1)^2 k(t_2)^2}} = \frac{v_1}{v_2},$$

$$\frac{k(t_1)^2}{kt_1 t_2} = \frac{v_1}{v_2}, \text{ or, } \frac{v_1}{v_2} = \frac{t_1}{t_2},$$

which means the instantaneous velocity is directly proportional to time.

In "The Two New Sciences" of 1638 Galileo states this fundamental principle as follows. He says of a moving object with uniform acceleration that he will,

"investigate and demonstrate some attributes (whatever be the cause of its acceleration) that the momenta of its speed go increasing after its departure from rest, in that simple ratio with which the continuation of time increases, which is the same to say that in equal times, equal additions of speed are made. And if it shall be found that the events that then shall have been demonstrated are verified in the motion of naturally falling and accelerated bodies, we may deem the definition assumed includes the motion of heavy things, and that it is true that their acceleration [he means change in velocity] goes increasing as the time and duration of motion increases."

In this verbose statement of the principle, notice that it is not enough for Galileo to have a theory of uniform acceleration where velocity is directly proportional to time, but the theory must be verified by 'naturally falling bodies', bodies in free fall. The mathematics which derives the result that instantaneous velocity is directly proportional to time starting with the experimentally determined relation that distance is directly proportional to time squared, ought to be true for naturally occurring motions, when impediments to motion are insignificant.

In the manuscript notes from 1604 Stillman Drake did find Galileo had the idea that in considering two distances traversed in motion with uniform acceleration the ratio of the first distance to the mean proportional distance between the two distances equals the ratio for the time for the first distance to the time for the second distance.   In Figure 43-3 and Table 1we found distance AB = 4, AC = 9, and the mean proportional distance between them is AD = 6 units distance.   The time for AB = 4 units of time, AC = 6 units of time .

$$\frac{dis\tan ce\ AB}{dis\tan ce\ AD} = \frac{4}{6} = \frac{time\ for\ AB}{time\ for\ AC} = \frac{4}{6}.$$

In "The Two New Sciences" Galileo gave this relation as a corollary to his law of free fall in the following words,

"If at the beginning of motion there are taken any two distances, run through in any two times, the times will be to each other as either of these distances is to the mean proportional distance between the two distances".

This principle can also be written,

$$\frac{s_1}{\sqrt{s_1 s_2}} = \frac{t_1}{t_2}.$$

The principle can be derived mathematically from the relationship that distance is directly proportional to the square of time.

$$\frac{s_1}{s_2} = \frac{(t_1)^2}{(t_2)^2}, \textit{ taking square roots, } \frac{\sqrt{s_1}}{\sqrt{s_2}} = \frac{t_1}{t_2}.$$

Multiply the left side by $\sqrt{s_1}/\sqrt{s_1}$ which equals 1.

$$\frac{s_1}{\sqrt{s_1 s_2}} = \frac{t_1}{t_2}.$$

Let us now take stock of our work with Galileo's study of uniformly accelerated motion and review the principles we have been studying.

11. $s = kt^2$, distance is directly proportional to the square of time. Experimentally derived from free fall experiments with pendulums and Galileo's water clock, and confirmed by the inclined plane experiments.

12. $\sqrt{s} = (k)(average\ velocity)$, the square root of distance is directly proportional to the average velocity. This discovered in spite of faulty reasoning, by Galileo's work with arbitrary ratios of numbers representing average velocities.

13. *average velocity* = $\frac{1}{2}v$, where *v* is the instantaneous velocity. Originally known as the Merton Rule from the 14th century, but derived independently by Galileo by using arguments involving infinitesimals in "The Two New Sciences" of 1638.

14.

$$\frac{S_1}{\sqrt{S_1 S_2}} = \frac{V_1}{V_2},$$

if we consider two distances traversed by a body moving with uniform acceleration the ratio of the first distance to the mean proportional distance between the two distances is equal to the ratio of the instantaneous velocity at the end of the first distance to the instantaneous velocity at the end of the second distance.   Discovered by the mathematics of proportion.

15. $s = kv^2$, or $\sqrt{s} = kv$, the square root of the distance traversed is directly proportional to the instantaneous velocity at that distance.   This relation could be derived mathematically, but could not be confirmed experimentally by Galileo, because he had no means to experimentally measure instantaneous velocity. Stillman Drake points out that Galileo was not certain this principle was valid in 1604.   However, by 1607 Galileo had written, *"the momenta of velocita [instantaneous velocity] of things falling from a height are to one another as the square roots of the distance traversed"*.   This relationship indicates that distance increases in a parabolic manner as the instantaneous velocity increases as shown in Figure 43-4.

16.

$$\frac{S_1}{\sqrt{S_1 S_2}} = \frac{t_1}{t_2},$$

the ratio of the first distance to the mean proportional distance between the first and the second distances is equal to the ratio of the time for the first distance to the time for the second distance. This was derived mathematically in 1604.

17. $v = kt$, instantaneous velocity is directly proportional to the time of motion. Derived mathematically by 1607 and the cornerstone of motion with uniform acceleration in "The Two New Sciences" of 1638.

Prior to Stillman Drake's researches into the unpublished manuscript notes of 1604 the sequence of discovery for the laws of motion with uniform acceleration was assumed by historians to be in the order given by Galileo in "The Two New Sciences". In this book Galileo started with a definition of uniform acceleration that occurs to naturally falling objects with no significant impediment to their motion. He stated, *"I say that motion is equable or uniformly accelerated which, abandoning rest, adds on to itself equal momenta of swiftness in equal times".*

In other words instead of starting with the idea that the acceleration is constant (uniform), and distance is proportional to the square of time, he starts with a definition that instantaneous velocity increases in direct proportion to time, $v = kt$, in uniformly accelerated motion, i.e. $a = k$. He then derived the other relations given in steps 1 through 7 by mathematical arguments.

In reviewing the work of Stillman Drake, which has showed us the true sequence of discovery, I. Bernard Cohen, a noted historian of 17th century science, wrote in his book, "The Birth of a New Physics" (1985 edition),

"It has been a cause of real astonishment to find that Galileo had in fact made his discovery of the laws of motion in a manner quite different from the public presentation he gave in his last treatise, the 'Two New Sciences'. His secret was well kept for over three and a half centuries, until Stillman Drake found and drew attention to Galileo's work sheets, which seem unquestionably to be the record of experiments on moving bodies somehow related to the laws of motion he had found. This is one of the great discoveries of the history of science of our times".

To conclude our work on uniformly accelerated motion we turn to "The Two New Sciences" where Galileo gives a rule known today as his 'double distance' rule. The rule states that a body moving with uniform acceleration a distance s_1 in time t_1, and then deflected in its motion so as to move at constant velocity at a value equal to the maximum velocity obtained at the end of the motion with uniform acceleration, will move with constant velocity in a time equal to t_1 twice the distance s_1. If s_2 is the distance moved with constant velocity then the rule says, $s_2 = 2s_1$. We can demonstrate the rule mathematically by using our equations of free fall in the metric system.

$s_1 = 4.9(t_1)^2$, and, $v_1 = 9.8t_1$, where v is the instantaneous velocity at t_1, which is the constant velocity which is the object moves after deflection through distance s_2 in a time equal to t_1. Thus, $s_2 = v_1t_1$, since v_1 is constant for this part of the motion. We are to demonstrate that, $s_2 = 2s_1$.

Since, $v_1 = 9.8t_1$, we substitute this value in the equation for s_2.

$s_2 = 9.8t_1t_1 = 9.8\ (t_1)^2$.

We now set up the proportion between s_1 and s_2.

$$\frac{s_1}{s_2} = \frac{4.9(t_1)^2}{9.8(t_1)^2} = \frac{1}{2}, \text{ or, } s_2 = 2s_1.$$

Galileo's argument found in "The Two New Sciences" uses geometry. We will give the essentials of his demonstration to illustrate his reasoning. He constructs Figure 43-5.

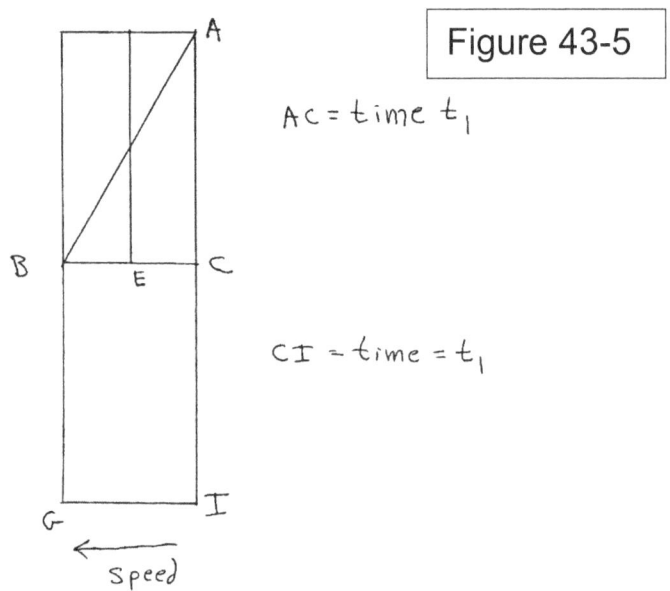

Figure 43-5

AC = time t_1

CI = time = t_1

In the figure the object starts from rest at point A, moving with uniform acceleration for time AC, which is measured as a vertical line. Let AC = t_1. Velocity is measured is the perpendicular line CB. At the end of time t_1 the object has the instantaneous velocity BC, which can also be called its acquired speed.   Line AB is connected.   As in the reasoning used by Galileo in Figure 43-2, the instantaneous velocity at any point in time from A to C is measured as the perpendicular distance from AC to line AB.

At the time represented by point C the object's motion is deflected so that it now begins to move without any acceleration at a constant velocity.   It moves with constant velocity for time CI, which is equal to time AC = t_1.   Thus line segment CI = AC.   Since the velocity is constant, then velocity CB = velocity IG.   Line BG is connected.   The velocity at any time from C to I is given by the perpendicular distance from CI to BG, which is always equal to BC.

Galileo argues that the area of triangle ACB equals ½ the area of rectangle CIGB.   As in the argument for Figure 43-2, the numerical values of the areas equal the distances moved in the two motions, AC with uniform acceleration and CI with constant velocity.   If the area of triangle ACB equals s_1, and the area of rectangle CIGB equals s_2, then Galileo states,

$s_2 = 2s_1$.

Area triangle ACB = ½(CB) (AC)

Area rectangle CIBG = (CB) (CI), and since CI = AC, area triangle ACB = ½ area rectangle CIBG, which establishes the 'double distance' rule by Galileo's method.

In our work with Galileo's law of uniform acceleration we have made use of the extensive work of

Stillman Drake and his analysis of Galileo's unpublished work sheets. We used this work rather than the demonstrations as given in "The Two New Sciences" except in a few cases. We found Galileo used the mathematics of proportion, but in addition Galileo introduced new mathematical concepts such as dividing a time interval into infinitesimal instants with an instantaneous velocity for each instant. This allowed him to represent the distance traversed a geometric area, which could be calculated by the formulas of Greek geometry.

With this method he points the way to the mathematics of the calculus so useful in problems of moving objects. The law of uniform acceleration is one of Galileo's great discoveries and the reader is encouraged to pursue its full development by reading the chapter in "The Two New Sciences" entitled "The Third Day", as translated by Stillman Drake listed in the bibliography to this chapter.

Bibliography for Chapter 43

1. Galilei, Galileo, "Discourses and Mathematical Demonstrations Concerning Two New Sciences", translated with a new introduction and notes by Stillman Drake, Second Edition, Wall & Thompson, Toronto, 1989. See Chapter, "The Third Day".
2. Drake, Stillman, "Galileo: Pioneer Scientist", University of Toronto Press, 1990.

3. Drake, Stillman, "Galileo's Discovery of the Law of Free Fall",   Scientific American, 228, Number 5, pages 84-92, May 1973.

4. Drake, Stillman, "Galileo at Work, His Scientific Biography", The University of Chicago Press, 1978.

5. Cohen, I. Bernard, "The Birth of a New Physics", Doubleday and Co., New York, 1985.

Chapter 44

Galileo's Principle of Inertia and the Parabolic Path

of Projectiles

In the last part of Galileo's treatise, "The Two New Sciences", he discusses the motion of projectiles and the law of horizontal inertia. As with the law of motion with uniform acceleration, Galileo's principle of inertia, as seen in his study of projectile motion, is one of the great discoveries of science, and was the prelude to Newton's law of inertia. In Galileo's presentation, found in the chapter entitled "The Fourth Day", he does not give the dates of his discoveries, nor the experiments he performed to support his theory. As we found in the preceding chapters, it is Stillman Drake's research into Galileo's work sheets that will help us understand how Galileo discovered the principle of inertia and its mathematics. We now know that Galileo did his work on this subject in the period of 1607 to 1609.

The first statement of a principle of inertia by Galileo goes back to 1590 when Galileo wrote a treatise on motion where he theorized without experimental confirmation, that if spherical balls were started in motion with constant velocity on a smooth horizontal table, they would continue to move in a straight line for an indefinite time; that is until they reached the end of the table, or were eventually slowed by the resistance of friction.

In contrast the Aristotelian philosophers had taught that the only thing that kept an object in motion is a continuously acting force. The source of this force was taken to be the air, which supposedly pushed the object along.   Once this force ceased to act, or if another force opposed it, the object ceased to move, and obtained its natural state of rest.

Galileo rejected these notions and felt that if there were no impediments to oppose motion, the idealized situation of continuous motion along a straight line would be realized. To quote from the "Two New Sciences",

"Hence to deal with such matters [impediments to motion] scientifically it is necessary to abstract from them. We must find and demonstrate conclusions abstracted from the impediments, in order to make use of them in practice under those limitations that experience will teach us... distances and speeds will for the most part not be so exorbitant that they cannot be reduced to management by good accounting".

Thus, Galileo sought the essential principle of motion in motion along a horizontal plane for an object with constant velocity. Galileo's principle of inertia was motion with uniform velocity which continues indefinitely along a line perpendicular to the earth's radius.

Of course, the 'line' along the earth's surface which is perpendicular to the radius from the center of the earth is not a straight line, but it becomes the circular circumference of the earth. However, for ordinary distances that can be experimentally studied by Galileo's methods, the line perpendicular to the earth's radius is on a horizontal plane on the surface of the earth, so that an object once in motion with constant velocity would continue to move in a straight line on the observer's horizontal plane.

Galileo envisioned, for example, that if a sailing ship were in motion with constant velocity, and the sea offered no impediment to its motion, the ship would continue to move along a straight line as far as the eye could see, and Galileo hints that the ship would even follow the circumference of the earth. Historians have debated the point as to whether Galileo meant to extend his principle of inertia in such a way as to involve motion around the earth's circumference. In his book, "Dialogue Concerning the Two Chief World Systems", published in 1632, he argues that motion started on the observer's horizontal plane with uniform velocity will continue perpetually. He states,

"But motion in a horizontal line which is tilted neither up nor down is circular motion about the

center; therefore circular motion is never acquired naturally without straight motion to proceed it; but being once acquired, it will continue perpetually with uniform velocity".

In "The Two New Sciences" in the chapter entitled "The Third Day" he discusses the principle in these words,

"But motion in the horizontal plane is equable [motion with constant velocity along a straight line], as there is no cause of acceleration or retardation...It may also be noted that whatever degree of speed is found in the moveable, this is by its nature indelibly impressed on it when external causes of acceleration or retardation are removed, which occurs only on the horizontal plane, for on declining planes there is cause of more acceleration [increases in velocity due to uniform acceleration of the objects moving down an inclined plane], and on rising planes of retardation. From this it likewise follows that motion in the horizontal is also eternal, since if it is indeed equable it is not even weakened or remitted, much less removed".

In these words, we see the idea of the horizontal plane in view of the observer, but he also considers that motion can be 'eternal',or extended indefinitely in a straight line. Thus, we have the assertion that an object once in motion with constant velocity with no impediment to its motion would continue moving with constant velocity in a straight line forever.

Here the principle seems to mean motion in a truly straight line extended outward away from the earth's surface. Of course, once the observer's plane is extended beyond the earth's surface the force of gravity causes an acceleration to retard the uniform velocity of the object as shown in Figure 44-1.

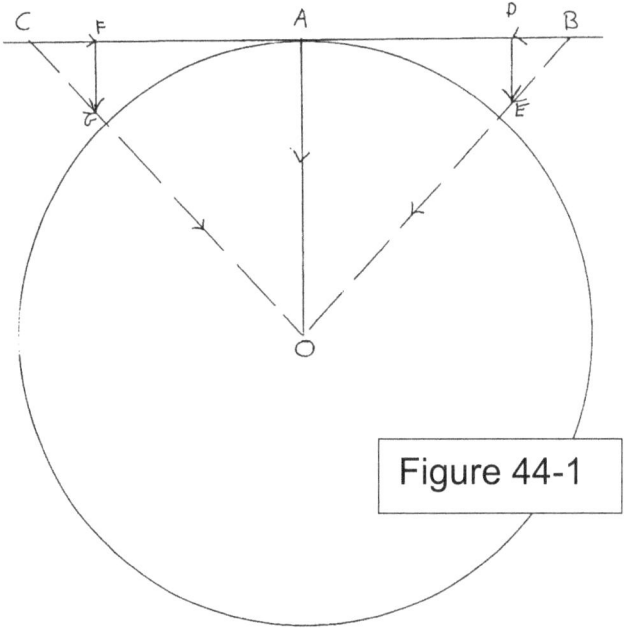

Figure 44-1

When the object is at A, or in the immediate vicinity of A, on the horizontal plane, the force of gravity is perpendicular to the plane and the acceleration of gravity will not retard motion along the plane so the object will move with uniform velocity. When the object is at B or C on the plane, we see the force of gravity acts along CO or

BO toward the center of the earth. We recall our work with the inclined plane in Figure 42-1, and draw the perpendicular components of the gravitational force for the object at positions C and B. Consider position B. The acceleration of gravity acts along CO and is represented by length CG. CG can be considered as having two perpendicular components, CF, and FG. Component FG acts to keep the object on the plane, while CF causes acceleration toward A, and thus retards the uniform velocity of the object. The same argument can be given to show the object's motion is retarded at B, or any other position on the plane except at A. Galileo did not use such an argument in his work, and we give this argument using the ideas later introduced by Newton, which give us a fuller appreciation of the principle of inertia as applied to planes near the earth's surface. The principle of inertia also means if a body is at rest, it has a velocity of zero, and the body will remain at rest unless a force changes its motion by giving acceleration to the body.

The conclusion of the principle of inertia is that no external force is required to maintain a body in motion with constant velocity along a straight line. Such motion is as natural as a body which is at rest. This concept of inertia was a complete change from the view held by Aristotle and the students of motion up to the work of Galileo. Both

Newton and Descartes would adopt the principle of inertia as a fundamental physical law with application throughout the universe, whereas Galileo studied it with experiments where motion was directly observable. The principle in its universal form states that any body will remain in its state of rest or uniform motion at constant velocity in a straight line unless compelled to change its state of rest or motion by forces impressed upon it.

For short distances of several meters to several hundred meters Galileo's principle of inertia meant an object moving with constant velocity with no impediments to motion would move in a straight horizontal line, horizontal to the earth's surface, as seen by an observer, and would move equal distances in equal intervals of time.

Galileo's original work with the principle included studies with projectile motion. This work was discovered by Stillman Drake among Galileo's work sheets. Drake has analyzed this work so we can understand the key experiments which confirmed the principle of inertia. In addition Galileo also discovered that projected bodies follow a parabolic path when there are no impediments to the motion.

To do these experiments Galileo made a grooved wooden plane which he polished. He placed the plane in a position horizontal to the earth's surface without any incline so no acceleration would occur.   He took a spherical ball and gave it a push and then measured the distance traveled in equal time intervals.   As in his previous work the unit for distance was the punti.   We recall 1 punti equals .00094 meters.   Galileo's notes indicate the ball moved 196, 155, 121, and 100 punti in equal time intervals.   Now the theory of horizontal inertia states that if the ball had no impediment to its constant velocity given to it at the start of its motion, then the distance moved in equal time intervals would be equal.   Clearly this experiment did not confirm the principle of inertia, because Galileo realized even the polished grooved plane was retarding the motion.

Galileo then had the idea of tilting the grooved plane so the ball would accelerate and acquire increasing velocity as in his previous inclined plane experiments, and then allow the ball to roll off the inclined plane into the air after being deflected by a deflector at the end of the plane. See Figure 44-2.

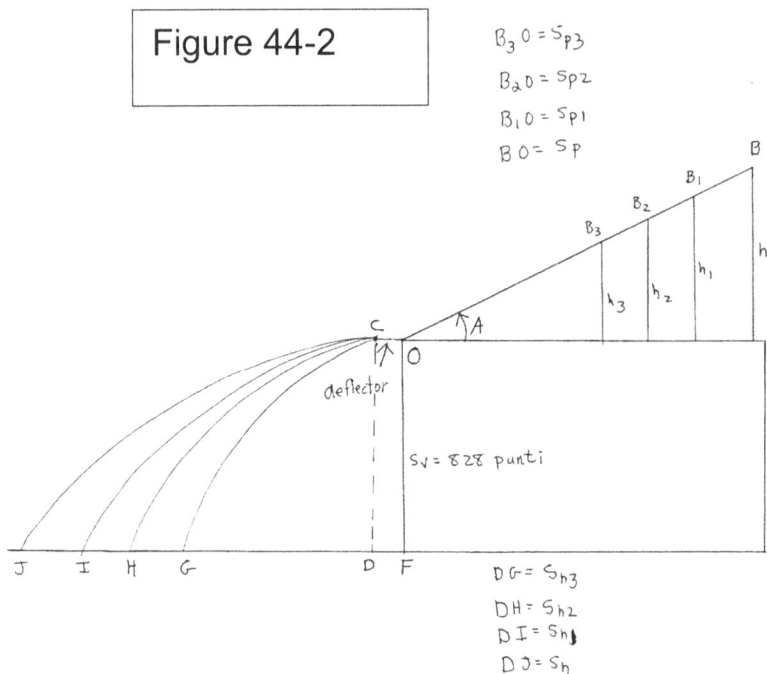

Figure 44-2

The purpose of the deflector was to cause the ball to leave the inclined plane with the direction of its motion horizontal to the floor (the earth's surface).

As soon as the ball left the plane the law of free fall would operate so the ball would fall toward the floor. But it would also continue to move away from the inclined plane. Its motion, Galileo reasoned, would be determined by two principles operating simultaneously. The principle of horizontal inertia should cause the ball to move equal distances horizontally in equal intervals of time, and the principle of free fall would result in the vertically downward distances to increase as the square of time according to the law of free fall (uniform acceleration).

618

In Figure 44-2 the plane is seen as inclined at angle A and placed on a table located distance s_v = 828 punti from the floor. The spherical ball is first released from position B which is distance h from the table as shown. The ball rolls down the plane, hits the deflector which changes its direction so that it leaves the plane moving parallel to the floor. This means it has a component of velocity in the horizontal direction when it is projected off the plane.

After leaving the deflector the motion of the ball is analyzed as having two components, (1) the horizontal component which should obey Galileo's principle of inertia such that motion in the horizontal direction away from the deflector occurs with constant velocity equal to the instantaneous velocity at the bottom of the plane, and (2) the vertical component which should obey the law of free fall and move in the vertically downward direction with uniformly accelerated motion. The two components of motion acting simultaneously result in the ball taking a curved path, which Galileo sketched in his work sheets much as shown in the figure. We will use mathematics to deduce the curve which the ball traces after it leaves the deflector.

Note that the higher the initial height of release of the ball on the plane, the farther away from the table is the arc of travel. For example, if the ball is released from position B, it moves through arc CJ, and if it is released from position B_1, it moves through arc CI. The horizontal distance the ball moves after leaving the deflector and striking the floor is DJ = s_h, when the ball is released from position B, and the horizontal distance the ball moves after leaving the deflector is DI = s_{h1}, when the ball is released from position B_1.

1. Using the law of free fall in the metric system, $s_v = 4.9t^2$, where $t = 0$ at point C the end of the deflector where the free fall component of motion begins.

2. If the ball is released from position B then,

DJ = s_h = vt, where v is the magnitude of the velocity at point C, and $t = 0$ at point C.

Galileo used the mathematics of proportion to calculate his predicted results based on the law of free fall and on the principle of horizontal inertia. He then could compare his predicted results with the results he obtained by doing the experiment to see if the law of horizontal inertia was verified.

For motion down the inclined plane in Figure 44-2 the distance the ball rolls is BO, designated by s_p, when the ball is released from position B. The figure indicates the designated distances the ball rolls down the plane when released from the other positions B_1, B_2, etc. Using the results from the last chapter that in motion with uniform acceleration the distance is directly proportional to the velocity squared, and the fact that motion down an inclined plane is a case of motion with uniform acceleration, we have,

3.

$$\frac{s_p}{s_{p1}} = \frac{(v)^2}{(v_1)^2},$$

where s_p = BO, s_{p1} = B_1O, and v, v_1 are the respective velocities at C.

4. $\sin A = h/s_p$, or, $s_p = h/\sin A$, and by the same reasoning

$s_{p1} = h_1/\sin A$.

We substitute these values into step 3.

5.

$$\frac{h/\sin A}{h_1/\sin A} = \frac{(v)^2}{(v_1)^2}, \ or, \ \frac{h}{h_1} = \frac{(v)^2}{(v_1)^2}.$$

6. The principle of horizontal inertia states the horizontal distances from position C are equal to the horizontal velocity at C multiplied by the time it takes for the ball to reach the floor. Thus,

$$\frac{s_h}{s_{h1}} = \frac{vt}{v_1 t_1},$$

where t is the time from point C to the point at which the ball reaches J when released from B, and t_1 is the time from C until the ball reaches I when released from B_1.   But, t must equal t_1 because the ball falls through the same vertical height $sv = 828$ punti, and according to the law of free fall, $sv = 4.9t^2$, and also, $sv = 4.9\ (t_1)^2$, so $t = t_1$. Thus,

$$\frac{s_h}{s_{h1}} = \frac{v}{v_1}, \text{ squaring, } \frac{(v)^2}{(v_1)^2} = \frac{(s_h)^2}{(s_{h1})^2}.$$

We use this value in step 5.

7.

$$\frac{h}{h_1} = \frac{(s_h)^2}{(s_{h1})^2}.$$

In words this means the square of the horizontal distance moved after the ball leaves the deflector is directly proportional to the height of the ball above the table on the plane when the ball is released.   Galileo can readily measure the distances involved and see if they are in this proportion.

In Galileo's experiments Stillman Drake found the value of angle A was about 60°. When $h_1 = 300$ punti, and $h = 600$ punti, Galileo found the value of distance DI $= s_{h1}$ to be 800 punti. Using the proportion in step 7 we can calculate the predicted value of distance DJ $= s_h$.

8.

$$(s_h)^2 = \frac{(s_{h1})^2 (h)^2}{h_1} = \frac{(800)^2 (600)}{300}$$

$$= 1{,}280{,}000, \quad s_h = 1{,}131 \; punti.$$

Galileo's measured value for s_h was 1,172 punti, a difference of 41 punti, or $(41)/(1,131)(100\%) = 3.63\%$ error.

We will give a table of values comparing measured values to predicted values for horizontal distances after the ball has been released from various positions on the inclined plane. We will designate the horizontal distances as s_h, and the height above the table at which the ball is released as h. The value of h_1 used in the calculations to find the predicted values of s_h was 300 punti, and the value of s_{h1} used was 800 punti. All distances are given in Galileo's units of punti.

h	s_v	Predicted s_h	Measured s_h	Error
600	828	1,131	1,172	3.6%
800	828	1,306	1,328	1.7%
828	828	1,329	1,340	.8%
1,000	828	1,460	1,500	2.7%

We find from the table Galileo had good agreement between the predicted values using the law of free fall and the principle of horizontal inertia, and the experimentally measured values. This convinced him that the principle of inertia was valid near the earth's surface over observable distances. We note that in using the mathematics of proportion Galileo did not have to know the numerical values of the velocities as the ball leaves the deflector. This is because the law of free fall applies so that the times to reach the floor (step 6) are equal, no matter what the value of the horizontal velocity at point C.

We can derive more information from Galileo's data using our formulas in the metric system. We will study the case where h is 828 punti, sv = 828 punti, and s_h = 1,329 punti as predicted by Galileo.

9. h = (828 punti)(.00094 meters) /(1 punti)

= .7783 meters.

sv = 828 punti = .7783 meters.

Predicted s_h = (1,329 punti)(.00094 meters)/(1 punti) = 1.249 meters.

In Figure 44-3 we indicate these values.

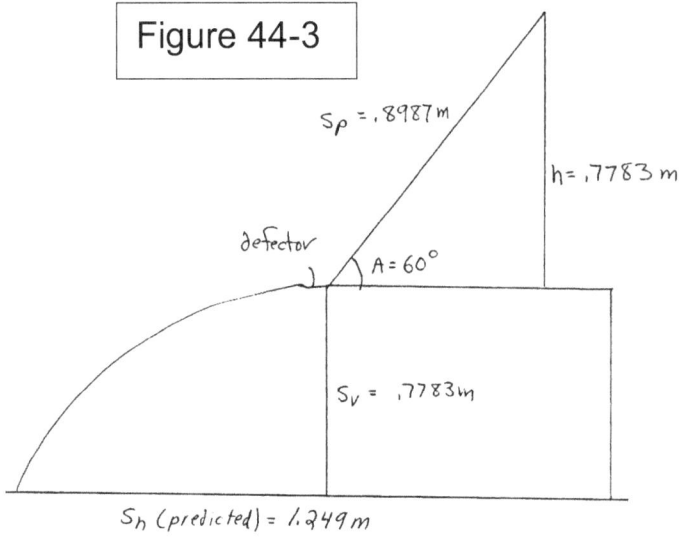

Figure 44-3

S_p = .8987 m

h = .7783 m

defector

A = 60°

S_v = .7783 m

S_h (predicted) = 1.249 m

10. sin 60° = h/s_p, or, s_p = h/sin 60°

= .7783/.8660 = .8987 meters.

If the ball had rolled down the inclined plane with no impediment to its motion, we can use the formula we derived for uniformly accelerated motion down an inclined plane, which we derived in chapter 42: $s = 4.9t^2(sin\ A)$.   We wish to calculate the time t it takes for the ball to reach the deflector after it is released from its starting position.

   11. $t^2 = (s_p)/(\ 4.9sin\ A)$

      $= (.8987)/(4.9)(.8660) = .2118$,

      or. $t = .4602$, or .46 seconds.

   12. The formula for the instantaneous velocity

      at the deflector is,

      $v = ds/dt = (2)(4.9)(t)(sin\ A) = 9.8t\ sin\ A$

      $v = (9.8)(.46)(.8660)$

   $v = 3.90$ meters/second.

   This is the value for the horizontal velocity as it leaves the deflector at C in the idealized situation when the deflector only changes the direction of motion and not its magnitude.

   Before proceeding we will give another method of finding the value of $v = 3.90$ meters/second.   In chapter 42 we also found that the instantaneous velocity of an object starting from rest and moving down an inclined plane of height h is given by,

   13. $v = 4.426\sqrt{h}$, or, $v = (4.426)(\sqrt{.7783}\ )$

      $v = 3.90$ meters/second.

   We recall our formulas for the motion beginning at point C in Figure 44-2 given in steps 1 and 2; $s_v = 4.9t^2$, and, $s_h = vt$, when $t = 0$ at point

C.   We solve for t, and then calculate what s_h should be.

14. $t^2 = (s_v)/ (4.9) = .7783/4.9$

$= .1588$ seconds squared,

$t = .399$, or, $.40$ seconds.

$s_h = vt = (3.90)(.40) = 1.56$ meters.

This differs from the value predicted by the method of steps 7 and 8, which we recall is $s_h = 1.249$ meters, and differs from the measured value of $s_h = 1,340$ punti, or 1.260 meters. The error is $(1.56-1.26)/(1.56) \times 100\%$, or, 19% in the latter case.

Stillman Drake in his book, "Galileo:Pioneer Scientist", (1990), feels the table height may in reality have been 800 punti, which is 800 punti x (.00094 meters) / 1 punti   = .752 meters, instead of the value presented in Table 1 of 828 punti = .7783 meters.   In addition, we recognize several other reasons to explain the difference between the predicted value using calculations based on the inclined plane and the experimentally determined value.   Friction between the ball in the plane does impede the motion so that the formula used for our calculation is not exact. The inclination of the plane may not have been exactly 60°. Also the force of gravity causes the ball to rotate as it moves down the plane which alters the value of the acquired velocity.

The first method of predicting the value for s_h given in steps 7 end 8 is more accurate, because we are comparing two experiments by the method of proportion thus resistance and rotation effects tend to cancel out.

Galileo also determined the curved path the ball follows after it leaves the deflector.   He found the curve was parabolic which he called a semiparabola.   Using the methods we learned in our study of coordinate geometry, we can show the path is parabolic from our equations in the metric system.   After the ball leaves the deflector the equations are, $s_h = vt$, and $s_v = 4.9t^2$.

We construct Figure 44-4 placing the origin of the coordinate system at the point where the ball leaves the deflector.

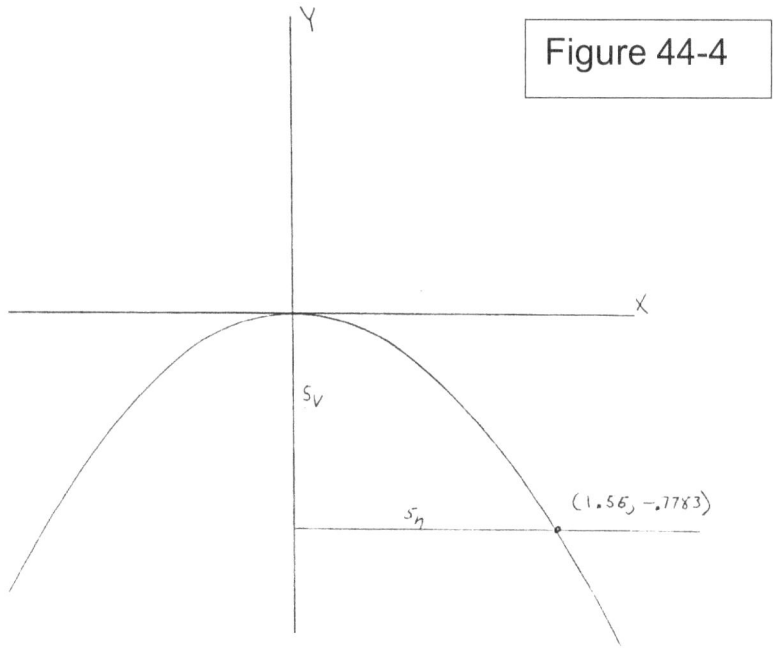

Figure 44-4

We also rotate the curve so s_h will become the positive x coordinate, and s_v the negative y coordinate. This means the path of the ball is represented in the fourth quadrant in Figure 44-4. The equations then become,

15. $x = vt$, and $y = -4.9t^2$.

We solve the first equation for t, and use this value in the second equation.

16. $t = x/v$, and then, $y = -4.9 [x/v]^2$, or,

$y = [(-4.9/v^2)]x^2$.

We recall from chapter 32 and Figure 32-7 that if a parabola is oriented as in Figure 44-4 it equation is, $y = - [1/4a] x^2$, where $4a$ is the length of the latus rectum, and a is the distance from the focus to the apex of the parabola.   The equation in step 16 is of this form. For example, if we use the value of v we determined for the case in step 12 we have,

$v = 3.90$ meters/second, and,

17. $y = [(-4.9)/(3.9)^2] x^2 = -.322x^2$, with a latus rectum of $1/.322 = 3.11$ meters.

Using the formula, $y = -.322x^2$, when $y = -.7783$ meters, the height of the table, then the value of x is found by,

18. $x^2 = -y/.322 = .7783/.322 =$ 2.417, and,

$x = sh = 1.56$ meters,

just as we found in step 14.

Galileo called the path a semiparabola since only part of a parabola is involved in the physical motion of the ball.   To establish the path of the projected ball Galileo did not have the above equations or the methods of coordinate geometry. Instead he used the mathematics of proportion. He first established the fundamental geometric property of the parabola, and then showed that his principle of inertia combined with the law of free fall results in the same proportional relationships that exist for the parabola.

See Figure 44-5.

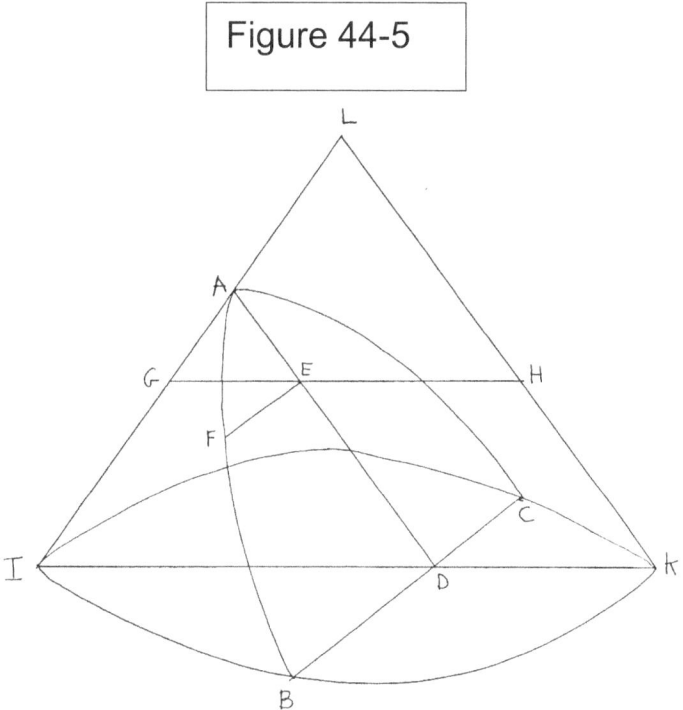

Figure 44-5

This figure shows a right circular cone with base IBKC, which is perpendicular to the axis of the cone, the axis is not drawn in the figure. Lengths IL and KL are equal because the axis of the cone is perpendicular to IK, which is a diameter of the

629

cone's circular base. The angle ILK does not have to be a right angle. We recall that if a right circular cone is cut by a plane such as BAC, parallel to a line from the apex to the base such as LK, then the curve BAC, formed by the intersection of the surface of the cone and the plane, is a parabola. We can imagine that the diameters of the base of the cone BC and IK can be any magnitude we wish so that the arcs of the parabola AB and AC can be any length desired.

BC is perpendicular to KI and AD, the axis of the parabola. AD is parallel to LK, since AD is in the plane of the parabola BAC. F represents any point on the parabola other than the apex A, or points B or C. From F we draw FE such that FE is perpendicular to AD and parallel to BD. Note if F were placed on arc AC of the parabola, then FE would be perpendicular to AD and parallel to CD. GEH is drawn parallel to diameter IDK of the base of the cone. H is on LK, and G on LI.

The fundamental proportional geometric relationship of the parabola is,

$$\frac{(BD)^2}{(FE)^2} = \frac{AD}{DE}.$$

Gallileo had to demonstrate this by means of the mathematics of proportion.

If we imagine AD as the Y axis, and a line perpendicular to AD as the X axis, then this proportion is equivalent to the relation, $y = kx^2$, where *k* is the proportionality constant between *x* and *y*, AE is the *y* coordinate, and FE is the *x* coordinate.   The proportion states the *y* coordinate is directly proportional to the square of the *x* coordinate, which we know from coordinate geometry is the relationship for a parabola.

To demonstrate the relation for the parabola Galileo imagines a plane parallel to the base of the cone through line GEH in Figure 44-5. This plane is shown in Figure 44-6.

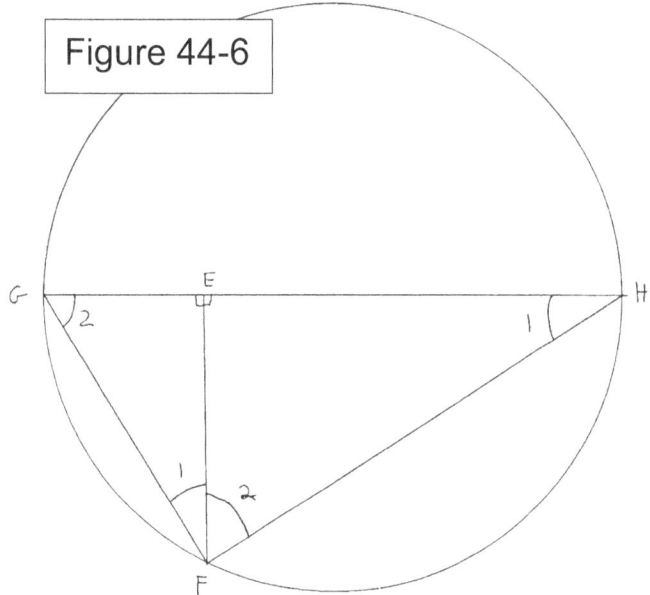

Figure 44-6

We see this section of the cone contains GEH as a diameter with FE perpendicular to GEH. Angle GFH is an angle inscribed in a circle which intersects the circle at the endpoints of a diameter and is thus a right angle.

19.   Angle GFE is marked as angle 1, and angle EFH as angle 2. We see, 1 + 2 = 90°.

Angle FGE is a complement to 1 since FEG is a right angle. We mark FGE as 2. Likewise angle FHE is a complement to 2 and is thus marked 1. This means triangles GEF and FEH are similar so that corresponding sides are in proportion,

20. GE/FE = FE/EH, or, $(FE)^2 = (GE)(EH)$.

We can also consider the circular base in Figure 44-5, namely IBKC, and imagine connecting BI and BK so that the two similar right triangles so formed BDI and BDK, have corresponding sides in proportion,

21. ID/BD = BD/DK, or, $(BD)^2 = (ID)(DK)$.

We divide this result by the result of step 20.

22. $(BD)^2/(FE)^2 = (ID)(DK) / (GE)(EH)$.

When we inspect Figure 44-5 we see DEHK is a parallelogram, since EH is parallel to DK, and DE is parallel to HK. This occurs because to form a parabola the plane that intersects the cone contains DE, and this must be parallel to the edge of the cone which contains HK. Then, EH = DK.

23. We can rewrite step 22 as,

$$\frac{(BD)^2}{(FE)^2} = \frac{ID}{GE}, \text{ since}, \frac{DK}{EH} = 1.$$

From Figure 44-5 we also see triangles AGE and AID are similar, since their bases GE and ID are parallel. Their corresponding sides are in proportion, so we have,

24. ID/AD = GE/AE , or, ID/GE = AD/AE , which we substitute into step 23.

25.

$$\frac{(BD)^2}{(FE)^2} = \frac{AD}{AE}.$$

This is the result Galileo sought and is his fundamental geometric proportion for the parabola. Galileo then uses this result to show that an object moving simultaneously with uniform horizontal velocity, and uniformly accelerated vertical motion will follow the path of a semiparabola.   In Galileo's words,

"the line described by a heavy moveable, when it descends with a motion compounded from equable horizontal and natural falling motion, is a semiparabola".

The figure used by Galileo for his demonstration is found in the chapter entitled, "The Fourth Day" in "The Two New Sciences", as well as in the work sheets from 1608 analyzed by Stillman Drake.   However, in the 1608 work sheets the demonstration is not written out.   We will give Galileo's arguments from "The Two New Sciences". See Figure 44-7.

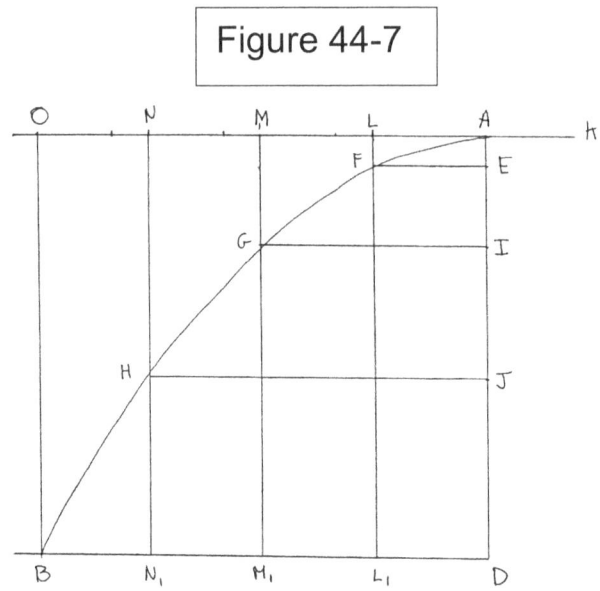

Figure 44-7

Galileo states the object moves from K to A with uniform horizontal velocity. At A the object leaves the horizontal plane KA and then experiences the simultaneous effects of free fall and uniform horizontal velocity. This results in the object following the curved path AFGHB, which Galileo demonstrates to be a semiparabola.

He first extends KA as a straight line to 0. The line AO is divided into equal segments, AL = LM = MN = NO. These segments represent equal time intervals. From points A, L, M, N, and O, perpendicular lines are dropped to line DB, which represents the floor in Galileo's experiments. Line DB represents the horizontal distance moved by the object after it leaves point A.

634

Because the velocity in the horizontal direction is uniform, Galileo can represent the segments of horizontal distance DL_1, L_1M_1, M_1N_1, and N_1B as equal to the segments representing time intervals, since in equal time intervals equal distances are moved in the horizontal direction. Thus, distance DL_1 corresponds to time interval AL, etc.

Motion is also taking place in the vertical direction and line AD represents the vertical distance.   After time interval AL the object has moved vertically downward a distance equal to AE. In time interval LM the object moves vertically downward distance EI, which we note is greater than distance AE, since the vertically downward motion is uniformly accelerated motion according to the law of free fall.   Galileo connects lines EF, IG, and JH. These lines are perpendicular to AD. In this way the path of motion AFGHB is traced out.

Galileo uses the law of free fall to assert,
26. distance LF / distance MG = (time AL)^2 / (time AM)^2.

This proportion is the law of free fall where vertical distances traversed are proportional to the square of the times. If we let time AL equal 1 unit of time then,
27. distance LF / distance MG = 1^2/ 2^2 = ¼.
If LF is designated as 1 unit of length then MG = 4 units of length.

28. By the same reasoning,

distance LH / distance NH = (time AL)^2 / (time AN)^2.

Because the time intervals AL, LM, MN, NO are equal, then   AN = 3(AL) = 3 units of time. Then, distance LF / distance NH = 1^2 / 3^2 = 1/9, or, NH equals 9 units of length.

29. Likewise, distance LF / distance OB = (time AL)^2 / (time AO)^2 = 1^2 / 4^2 = 1/16.

Note distance OB equals 16 units of distance, while time AO equals 4(time AL) = 4 units of time.

Recall lines EF, IG, JH, and DB all represent the horizontal distances traversed, and these distances are assigned the same units of measurement as the vertical distances LF, MG, NH, and OB.   Also, distances LF, MG, NH, and OB are equal respectively to distances AE, AI, AJ, and AD. Thus, for steps 26 through 29 we can write,

30.

$$\frac{AE}{AI} = \frac{(time\ AL)^2}{(time\ AM)^2},\ \frac{AE}{AJ} = \frac{(time\ AL)^2}{(time\ AN)^2},$$

$$\frac{AE}{AD} = \frac{(time\ AL)^2}{(time\ AO)^2}.$$

For the horizontal motion of the ball equal distances are passed in equal times.   This means,

31. EF/IG = AL/AM . We can then write using the first proportion in step 30,

$$\frac{AE}{AI} = \frac{(EF)^2}{(IG)^2}.$$

32.   Likewise, EF/JH = AL/AN, and using the second

proportion in step 30,

$$\frac{AE}{AJ} = \frac{(EF)^2}{(JH)^2}.$$

33. And also, EF/DB = AL/AO . Using this in the third proportion of step 30,

$$\frac{AE}{AD} = \frac{(EF)^2}{(DB)^2}.$$

We can invert this last proportion and rewrite the terms, so it reads,

34.

$$\frac{(BD)^2}{(FE)^2} = \frac{AD}{AE}.$$

If we compare this to step 25, and compare Figure 44-7 to Figure 44-5, we see they are the same, and that this establishes that the path of the ball in Figure 44-7 is that of the parabola of Figure 44-5. In order to do this demonstration Galileo had to treat the lengths AL, AM, AN, and AO as time intervals equal to their respective horizontal distances in order to write, AL/AM = EF/IG, etc. This requires the validity of the principle of inertia operating in the horizontal direction, while at the same time the ball is falling vertically according to the law of free fall. Galileo's genius was in being able to resolve the motion of the projected ball into two separate motions each motion acting according to a mathematical law. Galileo concluded his demonstration by stating.

"And it is similarly demonstrated, assuming any equal parts of time, of any size whatever, that the places of moveables carried in like compound motion will be found at those times in the same parabolic line."

Following the demonstration of the parabolic path for projected objects Galileo discusses objections to the postulate that such motion should be studied by removing all impediments such as the retarding effects of the medium through which the object moves. These objections might be stated as follows. The projectile's horizontal motion is always slowed somewhat by the medium through which it moves, so that in reality there is no uniform horizontal velocity. The surface of the earth in reality is not a horizontal plane, but it is curved so that the downward uniform acceleration of gravity changes direction as a projected body moves through the air.

Galileo admits the validity of these objections. Newton was to show that when the earth's curvature is taken into account the actual paths of projected bodies are elliptical rather than parabolic. Galileo points out that over commonly observed distances by a stationary observer on the earth, an observer will find the measurements from experiments agree closely with the laws of motion which Galileo proposed, and with his mathematical method of considering projectile motion as compounded of two motions, uniform horizontal motion according to the principle of inertia, and uniformly accelerated motion vertically toward the earth according to the law of free fall.

Galileo then takes up a long discussion in "The Two New Sciences" of the parabolic trajectory of cannon balls, where the cannon is aimed with different inclinations to the horizon plane.

We will study the motion of a projectile fired at an angle from the horizontal as shown in Figure 44-8.

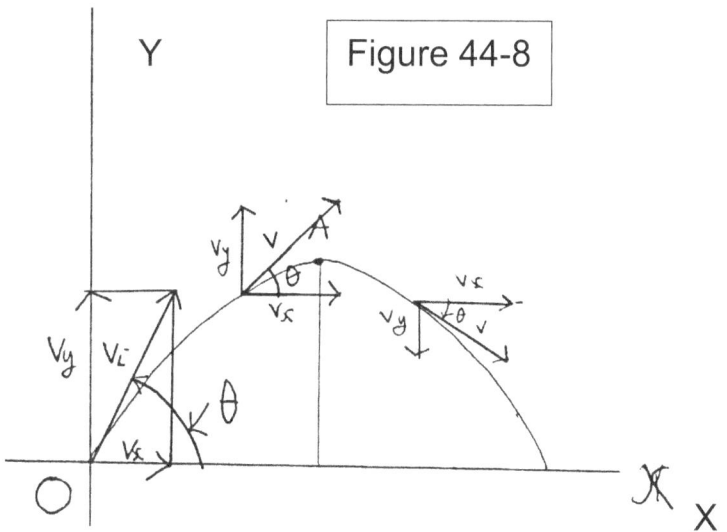

Figure 44-8

In the figure the origin of a Cartesian coordinate system is marked as the point of projection of a presumed cannon ball. The angle of projection from the horizontal plane of the earth's

639

surface, which is the X axis, is marked as angle θ (Greek letter theta).

We will work an example in the metric system where the initial velocity of the cannon ball is 300 meters/second. The direction for this initial velocity is not horizontal as in the inclined plane experiments done by Galileo in 1608, but at angle θ from the earth's surface measured counterclockwise. Throughout the trajectory of the cannon ball the acceleration due to gravity acts on the ball in the negative y direction. We call the value of this acceleration ay. From our previous work we know,

$ay = 9.8 \ meters/(second)^2$.

There is no acceleration imparted to the cannon ball in the horizontal or x direction. The explosion of the gunpowder gives the cannonball an initial velocity in the direction determined by θ. This means the cannon ball's motion in the positive x direction follows Galileo's principle of inertia wherein the motion of the cannon ball proceeds with uniform velocity. We see from the figure,

35. $\cos θ = v_x / v_i$, where v_x is the constant velocity of the cannon ball in the positive x direction, and v_i is the magnitude of the initial velocity at angle θ from the X axis.

We can represent the velocity v_i as having two components, v_x, which is the constant horizontal velocity, and v_y which is the changing

640

vertical velocity.   We have indicated the direction and relative magnitudes of these two components at three points of the cannon ball's trajectory.   We should remark that Galileo did not use this method of analysis.   Newton later in the 17th century did recognize that physical quantities such as velocity and acceleration have both magnitude and direction, either of which, or both of which, may vary in the course of an object's motion.   Further, he recognized that both the magnitude and direction of a physical quantity can be resolved into components such as those we have used in resolving v_1.   In this case the two components of the velocity at each point of the trajectory are resolved into components v_x and v_y perpendicular to each other.

We can write step 35 as,

36. $v_x = v_i \cos \theta = 300 \cos \theta$, since $v_i = 300$ meters/second.

We noted previously there is no acceleration in the x direction, so $a_x = 0$, where a_x is the acceleration in the x direction.   From the calculus we recall the derivative of the velocity is the acceleration so.

37.

$$\frac{dv_x}{dx} = a_x = 0,$$

or using differentials, $dv_x = 0\ dt$.   We then find the antiderivative, or integral, $\int dv_x = \int 0\ dt$,

$v_x = 0 + C$, where C is the constant of integration.

In integrating *0 dt*, we are finding the antiderivative whose derivative equals 0. The result must be a constant or 0, because the derivative of a constant or of 0 is always 0, no matter what values the constant may have.

Since we know $v_x = 300cos\ \theta$, then, C = 300cos θ. Thus we confirm with the calculus that v_x is constant for the motion of the cannon ball.   If we call the horizontal distance moved s_x, then, since the derivative of the distance is the velocity,

38.

$$\frac{ds_x}{dx} = v_x = 300\cos\theta,$$

or using differentials, $ds_x = 300cos\ \theta\ dt$.

Integrating, $\int ds_x = \int 300cos\ \theta\ dt$

$= [300cos\ \theta]\int dt,$

$s_x = 300(cos\ \theta)t + C$. When $t= 0$, $s_x = 0$, so C = 0.

$s_x = 300tcos\ \theta$.

This is the equation for the distance moved horizontally in the positive x direction. We next turn to the motion in the y direction. From Figure 44-8 when $t = 0$ the cannon ball is at the origin, and

39. sin $\theta = v_y / v_i$, or, $v_y = v_i$ sin $\theta = 300$ sin θ.

We know the acceleration due to gravity is in the negative y direction and equals, $a_y = -9.8$ meters/second^2. This acceleration is also equal to the derivative of v_y.

40.

$$\frac{dv_y}{dt} = a_y = -9.8, \, or, \, dv_y = -9.8dt.$$

Integrating, $\int dv_y = -9.8 \int dt,$

$v_y = -9.8t + C,$

where C is the constant of integration.

When $t = 0$, we know, $v_y = 300 \sin \theta$, so

$300\sin\theta = (-9.8)(0) + C$, or,

$C = 300 \sin \theta$.

Therefore, for any point of the trajectory,

$v_y = -9.8t + 300 \sin \theta$.

We can also interpret v_y as the rate of change of the vertical distance, or the derivative of the vertical distance s_y. We see from the figure v_y has a positive value as long as the cannon ball continues to move upward away from the earth or X axis. At the instant the cannon ball reaches the apex of its trajectory, point A, the cannon ball is not moving up or down, and thus $v_y = 0$ at that instant in time. Point A is also the maximum height the cannon ball reaches above the earth. We will find the value of the time when the cannon ball is at point A, and $v_y = 0$.

41. $v_y = -9.8t + 300 \sin \theta$,

$0 = -9.8t + 300 \sin \theta$,

$9.8t = 300 \sin \theta$, or,

$t = (300/9.8)\sin \theta = 30.61 \sin \theta$.

We next find the height of point A above the X axis. To do this we must find y coordinate of A when $t = 30.61 \sin \theta$. We use the fact the the derivative of the vertical distance equals v_y.

42.

$$\frac{ds_y}{dx} = v_y = -9.8t + 300 \sin\theta.$$

Using differentials, $ds_y = -9.8t + 300\sin\theta\ dt$.

Integrating, $\int ds_y = -9.8\int t\ dt + (300 \sin \theta)\int dt$.

$s_y = [-9.8 / 2]\ \ t^2 + (300\sin \theta)t\ \ + C$.

$s_y = -4.9t^2 + (300\sin\theta)t + C$.

When $t = 0$, $s_y = 0$, so we see $C = 0$. To find the maximum height, or the y coordinate of point A we substitute the value, $t = 30.61 \sin \theta$.

43.　$s_y = -4.9(30.61 \sin \theta)^2 + 300(30.61 \sin \theta)\sin \theta$,

$s_y = -4,591(\sin \theta)^2 + 8,183(\sin \theta)^2$

$s_y = 4,592(\sin \theta)^2$.

We see from this result that the maximum height the cannon ball reaches depends on the value of θ, the angle of inclination of the cannon when it is fired. If the cannon were pointed straight up, i.e., $\theta = 90°$, then $[\sin \theta]^2 = [\sin 90°]^2 = 1$, and the maximum height the cannon reaches would be 4,592 meters when the initial velocity v_i is 300 meters/second in magnitude.

We will now determine the equation of the trajectory and confirm that it is parabolic when $\theta > 0°$, but, $< 90°$.

644

44. We have found s_x and s_y as functions of the time t after firing the cannon. The values of s_x and s_y are the x and y coordinates of the trajectory so we can designate s_x by x, and s_y by y. We then have for our equations,

$x = 300t \cos \theta$, and, $y = -4.9t^2 + 300t \sin \theta$.

We can solve the first equation for t, and use the result in the second equation. This will give us y as a function of x, and we will show that this is the equation of a parabola. When the variables x and y are given in terms of a third variable, in this case t, mathematicians call the equations for x and y in terms of the third variable, t, the parametric equations of x and y. We have seen it is easier for us to derive the parametric equations using Galileo's principle of inertia and the law of free fall than it would be to find the equation of y in terms of x directly from Galileo's laws. First we solve the parametric equation involving x and t for t.

45. $t = x / 300 \cos \theta$. We substitute this value of t into the parametric equation relating y and t.

46. y

$= -4.9[x / (300 \cos \theta)]^2 + (300 \sin \theta) [x / 300 \cos \theta]$

$$y = \frac{-4.9x^2}{90,000\cos^2 \theta} + \frac{\sin\theta}{\cos\theta}x,$$

$$y = \frac{(-5.444)(10^{-5})}{\cos^2 \theta}x^2 + (x)\tan\theta.$$

We see that this equation, giving y as a function of x, is not in the form of the equation of a parabola that we have learned, namely, $y = ax^2$, or, $y = -ax^2$. The reason is that when we derived the equation for a parabola from Galileo's inclined plane experiments, we chose the coordinate axes so that the apex of the parabola was at the origin of the coordinate system as in Figure 44-4. In Figure 44-8 the origin is away from the apex A on one side of the parabola. It is possible to get the equation in step 46 into the form, $y = ax^2$, where a can be either positive or negative, by a process known as translation of the coordinate axes.

The equation in step 46 is in the form, $Ax^2 + Dx + Ey = 0$, and since we can regard E = -1, we can write A and D from the equation in step 46 as,

$A = (-5.444)(10^{-5})$ /cos θ, and D = tan θ.

We wish to have the equation in the form,

$$y = \frac{-A}{-E}x^2, where, \frac{-A}{-E} = a, \text{ so that, } y = ax^2.$$

We also want $\theta = 0^0$, so D = 0.

We see, $y = (-A/-E)$ x2, is equivalent to,

$Ax^2 + Ey = 0$, if both sides are multiplied by E, followed by adding Ax^2 to each side of the equation. Our task is then to learn how to convert an equation of the form, $Ax^2 + Dx + Ey = 0$, into an equation of the form, $Ax^2 + Ey = 0$.

To do this we relocate the coordinate axes in Figure 44-8 to the position shown in Figure 44-9.

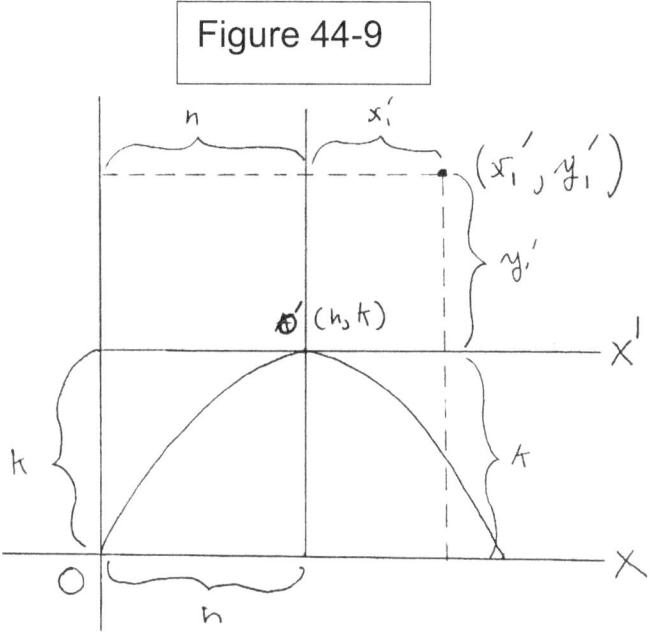

Figure 44-9

We do this because we know if the apex of a parabola is at the origin of the coordinate axes, then the parabola will have an equation of the form, $y = ax^2$, or the equivalent form. $Ax^2 + Ey = 0$. The new coordinate axes in Figure 44-9 are marked X' and Y'. Point O' is the origin of the X'Y' coordinate system, and has coordinates (h, k) in the XY coordinate system, and coordinates (0, 0) in the X'Y' coordinate system. In general we call the coordinates in the X'Y' system x' and y'. To convert coordinates from the X'Y' system into the XY system we use the conversion equations,

$x = x' + h$, and, $y = y' + k$.

This makes sense, since we can see from Figure 44-9 that a point marked with coordinates

647

(x_1', y_1') in the X'Y' system has a horizontal distance of $h + x_1'$ from point O, the origin of the XY system, and a vertical distance of $k + y_1'$ from point 0.

For example, when $x' = 0$, and $y' = 0$, in the X'Y' system, then by the conversion equations, $x = x' + h$, or, $x = h$. Likewise, $y = y' + k$, or, $y = k$, which we know are the coordinates of the origin O' in the X'Y' coordinate system.

If we start with the equation,

$Ax^2 + Dx + Ey = 0$, in the XY coordinate system, then with the proper choices for h and k, which should be the coordinates of O', then we should have an equation of the form, $Ax'^2 + Ey' = 0$, in the X'Y' system. We then substitute for x in the first equation $x' + h$, for x, and for y the value, $y' + k$.

47. $A(x' + h)^2 + D(x' + h) + E(y' + k) = 0$,

$A(x'^2 + 2x'h + h^2) + Dx' + Dh + Ey' + Ek = 0$,

$Ax'^2 + 2Ahx' + Ah^2 + Dx' + Dh + Ey' + Ek = 0$,

$Ax'^2 + (2Ah + D)x' + Ey' + G = 0$, where

$G = Ah^2 + Dh + Ek$.

We see G is a constant term.

We wish to find values for h and k in the last equation relating x' and y' such that, $G = 0$, and $(2Ah + D)$, the coefficient of x', also equals 0, for then the equation will be of the form, $Ax'^2 + Ey' = 0$, the equation of a parabola with the apex at A the origin of the X'Y' system.

648

48. We let $(2Ah + D) = 0$, so the coefficient of x' is 0.

   Then, $h = -D/2A$.

   We let $G = 0$, so $Ah^2 + Dh + Ek = 0$. Then,

   $Ek = -Ah^2 - Dh$, or

$$k = \frac{-Ah^2 - Dh}{E}.$$

We substitute the value of h,

$$k = \frac{-A(-D/2A)^2 - D(-D/2A)}{E} =$$

$$\frac{-A(D^2/4A^2) + (D^2/2A)}{E},$$

$$k = \frac{-D^2/4A}{E} + \frac{D^2/2A}{E} = -\frac{D^2}{4AE} + \frac{D^2}{2AE}.$$

   We recall $h = -D/2A$.

   When h and k have the above values, then,

   $2Ah + D = 0$, and

   $G = 0$. Thus,

49. $Ax'^2 + 0x' + Ey' + 0 = 0$

   $Ax'^2 + Ey' = 0$

   $Ey' = -Ax'^2$

   $y' = -(A/E)\ x'^2$.

   This equation is of the form, $y = -ax^2$, where, $a = A/E$, so we know the curve is parabolic.   We recall from step 46 above the equation in the XY system for the path of the cannon ball is,

$$y = \frac{(-5.444)(10^{-5})}{\cos^2 \theta} x^2 + (x)\tan\theta.$$

This means A = the coeficient of x^2, D the coefficient of x, and E =-1, in the general equation, $Ax^2 + Dx + Ey = 0$, which we now know can be converted into the equation for a parabola by choosing h and k according to step 48.   To find the numerical values of h and k we have to know the value of θ so we can calculate A and D.   Thus we know the trajectory of the cannon ball is parabolic in the ideal case where there is no impediment to its motion by the air.

Using the equation for the parabolic trajectory in the XY system we can ask what is the range of the cannon ball?   The range is the horizontal distance x at the time the cannon ball returns to the ground.   This occurs when $y = 0$.
50.

$$0 = \frac{(-5.444)(10^{-5})}{\cos^2 \theta} x^2 + (X)(\tan \theta),$$

$$\tan \theta = \frac{(5.444)(10^{-5})}{\cos^2 \theta} x,$$

after dividing both sides by x, and subtracting the first term from both sides,

$$x = (\tan \theta)(\cos^2 \theta)\frac{1}{(5.444)(10^{-5})},$$

$$x = \frac{\sin \theta}{\cos \theta}(\cos^2 \theta)(1.8367)(10^4),$$

$$x = 18{,}367 \sin\theta \cos\theta.$$

In chapter 13 we derived several trigonometric identities one of which we shall use now, namely, 51. sin 2A = 2(sin A)(cos A), or using θ instead of A, and dividing by 2, (sin θ) (cosθ) = sin 2θ /2 . We use this in step 50.

52.

$$x = \frac{18{,}367}{2} \sin 2\theta = 9{,}184 \sin 2\theta.$$

This equation gives us the range of the cannon ball. We see that it depends on the value of θ, as we would expect.   When θ= 45°, then sin 2θ = sin 90° = 1.   This is the largest value the sine of an angle may have, so that when θ = 45° we have the maximum value x attains, or the maximum range of the cannon ball. The value of x is 9,184 meters in our example where the initial velocity is 300 meters/second and θ = 45°.

The mathematical method we used can be used for any projectile, provided we understand air resistance is neglected in the calculation, and we know the initial velocity an angle of inclination of the projectile at the time of release.

We showed earlier (steps 41-43), that the time for the cannon ball to reach its maximum height was, 30.61(sin θ). Keeping this value in mind we will find the time it takes the cannon ball to reach the ground after being fired, i.e. the time it takes for the cannon ball to reach its range.   We start with the parametric equation, $y = -4.9t^2 + 300t$ sin θ.   We solve this equation for $y = 0$, the point at which the cannon ball returns to the earth.

53. $0 = -4.9t^2 + 300t$ sin θ. Dividing by t,

   $0 = -4.9t + 300$ sin θ,

   $4.9t = 300$ sin θ

   $t = 61.22$ sin θ, or, $t = 2(30.61)$sin θ.

   We note this is twice the time for the cannon ball to reach the apex or maximum height. This fits with the idea from inspecting Figure 44-9 that the parabolic trajectory is symmetric about the perpendicular line from the apex to the X axis.

   We next calculate the horizontal distance from the origin to the position perpendicular from point A in Figure 44-9. The parametric equation for the horizontal distance is,

54. $x = 300t$ cos θ.

   At the point below A, the maximum height, we found, $t = 30.61$ sin θ. Then,

   $x = (300)(30.61)($sin θ$)($cos θ$)$.

   Since, (sin θ) (cos θ) = sin 2θ /2 , we have.

$$x = \frac{(300)(30.61)}{2}\sin2\theta = 4{,}592\sin2\theta.$$

Since 4,592 is ½ of 9,184, then the distance from the origin to the point below the maximum height of the trajectory is ½ of the range, which we found in step 52 to be,9,184(sin 2θ).   This also confirms the symmetry of the motion about point A, the apex of the parabolic trajectory.

As we have noted several times Galileo's theory of free fall and the principle of inertia, we used to derive these equations neglect the air resistance.   The actual paths of cannon balls or other projectiles, such as golf balls, or hit baseballs are not true parabolas, but altered parabolas. Mathematicians in the 18th and 19th centuries were able to add into the basic equations the alterations which give a more accurate path for projectiles. Nonetheless, the approach of Galileo was a giant step in the understanding of projectile motion, and in fact may be considered the crucial step.

We will solve one more problem related to projectile motion. We will determine the magnitude of the velocity of the cannon ball at the time of impact with the earth.   Recall from steps 35 and 40, the initial velocity v_i , can be resolved into two mutually perpendicular components,

$v_x = v_i \cos \theta$, and, $v_y = -9.8t + v_i \sin 0$.

From Figure 44-8 we use the Pythagorean Theorem to state,

55. $v_i^2 = v_x^2 + v_y^2$.

This equation is true not only for the initial velocity, but also for the instantaneous velocity at all points on the trajectory of the cannon ball, including the point of impact with the earth. If we use v for the magnitude of the instantaneous velocity at any point on the path of the cannon ball and continue to use v_i for the initial velocity then,

56. $v^2 = v_x^2 + v_y^2$, where v_x and v_y have the values given above.

$v^2 = v_i^2 \cos^2 \theta + (-9.8t + v_i \sin \theta)^2$,

$v^2 = v_1^2 \cos^2 \theta + (9.8)(9.8)t^2 - 2(9.8)v_i\, t \sin \theta$
$+ v_i^2 \sin^2 \theta$,

$v^2 = v_i^2 (\cos^2 \theta + \sin^2 \theta] + (9.8)2t^2 - 2(9.8)v_i\, t \sin$
θ. We recall the trigonometric identity,
$\cos^2 \theta + \sin^2 \theta = 1$.

$$v^2 = v_i^2 + \frac{-2(9.8)^2}{-2}t^2 + (-2)(9.8)v_i\, t \sin\theta.$$

By rewriting the equation in this way we can factor the term, (-2)(9.8), from each of the three terms on the right side to get:

$$v^2 = (-2)(9.8)[\frac{v_i^2}{(-2)(9.8)} - \frac{9.8}{2}t^2 + v_i t \sin\theta].$$

The parametric equation for y in terms of t from step 44 is, $y = -4.9t^2 + 300t \sin \theta$. From step 39, $300 \sin \theta = v_i \sin \theta$, thus,

57. $y = -4.9t^2 + v_i\, t \sin \theta$. Since $-4.9 = -9.8/2$, we can substitute y for the last two terms in the brackets in the equation for v^2.

$$v^2 = (-2)(9.8)[\frac{v_i{}^2}{(-2)(9.8)} + y],$$

$$v^2 = v_i{}^2 + (-2)(9.8)y.$$

At the point of impact $y = 0$, so,

58. $v^2 = v_1{}^2$, **or, $v = v_i = 300$ meters/second.**

The magnitude of the velocity of the cannon ball at impact is the same as the initial velocity. However, from Figure 44-8 we see the direction of the cannon ball has changed 90° clockwise.

We have taken time to develop the mathematics of projectile motion at least in the idealized state where there is no air resistance. This involved much algebra and coordinate geometry and parametric equations, all of which increases our mathematical abilities to tackle the problems in subsequent chapters. Those readers who may wish to pursue the subject in more depth are referred to the bibliography of this chapter. Galileo completed his work on projectile motion in 1608 and published the theory in 1638 in "The Two New Sciences". He did not publish the work in 1609 or in the immediate period thereafter, because he began his researches with the newly invented telescope in 1609. In the following chapter we will review his telescopic discoveries and his defense of the Copernican system.

Bibliography for Chapter 44

1. Galilei, Galileo, "Discourses and Mathematical Demonstrations Concerning Two New Sciences", 1638, translated with a new introduction and notes by Stillman Drake, Second Edition, Wall and Thompson, Toronto, 1989.

2. Drake, Stillman and MacLachan, James, "Galileo's Discovery   of the Parabolic Trajectory", in "Scientific American", Vol. 232, March 1975, 101-110.

3. Drake, Stillman, "Galileo's Experimental Confirmation of Horizontal Inertia: Unpublished Manuscripts", in "Isis", 64(1973), 291-305.

4. Drake, Stillman, "Galileo: Pioneer Scientist", University of Toronto Press, 1990.

5. Kline, Morris, "Calculus: An Intuitive and Phsysical Approach", Part Two, Chapter 18:   "Rectangular Parametric Equations and Curvilinear Motion", John Wiley and Sons, Inc., 1967.

Chapter 45

Galileo and the Copernican System

In 1609 two Dutch lens makers, Hans Lippershey and Zacheria Jansen, constructed the first useful telescope. They used it to magnify distant objects on the earth's surface. Galileo heard of their invention and began work to make superior telescopes which he intended to use to study celestial objects. Not only did he improve upon the magnification, but also he improved the resolution of the telescope, that is the ability to distinguish two nearby points of light. He achieved a magnification of about 30 times with much better resolution than previous models.

Galileo made many observations of the night sky and wrote a brief treatise published in March of 1610 called, "*Sidereus Nuncius*", or "The Starry Messenger". He studied the surface markings of the moon, and he made the first observations of the moons of Jupiter which he named the Medicean Stars.

He also observed stellar phenomena and saw the phases of Venus which were very similar to those of our own moon. All of these observations began to convince Galileo that the Ptolemaic system was incorrect, and that the Copernican system represented the true reality of the heavens. He formed this opinion by 1613.

From 1613 to 1616 he tried to convince church authorities that the Copernican theory was a valid hypothesis, so that it could be discussed openly and not banned. In 1615 he wrote his "Letter to the Grand Duchess Christiana" where he states clearly his view. The letter wasn't published until 1636, although many scholars were aware of it. It is useful to quote from this letter to show the reader Galileo's belief in the Copernican system.

"I hold the sun to be situated motionless in the center of the revolution of the celestial orbs while the earth rotates on its axis and revolves about the sun...I support this position not only by refuting the arguments of Ptolemy and Aristotle, but by producing counterarguments..."

By 1616 Church authorities in Rome told Galileo to abandon teaching the theory of Copernicus and condemned it as heretical. However, Galileo continued to believe in the theory of Copernicus, and by 1626 he began work on a large treatise in which he argued in favor of the earth's motion about its axis and around the sun. The treatise is called the "Dialogue Concerning the Two Chief World Systems - Ptolemaic and Copernican", and it consisted of many arguments by interlocutors, one an Aristotelian, one an intermediary, and one a Copernican who represented Galileo's views. By structuring the book in this way and writing in his native Italian most educated people could follow the debate and understand the reasons why Copernicus had a valid system.

The book was published in 1632. It was very popular and all copies printed seemed to have been sold. Galileo became the most known of all astronomers because of his discoveries with the telescope.

However, when authorities from the church in Rome read the book they realized the book favored Copernicus, and the book was banned. Galileo was taken to the Inquisition and tried for heresy in 1633. He was found guilty of violating his admonition not to teach the Copernican theory from 1616. He was forced to recant his views and live in his house in Arcetri near Florence for the remainder of his life.

In our study of Galileo's support of Copernicus we will see how mathematics and the principle of inertia can be used to argue that the earth can be in motion, although we on earth do not perceive this motion.

Aristotelian philosophers developed the following argument against the motion of the earth. If a heavy object is dropped from a building, and if the earth is rotating about its axis, then during the time interval that the object falls the surface of the earth, which initially is directly below the dropped object, should move so that the object ought to land

659

some distance from the position directly below the point where the object was released. For example, if the object falls 200 meters, and obeys Galileo's law of free fall then,

1. $s = 4.9t^2$, or, $t^2 = 200/4.9$, and
    $t = 6.39$ seconds.

If we are at 36° north latitude, we see from Figure 45-1

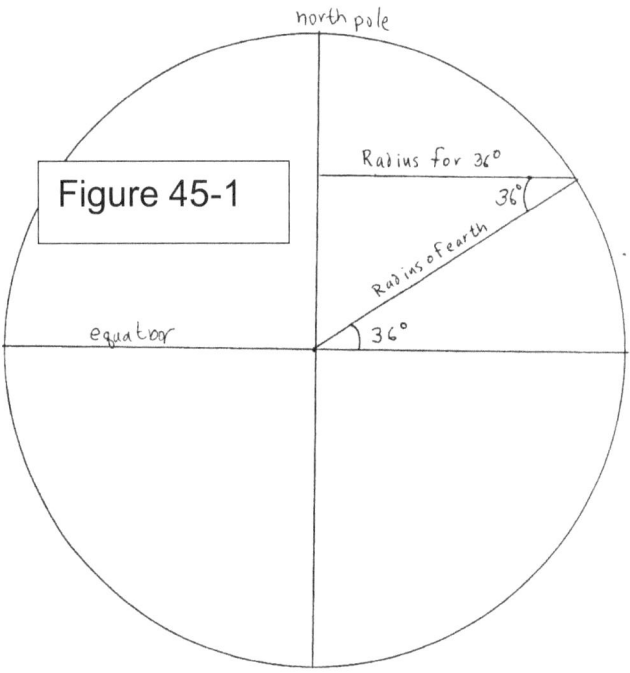

that the radius of the 36° latitude circle in terms of the earth's radius can be found by,

2. $\cos 36^0$ = radius for 36° / radius of earth,
   radius for 36° = R(cos 36°), where R is the radius of the earth.

3. If we take the circumference of the earth as 24,000 miles, or (24,000 miles)(1,609 meters/mile)

660

= 38,616,000 meters, then, $2\pi R$ = 38,616,000 meters. R = (38,616,000) / (2)(3.1416) = 6,360,000 meters.

4. radius for 36° = (6,360,000)(cos 36°) = 5,140,000 meters,

5. circumference for 36° latitude = 2π(5,140,000) = 32,300,000 meters.

6. The earth takes 24 hours to rotate through this
circumference.

Thus,

$$\frac{32{,}300{,}000\ meters}{24\ hours} = \frac{32{,}300{,}000\ meters}{86{,}400\ sec\,onds} = 374\ meters\ /\ sec\,ond.$$

This means if the Aristotelian philosophers are correct in the 6.39 seconds that the object falls 200 meters the earth moves,

7. (6.39 seconds)(374 meters/second) = 2,390 meters.

Everyone knew that objects dropped from buildings do not land a great distance from the building much less 2,390 meters. Because of this the Aristotelians argued that the earth could not be moving, that is rotating around its axis once every 24 hours.

Galileo refuted this type of argument in his book. His reply does not prove the earth rotates, but it does show that it is possible the earth could be rotating, and that the Aristotelian argument does not prove the earth is stationary. In our presentation of Galileo's argument we will paraphrase Galileo and make use of our knowledge of the metric system of measurement.

Galileo envisioned the following experiment. Imagine a sailing ship traversing along the shore of a large lake at uniform velocity as seen by an observer sitting on the shore. Let us say the ship is moving at 5,000 meters per hour, which is 1.40 meters/second. Let us imagine the observer can also see behind the ship and make out the opposite shore of the lake behind the ship.

Another observer is on the ship itself positioned near the bottom of the ship's mast. A sailor climbs up the mast to release a heavy brass ball from the top of the mast. The observer on board the ship sees the ball fall in a straight line and land below the point of release. Galileo maintains the observer on board the ship will see the ball move in a straight line from the top of the mast to the point directly below as long as the ship is moving at uniform velocity relative to the shore no matter how great the magnitude of the uniform velocity. The observer on board the ship records the motion of the ball as an example of the law of free fall, i.e. motion in a straight line with uniform acceleration even though the ship is moving at constant velocity as seen by the observer on the shore. Galileo confesses in the "Dialogue

Concerning the Two Chief World Systems" that he has not actually carried out this experiment, but argues the result must be as he describes, because the motion follows the same principle of free fall motion he has studied combined with motion with uniform horizontal velocity. Here is an example of Galileo's plan of deducing new knowledge from fundamental principles, the law of free fall and the principle of inertia, discovered in his previous experiments, and carrying out a thought experiment based on his principles.

The Aristotelian philosophers would argue that in this experiment the brass ball would only land at the bottom of the mast if the ship were stationary relative to the observer on the shore. When the ship is in motion relative to the shore, then the released ball must land, according to the Aristotelians, some distance behind the point of release.

When we explain this thought experiment today we say the observer on shore has a different reference system of measurement compared to the reference system of measurement for the observer on the ship. A reference system is a system of coordinate axes set up for determining the horizontal and vertical distances involved in motion as we show in Figure 45-2.

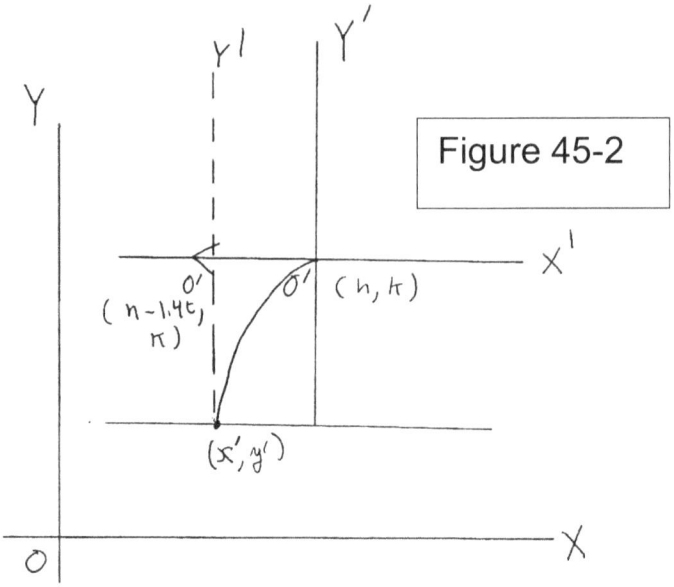

Figure 45-2

The reference system for the observer on the shore is the XY coordinate system. The reference system for the observer on the ship is the X'Y' coordinate system, which is stationary for the observer on the ship, but which is moving in this case in the negative x direction with uniform velocity as seen by the observer on the shore.

When two reference systems differ because one is moving with uniform linear velocity compared to the other, then we say the two reference systems are Galilean reference systems in honor of Galileo's famous thought experiment which we are describing.

In Figure 45-2 the entire X'Y' system and all its points are moving with linear (straight line) uniform velocity in the horizontal direction toward negative values for x. At the point the ball is released from the top of the mast the origin of the

X'Y' coordinate system has coordinates (h, k) as seen by the observer on shore, although the coordinates in the X'Y' system are $(0, 0)$ for the observer on the ship.   At the time the ball hits the deck below the mast a new position of the X'Y' system as seen by the observer on the shore is indicated by the broken line for the Y' axis.

The observer on the ship sees the ball as starting from point O', the origin of the X'Y' system, and following the Y' axis until it hits the deck at point (x', y').   Note that for the observer on the ship the value of x' is 0, since the ball follows the negative Y' axis, and the value of y' is a negative number determined by the law of free fall and the height of the mast.

The observer on shore sees a different path for the motion of the ball.   For him the ball follows the curved path from (h, k) to (x', y') in the XY coordinate system.   Since the mast, represented by the Y' axis, is moving at 1.4 meters/second in the negative x direction, after time t, the time it takes for the ball to fall from the top of the mast to the deck, the coordinates of 0' will be ([h - 1.4) t], k).   The y coordinate of the origin, of course, does not vary with linear horizontal uniform velocity in the negative direction.

Galileo argues, as in his work with projectile motion, that the ball for the observer on the earth has motion compounded by two components, the

665

horizontal motion of the ball given to it because it shares in the horizontal motion of the ship as seen from the shore, and the vertical motion in the negative direction due to the uniform acceleration of gravity after it is released from the top of the mast.   The horizontal motion obeys the principle of inertia because the ship and ball are moving with uniform horizontal velocity as seen by the observer on the shore.   The vertical motion obeys the law of free fall which is towards the center of the earth. This is the same situation as in projectile motion, so we can predict that the path of the ball as seen by the observer on the shore is that of a semiparabola.

The observer on the ship is unaware of the ship's motion relative to the shore as he watches the ball fall, and therefore sees the motion as an example of vertical motion in the negative direction with uniform acceleration.

We will derive the equations for the motion of the ball using the value of 1.4 meters/second for the horizontal motion of the ship toward the negative x direction.   The x coordinate of the ball in the XY system is:

8. $x = h - 1.4t$, and the y coordinate is,
9. $y = k - 4.9t^2$.

In these equations, $t = 0$ at the time the ball is released from the mast.   For the values of h and k we choose, $h = 10$ meters, and $k = 0$. This means in determining our equations the origin O' of the

666

X'Y' system lies on the X axis of the XY system instead of the position shown in Figure 45-2. We make this choice so the mathematics will be easier to follow. The top of the mast at the time of release is on the X axis 10 meters to the right of the origin 0 of the XY system.

10. Equation 8 becomes, $x = 10 - 1.4t$, or,

$$t = \frac{x - 10}{-1.4}.$$

11. Equation 9 becomes, $y = -4.9t^2$.

We then substitute the value of t from step 10 into this equation.

12.

$$y = -4.9(\frac{x - 10}{-1.4})^2 = \frac{-4.9}{1.96}(x^2 - 20x + 100),$$

$$y = -2.5x^2 + 50x - 250,$$

$$or, \ -2.5x^2 + 50x - y - 250 = 0.$$

This equation represents the equation for the path of the ball as seen by the observer on the shore at the origin of the XY coordinate system. This is the equation of a parabola of the form, $Ax^2 + Dx + Ey + G = 0$. We learned an equation of this form is a parabola in chapter 44.

After explaining the thought experiment of the ball dropped from the mast of a moving ship, Galileo can reason by analogy that the earth itself can be considered as a large ship sailing through space with an imaginary observer on the stationary sun. Such an observer would see the earth move at essentially uniform velocity, at least over short periods of time, and if objects could be seen dropped from buildings on the earth, they would have the compounded motion of uniform horizontal velocity because they are carried along with the earth, and the motion of uniform vertical acceleration towards the earth's center, and thus follow parabolic paths as seen by the observer on the sun, because the dropped objects would be obeying the principle of inertia and the law of free fall. Meanwhile observers on earth would see objects dropped from buildings as having motion with free fall, just as the observer on the ship in Galileo's thought experiment. Although this argument does not prove the earth is moving through space with a stationary sun, it does show that it is possible the earth is moving, and therefore the Aristotelian argument that the earth must be stationary is refuted.

Galileo's reasoning also establishes the importance of establishing reference frames or coordinate systems when studying the motion of objects. Later in our work when we study curvilinear motion under variable acceleration, we will see that it is crucial to define the reference frame and coordinate system for the motion.

Another argument of the Aristotelian philosophers involved considerations of a rotating sphere. The Aristotelians argued that if the earth were rotating about its north south axis daily as Copernicus claimed, then objects on the earth's surface, especially those on the equator where the rotational velocity is the largest, would be thrown off into space. This argument is not found in the works of Aristotle, but was developed as considerations in Ptolemy's theory. Copernicus discusses it in "*De revolutionibus*", and Tycho Brahe also referred to it.

Aristotle himself made the distinction, as we have mentioned, between natural and violent motion. Natural motion was the movement of objects towards the earth's center, which was considered the center of the universe. This explained the motion of dropped balls and movements down inclined planes. Violent motions were motions caused by an interfering agency. Examples are objects thrown into the air, or objects pushed or pulled along horizontal surfaces.

A motion such as rotation of the earth would have to be considered a violent motion in Aristotle's scheme, since the result of the motion is to cause objects to rotate around the center of the earth and not move towards its center. Aristotle rejected a rotation of the earth since it was not natural in his thinking.

Copernicus in Book I, Chapter 7 of "*De Revoltionibus*" paraphrases what he thought the ancient Greeks felt about a rotating earth,

"Now things which are suddenly and violently whirled around are seen to be utterly unfitted for reuniting... unless some constant force constrains them to stick together".

It turns out gravity itself is just such a force which holds objects on the earth even though the earth itself is rotating.   In the "Dialogue Concerning the Two Chief World Systems", Galileo considers this argument.   Galileo speaks through the interlocutor called Salviati.   He states everyone is familiar with what happens when a sling is whirled around and around and then let go.   When the sling is released, it follows a straight path leaving the circumference of its previously circular path following a tangent line to the circumference. Salviati says,

"And since if the earth revolved upon itself, the motion of its surface (especially near the equator) would be incomparably faster than the objects mentioned (stones in a sling], it would necessarily throw everything into the sky".

Galileo thus introduces us to the argument which he must refute if he is to convince his readers that the Copernican system is tenable. Today we say that a rotating object attached to a sling or rope rotated around and around experiences a center seeking or centripetal force which is transmitted to the object by the sling or rope.   The object is held in its circular path by the force pulling it along the sling or rope toward the center of rotation. If the sling or rope is released, the center seeking force no longer exists and the principle of inertia accounts for the linear motion along the tangent to the circle of rotation at uniform velocity. Of course, if the object is released near the surface of the earth it also experiences the force of gravity, which causes uniform acceleration toward the earth as we have studied in the case of projected objects.

We can explain why objects on the earth's surface are not significantly affected by the earth's rotation by using the mathematics of the calculus and by using parametric equations to represent the rotational motion. In addition we will introduce the concept of vectors to explain the direction of the velocity and the acceleration for objects experiencing rotational motion. Galileo did not have the necessary mathematics to explain rotational motion in full. We will give the modern explanation and then give Galileo's argument.

We consider Figure 45-3 which is a cross section of the earth through its center and the equator as viewed from the North Pole.

671

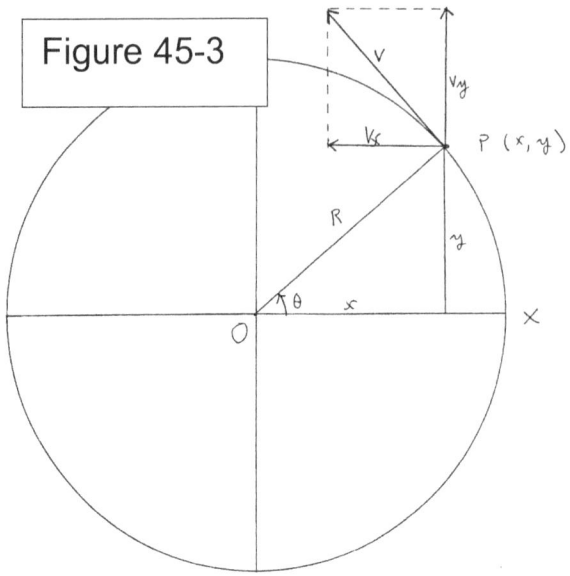

Figure 45-3

The earth's center is at O, and the radius is R = OX = OY. An object is located at point P on the equator and is rotating with the equator covering the distance of one circumference every 24 hours. We choose our coordinate reference frame so O is the origin of the XY coordinate axes. The coordinates of point P are designated (x, y). We draw radius OP = R, and mark angle XOP as angle θ, which is measured counterclockwise from the X axis. We see from the figure,

13. cos θ = x/R , or, x = R cos θ.

Also sin θ = y/R, or,

y = R sin θ.

We would like to express angle θ in terms of the parameter for time, t. The angular velocity of point P is designated by the Greek letter omega, ω. We know ω is equal to 360°/24 hours. However, it

672

is more convenient in working with the calculus to measure angles in the radian system which we learned in chapter 2. We recall the value of an angle in radians is equal to the length of the arc inscribed by the angle divided by the length of the radius, when the arc and radius are measured in the same system of units. For one revolution of the earth the arc length is equal to the circumference of the circle, in this case the equator, which equals $2\pi R$. Thus for one revolution,

14. θ = arc length / radius = $2\pi R / R = 2\pi$.

The angular velocity ω is equal to the circumference divided by the time for one revolution.

15.

$$\omega = \frac{2\pi \; radians}{24 \; hours} = \frac{2\pi \; radians}{86,400 \; seconds} =$$

$$(7.27)(10^{-5}) \; radians / second.$$

The instantaneous rate of change of θ with respect to time is the

$$\lim_{t \to 0} \frac{\Delta\theta}{\Delta t} = \frac{d\theta}{dt},$$

the derivative of θ with respect to time. This derivative is constant, since the earth rotates with uniform angular velocity equal to ω.

16. Thus,

$$\frac{d\theta}{dt} = \omega = (7.27)(10^{-5}) \; radians / second.$$

Using differentials, $d\theta = \omega\, dt$. To find the value of θ in terms of t and ω we integrate this equation.

17. $\int d\theta = \int \omega dt$

$\theta = \omega dt = \omega t + C$, where C is the constant of integration.

When P is on the positive X axis, then $t = 0$. This occurs when $\theta = 0$.   Then, $0 = (\omega)(0) + C$, or, $C = 0$, so, $\theta = \omega t$.   For example, if $t = 12$ hours, or, 43,200 seconds, then

18. $\theta = \omega t$

$= (7.27)(10^{-5})$ radians/seconds)(43,200 seconds)

$= 3.14$ radians $= \pi$ radians, or, ½ of one revolution as we expected.

We can now write the parametric equations of step 13 as follows:

19. $x = R \cos \omega t$, and $y = \sin \omega t$.

We next represent the velocity of the object at point P in reference to the origin, or center of the earth as *v*.   When using the symbol *v* in boldface type we are using the mathematical concept of vectors.   Although the concepts of vector algebra and calculus were not developed until the 19th century, both Galileo and Newton were aware that velocity and acceleration had magnitude and direction, and thus fit the modern meaning of vectors.

Vectors are defined mathematically as directed line segments, which can be used to

674

represent physical quantities which have both magnitude and direction.   The length of the vector is proportional to the magnitude of the vector, and the orientation of the vector in relation to the coordinate system gives the direction of the vector. In Figure 45-3 we show three directed line segments, or vectors, v, v_x, and v_y.   Vectors are indicated in print by symbols with boldface type, as we mentioned. In the figure the vector v is shown as a line segment of length $|v|$. We use the absolute value notation to represent the magnitude of the vector, since the vector v may be pointing in the negative or positive direction, but its magnitude is taken to be a positive quantity.

Vectors can be resolved into two mutually perpendicular components.   We show vector v resolved into the two perpendicular vectors v_x and v_y.   We use the Pythagorean theorem to find the magnitude of v, ($|v|$), if we know the values of the magnitudes of v_x and v_y.   Thus,

20. $|v|^2 = |v_x|^2 + |v_y|^2$, or,

$$|v| = \sqrt{|v_x|^2 + |v_y|^2}.$$

In Figure 45-3 vector v_x is parallel to the X axis, while vector v_y is perpendicular to the X axis. Vector v makes an angle with the X axis which is found by extending vector v to the X axis as shown in Figure 45-4.

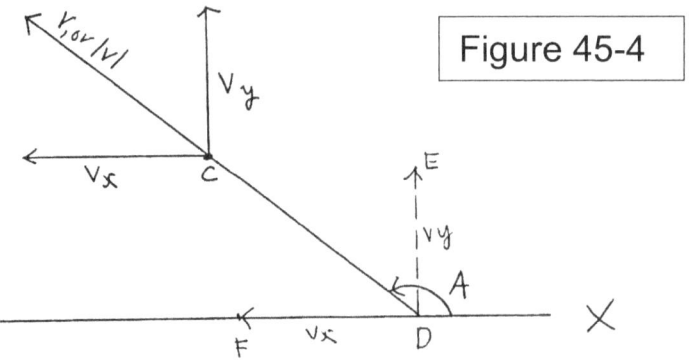

Figure 45-4

We see the angle of inclination of *v* is angle A.   If we mark off lengths DE and DF on the X and Y axes in Figure 45-4 equal to $|v_x|$ and $|v_y|$ respectively, we can show from trigonometry that,

21. $\tan A = v_y / v_x$.

The value is negative for the position shown in Figure 45-3 and 45-4, because angle A is in the second quadrant.   However, this equation is valid for all four quadrants.   Note we did not use the absolute values for *v_x* and *v_y* because in determining tan A we must take into account not only the magnitude of the vectors, but also whether the vector is pointing in the negative or positive direction.   When *v_x* is pointing in the negative direction, we must use a negative value for the magnitude of *v_x* in the equation determining tan A.

We can use the value of A to give us the direction of *v*. We know from our work with the calculus that the instantaneous rate of change of of the *x* coordinate, or the distance of P from the Y axis, is equal to the derivative of *x* with respect to *t*,

676

namely, *dx/dt* . But *dx/dt* is also equal to the velocity in the *x* direction, **v**_x. Since *x* = R cos ω*t*,

22.

$$v_x = \frac{dx}{dt} = \frac{d(R\cos\omega\, t)}{dt} = R\frac{d(\cos\omega\, t)}{dt}.$$

The instantaneous rate of change of the distance of P in the

y direction is **v**_y, which is equal to *dy/dt* , and since, *y* = R sin ω*t*,

23.

$$v_y = \frac{dy}{dt} = \frac{d(R\sin\omega\, t)}{dt} = R\frac{d(\sin\omega\, t)}{dt}.$$

We now face a new problem: how do we determine the derivative of the sine and cosine of ω*t* with respect to *t*, where *t* is the independent variable?

The derivatives and antiderivatives of the trigonometric functions, sine and cosine, are very important in problems of circular and other curvilinear motions such as the ellipse. We shall now take the time to learn how to find the derivatives of the sine and cosine functions.

We start by writing the sine function in terms of variables *x* and *y*. The sine function can then be written as, *y* = sin *x*, where *y* is the dependent variable, and *x* is the independent variable. We wish to find,

$$\frac{dy}{dx} \text{ , or, } \frac{d(\sin x)}{dx} .$$

677

The graph of the sine function, $y = \sin x$, is shown in Figure 45-5.

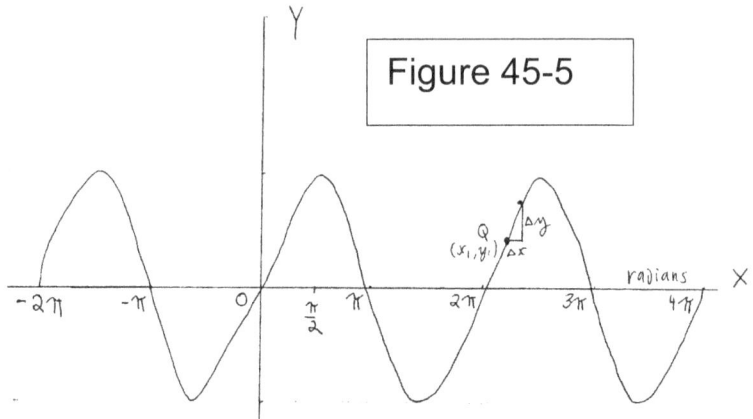

Figure 45-5

We see that it is a periodic function, that is, the graph from $x = 0$ to $x = 2\pi$ radians is repeated over and over for every interval of x equal to 2π radians (remember in the calculus the angle is measured in radians). We say the period of the function, $y = \sin x$, is 2π radians.

The function has a maximum value when $x = \pi/2$. The value of y when $x = \pi/2$ is 1. (Recall $\pi/2$ radians equals 90°, and $\sin 90° = 1$). If we have the function, $y = 2 \sin x$, as shown in Figure 45-6, we see again a periodic function with a period of 2π radians.

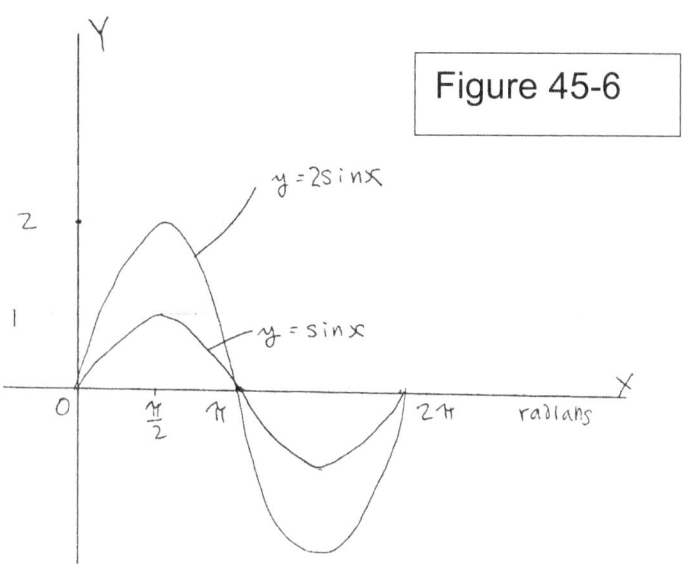

Figure 45-6

However, the maximum value of the function is 2 instead of 1, and occurs at x = π/2, just as in, y = sin x, which we graph along with the function, y = 2 sin x.

In general, if we have, y = a sin x, where a is a constant, then a is the value of the amplitude of the function, or the maximum value of the function in the interval of 1 period between 0 and 2π radians.

Now consider the function, y = sin 2x, which is shown in Figure 45-7 along with, y = sin x.

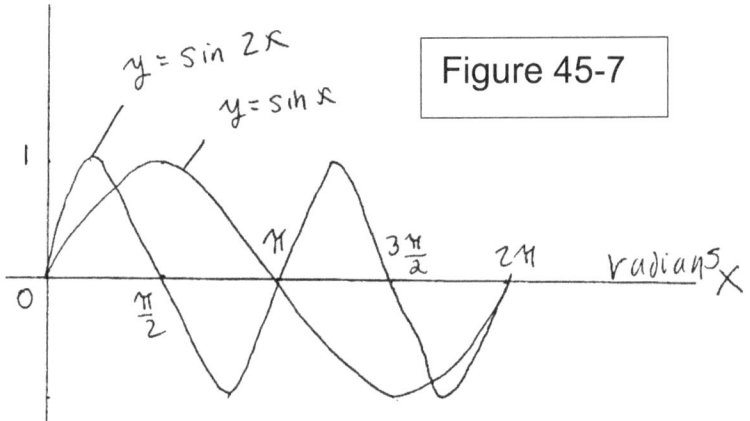

Figure 45-7

We see that the period of, $y = \sin 2x$, is ½ the period of, $y = \sin x$. In other words the graph of, $y = \sin 2x$, has traced out 1 period in the x interval from 0 to π radians, and 2 periods in the interval from 0 to 2π radians. The period of, $y = \sin x$, is 2π radians, and the period of, $y - \sin 2x$, is π radians. In general if we have, $y = \sin ax$, where a is a constant and x is the independent variable, then the period, designated by T, for the function is equal to $2\pi/a$, where a is the coefficient of x, when x is measured in radians.

We now turn to the problem of finding the derivative of, $y = \sin x$. We will use the method of increments. At point Q on the curve on Figure 45-5 , $y = \sin x$, the coordinates are designated (x_1, y_1). If x changes by an incremental amount Δx, then the new value of x is, $x_1 + \Delta x$. The corresponding change in y is Δy, so the new value of y is, $y_1 + \Delta y$. We then have,

24. $y_1 + \Delta y = \sin (x_1 + \Delta x)$.

680

We subtract, $y_1 = sin\ x_1$, from this equation, the left side from the left side and the right side from the right side.

25. $\Delta y = sin\ (x_1 + \Delta x) - sin\ x_1$.

We divide the equation by Δx.

26.

$$\frac{\Delta y}{\Delta x} = \frac{\sin(x_1 + \Delta x) - \sin x_1}{\Delta x}.$$

To find derivative means finding the limit of both sides of this equation as Δx approaches 0. The limit of the left side is the derivative dy/dx. However, the limit on the right side as Δx approaches 0 is not evident. We shall have to do some mathematical manipulation to find it.

In chapter 13 we learned the trigonometric identity.

27. $sin\ (A + B) = sin\ A\ cos\ B + cos\ A\ sin\ B$.

We apply this identity to $sin\ (x_1 + \Delta x)$.

28.

$$\frac{\Delta y}{\Delta x} = \frac{\sin x_1 \cos \Delta x + \cos x_1 \sin \Delta x - \sin x_1}{\Delta x},$$

$$\frac{\Delta y}{\Delta x} = \frac{(\sin x_1 \cos \Delta x - \sin x_1) + \cos x_1 \sin \Delta x}{\Delta x},$$

$$\frac{\Delta y}{\Delta x} = \sin x_1 (\frac{\cos \Delta x - 1}{\Delta x}) + \cos x_1 (\frac{\sin \Delta x}{\Delta x}).$$

It is true in the calculus that the limit of the sum of two terms is equal to the sum of the limits of each term.   We would then like to find as a first step,

29.

$$\lim_{\Delta x \to 0}(\sin x_1)[\frac{\cos \Delta x - 1}{\Delta x}].$$

Because we are finding the limit at the point Q where $x = x_1$, then sin x_1 is a constant term which can be multiplied into the limit of the second term in step 29.   We rewrite the limit in step 29 as,

30.

$$(\sin x_1)\lim_{\Delta x \to 0}(-1)[\frac{1-\cos \Delta x}{\Delta x}]$$

$$= -(\sin x_1)\lim_{\Delta x \to 0}\frac{1-\cos \Delta x}{\Delta x}.$$

In chapter 13 we learned another trigonometric identity, which was, 2 (sin² (A/ 2) = 1 - cos A.   We then seek to find,

31.

$$\lim_{\Delta x \to 0}\frac{2\sin^2 \Delta x/2}{\Delta x} = \lim_{\Delta x \to 0}(\sin \frac{\Delta x}{2})(\frac{\sin \Delta x/2}{\Delta x/2}).$$

By writing the limit in this way we see that as Δx approaches 0, then the term sin Δx must also approach 0, since the sine of 0 radians equals 0. Also the limit of sin $\Delta x/2$, as Δx approaches 0, must be 0 as well as sin 0 radians is 0. Then the limit of the step 30 is 0, which in turn means that in step 28, the limit of the first term to the right of the equal sign is 0. We then have to find the limit of the second term in step 28, namely,

32.

$$\lim_{\Delta x \to 0} \cos x_1 (\frac{\sin \Delta x}{\Delta x}).$$

Since cos x_1 is a constant we can write the limit we seek,

33.

$$(\cos x_1) \lim_{\Delta x \to 0} \frac{\sin \Delta x}{\Delta x}.$$

To aid us in finding this limit we construct Figure 45-8.

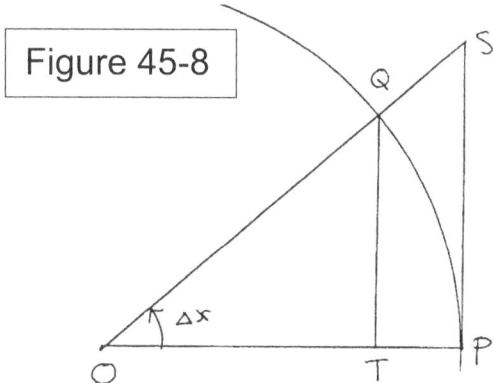

Figure 45-8

We see Δx as the angle with vertex at O. The circular arc PQ is drawn with center at O and radius OQ = OP = 1 unit.   We draw QT and SP perpendicular to OP.   S is on OQ extended.   We see,

34. area triangle OTQ < area sector OPQ <
    area triangle OPS.
    area triangle OTQ = ½(OT){TQ)

683

cos Δx = OT/OQ = OT, since OQ = 1 unit.

sin Δx = TQ/OQ = TQ.

area triangle OTQ = ½ cos Δx sin Δx.

The area of a sector of a circle is in proportion to the area of the circle as the central angle of the sector is to the central angle of the circle. The central angle of a circle is 2π in the radian system.

35.

$$\frac{area \ sector \ OPQ}{area \ circle} = \frac{central \ angle \ of \ sector}{2\pi} =$$

$$\frac{area \ sector \ OPQ}{\pi(OQ)^2} = \frac{\Delta x}{2\pi}, and \ since \ OQ = 1 \ unit,$$

$$area \ sector \ OPQ = \frac{(\Delta x)(\pi)}{2\pi} = \frac{\Delta x}{2}.$$

36. Area of triangle OPS

= ½(PS)(OP) = ½PS,

since, OP = 1 .

tan Δx = PS/OP = PS, as OP = 1.

area triangle OPS = ½tan Δx

= ½(sin Δx/cos Δx

We use these results in the first equation of step 34.

37. area triangle OTQ < area sector OPQ <
area triangle OPS,

½cos Δx sin Δx < Δx/2 < ½(sin Δx/cos Δx

We divide this inequality by the positive quantity, sin Δx /2.

38.   cos Δx < (Δx / sin Δx) < (1/cos Δx).

684

As Δx approaches 0, cos Δx approaches 1, since the cos of 0 radians equals 1. This means both terms in the inequality, cos Δx, and $(1/\cos \Delta x)$ approach 1 as Δx approaches 0. This means also that, $(\Delta x/\sin \Delta x)$ must approach 1 as Δx approaches 0, since this term lies between two terms which both approach 1. Since in the limit we can say, $(\Delta x/\sin \Delta x) = 1$, then also, $(\sin \Delta x/\Delta x) = 1$ in the limit as Δx approaches 0.

Going back to step 28 we can now take the limit of the entire equation as Δx approaches 0.

39.

$$\frac{\Delta y}{\Delta x} = \frac{\sin x_1 \cos \Delta x + \cos x_1 \sin \Delta x - \sin x_1}{\Delta x},$$

$$\frac{\Delta y}{\Delta x} = \frac{(\sin x_1 \cos \Delta x - \sin x_1) + \cos x_1 \sin \Delta x}{\Delta x},$$

$$\frac{\Delta y}{\Delta x} = \sin x_1(\frac{\cos \Delta x - 1}{\Delta x}) + \cos x_1(\frac{\sin \Delta x}{\Delta x}).$$

The limit of the left side is the derivative, dy/dx, for $y = \sin x$. The limit of the first term on the right side in the last equation is 0, as we determined. The limit of the second term on the right side must be cos x_1, since we showed the limit of sin $\Delta x/\Delta x$ is 1. Since x_1 can represent any value of x for the function, the derivative of sin x is cos x.

40. For $y = \sin x$, $dy/dx = \cos x$.

We know the derivative for any point (x, y) on the curve where y is graphed as a function of x is the slope of the tangent line to the curve at the point.   The value of the slope is then the value of the cosine of the x coordinate when the *x* coordinate is in radians.

We will now find the value of the derivative of the function, $y = \cos x$, where *x* is measured in radians.   In order to do this we will introduce another important concept in the calculus known as the chain rule.   Often in our work with the calculus we wish to find the derivative of what is called a composite function.   For example, consider the function,

41. $y = \sin x^2$.

In this function x^2 has replaced *x* in the function, $y = \sin x$.   Mathematicians consider x^2 to be a function of *x*, namely the function, $u = x^2$, where, *u* is a function of *x*.   We can then write step 39 as,

42. $y = \sin u$.

We think of *y* as function of *u*, and in turn *u* as a function of *x*. The function, $y = \sin x^2$, then is considered as a composite function, where the two composite functions are, $y = \sin u$, and $u = x^2$. In finding derivatives this idea is very useful.   We know the derivatives of, $y = \sin u$, and, $u = x^2$.

43.

$$\frac{dy}{du} = \frac{d(\sin u)}{du} = \cos u,$$

$$\frac{du}{dx} = \frac{d(x^2)}{dx} = 2x.$$

We now ask what is the derivative of *y* with respect to *x*, *dy/dx*?

The answer is the rule called the chain rule, which states,

44.

$$\frac{dy}{dx} = [\frac{dy}{du}][\frac{du}{dx}] = (\cos u)(2x), \text{ since } u = x^2,$$

$$\frac{dy}{dx} = (2x)(\cos x^2).$$

We have found *dy/dx* by multiplying *dy/du* by, *du/dx*. The rule is called the chain rule because a 'chain' of derivatives is multiplied to find the result.

To demonstrate the validity of the chain rule we recall our work with differentials.

45.

$$\frac{\Delta y}{\Delta x} - \frac{dy}{dx} = \varepsilon,$$

where Δy and Δx are the incremental changes in y and x, and ε (episilon) is the value of difference between the ratio of the increments and the derivative. ε approaches 0 as Δx approaches 0, since we have defined the limit $\Delta x/\Delta y$ as equal to the derivative as Δx approaches 0. We multiply both sides by Δx, although approaching 0 is not 0 but a very small quantity, which the 17th century mathematicians would call infinitesimal.

46.

$$\Delta y - \frac{dy}{dx}\Delta x = (\varepsilon)(\Delta x).$$

If y is a function of u and u is a function of x, we have,

47.

$$\Delta y - \frac{dy}{du}\Delta u = \varepsilon\,\Delta u, \text{ and, } \Delta u - \frac{du}{dx}\Delta x = \varepsilon\,\Delta x.$$

We rewrite these equations as,

48.

$$\Delta y = \frac{dy}{du}\Delta u + \varepsilon\,\Delta u, \text{ and, } \Delta u = \frac{du}{dx}\Delta x + \varepsilon\,\Delta x.$$

We divide both sides of both equations by Δx,

49.

$$\frac{\Delta y}{\Delta x} = [\frac{dy}{du}][\frac{\Delta u}{\Delta x}] + \varepsilon\frac{\Delta u}{\Delta x}, \text{ and, } \frac{\Delta u}{\Delta x} = \frac{du}{dx} + \varepsilon.$$

We then take the limit of the terms of both equations as Δx approaches 0. In the second equation we get, $du/dx,$ on both sides of the equation, since we have learned the limit of $\Delta u/\Delta x$ approaches du/dx, the derivative, and we realize ε approaches 0 as Δx approaches 0 from step 43.

In the first equation we get,

50

$$\frac{dy}{dx} = (\frac{dy}{du})\,(\frac{du}{dx}),$$

which is the chain rule.

We get the chain rule because ε approaches 0 as Δx approaches 0, so the second term in the first equation becomes 0 in the limiting process.

We can apply the chain rule to find the derivative of, $y = \cos x$. We recall that $\cos x$ = sine of the complement, $90° - x$, or in radians,

51. $\cos x = \sin (\pi/2 - x)$, since 90^0

$\qquad = \pi/2$ radians.

We define the function, $u = \pi/2 - x$. Then, we have the composite function, $y = \sin u$, where, $u = \pi/2 - x$. We apply the chain rule.

52.

$$\frac{dy}{dx} = [\frac{dy}{du}][\frac{du}{dx}],$$

$$\frac{dy}{du} = \frac{d(\sin u)}{du} = \cos u,$$

$$\frac{du}{dx} = \frac{d(\frac{\pi}{2} - x)}{dx} = \frac{d(\frac{\pi}{2})}{dx} - \frac{dx}{dx}.$$

We know the derivative of $\pi/2$, a constant, is 0, and the derivative of x is 1, so, $du/dx = -1$, as a minus sign proceeds dx/dx .

$$\frac{dy}{dx} = (\cos u)(-1) = -\cos(\frac{\pi}{2} - x).$$

Since the cosine of an angle equals the sine of the complement, and the complement of, $\pi/2 - x$, is x, then,

$$\frac{dy}{dx} = -\sin x, \textit{when, } y = \cos x.$$

Knowing the derivatives of the sine and cosine and the chain rule allows us to take up again the problem of the object rotating on the equator of the earth. We had found the parametric equations

689

for the coordinates of the moving object in step 20, and the equations for the two components of the velocity vector in steps 22 and 23. We place these steps below:

20. $|v|^2 = |v_x|^2 + |v_y|^2$, or,

$$|v| = \sqrt{|v_x|^2 + |v_y|^2}.$$

22.

$$v_x = \frac{dx}{dt} = \frac{d(R\cos\omega t)}{dt} = R\frac{d(\cos\omega t)}{dt}.$$

23.

$$v_y = \frac{dy}{dt} = \frac{d(R\sin\omega t)}{dt} = R\frac{d(\sin\omega t)}{dt}.$$

We will now calculate the derivatives $dx/dy = v_x$ and, $dy/dt = v_y$. We let $u = \omega t$ in steps 21 and 22, so $x = R\cos u$, and $y = R\sin u$.

The chain rule states,

53.

$$\frac{dx}{dt} = (\frac{dx}{du})(\frac{du}{dt}), \text{ and, } \frac{dy}{dt} = (\frac{dy}{du})(\frac{du}{dt}),$$

$$\frac{dx}{du} = \frac{d(R\cos u)}{du} = -R\sin u,$$

$$\frac{du}{dt} = \frac{d(\omega t)}{dt} = \omega,$$

$$\frac{dx}{dt} = -R(\sin u)(\omega) = R\omega\sin\omega t = v_x.$$

We find dy/dt by a similar process.

$$\frac{dy}{dt} = [\frac{dy}{du}][\frac{du}{dt}],$$

$$\frac{dy}{du} = \frac{d(R\sin u)}{du} = R\cos u,$$

$$\frac{du}{dt} = \omega,$$

$$\frac{dy}{dt} = R(\cos u)(\omega) = R\omega\cos\omega t = v_y.$$

We now have the values for the two components of the velocity vector v in Figure 45-3 for an object at point P moving on the earth's equator with uniform angular velocity with the reference frame set so the earth's center is stationary.   We now calculate the magnitude of the velocity vector with the Pythagorean theorem.

54. $|v|^2 = |v_x|^2 + |v_y|^2,$

$|v|^2 = (-R\omega \sin \omega t)^2 + (R\omega \cos \omega t)^2 ,$

$|v|^2 = \omega^2 R^2(\sin^2 \omega t + \cos^2 \omega t),$ and

since,

$\sin^2 \omega t + \cos^2 \omega t = 1$, then,

$|v|^2 = \omega^2 R^2$, or,

$|v| = \omega R,$

where v is measured in meters/second, ω is measured in radians/second, and R is measured in meters. Today we know the radius of the earth is 6,378,000 meters.   We had found ω equaled $(7.27)(10^{-5})$ radians/second. Thus,

55. $|v| = (7.27)(10^{-5})(6,378,000)$

$= 463.7$ meters/second.

This is the magnitude of **v** in Figure 45-3. We next seek an expression for the direction of **v** for any position of P on the earth's equator.   We should establish that the direction of **v** is always in the direction of the tangent line to the circle for any position of P.

We know the parametric equation for the circle of the equator in terms of the parameter θ are, x = R cos θ, and, y = R sin θ.   We know if we have y as a function of x, then the derivative, dy/dx, is equal to the magnitude of the slope of the tangent line at any point (x, y) where the derivative exists. Furthermore, the slope of the tangent line is equal to the value of the tangent of the angle of inclination that the tangent line makes with the X axis. See Figure 45-9.

Figure 45-9

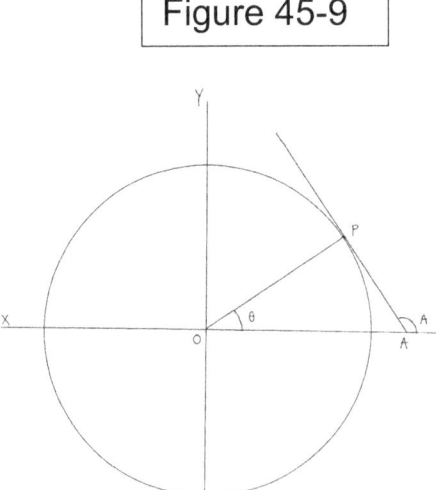

We see angle A is the angle of inclination which the tangent line to the curve at P makes with the X axis.   Note θ is the angle which the radius of the equator circle to the point P makes with the X axis. We then have,

56. $dy/dx = \tan A$.

We will also determine dy/dx by using the parametric equations of the circle. We note an incremental change in θ, $\Delta\theta$, results in corresponding incremental changes in x and y, Δx and Δy. We can write,

57.

$$\frac{\Delta y}{\Delta x} = \frac{\dfrac{\Delta y}{\Delta \theta}}{\dfrac{\Delta x}{\Delta \theta}},$$

by dividing numerator and denominator by $\Delta\theta$.

As $\Delta\theta$ approaches 0, so too do Δx and Δy. We can take the limits of both sides of the equation as $\Delta\theta$ approaches 0. It can be shown in the calculus that the limit of a quotient is equal to the quotient of the limit of the numerator divided by the limit of the denominator provided that the limit of the denominator is not 0. The limits in the equation are the derivatives. Thus,

58.

$$\frac{dy}{dx} = \frac{\dfrac{dy}{d\theta}}{\dfrac{dx}{d\theta}},$$

provided $dx/d\theta$ does not equal 0.

We calculate $dy/d\theta$, and, $dx/d\theta$.

59.

$$\frac{dy}{d\theta} = \frac{d(R\sin\theta)}{d\theta} = R\cos\theta,$$

$$\frac{dx}{d\theta} = \frac{d(R\cos\theta)}{d\theta} = -R\sin\theta.$$

Then,

60.

$$\frac{dy}{dx} = \frac{R\cos\theta}{-R\sin\theta} = -\cot\theta.$$

In Figure 45-9 we also know, dy/dx = tan A. Then,

61.

$$\frac{dy}{dx} = \tan A = -\cot\theta.$$

This expression means the slope of the tangent line to the circle of the equator at point P is equal to the value of - cot θ, where θ is the angle from the X axis to the radius from the center to P. We also argue this shows that radius OP must be perpendicular to the tangent line AP.   To see this we construct Figure 45-10.

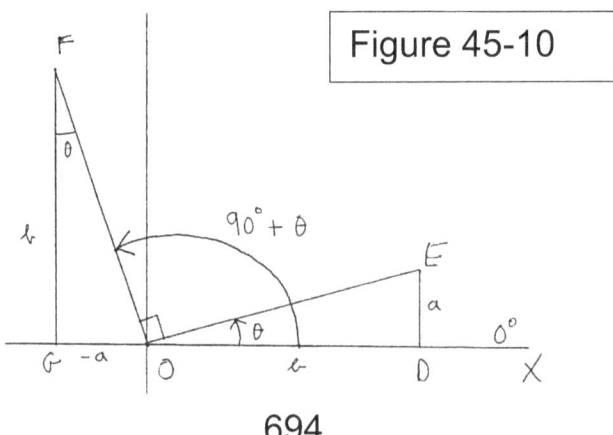

Figure 45-10

Here we have drawn θ < π/2 radians (90°). We have constructed the angle marked, 90° + θ. We recall from our study of trigonometry,

62. tan θ = a/b, and cot θ = b/a , where a and b are the lengths of the sides of right triangle ODE as   marked in the figure.

By constructing right triangle OGF with OG = $-a$, and GF = b, and angle EOF = 90°, we see,

63. tan (90° + θ) = $b/-a$ = $-b/a$ = - cot θ.

This is correct because the trigonometric functions of 90° + θ, when θ < 90°, are defined to be the trigonometric functions of angle FOG in triangle OGF as we have constructed the triangle.

Going back to Figure 45-9 we see in triangle OPA that,

64. θ + angle OPA = angle A.

This holds because the external angle of a triangle, A, equals the sum of the two opposite interior angles, θ and angle OPA.

65. Since, tan (90° + θ) = - cot θ, and since,
     tan A = - cot θ, from step 61,
     tan (90° + θ) = tan A, and 90° + θ = A.

From steps 64 and 65, angle OPA must equal 90°(θ + angle OPA = angle A = 90° + θ), so, θ + angle OPA = 90° + θ, subtract θ from both sides). If this is true then AP is perpendicular to OP, or the radius of the circle to any point P is perpendicular to the tangent line which touches the circle at P.

We found earlier,

$$v_x = -\omega R \sin \omega t, \text{ and,}$$

$$v_y = \omega R \cos \omega t. \quad \text{Since } \theta = \omega t, \text{ then,}$$

66. $$v_x = -\omega R \sin \theta, \text{ and,}$$

$$v_y = \omega R \cos \theta.$$

From Figure 45-4,

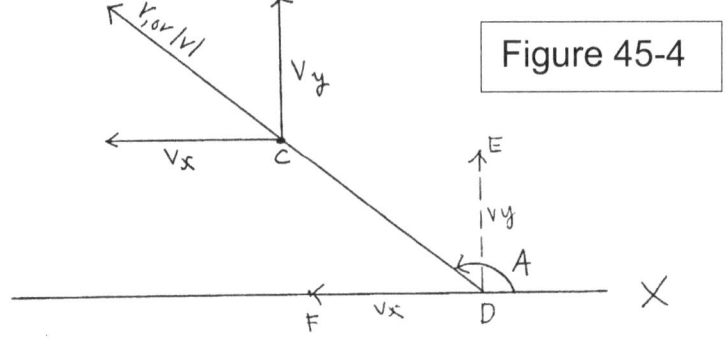

Figure 45-4

67.

$$\tan A = \frac{v_y}{v_x} = \frac{\omega R \cos \theta}{-\omega R \sin \theta} = -\frac{\cos \theta}{\sin \theta} = -\cot \theta.$$

This is the same result as step 61, but found using the values of the velocity vectors in Figure 45-4. Because vector v_y points in the positive y direction, and vector v_x points in the negative x direction at the position of point P shown in Figure 45-3,

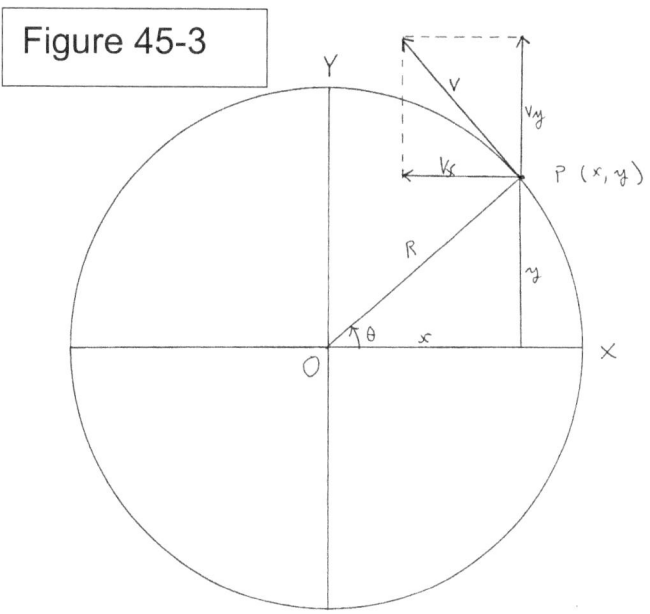

Figure 45-3

as point P rotates with the equator its direction of motion is in the direction of vector **v**, which corresponds to the counterclockwise rotation of the earth seen from the north pole in Figure 45-3. We also just learned, - cot θ = dy/dx = slope of the tangent line vector **v**.

This confirms the velocity of P on the equator of the earth has the magnitude of 463.7 meters/second, and a direction perpendicular to the radius OP along the tangent to the equator at P. We now have a complete description of the velocity of objects on the earth's equator. We now ask what is the acceleration of an object at P?

From our work with Galileo we would answer that the acceleration of objects on the earth's surface is directed toward the center of the earth along radius OP with a magnitude of about 9.8 meters/second^2. However, because the object at P is rotating with the earth's equator in Figure 45-3, the velocity, although constant in magnitude at 463.7 meters/second, is continually changing directions. A change in the velocity vector whether it be in magnitude or in direction is always due to an acceleration. There must be some part of the acceleration responsible for the change in the direction of the velocity vector.

The source of the acceleration causing the change in the velocity vector is the earth's gravity. After the work of Isaac Newton scientists realized that the earth's gravity had two actions for objects rotating with the earth. The earth's gravity is responsible for the acceleration of objects toward the center of the earth, and it is responsible for the change in direction of the velocity vector as a result of the earth's rotation. Note that the earth's gravity does not cause the earth to rotate, but it acts to keep those objects on the earth's surface rotating with the earth.

Thus today we think of the earth's gravity as causing the weight of objects on the earth's surface or the acceleration toward the earth's center at about 9.8 meters/second^2 for objects near the earth's surface. This acceleration is responsible for Galileo's law of free fall. But in addition to this effect of the earth's gravity there is also the effect of the earth's gravity to keep objects on the earth's surface rotating with the earth in a circular path. The earth's atmosphere and objects in it are also rotating with the earth and also experience this part of the acceleration due to the earth's gravity.

In the case of objects like those at point P on the equator in Figure 45-3 the center of the circle of rotation is the earth's center. For objects at latitudes north or south of the equator, the center of the circle of rotation lies along the axis of the earth from the north to the South Pole. Only two points on the earth's surface do not participate in the rotation of the earth, the North Pole and South Pole. At these two points the only action of the earth's gravity is to cause acceleration towards the earth's center.

These concepts were not known to Galileo so he could not use them in his arguments in favor of the Copernican theory of a rotating earth. We are developing these concepts at this time so we see more clearly why objects on the earth's surface are not thrown out into space along a tangent line to the circle of rotation.

We will now proceed to determine the value and direction of that part of the earth's gravitational acceleration which is responsible for the change in the velocity vector *v* in Figure 45-3.

We show the components of the acceleration vector a in Figure 45-11 for different positions of P on the earth's equator.

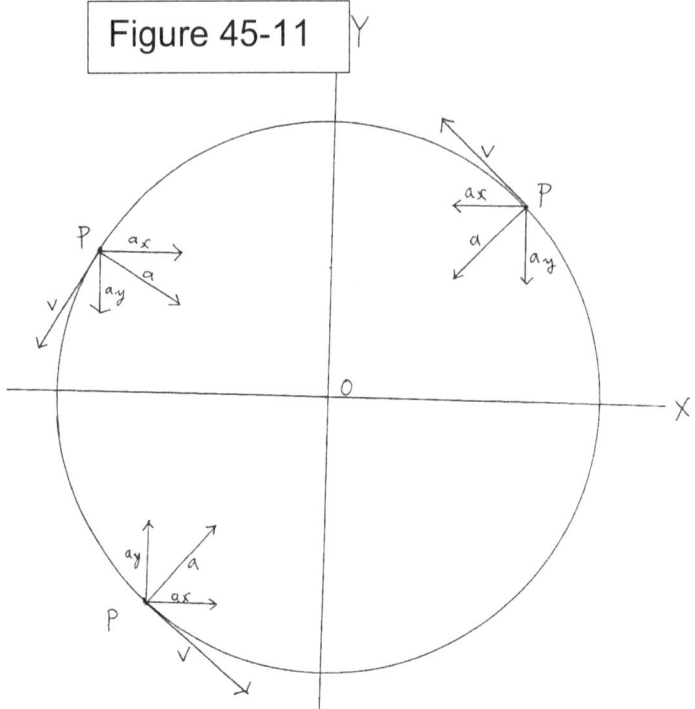

Figure 45-11

Also shown are the velocity vectors, **v** , for these points. In drawing these vectors it is essential to choose the appropriate reference frame for the motion of P.   Since the stationary point in considering the rotation of the earth's equator is the earth's center 0, we chose O as the origin of our coordinate system in figure 45-11 just as in figure 45-3.

We see the acceleration vector a can be resolved into two mutually perpendicular components a_x and a_y, just as in the case of the velocity vector.   We designate the magnitude of the acceleration vector as |**a**|.   We calculate the

700

components of the acceleration vector as follows. We know the instantaneous rate of change of the velocity in a given direction is the acceleration in that direction. We also know the instantaneous rate of change of one quantity with respect to another is the derivative. Hence,

68.

$$a_x = \frac{dv_x}{dt} = \frac{d(-R\omega \sin \omega t)}{dt}, \text{ and,}$$

$$a_y = \frac{dv_y}{dt} = \frac{d(R\omega \cos \omega t)}{dt}.$$

We let, $u = \omega t$, so we have composite functions in order to apply the chain rule.

$$a_x = [\frac{d(-R\omega \sin u)}{du}][\frac{d(\omega t)}{dt}] =$$

$$-R\omega(\cos u)\omega = -R\omega^2 \cos \omega t,$$

$$a_y = [\frac{d(R\omega \cos u)}{du}][\frac{d(\omega t)}{dt}] =$$

$$R\omega(-\sin u)\omega = -R\omega^2 \sin \omega t.$$

We find the magnitude of the acceleration vector by the Pythagorean Theorem.

69. $|a|^2 = |a_x|^2 + |a_y|^2,$

$|a|^2 = (-R\omega^2 \cos \omega t)^2 + (-R\omega^2 \sin \omega t)^2,$

$|a|^2 = R^2\omega^4 \cos^2 \omega t + R^2\omega^4 \sin^2 \omega t.$

$|a|^2 = R^2\omega^4(\cos^2 \omega t + \sin^2 \omega t).$

Since, $\cos^2 \omega t + \sin^2 \omega t = 1,$

$|a|^2 = R^2\omega^4,$

$|a| = R\omega^2.$

We found, $|v| = R\omega$, or, $|v|^2 = R^2\omega^2.$

701

70. We can write for |**a**|.

$$|a| = \frac{1}{R}R^2\omega^2.$$

We then substitute, $|v|^2$, for, $R^2\omega^2$

$$|a| = \frac{|v|^2}{R}, or, a = \frac{v^2}{R}.$$

This is the equation for the magnitude of the acceleration due to the earth's gravity which keeps the object at P rotating with the earth on the equator. We previously determined that, $|v|$ = 463.7 meters/second, and, R = 6,378,000 meters. Using these values we calculate |**a**|.

71.

$$|a| = \frac{|v^2|}{R} = \frac{(463.7)^2}{6,378,000} =$$

.0337 *meters* / sec*ond*^{2}.

We see this is a much smaller value than 9.8 meters/second^2, which is the value determined for the acceleration of gravity for free fall directed towards the earth's center. This means very little acceleration is required to carry objects with the earth as it rotates at the equator, compared to the acceleration which pulls objects to the center of the earth. If we consider an object at either the north or South Pole of the earth, the acceleration due to gravity, which causes rotation, is 0 since at the poles there is no rotation. In other words at the poles all of the earth's gravitational force is used to produce attraction of objects towards the earth's center, and no gravitational force is required for rotation.

We will learn in our work with Newton that the gravitational force that the earth produces is related to the quantity of matter in the earth. For objects at the equator most of the gravitational force results in an attraction of objects towards the earth's center, and very little of the earth's gravitational force is required to keep objects moving with the earth's rotation. The acceleration of 9.8 meters/second^2 is a measure of the force that causes attraction towards the earth's center, and the acceleration of .0337 meters/second^2 is a measure of the force needed to keep objects on the earth's equator rotating with the earth.

If the earth were a perfect sphere and all the matter of the earth were the same we would expect the value of the acceleration for free fall would be greater at the north and south poles and less at the equator by an amount equal to .0337 meters/second^2. The actual values found in the modern age for the acceleration causing free fall due to gravity at sea level at the north and south poles is 9.83217 meters/second^2. At the equator the acceleration causing free fall is 9.78039 meters/second^2. The difference is .0518 meters/second^2, a bit greater than .0337 meters/second^2. It turns out that the earth is not a perfect sphere and the equatorial radius of the earth is 6,378,000 meters (the value we have been using), which is greater than the polar radius of 6,357,000 meters. When we consider the work of Newton we will find the value of the acceleration directed to the earth's center is not constant, but increases as the latitude from the equator to the poles increases. The main reason is effect of the earth's rotation as discussed above. The lower value for the acceleration at the equator accounts for the equatorial bulge of the earth, which also has the effect of reducing the value of the acceleration causing attraction to the earth, and is the reason the difference in value between the equator and the poles is greater than .0337 meters/second^2.

The reason objects on the earth's surface are not projected into space is now evident. The acceleration of gravity pulling objects to the earth's center, which we detect as an object's weight, is not significantly reduced by the earth's rotation as shown by our calculations. Our mathematics has shown the Aristotelian philosophers were wrong in maintaining the earth cannot be rotating by saying that if it were then we would be hurled into space.

Out of curiosity we may ask, how fast would the earth have to be rotating so that objects on the earth's surface would be left behind the spinning earth? That is, what is the value for ω, the angular velocity of rotation, at which all the acceleration

704

from the earth's gravitational force would be used to keep objects rotating with the earth so that there would be no acceleration left to attract objects towards the earth's center?   The answer is if the value for $|a|$ in step 71 were equal to 9.8 meters/second^2.

72. $|a| = |v|^2 /R,$

$|v|^2 = R|a| = (6,378,000)(9.8)$

$= 62,504,000,$

$|v| = 7,905$ meters/second.

The earth's circumference at the equator is, $2\pi R$, which is 40,074,000 meters. Then,

73.

$$(\frac{40,074,000 \ meters}{1 \ revolution})(\frac{1 \ sec ond}{7,905 \ meters})(\frac{1 \ m}{60 \ s})$$

$$(\frac{1 \ hour}{60 \ min utes}) = \frac{1.41 \ hour}{1 \ revolution}.$$

$\omega =$

$$(\frac{2\pi \ radians}{1 \ revolution})(\frac{1 \ revolution}{1.41 \ hours})(\frac{1 \ hour}{60 \ min utes})(\frac{1 \ m}{60 \ s})$$

$$= \frac{.0012 \ radians}{1 sec ond}.$$

The earth would have to be rotating so fast that 1 solar day would only last 1.41 hours.   If the earth were spinning that fast or faster, then objects on the surface would experience no acceleration towards the earth's center, and would be weightless.   The earth would be rotating past all objects on its surface, so in effect everything would be 'thrown from the earth', and the earth's atmosphere would disperse into outer space.

We have yet to consider the direction of the acceleration vector, which is responsible for keeping objects in rotation with the earth on the equator. In Figure 45-11 we have drawn the acceleration vector **a** from various positions of P as pointing towards the earth's center, the very same direction of the acceleration vector in free fall. To establish mathematically that this is correct we construct Figure 45-12.

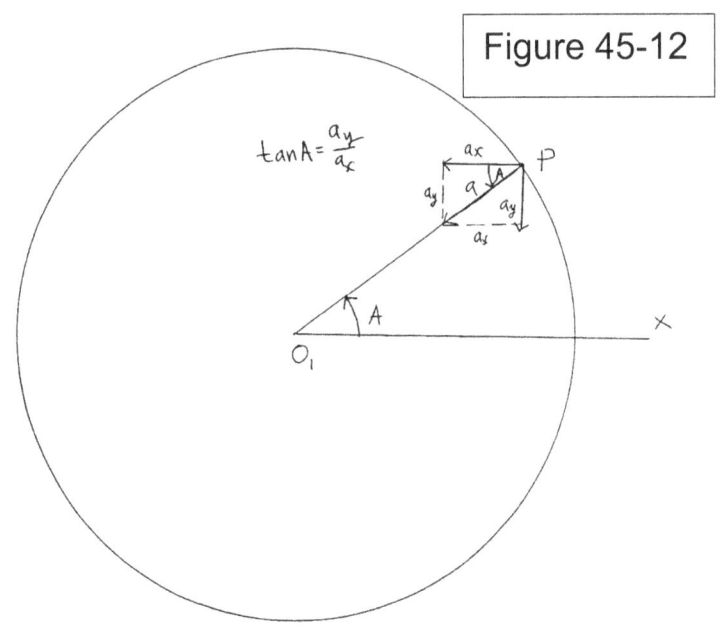

Figure 45-12

In this figure we indicate at a point P on the equator circle the components of the acceleration vector a_x and a_y, as in Figure 45-11. We see the acceleration vector **a** as the diagonal of a rectangle with sides a_x and a_y. We extend the line

706

containing *a* to its intersection with the X axis at point O_1. We indicate angle A as the angle of inclination of *a*.   We wish to demonstrate that O_1 is the coincident with the origin of the coordinate system of figure 45-11, and that angle A is equal to θ, the angle position P makes with the X axis as in Figure 45-9. Because a_x is parallel to the X axis,

74.

$$\tan A = \frac{a_y}{a_x} = \frac{-R\omega^2 \sin \omega t}{-R\omega^2 \cos \omega t}.$$

using the values from step 68.

Because both a_x and a_y have negative values when angle ωt is in the first quadrant, these vectors must be pointed in the negative x and y directions respectively as shown in Figure 45-12.   This means that *a* must be pointing in the direction of O_1, on the X axis.

75. $\tan A = \sin \omega t / \cos \omega t = \tan \omega t$, from step 70.

Since $\omega t = \theta$ in Figure 45-9, then, angle A = θ. This means line OP_1 makes an angle θ with the X axis, and this can only be if O_1P coincides with the radius of the circle, thus O_1 and O coincide, which means the acceleration vector *a* points to the center of the circle.

We now have shown that the two parts of the acceleration due to the earth's gravity both are directed towards the earth's center.   The first part, which causes free fall was that part studied by Galileo, and results in the law of free fall, and the parabolic trajectory of projected objects, is directed to the earth's center, and has a value of about 9.8 meters/second^2. The second part is the acceleration which keeps objects in rotation with the earth. This is directed towards the center of the circle of rotation as has a value of about .034 meters/second^2, if the circle of rotation is the equator. If the circle of rotation is at a latitude north or south of the equator, the acceleration is directed to a point on the north south axis of the earth, which is the center of the latitude circle. Note that at latitudes north or south of the equator the direction of the acceleration for free fall is directed to the center of the earth, while the acceleration keeping objects rotating with the earth is not directed to the earth's center, but to the center of the circle of rotation.

The discovery of the mathematical methods required to explain the value and direction of the acceleration responsible for keeping objects in rotation with the earth was a consequence of the work of Newton, and the use of concepts of the calculus and the vector qualities of velocity and acceleration.

The subject of the direction of the acceleration vector is so important in the solution of the problem of planetary motion, which we will consider in our work with Newton, that we will give one more demonstration of how the direction of the acceleration vector can be determined in the case of circular motion with uniform angular velocity. In this demonstration we will consider the acceleration vector *a* as the instantaneous rate of change of the velocity vector *v*.

75. $a = dv/dt$.

We construct Figure 45-13.

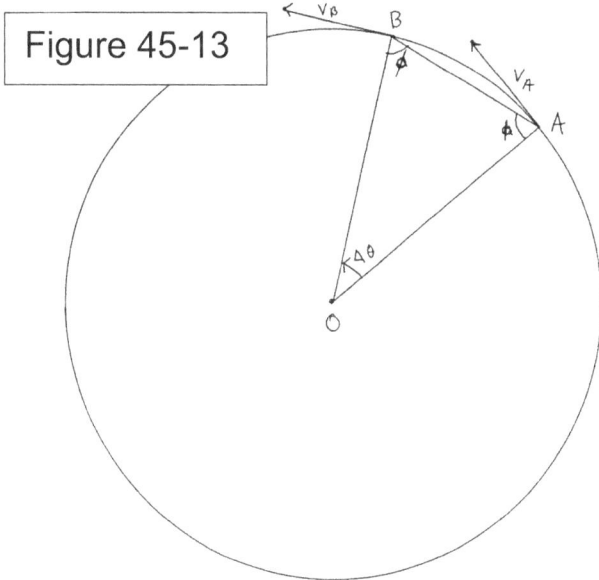

Figure 45-13

The circle is the earth's equator rotating counterclockwise as viewed from the north pole. Point A represents the initial position of an object on the equator. An increment of time passes in which the object moves from position A to position B. We complete triangle BOA and indicate the incremental angle of rotation as $\Delta\theta$, and indicate the other two angles of the triangle by φ (Greek letter phi).

Chord **AB** represents the change in position of the rotating object in moving from A to B. We can consider chord **AB** as a position vector, and designate it with boldface type as **AB**. Vector **AB** is a result of the change in position from A to B of the object associated with the change in the direction of the velocity vector from v_A to v_B. The magnitude of the velocity vector, $| v |$, remains constant during the motion, since the earth is rotating with uniform angular velocity, but the direction of the velocity vector has clearly changed from the direction of v_A to the direction of v_B . We have drawn the velocity vectors perpendicular to the radius OA and OB, as they should be from our work with Figure 45-3.

We would like to demonstrate that the angle between vector v_A and vector v_B is equal to angle $\Delta\theta$. To do this we construct Figure 45-14.

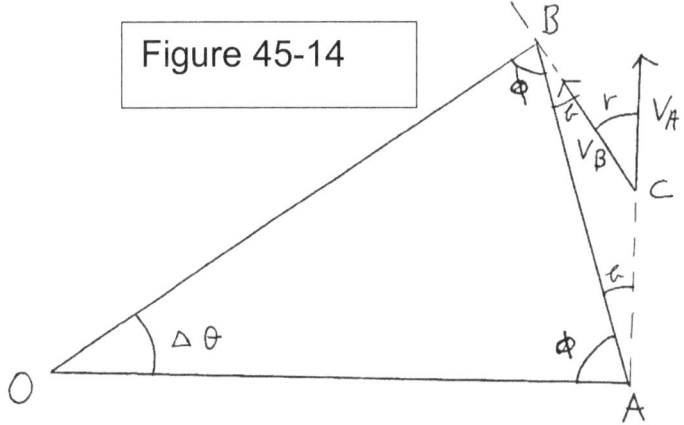

Here we have drawn triangle OAB from Figure 45-13 without including the circle of the equator.

We have moved vectors v_A and v_B so their points of origin are at point C. We moved v_A along the tangent to the circle at A, and v_B along the tangent to the circle at B. This means the vectors still have the same directions with v_A perpendicular to OA, and v_B perpendicular to OB.

76.   angle OAC = $\pi/2$ radians = 90°

angle OAC = φ + b = 90^0,

where angle b = angle BAC.

angle OBC = $\pi/2$ radians = 90°

angle OBC = φ + angle ABC = 90^0.

Then,

φ + angle ABC = φ + b, so

angle ABC = b, and is so marked.

Angle r is the angle between vectors v_A and v_B. Angle r is also the external angle of triangle ABC and is equal to the sum of the two opposite interior angles.

77. r = b + b. Since, φ + b = 90°, b = 90° - φ,

r = (90° - φ) + (90° - φ),

r = 180° - 2φ, or, 2φ = 180° - r.

In triangle OBA, $\Delta\theta$ + 2φ = 180°, so:

$\Delta\theta$ + (180° - r) = 180°,

$\Delta\theta$ - r = 180° - 180° =0, or,

$\Delta\theta$ = r.

We have demonstrated the angle between the velocity vectors is equal to $\Delta\theta$, the incremental angle of change when the equator has rotated to move an object from A to B.

We next wish to find the direction of the acceleration vector for position A, which can represent any point on the equator.   We construct Figure 45-15.

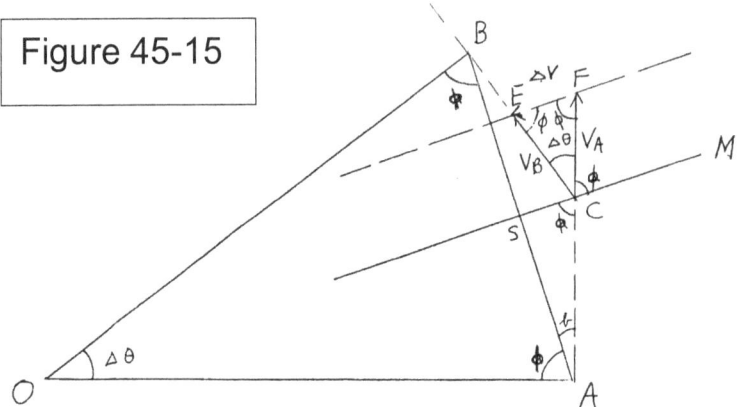

This figure is almost the same as figure 45-14.   We have indicated the angle between v_A and v_B as $\Delta\theta$, and have completed triangle EFC. Because the magnitudes of of v_A and v_B are equal, the lengths of the directed line segments representing their magnitudes must be equal. This means triangle EFC is an isosceles triangle similar to triangle OAB, since both triangles have two equal sides, and the angle between the corresponding equal sides is the same in both triangles. This in turn means the angles at E and F are equal to φ, and are so marked.

We draw MCS parallel to EF, which means angle EFC (φ) equals angle FCM, which equals angle ACS. Thus, both angles FCM and ACS equal φ, and are so marked.

712

The change in velocity vectors in going from A to B can be considered a vector, which we designate by Δ**v**, and geometrically is the directed line segment F to E. The instantaneous rate of change of the velocity vector at any point is the derivative of the velocity vector, and equals the acceleration vector. We can write this as a limit:

78.

$$\lim_{t \to 0} \frac{\Delta v}{\Delta t} = \frac{dv}{dt} = a.$$

This is true because as t, the time of the incremental rotation, approaches 0, then B is closer to A, and in the limit B is at A. In turn vector Δ**v** becomes the instantaneous rate of change of the velocity vector, or the acceleration vector **a**. If we determine the direction of Δ**v** in the limit, we then have determined the direction of the acceleration vector **a**.

We see in triangle ASC,

79. φ + b + angle ASC = 180°.

We recall, φ + b = 90°, so,

90° + angle ASC = 180°, or,

angle ASC = 90°.

Since SC is parallel to to EF, then EF must intersect AS extended (AB) at a right angle. This means Δ**v** which is represented by EF is perpendicular to chord AB. In the limiting process Δ**v** moves toward, and will coincide with OA, the radius of the equator at A, in the limit. In the limit

Δv will be pointing toward O the center of the equator, which is the center of the earth. Thus the acceleration vector **a**, which is what Δv becomes in the limit, points to the center of the rotating circle, just as we found in our first method of demonstrating the direction of the acceleration vector for uniform circular motion.

We have taken the time to find the direction of the acceleration vector by two methods. We did this because the result that objects in circular motion have an acceleration directed to the center, called by Newton a centripetal acceleration, is very important in Newton's study of the laws of motion. We also used these demonstrations to better acquaint ourselves with the concepts of vectors, and how they are used in the ideas of the calculus. We are realizing how important the calculus is in solving the problems of motion.

Galileo, however, did not have the calculus to explain his idea of why objects are not hurled into the air by a rotating earth. His argument has flaws, but nonetheless is important, because his thinking may well have inspired Newton to carefully consider rotating motion with the new methods of analysis which included the calculus.

In the "Dialogue Concerning the Two Chief World Systems", Galileo mentions in his marginal notes a comment on his principle of inertia. We will quote from the translation of Thomas Salusbury of 1661, since this version was probably read by Newton.

"The motion impressed by the projicient is onely in a right line".

The note means that if an object is projected it has a component of its motion along a straight line with uniform velocity. Another note states,

"The projectile moveth by the Tangent of the circle of motion precedent in the point of separation".

Here Galileo meant that if an object is projected from the circumference of a circle, such as stone whirled around one's head attached to a string, and then released, the object would proceed to move along the tangent line to the circle at the point of release. A third marginal note reads,

"A grave projectile, as soon as it is separated from the projecient, beginneth to decline".

This means that after an object is projected there is a component of motion towards the center of the earth according to Galileo's law of free fall.

One of Galileo's drawings is shown in Figure 45-16.

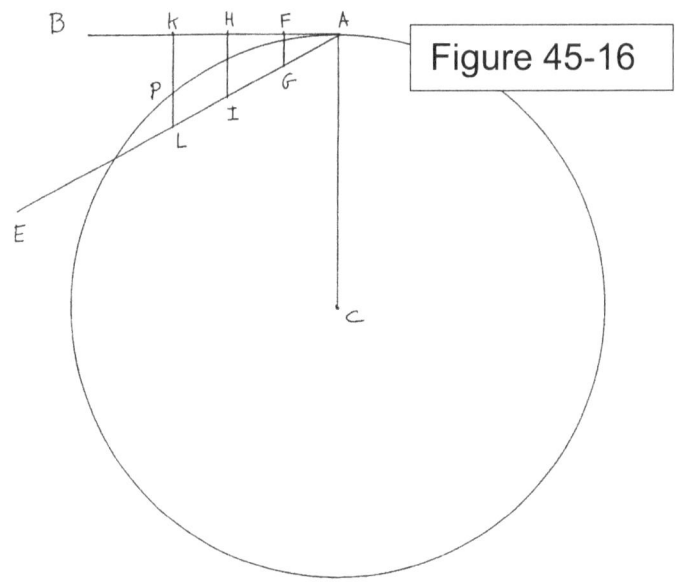

Figure 45-16

The earth is represented with center at C. If
objects were projected into space by the earth's
rotation, then, Galileo says, they would follow the
straight line AB, the tangent to the earth's
circumference. The projected object would have a
component of motion along AB with uniform
velocity, because of the principle of inertia. In
addition the law of free fall means the projected
object would have a component of motion with
uniform acceleration along lines FG, HI, and KL,
according to Galileo.

Although in Galileo's figure the lines FG, HI,
and KL are drawn perpendicular to the tangent AB,
and thus parallel to each other, we know Galileo
had established the direction of the motion of free
fall as being towards the earth's center. However,
the arc of the circumference in the immediate

vicinity of A is essentially a straight line to a good approximation.

In the figure Galileo has marked AF = FH = HK. He designates the lengths of these segments as equal time intervals. AE is drawn as any line such that angle BAE < 90°. He then draws the perpendicular lines FG, HI, KL from AB, which meet AE. He then states the lengths of these lines represent the acquired velocities from free fall of the projected objects after the respective time intervals, AF, AH, and AK.

This construction by Galileo plots a line representing time intervals with lines representing acquired velocities. This method was used again by Galileo in "The Two New Sciences" of 1638, but we see it first used here in the "Dialogue Concerning the Two Chief World Systems" of 1632. It is significant because it is the precursor to the mathematical idea of a function of two variables, where the variables represent different kinds of quantities, i.e. time and velocities.

Next Galileo sets up proportions involving the similar triangles, AKL, AHI, and AFG.

76. KL/AK = HI/AH = FG/AF .

He then argues that he can choose the time intervals, AF etc., so they become smaller and smaller down to the first instant of time that the object is presumed to have been projected into space by the earth's rotation at A. Because angle

BAE is always greater than the angle made by arc of the circumference and tangent BA, from the first instant of the presumed projection the object will have a downward component of motion from free fall so great that the object cannot reach a position between point F and the surface of the earth.   This means the object will never be able to leave the earth's surface because of the earth's rotation.

What Galileo did not realize was that the component of acceleration causing the object to move towards the center of the earth diminishes as the angular velocity of rotation increases, as we have shown by our study of the acceleration due to the earth's gravity. We know that if the earth's rotating angular velocity were great enough objects would become weightless, and they would be 'left behind' by the rapidly spinning earth. What happens to the object would depend on how the situation is envisioned. If we imagine some force causing the earth to increase its rotating angular velocity faster and faster with an object on its surface, then when the earth's angular rotation ω reached a value greater than .0012 radians/second, the object would no longer experience any acceleration due to gravity.   It would have achieved a large value for the velocity vector from the increasing rotation of the earth so as to move off along the tangent line AB in Galileo's Figure 45-16.

We present Galileo's argument not just to show why it is incorrect, but also to indicate its historical significance. Galileo used mathematical principles and his law of free fall. Stillman Drake has argued that Newton studied the "Dialogue Concerning the Two Chief World Systems", and was familiar with Galileo's arguments. It is likely that reading and studying Galileo's arguments influenced Newton to carefully think through the meaning of the earth's gravity which led him to the law of universal gravitation.

Galileo never attempted a theory of gravity which was truly universal. However, he did write in the "Dialogue Concerning the Two Chief World Systems",

"I say that what makes the earth move is a thing similar to whatsoever moves Mars or Jupiter. If [someone] will advise me as to the motive power of one of those moveable bodies, I promise I can tell him what makes the earth move."

When it is suggested that it is gravity which makes thing move toward the earth, Galileo replies,

"everyone knows it is called 'gravity'. What I am asking you for is not the name of the thing but its essence, of which essence you know not a bit more than you know the essence of whatever moves the stars around... But we do not really understand what principle or what force it is that moves stones downward, any more than we

understand what moves them upward after they leave the thrower's hand or what moves the moon around".

Notice the way Galileo links the idea of a force which is involved with free fall, the projection of stones, and the motion of the moon around the earth. Stillman Drake has speculated that when Newton read these words in the Salusbury translation, they may have planted the seed in Newton's brain that there is a gravitational force within the earth acting not only on earthbound objects, like the apple falling from a tree, but also on the moon in its orbit around the earth.

We know many years after the publication of his great book, "The Principia", Newton told the story that observing the fall of an apple in his orchard in 1665 got him to thinking that the motion of the moon was controlled by the same force which made the apple fall. If the moon was not under the control of the earth's gravity then it must obey the principle of inertia and move in a straight line along the tangent to its orbit. In this way the law of universal gravitation may have had its birth in Newton's mind.

Galileo also used arguments from his telescopic observations of Venus to argue in support of the Copernican hypothesis that the planets moved around the sun. Galileo had observed phases for Venus which were the same as the phases of the moon seen from the earth as it rotates around the earth.   We indicate the phases in Figure 45-17.

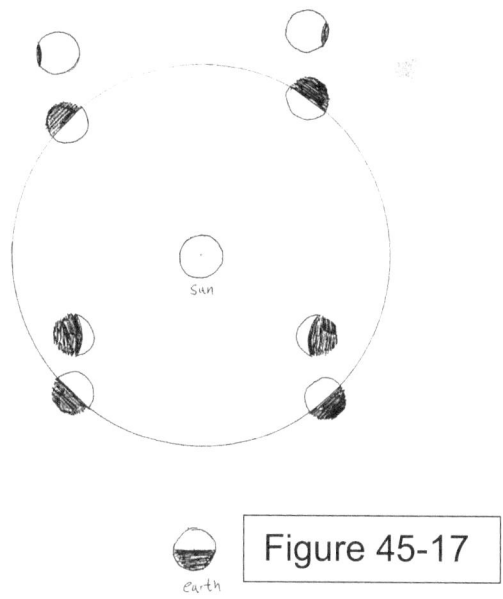

Figure 45-17

Galileo had made many observations of Venus since 1610 with his telescope.   He particularly noted that at maximal elongation from the sun Venus appeared as a crescent shape as shown in the elongations pictured at the bottom of Figure 45-17.

These elongations occur when Venus lies between the earth and the sun. The one on the circle bottom right is seen before sunrise, and the one on the bottom left on the circle is seen after sunset when Venus is at maximum elongation.   In the figure Venus as seen from the earth is shown above the drawing of Venus in its orbit illuminated by the sun.   We see that one half of Venus is

721

always illuminated, but what we see from earth depends on the relative positions of earth, Venus, and the sun.

If we review Ptolemy's theory in Figure 28-3 we see that Ptolemy maintained Venus would always be between the earth and the sun.

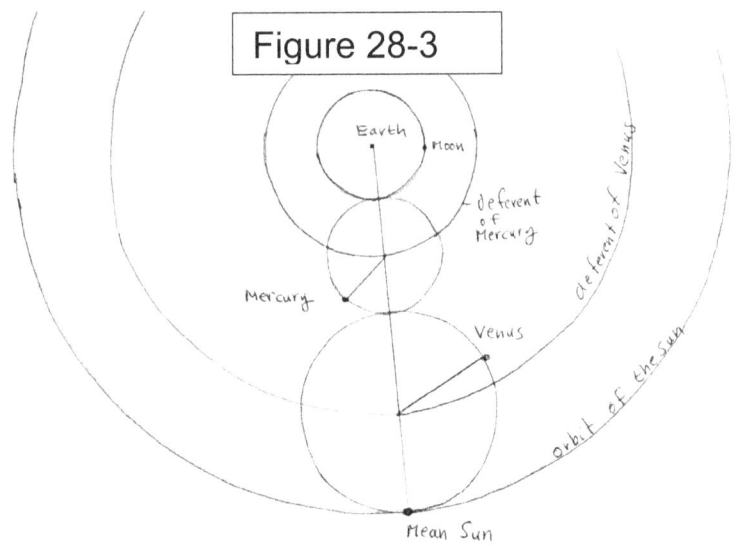

Figure 28-3

In Ptolemy's theory if Venus receives its light from the sun which is then reflected to the observer on earth, then the observer on earth would always see Venus with the telescope as having a horned or crescent shape when seen at maximum elongation, just as in the lower drawings of Figure 45-17, which are inside the circle representing Venus' orbit.

However, by 1610 Galileo had also observed with his telescope elongations of Venus with a gibbous shape, that is with more than half its visible surface illuminated. This phenomenon

722

corresponds to the drawings in the upper half of Figure 45-17 on the circle. This means that the sun has to be in between the earth and Venus. The observations of Venus by Galileo showing both crescent shapes and gibbous shapes for different elongations, indicates that Venus must be orbiting the sun as Copernicus had proposed.   Moreover, these observations show the Ptolemaic system cannot be correct, since Ptolemy's system cannot explain these observations without also proposing that Venus gives off its own light.

We should note that the observations do not prove the earth orbits the sun. The observations are also compatible with the Tychonic system, where the earth is stationary and the sun and planets orbit the earth as a system.

Galileo's accomplishment was to show that the Copernican system was possible and probable, and should be the working hypothesis for future mathematicians and astronomers.

Galileo also gave an account of the moons of Jupiter which he had observed.   He documented eclipses of these moons by Jupiter as seen from the earth.   He then could argue that these moons orbit Jupiter, and the earth was not unique in having a moon.

One of Galileo's last arguments for a rotating moving earth was the phenomenon of the ocean tides.   We will not give this argument because it turned out to have flaws.   Later with the law of universal gravitation Newton was to show the ocean tides were due to the combination of the gravitational forces of the moon and the sun acting on the earth's oceans and seas.

We have now completed our survey of the work of Galileo and his method of study.   He is epitome of the new scientist.   He was convinced nature was rational an understandable by the human intellect.   The scientist could perform experiments which would help reveal the behavior of moving objects.   The results would, moreover, be found to be best expressed in the language of mathematics.   The true scientist would be a student of both nature and mathematics.

We close this chapter with a quotation from one of Galileo's letters known to us as "The Assayer", which he wrote to a supporter of the Copernican system.

"Philosophy is written in this grand book, the universe, which stands continually open to our gaze.   But the book cannot be understood unless one first learns to comprehend the language and read the letters in which it is composed. It is written in the language of mathematics, and its characters are triangles, circles, and other geometric figures without which it is humanly impossible to understand a single word of it; without these one wanders in a dark labyrinth."

Bibliography for Chapter 45

1. Galileo Gallilei, "Dialogue Concerning the Two Chief World Systems - Ptolemaic and Copernican", translated by Stillman Drake, University of California Press, Berkeley and Los Angeles, 1962.

2. Galileus Gallileus, "The System of the world in Four Dialogues Wherein the Two Grand Systems of Ptolemy and Copernicus Are Largelly Discoursed Of", Englished from the Original Italian Copy by Thomas Salusbury, London, 1661.

3. Galileo, "Discoveries and Opinions of Galileo", Translated with an Introduction and Notes by Stillman Drake, Doubleday and Company, Inc., Garden City, New York, 1957.

4. Kline, Morris, "Calculus: An Intuitive and Physical Approach", Part Two, John Wiley and Sons, Inc., New York, 1967.

5. Drake, Stillman, "Galileo: Pioneer Scientist", University of Toronto Press, Toronto, 1990.

Chapter 46

Isaac Newton and the "Principia"

On Christmas day 1642 by the English calendar at the Manor House of Woolsthorpe in Lincolnshire Isaac Newton was born. Newton's father, also Isaac Newton, had inherited the house and had married Hannah Ayscough in April 1642, but Newton's father died soon after the marriage.   The Newton family had no tradition of schooling and Newton's father was illiterate having to sign his name with an 'x'.   However, the Ayscough family was noted for their educational achievements. Newton's uncle, William Ayscough, was educated at Cambridge University and became a cleric.   He would play a role in seeing that young Isaac Newton received a university education.

Many commentators on Newton have noted that Galileo died on 8 January 1642 at 80 years of age. Newton was to live until he was nearly 85 years of age. We should note that at the time of Newton's birth England used the Julian calendar while Italy had converted to the Gregorian calendar so the date of Newton's birth in Italy and the rest of continental Europe was 4 January 1643.

In 1646 Newton's mother married again, but she separated from young Isaac to live with her new husband.   Isaac was raised by his grandmother until 1653 when his mother returned home to Woolsthorpe Manor.   In 1655 Newton was sent to attend the Free

Grammar School in nearby Grantham. The most important part of his education at Grantham was acquiring knowledge of the Latin language. Later this ability would allow him to pursue an independent course of study of Latin books in philosophy and mathematics.

By 1659 Newton's mother recalled Isaac back to Woolsthorpe Manor with the idea of making Newton into a farmer. However, by this time the young Newton had an inclination for study, and he had learned of sundials making several for his rooms. He was also experimenting with kites and how to make them fly better using various attachments for the strings. He became known as a 'thinker' and solver of problems. William Ayscough realized Newton could become a scholar, and urged his sister to send Isaac back again to Grantham for more education. The headmaster of the Grantham school, Mister Stokes, also urged Isaac's return, so Newton returned to Grantham in 1660 to become the leading pupil in his class. This in turn led William Ayscough to help place Newton in Trinity College at Cambridge University in 1661, the same college Ayscough had attended.

Newton began his career at Trinity as a subsizar, a poor student who was expected to do menial chores for the fellows of the college, those who were full time at the college working as tutors or instructors. In spite of his duties Newton found life at the university filled with a vast new world of books and learning which he could pursue on his own after he had been there a year or two. Today scholars have asked what did Newton study, and how did he become interested in mathematics?

Several books have recently been written describing in detail the range of Newton's reading and studies, particularly in the years 1664 through 1666. Richard S. Westfall has written a biography entitled, "Never at Rest: A Biography of Isaac Newton", published by the Cambridge University Press in 1980. Here some details and references of Newton's mathematical and scientific work as well as the people and events in Newton's life are given. The details and sequence of Newton's mathematical researches are given in full in the massive eight volume series edited and annotated by Derek T. Whiteside: "The Mathematical Papers of Isaac Newton", the first volume of which was published in 1967 by the Cambridge University Press. Volume 1 covers the years 1664 through 1666, the years in which Newton discovered mathematics.

Newton's first years at the university consisted of the standard course of learning the work of Aristotle and other philosophers who wrote on ethics, metaphysics, and theology. But Newton was distracted from this course of study when he became aware of the new philosophy and mathematics, particularly the work of Rene Descartes. He also read of the atomistic philosophy of Gassendi who had argued that matter was made up of small particles called atoms, and that there were empty spaces between the atoms. This was in contrast to Descartes who had argued that all matter is

728

extension, that is, there is no such thing as empty spaces or a vacuum where matter does not exist.

In studying the work of Descartes in his private reading Newton was led to a translation of Descarte's "*Geometrie*" into Latin by Frans von Schooten who had annotated and elucidated the text. This book was published in 1661. It was in the spring of 1664, Newton's last undergraduate year at Cambridge, that his mathematical studies began in earnest. We must note that Newton was self-taught in mathematics.   He did not attend lectures or had tutoring in mathematics, but he purchased his own books for private study. Years later Newton told the mathematician Abraham De Moivre how he proceeded. The story was recorded by John Conduitt in 1727.

"In 63 being at Sturbridge fair bought a book of Astrology, out of a curiosity to see what there was in it. Read in it till he came to a figure of the heavens which he could not understand for want of being acquainted with Trigonometry. Bought a book of Trigonometry, but was not able to understand the Demonstrations. Got Euclid to fit himself for understanding the ground of Trigonometry. Read only the titles of the propositions, which he found so easy to understand that he wondered how any body would amuse themselves to write any demonstrations of them. Began to change his mind when he read that Parallelograms upon the secundi generis. In 65 & 66 began to find the method of Fluxions, and writ several curious problems relating to

that method bearing that date which were seen by me above 25 year ago."

Apparently Newton's first work with Euclid was superficial. Later he obtained Isaac Barrow's Latin translation of Euclid of 1655 in which he made many annotations and eventually mastered. Another book by William Oughtred entitled, "*Clavis Mathematicae*", contained the explanation of proportions and more advanced mathematics pertaining to the solution of algebraic equations. The "*Arithmeica Infinltorum*" by John Wallis was Newton's introduction to the mathematics of the binomial theorem and the infinitesimal, which along with Descartes' "*Geometrie*" annotated by Frans von Schooten, encouraged Newton to proceed on his own. The method of fluxions is Newton's version of the calculus.

We wish to note Newton's method of studying mathematics. He reads a number of pages and finds he does not completely comprehend the argument. He goes back and starts over, sometimes again and again until the demonstrations become clear. This shows the perseverance necessary to become a mathematician which holds true to this day. It also shows the edifice of mathematics begins with geometry, trigonometry, and algebraic equations as we too have seen with our historic approach in this book.

The most important of the books studied and mastered by Newton was the Latin edition of Descartes' "*Geometrie*" edited by von Schooten.   Richard Westfall has called this two volume edition of 1661 with its commentary, the 17th "century's most influential mathematical publication", since it gave to Newton and other 17th century mathematicians the ability to analyze geometric constructions with algebraic relations as well as contributing insight into the problems of the calculus such as a means of finding tangents to curves.

Newton also studied Galileo's work, "The System of the World in Four Dialogues Wherein the Two Grand Systems of Ptolemy and Copernicus are Largely Discoursed Of", that is the English translation of Galileo's, "Dialogue Concerning the Two Chief World Systems" of 1632 translated and published in 1661 by Thomas Salusbury. This introduced him to Galileo's laws of motion with uniform acceleration and the idea of inertia.

1664 was Newton's last undergraduate year at Trinity College, and in January 1665 he received his Bachelor of Arts Degree. Although Isaac Barrow was the Lucasian professor at Cambridge since 1663, when the chair was first created, Newton had little contact and knowledge of Barrow's work until around 1669. It is of interest that in 1669 when Newton was a graduate student Barrow examined Newton on mathematics and found Newton wanting in a thorough knowledge of Euclid. Barrow had printed his Latin edition in 1655 and an English translation in 1660. Newton had studied Euclid as noted above, but somehow did not impress Barrow with his knowledge. Newton was so shy he neglected to tell Barrow he had mastered the works of Descartes, von Schooten, Wallis, and Oughtred.   When Barrow finally did see the depth of Newton's independent mathematical work, he recognized Newton as the proper successor to the Lucasian chair of mathematics.   Newton was elected to that chair in October 1669.

Going back to the years 1665 and 1666 we find Newton continuing his studies at Trinity College. However, in August 1665 the university was closed because of the threat of the bubonic plague which was rampant in London.   Newton returned to Woolsthorpe Manor until March of 1666 when he could reenter Cambridge, but only until June when again the university was closed because of the threat of plague.   He returned to Woolsthorpe Manor again from June of 1666 until April of 1667 when he finally could live at Trinity college full time.

The periods at Woolsthorpe Manor were nonetheless extremely productive for Newton's studies and original work in mathematics.   His great mathematical tract on the calculus was created in this period, and was later read in 1669 by Barrow and other mathematicians at Cambridge.   The tract as written in 1669 was in Latin entitled, "*De analysi per aequationes numero terminorum infinitas*" (On Analysis by Equations Unlimited in the Number of Their Terms).   The tract explains the rules of finding derivatives and integrals by the use of infinite series, although the terms used by

Newton were different. It was the most advanced tract on the calculus yet written, and it had within it a general method of finding derivatives and integrals, the areas or quadrature of curvilinear figures. It was the paper on which Newton would base his claim that he had discovered the general method of the calculus. Despite the great importance of this paper Newton failed to have it published. The tract of 1669 was an amplified version of the 1666 papers he had written. It was not published for the mathematical world to see until 1711.

By October 1667 Newton was elected a minor Fellow of Trinity College, and in 1668 a major Fellow. He received his Master of Arts Degree in July 1668.

1669 was the year of his accession of the Lucasian chair of mathematics when Barrow resigned to pursue studies in theology. Nonetheless, Barrow continued to support Newton in his mathematical studies, and Newton on his part became familiar with Barrow's mathematical work. In 1671 Newton wrote that partly upon Isaac Barrow's persuasion he undertook an improvement of his method of the calculus and wrote a fifty page tract called in Latin, "*De methodis fluxionum et sererum infinitorum*" (A Treatise of the Method of Fluxions and Infinite Series). The tract again explains his calculus more extensively and more definitively. He regards the variables as generated by the continuous motion of points, lines, and planes. A variable quantity is called a fluent (that which flows), and its instantaneous rate of flow a fluxion (what we call the derivative). He also explains the fundamental problem of the calculus: given the relation between two variables, the fluents, find the relation between their fluxions, and conversely. Once again Newton failed to have this remarkable tract published, and it was only known to his intimates at Cambridge. It was not published until after his death in 1736.

Also in the 1670's Newton continued his work on optics which he had studied extensively in the two years at Woolsthorpe Manor, 1665 and 1666. He also invented a reflecting telescope which produced sharper images to the improvement of astronomy of his day. He began to correspond with the Royal Society of London for Improving Natural Knowledge, and became a member of the society.

The Royal Society was chartered in 1663 by King Charles II for the purpose of regular meetings of its members to discuss achievements in natural philosophy (science), mathematics, architecture, biology, and medicine. A society separate from the universities at Oxford and Cambridge was necessary to the pursuit of scientific and mathematical work, because the universities had so few courses in science and mathematics in this period, being concerned far more with philosophy and theology, and the ancient classics in Greek and Latin.

A record of the activities and papers presented to the Royal Society was kept by its secretary and known as the, "Philosophical Transactions of the Royal Society of London".   The papers often included the results of experiments or of observations under controlled conditions in order that other investigators could reproduce the results.   It was believed that nature was ordered and predictable under the controlled conditions of experiment. The secretary also kept a record of works received in its Register.   For scientists and mathematicians of Newton's time it was a great honor to be chosen for membership and to have works recorded in the Society's register and published in the transactions.

Newton became a member in 1675, and during the 1670's presented a number of papers to the society (but not his important mathematical discoveries), although at this time he attended few of the meetings which were held at Gresham College in London.

In November 1679 Newton corresponded with Robert Hooke, a prominent member of the Royal Society who had done much work on motion, forces, and the compression of springs.   Hooke had been interested in planetary motion and the forces responsible for the motion.   Hooke introduced the idea to Newton of a force directed to the sun decreasing by the distance squared of the planet from the sun.   He wondered if the motion of the planets were caused by a combination of this force and the planets inertia or tendency to move off in a straight line.

The popular story that Newton had discovered the law of universal gravitation in 1665 or 1666 while at

Woolsthorpe Manor after watching the fall of an apple from a tree in his orchard is not entirely correct. Newton had really not developed the concept of universal gravitation nor the concept of a centripetal force from the sun combined with planetary inertial motion until after the correspondence with Hooke. The correspondence with Hooke only lasted a few months, but it stimulated Newton's thinking and analysis of the problem of planetary motion. Derek Whiteside has pointed out that after the correspondence with Hooke, Newton most likely first worked out a demonstration that if a planet orbited in an ellipse, then the force from the sun had to vary inversely as the square of the distance. Once again Newton failed to communicate his work, and it wasn't until 1684 under the urging of Edmund Halley that he again took up the problem.

As a full time fellow of Trinity College and holder of the Lucasian chair of mathematics Newton was required to take Holy Orders in the Anglican Church. However, Newton's faith was more Unitarian and he refused to take orders in the Anglican Church. This could have led to his dismissal and loss of income, but in 1675 he received a patent granted by the Crown to excuse him from taking Holy Orders, thus allowing him to continue his position at the university.

It may seem somewhat strange to us that as the Lucasian professor of mathematics Newton had to give very few lectures or tutor many students. Moreover, his lectures were poorly attended by the students, since he was not noted as an inspiring speaker, and few knew of his mathematical genius. However, this gave Newton much time for his personal researches and study.

In the early 1680's Newton made numerous astronomical observations of comets, and corresponded with the leading English astronomer, John Flamsteed who became the first Astronomer Royal of England appointed by King Charles II. Flamsteed saw to the construction of the observatory at Greenwich near London, which received the designation of 0° longitude.

In August 1684 Halley visited Cambridge to ask Newton about the elliptical paths of the planets and how an inverse square law explained the motion. Newton stimulated by the problem, began work in earnest on what was to become his masterpiece, "*Philosophia Naturalis Principia Mathematica*" (The Mathematical Principles of Natural Philosophy), known as the "Principia". Newton sent a paper, "*De motu corporum in gyrum*" (On the Motion of Revolving Bodies), to Halley in London which was acknowledged, and entered on the Register of the Royal Society of London on 23 Febuary 1685.

Newton had succeeded where Hooke, Wren, and Halley had failed: to demonstrate that if a planet moves in an elliptical orbit due to an attractive force from the sun at one focus combined with an inertial tendency to move along the tangent to its orbit, then the force from the sun varies inversely in intensity as the square of the distance from the sun. The demonstration was based on

737

physical laws of motion and the use of mathematical analysis. Halley encouraged Newton to write a complete treatise on his great demonstration and the laws of motion. This treatise became the "Principia".

Newton worked on this project often day and night. Originally there were to be two books, but Newton expanded the "Principia" to three books. The first book demonstrated mathematical principles of forces acting on points without claiming to represent physical reality. The second book continued the mathematical demonstrations, and considered motions in resisting media. The third book was entitled, "*De mundi systemate*" (The System of the World). It explained whether or not the propositions in the first two books could be applied to the observed phenomena of planetary and lunar motion. Here for the first time was revealed Newton's most famous discovery, the law of universal gravitation, wherein every object has a gravitational attraction to every other object so that in reality the sun attracts the planets and the planets attract the sun and each other, such that the entire system moves about a center of gravity. This meant the sun was not truly stationary at the focus of an ellipse as Kepler had contended.

On 19 May 1686 the Royal Society announced its intention to publish the "Principia". Halley and Newton corresponded on the "Principia", and Newton learned that Hooke was claiming priority in the discovery of the inverse square law and its association with elliptical orbits.   By 2 June 1686 Halley was instructed by the Royal Society that the "Principia" would have to be published with Halley's private funds since the treasury of the Royal Society was depleted.   Newton was so upset with Hooke's claims that he told Halley he would not have the third book of the "Principia" published. Fortunately for science Halley convinced Newton by 14 July to restore the third book.   By July of 1687 the "Principia" was published under license of the president of the Royal Society, Samuel Pepys.   Figure 46-1 shows the title page of the first edition.

Figure 46-1

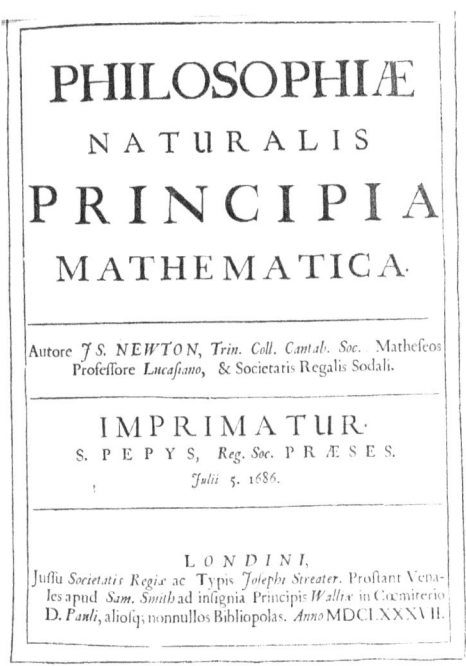

Since 1631 in England all printed books had to be licensed according to Acts of the King.   Licenses could only be granted to certain important individuals such as the Archbishop of Canterbury, or as in the case of the "Principia" the President of the Royal Society.

It is estimated there may have been only 300 to 400 copies of the first edition published. Original copies of the first edition are very rare. The most famous copy is Newton's own copy now at the Wren Library at Trinity College Cambridge. It contains notes in Newton's hand which he used years later for the second edition of 1713.

Newton continued to work on problems in the "Principia", since he received letters and criticisms from those who read it. Eventually this led to a second edition with numerous changes. He also worked on the calculus, and by 1693 another major tract, "*De quadratura curvarum*" (Quadrature of Curves) was completed. Publication did not occur until 1704. Newton also had an interest in many other endeavors including alchemy and the chronology of the Bible on which he was to write extensively.

By 1696 he was offered the post of Warden of the Mint in the Tower of London which he accepted, therefore leaving Cambridge.

After 1700 a dispute on the priority of discovery of the calculus began between Newton and his followers and Leibniz and his supporters. The dispute arose because Newton, as we have noted, did not publish his works on the calculus, but suppressed them so that only a few mathematicians were aware of his accomplishment. Meanwhile Leibniz had begun his own researches on the calculus in 1674, and had his work published before Newton.   Leibniz gave us the notations dy/dx, and the sign for integration, $\int$.   These notations became standard everywhere except in Great Britian, where Newton's notation was used until finally in the 19th century British mathematical works adopted the Leibnizian notation.   The priority dispute was long and often bitter.   It was never fully settled in the lifetimes of Newton and Leibniz, but today historians acknowledge that both mathematicians discovered the fundamental theorems and procedures of the calculus independently of each other and both are today honored as the chief developers of the calculus.   We have noted the contributions of Barrow, Fermat, Wallis, and Descartes as well.

In 1703 Hooke died and Newton was elected President of the Royal Society.   By then Newton was recognized as the greatest scientist and mathematician England had ever produced and he was knighted by Queen Anne at Cambridge in 1705.

In 1709 Newton began correspondence with Roger Cotes about a second edition for the "Principia". As noted above the second edition was not published until 1713.   It contained a preface by Cotes elaborating on the concept of universal gravitation.   In this edition 397 pages of the 511 pages of the first edition were modified by Newton.   He added *scholiums* where he argues that universal gravitation really exists, although he could not give a physical explanation of its cause. He was asking his readers to accept that universal gravitation is a continuously acting force with its action apparent across the vast distances of the solar system even though there is no material connection between the sun and the planets.

There was to be one more edition of the "Principia" before Newton died, the third edition of 1726. This edition did not contain major changes.   We note the second edition had 750 copies and the third edition 1,250 copies.   All three editions were in Latin. The first English translation was the 1729 translation by Andrew Motte based on the third Latin edition. Up to 1999 in the 20th century the most widely available translation for English readers is the 1929 revision of the Motte translation by Florian Cajori.   In 1999 a new modern translation by I. Bernard Cohen, Anne Whitman, and Julia Budenz was published.

By 1727 Newton was 84 years old.   In his eighties he gave some reminiscences of his past many of which are suspect, but have perpetuated legends such as Newton and the apple.   Newton attended his last meeting of the Royal Society on 2 March 1727 and died on 20 March 1727.   He was buried with great honor and ceremony at Westminster Abbey.

Our goal in this book is not to give an account of the vast mathematical and physical discoveries in Newton's long and amazing career, but rather to focus on those elements of his work that pertain to the study of motion and the laws of force that cause motion, particularly planetary motion.   We will then see how mathematically we can demonstrate using the calculus that an elliptical orbit means an inverse square law is present.   We will also undertake to demonstrate that from Newton's laws of motion and the law of universal gravitation an orbiting planet or comet must follow the path of a conic section.   If the body continues to orbit the sun, it must move in an ellipse, or if conditions are just right a circle, otherwise it must follow a parabola or hyperbola and thus leave the solar system such as occur with certain comets.

We will point out how modifications in Kepler's Laws must be made because the law of universal gravitation also means the planets act on the sun, so that in reality the sun is in motion as well.   We will be using the calculus which we have already developed in our work with Galileo.   We will have to expand our knowledge of the calculus and analytic geometry before we can accomplish our demonstrations.   We will also be using the work of Newton in the "Principia", and discuss Newton's own methods in contrast to the modern demonstrations of his propositions.

At this point we point out that both in the paper, "*De motu corporum in gyrum*", and in the first edition of the "Principia", Newton had argued that if an object, presumably a planet, is following an elliptical path it moves under the influence of a centripetal (center seeking) force originating from one focus of the ellipse, presumably the locus of the sun.   This force is demonstrated to vary inversely as the square of the distance from the focus to the orbiting object. In Newton's demonstration he does not invoke the law of universal gravitation, and assumes the object is a dimensionless particle just as is the focus of the ellipse. The force acts continually and at a distance with no material contact. The demonstration is mathematical and must be considered only an approximation to the situation in the solar system, since the sun is a huge object made of much matter, and though the planets are smaller they too contain considerable matter. The demonstration is again found in the "Principia" in Book I proposition 11. In propositions 12 and 13 of the "Principia", Newton demonstrates that if an object is moving in the paths of either an hyperbola or a parabola, it moves under the influence of a force from one of the foci of the hyperbola, or the focus in the case of the parabola, which also varies inversely as the square of the distance from the focus to the object.

Newton did not give a complete demonstration that if there is an attractive force from a point which varies inversely as the square of the distance from the point, then an object put in motion must follow the path of a conic section. The particular conic section the object follows depends on the objects initial velocity. Newton did state that such a demonstration is possible in a corollary to proposition 13, however, the mathematical argument is not given there. We note in the preface to this book it was said when Halley went to visit Newton at Cambridge in August 1684, he asked

Newton what path would a planet follow under the influence of an attractive force from the sun that varies inversely as the square of the distance. Newton is said to have replied that it must be an ellipse, and that he had demonstrated this mathematically. Although a demonstration starting with a centripetal force obeying an inverse square law originating from a point, and then showing that a body in orbit about the point must move in a conic section, one of which is an ellipse in the case of the planets, is not found as such in the "Principia". A demonstration in proposition 41 is given of the general problem of how to determine the curve which a body follows given a centripetal force of any kind. The demonstration is also found in a revised version of *De motu corporum in gyrum*" from 1685. The "Principia" in proposition 11 starts with the elliptical orbit and derives the inverse square law. The demonstration that an inverse square law implies an elliptical orbit is not equivalent, since we could argue that other paths of motion other than the ellipse or other conic sections are possible, if we start with the inverse square law. Several scholars, including I. Bernard Cohen and Robert Weinstock in the works cited in the bibliography to this chapter, have discussed this point in some detail. We can't be sure why Newton omitted the complete demonstration that the inverse square law implies a conic orbit as a special case of proposition 41 in the "Principia".

We will look into both demonstrations in our work, and show how we can use the law of universal gravitation in our demonstrations. It is the mathematics of the calculus that makes these demonstrations possible and understandable. However, if we read the "Principia", we do not see the notations of the calculus or functions from analytic geometry. Yet the "Principia" is filled with concepts of the calculus given in a particular language, which Newton chose for the "Principia", which he calls *de methodo rationum primarum & ultimarum* (the method of prime and ultimate ratios). This method includes the idea of limits applied to the geometric constructions which Newton used in his propositions.

In the next chapter we will begin with new concepts applied to physical bodies, namely mass and quantity of motion (momentum). We will take up the three laws of motion given in the "Principia", and so associated with Newton's work they are now known as Newton's laws of motion.

Bibliography for Chapter 46

1. Westfall, Richard S., "Never at Rest: A Biography of Isaac Newton", Cambridge University Press, 1980.
2. Newton, Isaac, "The Mathematical Papers of Isaac Newton", Volume I, 1664-1666, edited by D.T. Whiteside, Cambridge University Press, 1967.
3. Gjertsen, Derek, "The Newton Handbook", Routledge & Kegan Paul, London and New York, 1986.

4. Cohen, I. Bernard, "The Newtonian Revolution", Cambridge University Press, 1980.

5. Newton, Sir Isaac, "The Mathematical Principles of Natural Philosophy", translated into English by Andrew Motte 1729, 2 vols, Dawsons of Pall Mall, 1968.

6. Newton, Is., "Philosophiae Naturalis Principia Mathematica", Londini Jussu Societatis Regiae Typis Josephi Streater, Anno MDCLXXXVII.

7.Newton, Sir Isaac, "Sir Isaac Newton's Principles of Natural Philosophy and His System of the World" 2 vols, edited by Cajori Florian, University of California Press, Berkeley, 1962.

8. Newton, Sir Isaac, "The Principia, Mathematical Principles of Natural Philosophy", translated by, I. Bernard Cohen, Anne Whitman, and Julia Budenz, University of California Press, 1999.

9. Weinstock, Robert, "Dismantling a centuries old-myth: Newton' 'Principia' and inverse-square orbits", American Journal of Physics, 50:610-617.

10. Cohen, I. Bernard, "The Birth of a New Physics", revised and updated, W.W. Norton & Company, New York and London, 1985.

11. Christianson, Gale E., "In the Presence of the Creator: Isaac Newton and His Times", The Free Press, A Division of McMillin Publishers, New York, 1984.

Chapter 47

The Concepts of Mass, Force, and the Laws of Motion

In the preface to the first edition of the "Principia" Newton explains the general plan for the work. He will begin with definitions and axioms of motions, then in Books I and II demonstrate mathematical propositions using the definitions and laws of motion. In Book III he will consider natural philosophy, that is, what today we call the physical science of motion, mechanics and dynamics (motion under the influence of force). Newton states this involves observing nature and, *"from the phenomena of motions to investigate the forces of nature, and then from these forces to demonstrate the other phenomena".* In the third book he will use the previously established mathematical propositions from Books I and II when they agree with the observed phenomena. When there are discrepancies, that is when the observations don't agree with the mathematical propositions, then a modification of the mathematical propositions must be considered. In the case of planetary motion the great modification was the law of universal gravitation. We will begin our study with the fundamental definitions and axioms given in the first edition of the "Principia".

The first definition introduces us to the concept of inertial mass. Newton says each object has a quantity of matter within it.   We call this quantity of matter the inertial mass of the object.   Newton says the quantity of matter is determined by the density and volume of the object.   In Latin Newton wrote,

"Quantitas Materiae est mensura ejusdem orta ex illisu Densitate & Magnitudine conjunctim".

The translation is *'the quantity of matter is the measure of the same, and arises from its density and magnitude (volume) in conjunction'.*   Newton is defining a fundamental physical quantity, inertial mass, by two other quantities, volume and density.   Volume is, of course, determined by the space the object occupies, and in the metric system is measured in the units of meters cubed.   Newton does not define density for his readers. Today we say density is the mass per unit volume, which in the metric system is the mass per cubic meter.   Newton has been criticized for using circular reasoning in his definition because density is defined in terms of mass, the quantity he wishes to define.

However, in Newton's time the density of water, called the specific gravity of water, had been defined as equal to unity.   Thus, the density of any object could be compared to water, and thus the mass of an object can be found by comparison of the density of the object with the density of water.   We can call such a definition an operational definition, since to find the mass of an object using Newton's definition, we must perform the operation of comparing densities.

For example, if an object has a volume of 3 cubic meters and is found to be twice as dense as water then: quantity of matter (inertial mass)

= (2)[1 unit mass /1cubic meter](3 cubic meters)

= 6 units of mass, or six times the quantity of matter that is in 1 cubic meter of water.

In the metric system the unit of mass is defined as 1 kilogram. The kilogram was originally to be defined such that the density of water under conditions where water has its maximum density, 4° Celsius, or 39° Fahrenheit, would be, 1,000 kilograms per 1 cubic meter, and not unity as in the time of Newton. However, the mass chosen as the standard kilogram missed this objective by 3 parts in 100,000. For practical purposes, however, we can say the specific gravity of water in the metric system is 1,000. Thus, in our example of an object with a volume of 3 cubic meters and twice as dense of water, we should write: quantity of matter (inertial mass)

= (2)[1,000 kilograms /1 cubic meter](3 cubic meters)

= 6,000 kilograms of inertial mass.

Today the reference standard for a mass of 1 kilogram is kept at the International Bureau of Weights and Measures at Sevres, France near Paris. It is made of platinum and iridium.

Students who study chemistry or physics today are taught that the mass of an object is directly proportional to its weight. An unknown mass can then be determined not by comparing its density and volume to that of water, but by weighing the object on a balance scale until it is exactly balanced by a known mass. For example, if an object of unknown mass is exactly balanced by known masses of 1.25 kilograms, then we say the mass of the unknown object is 1.25 kilograms.

However, when mass is determined in this manner we are determining the gravitational mass of the object. Newton says that the gravitational mass of an object is identical to the inertial mass of an object, when he says masses are proportional to their weights, but adds that this proportionality had to be experimentally determined by very carefully done experiments with pendulums. He describes these experiments in Book III, proposition 6 of the "Principia". Later we will discuss these experiments after we are more acquainted with Newton's concepts of force, and the methods of determining the inertial mass by experiments which do not involve weighing objects, that is, by methods not using the force of gravity.

To get an estimate of the kilogram, we can compare the kilogram with the common measure of English pounds. If we weigh a 1 kilogram mass, we will find its weight to be 2.2 pounds when the acceleration of gravity is 9.8 meters per second squared (on the earth's surface). A man who weighs 154 pounds on the surface of the earth has a mass of 154 pounds / 2.2 pounds per kilogram, or 70 kilograms. A motorcar that weighs 2,000 pounds on the earth's surface has a mass of 2,000 pounds / 2.2 pounds per kilogram, or 909.09 kilograms.

Newton wanted to separate the concept of inertial mass from the concept of mass determined by weight on the earth.   An object has weight because it experiences the continuously acting force of gravity which at the surface of the earth has a uniform acceleration of 9.8 meters per second as we have learned.   However, if we imagine weighing an object in outer space away from the acceleration of gravity, then an object's weight would be 0, and in fact all objects would have the same weight of 0.   However, the quantity of matter, or the object's inertial mass is the same whatever the value of the acceleration due to gravity.   We, therefore, need a method or operation to determine inertial mass independent of the acceleration of gravity.   That is, the method should be able to determine an object's mass when the acceleration of gravity is zero, or when the acceleration of gravity plays no role in determining an object's mass. The method of determining inertial mass where gravitational acceleration plays no role involves experiments where the motion of an object is changed by the action of an impressed force upon the object, as we shall see in our subsequent work.

At this point we refer back to Newton's definition of inertial mass as being the product of the density and volume of the object under consideration.   We said we would have to compare the density of the object to the density of water, which in the metric system has the value of 1,000 kilograms per cubic meter. How do we compare the density of an object to the density of water in practice? The answer to this question does involve weighing the object, and goes back to a famous experiment of Archimedes.

The story of Archimedes' method is given to us by the Roman writer Vitruvius. Archimedes was, as we have previously noted, the greatest mathematician of his time in the third century BC. He lived in Syracuse in Sicilly under King Hiero. The occasion arose that King Hiero wished to place a pure gold crown in a temple dedicated to the gods. A contractor was engaged to make the crown from a piece of pure gold of known weight given to him. However, after the crown was completed rumors reached the King that the contractor had taken part of the gold for himself, and replaced the weight of gold he stole with pure silver, and then made a gold and silver alloy into the crown, such that the crown looked like a gold crown, and had the same weight as the original piece of pure gold. The King could not determine if the rumor was true without having to melt the crown and thus destroy the workmanship, which he did not wish to do.

He then called on Archimedes to see if he could devise a method of determining whether or not the crown was an alloy without melting it or changing it. Initially Archimedes pondered the problem without any result. He then had occasion to take a bath at the public baths. He entered a tub, which happened to be filled to the brim, and as he immersed himself, he noticed as water overflowed the tub, he felt his body partly buoyed up by the water. As he thought about this phenomenon, he realized how he could determine whether or not the King's crown was an alloy. He was so transported with joy at his solution that he jumped out of the tub and ran through the streets of Syracuse shouting Ευρεκα (Eureka, Eureka !, which means, 'I got it, I got it !).

Here is what Archimedes did. He weighed the crown on a scale, and then asked the King for a piece of pure gold and a piece of pure silver each of which was equal in weight to the weight of the crown. He then filled a large vessel with water with a spout at the top until it was completely full so that any object placed in the water would cause the overflow to run out the spout where Archimedes could collect it and measure its volume. He knew from experiments that when an object is placed in a vessel of water the object displaces the water upward, or in this case to overflow the vessel in an amount equal to the volume of the immersed object. Notice that this is a physical law which fluids obey when in containers on the earth's surface. As a physical law it cannot be derived mathematically, but must be found by doing experiments with objects of diverse volumes placed in different types of fluids.

He first placed the piece of silver in the vessel and collected the overflow and measured its volume, and thus he knew the volume of the silver piece which was equal in weight to the crown.   He refilled the vessel to the level of the spout with water, and then immersed the piece of gold in the vessel.   The amount of water which overflowed was again measured, and he found the volume of the piece of gold.   Less water overflowed than in the case of the silver piece.   When he immersed the crown itself in the water after again refilling the vessel with water to the spout, he found the volume of water which overflowed to be in between the volume of water displaced by the silver piece and the gold piece. He then knew the crown was an alloy, and that the King's contractor was a thief.

The reason is that gold has a greater density than silver, which in turn has a greater density than water, all of which Archimedes knew.   That is, 1 cubic meter of pure gold has a greater mass than 1 cubic meter of pure silver.   Today we know the density of pure gold in the metric system is 19,300 kilograms per cubic meter, and the density of silver is 10,500 kilograms per cubic meter.

For example, if we have a piece of pure gold that has a mass of 2 kilograms, the piece would occupy a volume of,

(2 kilograms)[1 cubic meter/19,300 kilograms]

= .000104 cubic meters.

A piece of silver with the same mass of 2 kilograms would occupy a volume of,

(2 kilograms)[1 cubic meter/10,500 kilograms]

= .000190 cubic meters.

When these pieces of gold and silver are placed in water one at a time, and the volume of water displaced is measured by the same method as Archimedes used, we would find more water overflows in the case of the silver piece, namely .000190 cubic meters, than the gold piece, namely .000104 cubic meters, a difference of .0000860 cubic meters, because of the difference in density of the two metals.

If a crown of 2 kilograms mass made of an alloy of gold and silver were immersed in water and the displaced volume is measured, the volume displaced would be between .000104 and .000190 cubic meters.

We note that when Newton worked on the propositions for the "Principia" his secretary reported in later years that sometimes Sir Isaac would cry out in his studies, 'Eureka, Eureka!', at having found a demonstration.

Archimedes wrote a treatise entitled, "On Floating Bodies".   In it he deals with objects that float or sink when placed in a container of fluid, and he introduces the proposition known as 'Archimedes' Principle'.   It states that an object immersed in a fluid, or an object floating on top of a fluid in a container experiences a force in the opposite direction to the downward attraction of gravity.   In proposition 7 of the treatise Archimedes says,

"Solids heavier than the fluid [i.e., have a greater density], when thrown into the fluid will be driven down as far as they can sink [by the force of gravity], and they

will be lighter [in weight] in the fluid by the weight of a portion of the fluid having the same volume as the solid".

That is, the loss of weight the solid experiences when immersed is equal to the weight of the fluid displaced, and the solid displaces a volume of fluid equal to its own volume. This loss of weight is the buoyant force Archimedes experienced in the tub of water during his bath. This last principle is Archimedes' proposition 5, which we considered in the case of the crown. Objects that float also experience an upward force equal to the weight of fluid displaced, so they also lose weight when they are floating.

Once again 'Archimedes' Principle' is a physical law valid for objects placed in fluids in containers on the surface of the earth. The principle must be established by doing experiments on a variety of objects with different densities placed in different types of fluids. Although Archimedes gives it as a proposition in his treatise, today we realize it is a physical principle and not a mathematical proposition to be demonstrated.

We will concern ourselves with determining the mass of an unknown piece of metal which sinks in a vessel of water, using the comparison of the density of water to the density of the piece of metal, which is determined by 'Archimedes' Principle'. We note the upward force which a submerged object experiences, and which equals the amount of weight the object loses when it has sunk into the water, is called the buoyant force. To determine the mass of our unknown piece of metal, which in this example is suspected of being pure gold, we first have to determine its volume.   This can be readily done by the method of Archimedes, which he used to solve the problem of King Hiero's crown.   We place the piece of metal in a vessel of water filled to the brim and measure the volume of water displaced when the piece of metal is placed in the vessel.   Let us say we determine the volume to be .001 cubic meters.   Next we suspend the piece of metal when it is immersed in the water onto a spring scale as shown in Figure 47-1.

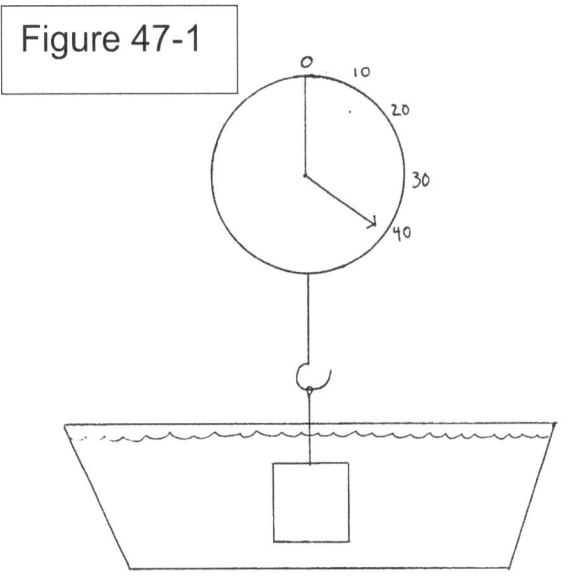

Figure 47-1

This allows us to determine the weight of the piece of metal in the water.   Our scale gives us the weight in pounds.   Of course, such a scale will have been calibrated to give accurate readings with known weights.   We find our piece of metal weighs 40.26 pounds when in the water.

From 'Archimedes' Principle',

Weight in water = Weight in air - buoyant force, or,

758

Weight in air = Weight in water + buoyant force

Weight in air = volume times density of the metal,

Density of metal = weight in air/ volume.

Then, density of metal

= weight in water + buoyant force/ volume.

We know the volume of the piece of metal is .001 cubic meters, and the buoyant force is equal to the weight of the water displaced.   We know the volume of water displaced is also .001 cubic meters.   To find the weight of the water displaced we must multiply its volume, .001, by the density of water expressed in the units of pounds per cubic meter.   From our previous work the density of water is 1,000 kilograms per cubic meter, and the weight of 1 kilogram is 2.2 pounds on the surface of the earth where the weights in this example are measured.   Thus, 1,000 kilograms weighs,

(1,000 kilograms)(2.2 pounds/kilogram)

= 2,200 pounds.

The density of water is then 2,200 pounds per cubic meter.   This in turn means the weight of .001 cubic meters of water is,

(.001 cubic meters)(2,200 pounds/cubic meter)

 = 2.2 pounds.

We can now find the density of the piece of metal.

Density of metal

= 40.26 pounds + 2.2 pounds / .001 cubic meters.

Density of metal = 42,460 pounds /1 cubic meter .

We can compare this value to the density of water: Density of metal / Density of water

= 42,460 / 2,200 = 19.3.

Thus, we have found the density of our unknown piece of metal is 19.3 times as dense as is water. Since in the metric system the density of water is 1,000 kilograms per cubic meter, then the density of our piece of metal is, (19.3)(1,000 kilograms/cubic meter) = 19,300 kilograms per cubic meter. From our work above we know this is the density of pure gold, so our piece of metal is pure gold.

The mass of our object is the product of its volume times its density or,

(.001 cubic meters)[19,300 kilograms/ 1 cubic meter] = 19.3 kilograms.

We note that because we used the weight of the metal in determining its density, and then comparing its density to that of water with a standard density of 1,000 kilograms per cubic meter, we have really determined the gravitational mass of the metal. As long as we are convinced the gravitational mass is equivalent to the inertial mass this method is valid.

Although the method of determining density and then mass is cumbersome, since we could just weigh the object on a scale calibrated in pounds, and then convert its weight in pounds to its mass in kilograms by using the conversion factor of 2.2 pounds per kilogram, nonetheless this method has taught us how to use 'Archimedes Principle' to find the density of objects, and it has shown us what is required in order to find the mass of objects using Newton's definition of density times volume for the quantity of matter or mass. As noted we also need a method of determining mass which does not involve the operation of weighing, that is using the force of gravity. Newton's subsequent definitions and concepts of motion will provide us with such a method.

Newton's second definition in the "Principia" is the definition of 'quantity of motion'. 'Quantity of motion' is measured by velocity and inertial mass conjointly. In practice this means the 'quantity of motion' is equal to the product of the mass of an object multiplied by the velocity vector for the object. This means that 'quantity of motion' has both magnitude and direction; it is a vector quantity just as are velocity and acceleration. Since Newton's time a special name has been given by scientists for the 'quantity of motion': momentum. The usual symbol for momentum is p, which we print in boldface type to indicate it is a vector quantity. We can then write,

p = mv, where m is the mass of the object in question, and v is its velocity vector.

When an object is in motion we must specify the reference frame at rest relative to the object's motion. For example, Galileo chose the surface of the earth to be the reference frame at rest in his experiments with pendulums, free fall, and projectile motion. For other discussions we may choose the earth's north south axis to be the reference frame at rest, as when we discuss the earth's diurnal rotation around its axis, and we may choose the sun to be at rest at the origin of a reference frame at rest when we speak of the motion of the planets. For our initial examples we will choose as our reference frame a smooth surface on the earth's surface perpendicular to the acceleration vector of gravity, so the force of gravity does not affect the motion of an object, at least to any measurable degree over the short distances of our examples. Furthermore, we will choose a surface such as an air hockey rink where an air hockey puck can be a body with a given mass which we place in motion relative to the rink with an impulsive force. The air hockey rink provides a thin cushion of air over which the puck travels, which offers minimal or immeasurable friction to retard the motion of the puck over short distances. Let us say we have a puck with a mass of .25 kilograms. From now on we will use the standard abbreviations for kilogram, i.e. 'kg', and the standard abbreviation for meters, i.e., 'm', and for seconds, 's'. If our puck is given an impulse such that it moves off due east with a velocity of 15 m/s, then its quantity of motion, or momentum is, $p = mv = (.25$ kg)(15 m/s) = 3.75 kg(m/s) in the due east direction, according to Newton's definition. Note that momentum has the physical units of kg(m/s) in the metric system, and that in order to determine the momentum of an object, we must have means of performing the operations needed to determine the object's mass and its velocity vector.

Newton then makes a third definition. He defines what he calls in Latin the *vis insita* for an object. This has often been translated as the innate or inherent force of matter. However, it is not a force as Newton will later define forces, nor is the *vis insita* determined by Newton's laws of motion. In Latin Newton writes, *"Materiae vis insita est potentia resistendi..."*. This means that the *vis insita* is really a potential quality that matter has of resisting a change in motion. Moreover, this quality of resisting a change in motion is directly proportional to the inertial mass of the object under consideration. The *vis insita* is the same whether the object is at rest or in motion. Newton states that a body, whether it is in motion or at rest, is with difficulty changed from its state of rest or motion, and the difficulty in making the change is in proportion to its mass. He also says we can call the *vis insita* by another name, *vis inertiae*, or the inertia of a body either at rest or in motion. Perhaps the best way to learn about this property of matter is to consider collision experiments such as we might perform on our air hockey rink. Newton says an object only manifests its *vis inertiae* when a force is impressed upon it. In theory the force impressed could be acting continuously, or just for a brief increment of time. In the collision experiments we will discuss the impressed force only acts for a very brief increment of time.

Let us imagine an experiment on our air hockey rink where an air hockey puck of mass .25 kg is moving due east at 15 m/s with uniform velocity, and then collides with another air hockey puck at rest on the rink with a mass of .125 kg. The impact of collision is equivalent to an impressed force on both hockey pucks. This means, according to Newton, the motion of both hockey pucks must change, and this is verified by experiment.

In Figure 47-2 we have shown a means whereby we can record the velocities of both hockey pucks before and after the collision.

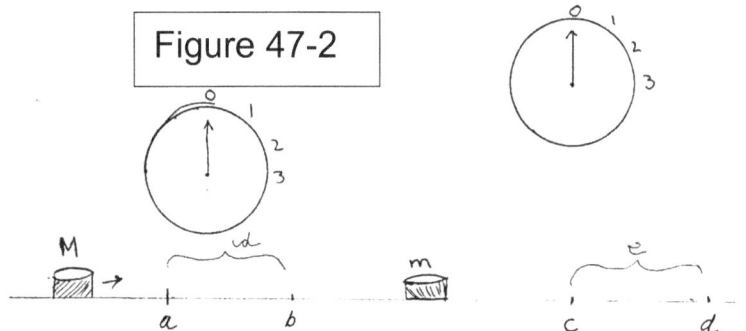

We call the hockey puck in motion M, and the hockey puck at rest m. Before the collision puck M moves over distance *d* from *a* to *b*. A stop-watch is activated to measure the time for M to move from *a* to *b*. If *d* is one meter we find the time to be 1/15 of a second so M is moving at 15 m/s due east.   Puck m has 0 velocity before the collision.

After the collision the two hockey pucks are observed to move off in the due east direction, and stop watches measure the time for each puck to move over distance *e* between points *c* and *d*.   We then can calculate the velocities of the two pucks after collision. Now experiments must be done to determine the results, and to find whether or not a physical law of motion is operative.

It is found that a mathematical relationship does exist relating the momentums before and after such collisions.   It is experimentally determined that the sum of the momentums for the two pucks before the collision equals the sum of the momentums of the two pucks after the collision. This means the total momentum of the two bodies before the collision equals the total momentum after the collision. This physical phenomenon is known as the law of conservation of momentum, and its basis is the many experiments done with colliding bodies.   It is also found that there are different types of collisions, as was discovered by doing experiments with pucks made of different materials, or altered before the experiment. For example, if both pucks are lined on the rims with putty which causes them to stick together after the collision, then they both move off together with the same uniform velocity in the due east direction. Such a collision is called an inelastic collision. In our case for the inelastic collision the momentum of M is, $p = mv =$ $(.25)(15) = 3.75$ kg(m/s), and the momentum of m is 0, since its velocity is 0 before the collision.   The total momentum before collision is, 3.75 kg(m/s) + 0 kg(m/s), and the total momentum after collision is, according to the law of conservation of momentum, the same, 3.75

kg(m/s) in the due east direction. The total mass is now, M + m = .25kg + .125 kg = .375 kg. We can calculate the velocity vector for this inelastic collision by,

$v = p /(M + m) = 3.75/.375 = 10$ m/s in the due east direction.

In fact if there were no significant friction or resistance to the motion, this would be the velocity we would obtain in our experiment using Figure 47-2. Of course, in real experiments small experimental errors occur, such as measuring the time accurately with stop watches, and measuring the distances accurately and small effects of friction affecting the results. Nonetheless, such experiments confirm the law of conservation of momentum to a remarkable degree of accuracy when pucks of various masses are used, and different velocities are given to puck M.

Now what would be the velocity of the two pucks moving off together if the mass of m were 4 times .125 kg, or .5 kg in an inelastic collision? Using the law of conservation of momentum, the total momentum before the collision is still 3.75 kg(m/s), since only M is in motion. After the inelastic collision, the mass of the two pucks moving as one is, M + m = .25 kg + .5 kg = .75 kg. The momentum after collision must be the same as before, or 3.75 kg(m/s), so,

$v = p /(M + m) = 3.75/.75 = 5$ m/s in the due east direction.

We see that the change in velocity is half the value found when the mass of m is .125 kg. This is what Newton means in saying a body has a resistance to a change in motion in proportion to its mass when a force is impressed upon it.   In both experiments M has the same momentum before the collision, and the total momentum before and after the collision is the same. The only difference being the mass of m is 4 times greater in the second experiment of the inelastic collision.   Yet this difference changes the subsequent motion of the two pucks after the inelastic collision. Newton would attribute the greater resistance to a change in motion in the second experiment to be due to the *vis inertiae* of m being greater in the second experiment.   Note the actual measured change in velocity is not 4 times less in the second experiment, even though the mass of m is four times greater.   The actual velocities found depend, as we have said, on the law of conservation of momentum.   In our experiments we are dealing with motion in a straight line, so the law of conservation of momentum in this situation is called the law of conservation of linear momentum.

We ourselves have probably sensed the *vis inertiae* of our own bodies at one time or another. Imagine an experience where a small child is running into our arms. When he lands we feel the impact, but we grab the child and maintain our position.   We easily resist the attempt of the child to change our position with his momentum due to our body's *vis inertiae*.   On the other hand, if we attempted to run into a child, we would knock him down, or send him flying, since his body's *vis inertiae* is so much less than ours due to his much lower mass.

There is another type of collision which was investigated by both Christian Huygens in his treatise on elastic impact published in 1703, and by John Wallis who published results on impact experiments in 1671. This type of collision is called a perfectly elastic collision. In this type of collision the differences between the velocities of the two pucks before the collision is equal to the differences between the velocities of the two pucks after the collision except for a change in sign.

The results of this work shows that if **U** is the velocity of M before the collision, **u** the velocity of m before collision, and **V** the velocity of M after collision, while **v** is the velocity of m after collision, then for perfectly elastic impacts,

$$U - u = - (V - v), \text{ or, } U - u = v - V.$$

We solve for **v**:

1. $v = U - u + V.$

The law of conservation of linear momentum also holds for perfectly elastic impacts, as has been determined experimentally. This means the total momentum before collision, MU + mu, equals MV + mv, the total momentum after collision. Using the value for **v** found in step 1 in this relation,

2. $M U + m u = M V + m (U - u + V),$

   $M U + m u = M V + m U - m u + m V.$

We solve this equation for V.

3. $M U + mu - m U + m u = M V + m V$

$= V (M + m)$, or

$V (M + m) = M\,U + 2m\,u - m\,U$

$V (M + m) = M\,U + m\,(2u - U\,)$

$$V = \frac{MU + m(2u - U)}{M + m}.$$

We see by inspecting this result, if we measure the velocities of the two pucks before they collide in a perfectly elastic collision, we can predict the velocity of puck M after the collision, provided, of course, we know the masses of the two pucks and the initial velocities of the pucks.   Let us say that puck M has a mass of .25 kg, and is moving at 15 m/s due east and collides in a perfectly elastic collision with puck m at rest with mass .125 kg. In reality no collisions are perfectly elastic, since when two objects collide they are slightly deformed by the collision, which prevents perfect elasticity, however, if materials are chosen of proper 'hardness'   and 'elasticity', results can be obtained that are very close to a perfectly elastic collision.   We then calculate V, the velocity of M after the collision.

    4.

$$V = \frac{(.25)(15) + (.125)[(2)(0) - 15]}{.25 + .125}$$

$$= \frac{3.75 - 1.875}{.375} = 5\ {m}/{s}\ due\ east.$$

We can find the value of v by solving step 1 for V, and substituting the result into the law of conservation of momentum, $MU + mu = MV + mv$.

5. $- V = U - u - v$, or,

$V = u + v - U$. Then,

$M U + m u - M (u + v - U) + m v$,

$M U + m u = M u + M v - M U + m v$,

$M U + m u - M u + M U = M v + m v$

$= v (M + m)$,

$V (M + m) = m u + 2M U - M u$

$= mu + M (2U - u$,

$$v = \frac{mu + M(2U - u)}{M + m}$$

$$= \frac{(.125)(0) + (.25)[(2)(15) - 0]}{.25 + .375}$$

$$= \frac{7.5}{.375} = 20m \, / \, s \; due \; east.$$

We see in the perfectly elastic collision that the motion of both pucks has been changed as a result of the collision.   M has had its motion slowed from 15 m/s to 5 m/s, while m has had its motion increased from 0 m/s to 20 m/s.

To show the role of the *vis inertiae*, let us calculate the result if the mass of m is increased to .5 kg, an increase of 4 times. We would expect from what we have learned of the *vis inertiae* that m would have its velocity increased far less than 20 m/s, and that M would have its motion slowed even more, in fact the calculation shows that after the collision M reverses direction or bounces backwards due to the *vis inertiae* of m.

6.

$$V = \frac{(.25)(15) + .5(0 - 15)}{.25 + .5} = \frac{3.75 - 7.5}{.75} = -10m / s,$$

which means M now moves at 10 m/s in the due west direction.

$$v = \frac{(.5)(0) + (.25) \, [\,(2)(15) - 0\,]}{.25 + .5} = \frac{7.5}{.75}$$

$$= 10 \; m / s \; due \; east,$$

significantly slower than 20 m/s, when the mass of m is .125 kg.

Actual experiments using proper hockey pucks and the techniques shown in Figure 47-2 confirm the calculations quite well.

We can go further with the equations for **V** and **v**. For example, we could have pucks made of material so that a perfectly elastic collision would occur. We would place a puck of unknown mass m at rest, and then cause a collision with M. We would know the mass of M, and we would be able to determine the velocities of M before the collision, and the velocities of both M and m after the collision by the techniques in Figure 47-2. Let us say that in this case M is 2 kg, and moves due east at uniform velocity of 10 m/s due to an impulse given to it. m starts out at rest and is found after the collision to be moving at 13.33 m/s due east, while the velocity of M after the collision is found to be 3.33 m/s due east after the collision. We wish to find the inertial mass of m. Note that we will truly be finding the inertial mass of m and not its gravitational mass, since in this method gravity is causing no change in the motion of the hockey pucks. But how can we be sure the value we use for mass M is its true inertial mass? It would have to have been determined by using an experimentally determined comparison to a standard mass.

First, we solve the equation in step 5 for m, then use values from the experiment.

8.

$$v = \frac{mu + M\,(2\,U - u\,)}{M + m}$$

$$v(M + m) - mu = M\,(2\ U - u),$$

$$vm - mu = M\,(2\,U - u) - vM,$$

$$m = \frac{M(2\,U - u) - vM}{v - u} =$$

$$\frac{[(2)[(2)(10) -\ \ 0]\ \ - (13.33)(2)}{13.33 - 0}$$

$$\cong 1kg.$$

Collision experiments allow the determination of inertial mass. Newton in Book III of the "Principia" described pendulum experiments where he shows the inertial mass is equal to the gravitational mass. However, before we take up these experiments we need to know more about the laws of force and motion, and how they affect the motion of a pendulum. We have at this point, at least in principle, shown an operational method of determining the inertial mass of an unknown mass by using an interaction with a known mass in such a way as to change the quantity of motion or momentum of both masses. We have also seen what Newton means by the *vis inertiae* of a body, and how a body of a given mass resists a change in motion after an interaction.

At the end of his discussion of the *vis inertiae* Newton states that many, presumably scientists working both before and during Newton's time, thought a body only possessed the power to resist motion (*potentia resistendi*) if the body were at rest.

Newton, however, states that motion and rest for a body are only relatively distinguished, so that a body in motion also has the power to resist motion, the *vis inertia*.

In the collision experiments we discussed we placed air hockey puck, m, at rest in each case. We could have described experiments with both hockey pucks in motion before the collision, and we .0

could have shown the same principles of conservation of momentum, and the effect of the *vis inertia*. The equations for **V** and **v**, the velocities of the pucks after a perfectly elastic collision, are valid when both pucks are moving with uniform velocity in a straight line before the collision. In such cases the value of **u**, the velocity of m before the collision, would not be 0 as it was in our examples.

In Definition IV of the "Principia" Newton defines what he means by an impressed force. Such a force is an action exerted upon a body which changes its state

773

of motion. If the body is at rest, its state changes to a state of motion with uniform velocity in a straight line; if the body is in motion with uniform velocity in a straight line, its state of motion is changed so that its velocity or direction or both are changed; but the result will be either a condition of rest or motion with uniform velocity in a straight line.

Although this statement is verbose, Newton's Latin is more compact: *"vis impressa est actio in corpus exercita, ad mutandum ejus statum quiescendi vel movendi uniformiter in directum"* (The *vis impressa* is the impressed force, *mutandum* is the change in motion).

The impressed force may only act for an increment of time as in collision experiments, and after the action of the force the body has a new state of motion in which it continues to maintain by its inertia, or as stated in Newton's Latin, *"Perseverat enim corpus in stata omni nova per solam vin inertiae"*. The idea of a body persevering in a new state of motion after the action of a force appears again in Newton's first law or axiom of motion.

Newton's laws are also called axioms, which reminds us of Euclid's axioms for the study of geometry. These laws had been found to be true in Newton's study of motion and in his assessment of the work of others, and Newton considered them valid throughout the universe. They are discoveries based on observations in experiments with colliding bodies, as well as observation of natural motions, or motions produced by applied forces, and by a study of the effects of gravity.

The first law is often called the law of inertia. The physics of Newton and Galileo has been called inertial physics, since they argued that complicated motions, such as the parabolic path of projectiles, or the elliptical orbits of the planets are due to the compound effects of inertial motion with motion caused by forces such as gravity acting simultaneously with inertial motion.   The first law states that if no force is acting on an isolated body, then the body will persevere either in a state of rest or in a state of motion with uniform velocity in a straight line, until such time a force is impressed on the body when its state of motion or rest will be changed.   We know a change in motion means a change in the quantity of motion, the momentum, mv, such that either the magnitude, direction, or both magnitude and direction of the velocity vector of the body changes.   Since the laws of motion are central to Newton's system, we will state them in Latin as given in the first edition of the "Principia".

"*Lex I*

Corpus omne perservare in statu suo quiescendi vel movendi uniformiter in directum, nisi quatenus a viribus impressis cogitur statum ilium mutare".   (Every body perseveres in a state of rest or uniform motion in a right line unless it is compelled to change that state by forces impressed thereon [Motte translation])."

I. Bernard Cohen in his book, "Introduction to Newton's 'Principia", has shown that when first considering this law Newton did not include the state of rest in the definition of inertia. The law in its completed form in the "Principia" states that a body can be in either of two conditions if a force is not acting on it: rest or rectilinear motion in a straight line.

Newton says in his explanation of the first law that a projectile, such as we studied in our work with Galileo, would move in a straight line with uniform velocity if it were not for the continuous action of the resistance of the air and the force of gravity pulling the projectile downward towards the earth. As we think about the first law it reminds us of Galileo's principle of horizontal inertia. In fact in Galileo's "Dialogue Concerning the Two chief World Systems" of 1632, and translated into English by Salusbury in 1661, which Newton had studied, we find Galileo had posed the problem, what would happen to a moveable if it were placed upon a surface with no slope? He states that the moveable would remain at rest, since there would be no tendency [force] to cause it to change its state of rest. Galileo says if the moveable were given an impetus in one direction, then the moveable would move off in a straight line on the surface, and further, if the surface were boundless, Galileo speculates its motion would be perpetual along the straight line with uniform velocity. Of course, Galileo realized the earth's surface curves, and thus his principle of horizontal inertia is not valid over long distances. He also ignored gravity. The object does not move because the surface exerts a force equal to gravity and in the opposite direction (this will be taken up in discussion of newton's third law of motion).   Then at one point he speculates a ship, if unimpeded and in motion with uniform velocity would keep circling the earth's circumference, which we recognize as circular motion due to acceleration towards the earth's center and not inertial motion.

Just as parts of Galileo's work appear in the law of inertia, so too does the work of Rene Descartes. Newton had studied Descartes' philosophical works as well as his work on geometry in the 1660's, and in particular he had studied Descartes' "*Principia Philosophiae*", written in Latin, and published in 1644. Although in the 1660's Newton used the principles in this book to explain the motion of the planets, later he came to reject Descartes' conception, and argued against it in the second edition of the "Principia".   Descartes had introduced the vortex theory, wherein there is no vacuum in space, and the extension of matter from the sun to the planets was whirling in a vortex, which by mechanical extension moved the planets.   However, Newton did accept as a valid hypothesis two of the articles, 37 and 39, from Part II of the "*Principia Philosophiae*".   Article 37 states the first law of nature is that each body always remains in the same state unless acted upon; if at rest it will not move again unless an external intervention causes a change in its state; if in motion it will continue in motion as far as itself is concerned; that is as long as no external force retards its motion.

In article 39 Descartes adds that the motion a body has on its own with no external force affecting its motion is motion along a straight line.

Newton also had studied the treatise by Chistian Huygens entitled, "*Horologium oscillatorum*" (The Pendulum Clock), published in 1673, after he was familiar with Descartes' formulation of the principle of inertia. In Part Two of his treatise Huygens introduced hypotheses on motion in order to discuss the motion of a pendulum under the influence of gravity. The first hypothesis is very similar to Newton's first law of motion. It states,

"If there were no gravity, and if the air did not impede the motion of bodies, then any body will continue its given motion with uniform velocity in a straight line". Huygens did not mention the situation of a body maintaining a state of rest as another possibility for a body not under the influence of gravity or other force.

We might wonder why we have found so many of the scientists and mathematicians of the 17th century, namely Galileo, Descartes, Huygens, and Newton, expounding the principle of inertia as so central to the study of motion. We must recall that before Galileo's pioneering work almost all philosophers accepted the Aristotelian viewpoint that all bodies have a natural place in the universe and that place is a state of rest at

778

the center of the universe, the earth. Celestial objects were attached to spheres which were in rotation around the earth. Consequently, it was not believed that any moving body would continue in motion on its own at uniform velocity in a straight line, if no force were not continually acting on it. With the work of Kepler and Galileo the Aristotelian views of motion and rest were shattered. The new scientists had to state and argue by their studies that the principle of inertia was real and central to a study of motion. The observations that objects in motion come to rest when left alone was explained by the presence of resistive forces such as friction and air resistance. Knowing the difficulty some readers would have with the idea of constant motion at uniform velocity when there are no forces acting, we can see why Newton developed this concept as Lex I, his first law of motion: the law of inertia.

The study of Newton's law of inertia has led physicists to introduce the concept of inertial reference frames. We have already designated a coordinate system, which is to be taken at rest or in motion with reference to another reference frame at rest, as a Galilean reference frame. Recall the explanation to Figure 45-2 in chapter 45. Inertial reference frames are also Galilean reference frames because they can be taken to be rest or in motion with reference to another reference frame. However, inertial reference frames have the particular property given in Newton's first law of motion: they have no force acting upon them.

Because here on earth a reference frame on a surface like our air hockey rink is in fact moving around the earth's north south axis once a day, it is experiencing motion caused by the acceleration of the earth's gravity, and hence is not truly at rest, or in motion with uniform velocity in a straight line. In addition it experiences the force of gravity which pulls it towards the earth's center. Therefore, we must conclude the air hockey rink, or any other reference frame attached to the earth's surface, is not a true inertial reference frame. On the other hand the gravitational force plays essentially no role in the motion of the hockey pucks over the short distances of the experiments, and since the time intervals of the experiments are small, the rotation of the earth during the experiments does not affect the experiment, we can conclude in practice that the reference frame attached to the air hockey rink and designated as being at rest is a close approximation to an inertial reference frame. It is close enough to demonstrate the validity of the law of inertia by experiments.

We note also that any reference frame either at rest or in motion with uniform velocity in a straight line relative to one inertial reference frame is also an inertial reference frame. If we attach a reference frame to an air hockey puck which is moving across the rink with uniform velocity in a straight line, we would conclude our hockey puck is at rest. If another hockey puck is moving along a parallel straight line with the same uniform velocity, but several meters to the east of the hockey puck with the reference frame, we would conclude this second hockey puck is also at rest.

Now let us imagine a reference frame attached to the rink itself.   We now propel an air hockey puck across the rink with a continuously acting air stream from an air compressor. This means the puck is experiencing a continuously acting force so its velocity is continually changing.   Its motion is accelerated as judged from the reference frame of the hockey rink, which we take to be inertial.   On the other hand if we attach a reference frame to the hockey puck experiencing the continuous force, it will not be an inertial reference frame because from the hockey puck, objects at rest on the rink will appear to be accelerating in the opposite direction.   But this would mean such objects are experiencing a force, which they are not. This means the reference frame attached to the accelerating hockey puck is not an inertial reference frame.

Newton did not discuss the concept of inertial reference frames in such language, but the ideas behind them stem as a consequence of Newton's first law.   For a body which is isolated, that is a body experiencing, no force so as to change its state of motion or rest, it is always possible in theory to find coordinate systems which are inertial reference frames so that the body will be judged to be either at rest or in a state of motion with uniform velocity in a straight line.

The study of inertial reference frames leads to what we call today the Galilean relativity principle.   It states that if Newton's laws of motion are valid in a given inertial reference frame, they are valid in any other reference frames moving with uniform velocity in a straight line relative to the given inertial reference frame. Fortunately, the earth's surface taken as a horizontal plane perpendicular to the earth's radius is very close to an inertial reference frame, if motions are restricted to short distances on this plane as in our air hockey experiments.

Newton wrote a scholium or explanation to his definitions in the "Principia" in which he introduces the ideas of absolute time, absolute space, and absolute motion.   Absolute time flows of its own nature equally without relation to anything external, and is called duration.   We measure time by clocks which involve moving parts relative to some reference frame, or in the case of sidereal time we measure time by the apparent motion of the stars which is really the motion of the earth around its north south axis. Consequently we do not measure absolute time since none of our clocks are based on a true inertial reference frame.

Absolute space by its own nature is never changing and immoveable throughout absolute time. However, we measure space by standards of distance such as rulers relative to some point we hold to be at rest.   We can never be sure any given point is truly at rest relative to absolute space, however.

Absolute motion is movement of a body from one point in absolute space to another point in absolute space.   Since we cannot be sure if a given point is at rest or in uniform motion in a straight line relative to absolute space, we cannot be sure if a body is in absolute motion.   What the scientist can do is determine if the laws of motion are valid in a given reference frame, or a plane in a given reference frame.   If they appear to be valid by experiment, he can conclude the reference frame he is using is close to an inertial reference frame so that he may make predictions of subsequent motions due to the action of forces and check them by experiments.   For example, our air hockey rink seemed to be an inertial reference frame for objects moving on the plane, since when a puck was placed in motion then left alone, it continued to move in a straight line at uniform velocity for a short period of observation.   If the puck flew off the rink and fell to the floor, we would see it follow a nearly parabolic path of a projectile, telling us the plane perpendicular to the hockey rink is not an inertial reference frame, since the puck experiences accelerated motion in that plane indicating a force is acting on the puck.

Later we will study Newton's law of universal gravitation and learn that all objects in the universe that have mass attract all other objects that have mass. Thus strictly speaking there can be no completely inertial reference frame, since no body is completely isolated from all other bodies, and hence all bodies experience some gravitational attraction no matter how far out in space and isolated we imagine them to be.

In the 20th century Newton's ideas of absolute space and time have been replaced by the relativity theory of Albert Einstein, where the fundamental quantity is the speed of light, which always has the same velocity in a vacuum for any reference system.

We now turn our attention to Newton's second law of motion. Historians, scholars, and physicists have given this law extensive study as its mathematical formulation is the means of understanding and applying Newton's dynamical system, dynamics being the study of motion under the influence of forces. Newton's intuitive genius was manifest when he started with a law which came from the study of impacts or collisions, and applied it to continuously acting forces, such as the gravitational force of the earth, which never ceases to act unlike the forces of impact which exist over brief time intervals.

Today most physics textbooks give the second law as an equation, $F = ma$, where F is the continuously acting force, which we note is a vector quantity having magnitude and direction, m is the mass of the body experiencing the force, and a is the acceleration vector which the body experiences from the continuously acting force. Since we are already familiar with the continuously acting acceleration vector in free fall experiments near the earth's surface from our work with Galileo, we can say that bodies in free fall experience a continuously acting force equal to the mass of the body times the acceleration vector. For example, a mass of 1 kg in free fall experiences a force of, $F = (1 \text{ kg})(9.8 \text{ m/s}^2) = 9.8 \text{ kg}(\text{m/s}^2)$, since the acceleration vector in free fall is 9.8 m/s² directed toward the earth's center. Note the units of the magnitude of the force vector are m(kilograms) x a(meters per seconds squared). These are not the units of momentum, $p = mv$ = kilogram x (meters per second).

The second law of motion is not given as an equation, nor does it mention acceleration. In Latin the law is:

"Lex II

Mutationem motus proportionalem esse vi motrici impressae, & fieri secundum lineam rectam qua vis ilia imprimitur. (The alteration of motion is ever proportional to the motive force impressed, and is made in the direction of the right line in which that force is impressed[Motte translation])". The term *mutationem motus*, although translated as 'alteration of motion' represents the change in the 'quantity of motion' from Definition 2, which we have defined as the momentum. Thus, the *mutationem motus* is the change in momentum which occurs to a body as a result of the action of an impressed motive force (*vi motrici impressae*). The impressed motive force has the direction of the straight line along which the impressed motive force acts.

As stated the law seems to be concerned with forces like impact or collision, which as we have seen results in a change of momentum. To better understand this law as it applies to impact we will consider the following experiment with our air hockey rink again taken as an inertial reference frame. See Figure 47-4.

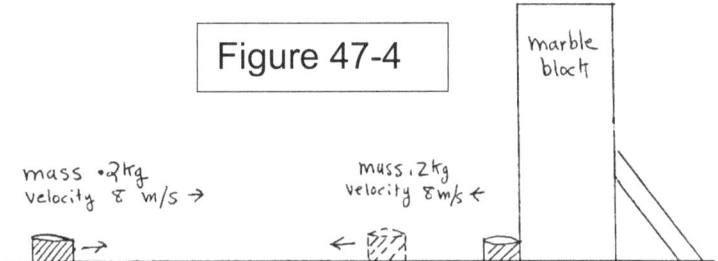

Figure 47-4

marble
block

mass •2kg
velocity 8 m/s →

mass .2kg
velocity 8m/s ←

A hockey puck of .2 kg mass is placed in uniform motion in a straight line in the due east direction at 8 m/s, so that it will collide with a marble block attached to the rink so it cannot continue to move in the same direction after the collision, and the block is oriented perpendicular to the path of the hockey puck. We find the hockey puck bounces off the marble block in an essentially perfectly elastic collision such that it moves due west at 8 m/s.

We envision the impact of the collision to take place over a very short increment of time Δt. It is during this short interval of time that the force acts. But what is the force? We say the marble block acts on the moving hockey puck to change its momentum mv. The momentum before the collision is,

p = (.2 kg)(8 m/s) = 1.6 kg(m/s) in the due east direction.

The momentum after the collision is,

p = (.2 kg)(- 8 m/s) = - 1.6 kg(m/s), since it is in the due west direction.

We designate the change in momentum as a result of the collision as, Δp = Δ(mv) = the difference between the momentum before collision and the momentum after collision.

Δmv = 1.6 kg(m/s) - (-1.6 kg(m/s) = 3.2 kg(m/s).

Now in our reading of Newton's second law we find the *mutationem mutis* (change in the quantity of motion) to be, Δmv = 3.2 kg(m/s).   Newton says this change in the quantity of motion is proportional to the motive force impressed (*vi motrici impressae*), which in this case occurs during the interval of collision, Δt, between the hockey puck and the immoveable marble block.   Can we designate the *vi motrici impressae* in mathematical terms with units of measurement?

The impressed force begins its action at the instant in time when the collision starts, then it acts during the time interval Δt, then ends its action at the instant in time the hockey puck separates from the marble block.   It also appears clear the force is an impulsive force of direct contact between the puck and the block.   If we plot what happens during the collision on a graph we have the situation shown in Figure 47-5.

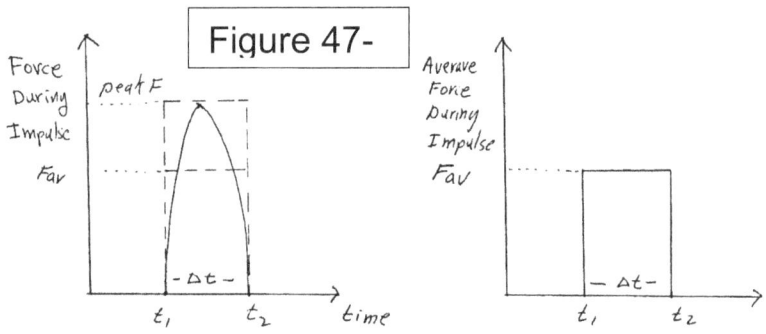

Figure 47-

In the plot on the left we note t_1 is the instant in time that the collision begins, and t_2 is the instant the collision ends. The interval Δt in such experiments is very brief, on the order of .001 seconds. The force that the marble exerts on the puck begins with a value of 0 at time t_1, and moves to a peak value during Δt, and then falls back to 0 at the time of separation, t_2. We can say the entire impulsive event of the collision is equivalent to the area under the curve of force plotted against time. On the plot to the right in the figure we have drawn the situation as if an average force, **F_{av}**, acted at one value for the entire interval. Then we can say the average force times the interval Δt is the area of the rectangle on the graph. The average force is such that the area, **(F_{av})** (Δt), equals the area under the curve in the graph to the right, and represents the 'impulse' given to the puck by the marble block. In physics the quantity, **(F_{av})**(Δt), is defined as the impulse of an impact between two objects. We can see from the second law of Newton, as stated in the "Principia", that Newton indicates the change in the quantity of motion ($\Delta m\mathbf{v}$) is proportional to the motive force impressed, and

788

he gives no indication that he conceived of impulse in modern terms.   Nonetheless, today we say Newton's second law means the *vi motrici impressae* is the impulse, $(Fav)(\Delta t)$, as we have defined it.   We make this change in Newton's formulation of the second law because of the necessity of having a consistent system of units of measurement for force, mass, distance and time, which for us is the metric system.

Newton did not concern himself with physical units in his statement of the second law as we do today. In order for the quantity of force to have the proper units, impulse must be defined as force multiplied by the increment of time.   It is the impulse which for us is proportional to the change in momentum as result of the collision.   Thus, we write,

$$(Fav) (\Delta t) = \Delta p = \Delta(mv).$$

This is the equation we obtain by our alteration of Newton's second law based on our analysis of collisions between objects. Therefore, we suggest that instead of translating *vi motrici impressae* as 'the motive force impressed', we translate the Latin as 'the impulse impressed'.   I. Bernard Cohen in his discussion of Newton's second law in his, "Introduction to the 'Principia'", has pointed out that Newton did not make a distinction between force and impulse as we have done. However, Newton was well aware the force impressed on objects during collisions was proportional to the change in momentum which results.   In symbols, F_{av} is proportional to $\Delta(mv)$. What we have done in the modern

789

definition of force and impulse is add Δt into the proportion to create our equation, $(\textbf{Fav})(\Delta t) = \Delta(mv)$. So doing gives us the proper units for the quantity of force. We do keep Newton's proportionality by placing Δt in the proportion, since for any collision Δt is constant.

In our experiment with the hockey puck let us say we find the value of Δt to be .001 s. We then can calculate the value of $\textbf{F}_{av}$, and determine the proper units for the force vector (force has magnitude and direction, and in this case the direction is due west).

$(\textbf{F}_{av})(\Delta t) = \Delta(m\textbf{v})$, we found, $\Delta(m\textbf{v}) = 3.2$ kg(m/s). $(\textbf{F}_{av}) = \Delta m\textbf{v}/\Delta t = 3.2$ kg(m/s) /.001 s $= 3,200$ kg(m/s^2) in the due west direction.

We have then found the average force acting on the hockey puck for the interval of collision. Note the units for the force vector: kilograms(meters per second squared). In the metric system the value for 1 kg(m/s^2) is called 1 Newton of force in honor of Sir Isaac Newton. Thus, our value for the average force of the collision is 3,200 N, where N is the symbol for Newtons of force.

For reference, 1 Newton of force in the English system of measurement is .224 pounds. For example, if an object of mass 10 kg is weighed on the earth's surface where the acceleration of gravity is 9.8 m/s2, it has a weight due to the force of gravity of, (10 kg)(9.8 m/s2) $= 98$ kg(m/s^2) $= 98$ Newtons, or, (98 Newtons)(.224 pounds/Newton) $= 21.95$ pounds $\cong 22$ pounds. We begin to see that a body's weight is really the value of the force of gravity acting on the body. If the 10 kg mass were in orbit around the earth in a satellite where the force of gravity is all taken up in keeping the satellite in its orbit, then the component of the acceleration of gravity toward the center of the earth is 0 m/s^2. The object has a weight of (10 kg)(0 m/s2) = 0 Newtons = 0 pounds.   (The subject of weightlessness is discussed in chapter 45).

In our collision experiment we found a large average force of 3,200 Newtons had acted, and in fact from Figure 47-5 the peak force would be even larger. However, the force acts for a very short interval to produce the change in momentum of 3.2 kg(m/s) of the hockey puck.

In his discussion of the second law Newton states that doubling the force will result in a doubling of the change in the quantity of motion, (mv), whether the force is impressed in one impulse or in successive impulses. This statement also suggests Newton was speaking of forces impressed on bodies during a collision or in a series of collisions.   But what about continuously acting forces where the momentum, (mv), of a body is continuously changing with time?   To find the answer we have to look at Definitions V, VI, VII, and VIII, which precede the statement of the laws of motion in the "Principia".

In these definitions Newton discusses and defines 'centripetal force', which is a continuously acting force.

"Definition V

A centripetal force is that by which bodies are drawn or impelled, or in any way tend, towards a point as to a centre".

Newton lists gravity and magnetism as continuously acting forces of this sort.   Further, he says, another centripetal continuously acting force is, *"that force, whatever it is, by which the Planets are perpetually drawn aside from the rectilinear motions, which otherwise they would pursue, and made to revolve in curvilinear orbits."*

Here we see that Newton is already thinking of planetary motion as a motion compounded of inertial motion along a straight line, and motion produced by a centripetal force from the sun, which draws the planets away from rectilinear motion.   This reminds us of our study of Galileo's discovery of the parabolic paths of projected bodies.   Note also what Newton says of the centripetal force moving the planets: *"whatever it is".* He will reveal later it is the force of universal gravitation, a force acting across the vast distances of the solar system, but the cause of which Newton does not know.

Newton says the quantity or magnitude of the centripetal force is of three kinds: absolute, accelerative, and motive.   Definition VI states that centripetal forces have a magnitude which propagates from the center of action.   Definition VII asserts the accelerative quantity of a centripetal force also has a magnitude which is proportional to the velocity it generates in a given time. This statement correlates with our work in the differential calculus where the acceleration vector is related to the instantaneous velocity vector by, $v = at$, and is derived from,

$dv/dt = a, dv = adt, \int dv = \int a\, dt,$

$v = at$ + C, where C is the constant of integration, and is 0 if v is 0 when t is 0).

Newton then states the magnitude of the earth's gravitational acceleration is dependent on the distance from the center of the earth; it is less on the tops of mountains and more in deep valleys, but constant at equal distances from the center.   He repeats Galileo's rule that the accelerative force of the earth's gravity equally accelerates all bodies in free fall no matter what their weights, if we eliminate the effects of air resistance.

In Definition VIII Newton defines the motive quantity of the centripetal force.   It is in the discussion of this definition that Newton gives the means of determining the magnitude of a continuously acting accelerative centripetal force.

"Definition VIII

The motive quantity of a centripetal force, is the measure of the same, proportional to the motion which it generates in a given time".

As we think about this definition we see it anticipates the second law, but with some differences. In the second law an impressed motive force also generates a change in motion (Δmv).   However, in Definition VIII we have the additional remark, "*in a given time*", meaning as time goes on the change in motion is

793

increasing. Thus, when centripetal force acts continuously, m**v** is continuously increasing over time. In discussing the definition Newton says,

*"For the quantity of motion [what we call momentum, **p**) arises from the celerity [**v**] multiplied by the quantity of matter [m] ; and the motive force [which we call **F**) arises from the accelerative force [in Latin: vi accelerative, which we call, **a**, the acceleration vector] multiplied by the same quantity of matter [m] . "*

We can then conclude that the magnitude of a continuously acting centripetal force can be found by a simple equation,

F = ma, where a is the acceleration vector given to a body of mass, m, by a force of magnitude F. The force acts in the same direction as the acceleration vector.

In speaking of gravity as a centripetal force Newton says that if the value of the acceleration of gravity were less than it is on the earth's surface, then the weight of the body would be less, because weight is the value of the gravitational force; weight = F (gravitational) = ma. Here Newton makes the distinction between mass and weight clear. The weight of a body is determined by its mass multiplied by the value of the acceleration due to gravity. Because the value of the acceleration can vary depending on location, the weight varies in proportion to the value of the acceleration due to gravity.

For example, the force of gravity on a 1 kg mass on the earth's surface is, F = ma, where m is 1 kg, and a is 9.8 m/s^2 directed to the earth's center. Thus, weight = F = (1 kg)(9.8 m/s^2) = 9.8 Newtons. The gravitational force on a 2 kg mass at the same point on the earth's surface is, weight = F = (2) (9.8) = 19.6 Newtons. If we imagine the masses to be at a distance from earth where a_g (directed to the earth's center) is only 5 m/s^2, then the weight of the 1 kg mass would be 5 Newtons, and the 2 kg mass would weigh 10 Newtons. We see the force of gravity is different for the masses, and in fact the force of gravity varies directly as the mass. Before Newton's work many thought the force of gravity was the same for all objects even though their weights differed. Weight was considered a primary attribute of a body, and was not considered as being determined by the force of gravity.

Newton showed that the proper way to understand gravity was to show that the weight of an object is equal to the force of gravity on the object, and that mass is the fundamental quantity of matter. Moreover, Newton had shown, as we shall discuss, the gravitational mass, determined by experiments involving the force of gravity, is identical to the inertial mass, as determined by impact experiments.

Before Newton, Galileo, Descartes, and Huygens had all related the inertia of a body to its velocity. Newton showed that mass was the quantity that determined inertia, a body's resistance to a change in its state of motion, as we have discussed.

Newton also recognized that a body in unhindered motion under the action of a continuously acting centripetal force, such as gravity, is constantly experiencing a change in its quantity of motion (momentum, mv).   The rate at which the momentum changes is equal to the product of the body's mass and its acceleration.   This concept leads directly into a consideration of how to express the rate of change of momentum in the language of the differential calculus.

The instantaneous rate of change of a quantity with respect to time is the derivative of the quantity with respect to time.

The instantaneous rate of change of momentum under the action of a continuously acting centripetal force is,

$$\frac{d(mv)}{dt} = m\frac{dv}{dt} = ma.$$

But the motive force, F, also equals ma.

Thus,

$$F = \frac{d(mv)}{dt} = m\frac{dv}{dt}.$$

In words this means a continuously acting centripetal force at any time of its action is equal in magnitude to the derivative of the momentum at that time, and the force acts in the same direction as the change in momentum at that time.   If we take the case of an object on the earth's surface where the acceleration vector is constant at 9.8 m/s^2, the force of gravity on a body of mass m is constant at that point and can be calculated by ma.

After all this discussion of how to determine the magnitude of forces, either centripetal or impulsive, we may well conclude that Newton interpreted motive force in the case of continuously acting force to be one quantity, and impressed force in the case of impact to be another quantity. In the study of the second law of motion we found the impressed force was proportional to the change in momentum Δ(mv) of the object on which the force was impressed. In the case of centripetal forces we found the motive force to be equal to the product of mass and acceleration, ma.

Newton has been criticized for not discussing these two concepts under the heading of his second law, and for not rewording the second law to clearly include both concepts. Since Newton's time physicists have brought both concepts under one second law of motion by using simple equations involving derivatives.

For impulsive forces:

Impulse = (F_{av}) (Δt) = Δ(mv), or,

$$F_{av} = \frac{\Delta(mv)}{\Delta t}.$$

But now instead of the average force over the interval of impact, we wish to find the value of the force at any instant of the interval Δt. We see on the left sided graph of Figure 47-5 that the force of impact does vary during the interval of action. In order to find the

797

value of the force at any instant of time from t_1 to t_2 we choose a value of t in the interval, say t_i, and let the incremental interval around t_i approach 0. We designate this incremental interval as Δt_i, the instantaneous force as F_i, and the instantaneous velocity at t_i as v_i. Our equation for the average force in the incremental interval Δt_i is,

$$F(averge \text{ in } t_i) = \frac{\Delta(mv_i)}{\Delta t_i}.$$

Taking the limit as Δt_i approaches 0 , gives, F_i, on the left and, $d(mv_i)/dt$, on the right side of the equation.

Then,

$$F_i = \frac{d(mv_i)}{dt} = ma,$$

where a is the acceleration vector at t_i.

This is the same equation we found for the motive force in the case of centripetal forces.

Because of this interpretation of impressed forces in the case of impact, today we see the second law of motion given essentially as follows:

The instantaneous rate of change of momentum, $d(mv)/dt$, is equal to the motive force active at any instant in which the force acts, and it is a vector quantity with a direction in the straight line along which the force acts.

Let's look at a geometrical representation of the instantaneous rate of change of momentum in the case of our collision experiment, where the .2 kg hockey puck

moving at 8 m/s due east collided with the marble block
and bounced back at 8 m/s in the due west direction.
See Figure 47-6.

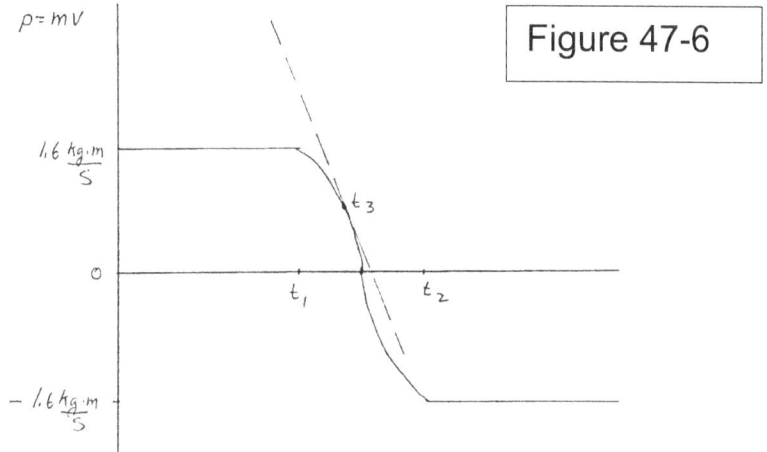

The momentum before the collision was 1.6
kg(m/s), and the momentum after the collision was -1.6
kg(m/s), since it moves in the opposite direction.   We
plot the relationship between the momentum and time.
The force causing the change of momentum of the
hockey puck occurs between time t_1 and time t_2, and is
due to the action of the marble block on the hockey
puck.   The value of the instantaneous rate of change of
momentum is, $d(mv)/dt$ is equal to the magnitude of the
force vector at any instant in the time interval Δt between
t_1 and t_2 .   We show in the figure the instant of time t_3 in
the interval Δt.   The derivative, $d(mv)/dt$, is the value of
the slope of the tangent line to the curve at t_3 . The
tangent line is shown as a broken line in the figure.
This experiment of an impact shows that even when a
force acts impulsively over a brief interval of time, we
can consider the magnitude of the force at an instant in

time as equal to the instantaneous rate of change of momentum, just as in the case of a continuously acting centripetal force.

Despite our success in using Definition VIII and Lex II of the "Principia" to derive mathematical expressions for forces acting on a body, we find Newton himself not entirely clear on the cause of centripetal forces such as gravity and magnetism. Newton, of course, believes these forces exist because of their observable effects on the motion of bodies under their influence. Yet he wrote at the end of his discussion of Defintition VIII,

"I refer the motive force to the body as an endeavor and propensity of the whole towards a centre, arising from the propensity of the several parts taken together... as endued with some cause, without which these motive forces would not be propagated through the spaces round about...For I here desire only to give a mathematical notion of those forces without considering their physical causes and seats".

Newton admits to uncertainty as to why centripetal forces occur as they do. There is for him an unresolved problem of how a force like gravity can cause an action like free fall. It is action at a distance, since the earth is not in contact with the object as it falls, yet it is the earth's gravity that is the source of the motion. Nor could Newton explain why the inertial mass of a body determined by collision experiments is exactly equal to the gravitational mass determined from the relation, $m = F/a$, when F is the force of gravity (the weight of the body), and a the value of the acceleration vector due to gravity. Two centuries after Newton, Albert Einstein wrote of these uncertainties which Newton had professed.

"Newton himself was better aware of the weaknesses inherent in his intellectual edifice than the generation of learned scientists which followed him. This fact has always aroused my deep admiration."

Newton's final axiom or law of motion indicates that forces never occur singly but always in pairs.

"Lex III

Actioni contrariam semper & aequalem esse reactionem: sive corporum duorum actiones in se mutuo esse aequales & in partes contrarius dirigi (To every action there is always opposed an equal reaction: or, the mutual actions of two bodies upon each other are always equal, and directed to contrary parts (Motte translation])."

Newton in his explanation of the law says that if we press down on a stone with our finger, our finger is also pressed upward by the stone with an equal force, even though no motion is observed. If a horse draws a stone tied to a rope, the horse feels a pull drawing it towards the stone of equal magnitude.

In the case where we push on a stone with our finger, note the forces: F_f, the force of our finger acting downward, and the force, F_s, of the stone acting upward, such that, **Ff = - Fs**, since the two forces act in opposite directions.

Newton's third law tells us that a single isolated force is an impossibility. If we have analyzed a situation as having only one force acting (as in the case of the hockey puck impacting with a fixed marble block), we have been incomplete in our analysis, since the third law states forces always occur in pairs which are equal to each other in magnitude and opposite in direction.

In our discussion of the collision of the air hockey puck with the marble block, we spoke of the force of the block on the puck as the cause of the change of momentum of the puck. The third law tells us the hockey puck must have acted on the marble block with an average force of 3,200 Newtons in the due east direction in Figure 47-4. If we call the force of the marble block which acts on the puck the 'action' force, then the force of the hockey puck on the marble block is called the 'reaction' force. The importance of the third law is the emphasis of a mutual simultaneous interaction of two forces. In Newton's Latin he emphasizes this with the words, "*corporum duorum actiones in se mutuo*" (mutual action of two bodies).

You may wonder whether the marble block acquires any change in momentum as the result of the collision, and the action of the force of the puck on the block. The physicist in analyzing this situation says the marble block is attached to the earth and the earth plus the marble block experiences the force of the hockey puck acting on the marble block. This force gives the earth an acceleration according to, $Fpm = m_e \, a$, where, Fpm is the force of the puck on the block, m_e, is the mass of the earth, and a is the acceleration vector of the earth in the due east direction in Figure 47-4. But the mass of the earth is so enormous, $(5.975) (102^{24})$ kilograms, that the acceleration of the earth and block is negligible and undetectable.

$$Fpm = m_e \, a,$$
$$3,200 = (5.975)(10^{24})a, \quad \text{or,}$$
$$|a| = (5.36) (10^{-22}) \text{ m/s}^2.$$

This is why we can neglect the force of the puck on the marble block in practice. However, to understand the phenomenon we should include this force. Consider that Newton speaks in the third law of two bodies; the action force acts on the first body, while the reaction force acts on the second body. This is an essential part of the law. If both action and reaction forces were to act on one of the bodies, we would have the situation where a body would have two forces of equal magnitude, but in opposite direction acting upon it. The body could never experience a change in motion, since the resultant force on the body (the sum of the action and reaction forces) would be zero. Thus when we analyze a situation where forces are acting we must be certain we assign the action force to one body and the reaction force to the other body. To further apply our knowledge of the third law let us analyze the situation of Newton's horse as it pulls on a rope attached to a stone.   We will see we have two separate pairs of forces acting, each pair of forces involving the two bodies. See Figure 47-7.

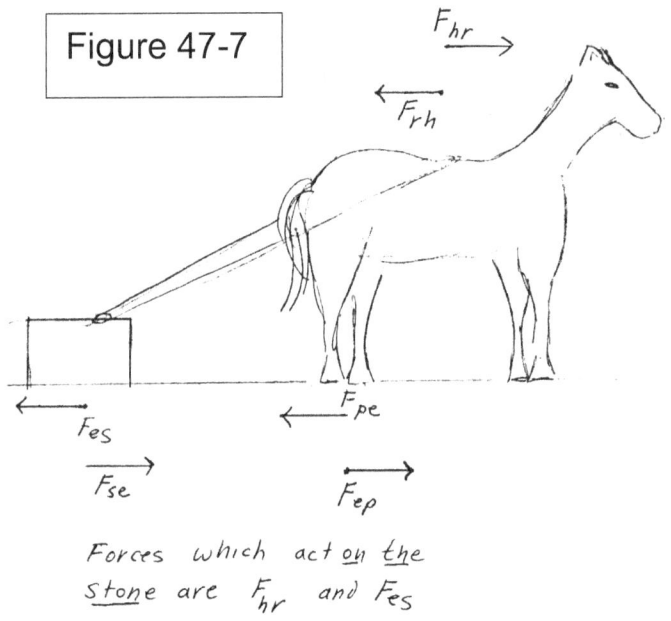

Figure 47-7

F_{hr}

F_{rh}

F_{pe}

F_{es}

F_{se}

F_{ep}

Forces which act on the stone are F_{hr} and F_{es}

     We see the horse pulls on the rope with an action force shown in the figure as **Fhr** (meaning the force of the horse acting on the rope).   The rope, considered as the second body in the pair, must in turn act with a reaction force shown in the figure as **Frh** (meaning the reaction force of the rope on the horse).

     When the horse starts to draw the stone to the right, the horse's hoofs push on the earth to the left in the figure with a force shown in the figure as **Fpe** (meaning the action force of pushing on the earth). Newton's third law says there must be a reaction force to this action of the horse's hoofs, which we designate as **Fep** (meaning the force of the earth pushing in the opposite direction to the horse's hoofs).

The rope is considered to be rigid at all times and of negligible mass in our calculation. As we inspect the figure, we conclude the external forces acting on the stone are **Fhr**, the force which the horse exerts on the rope in the direction to the right, and **Fes**, the force of friction of the earth acting on the stone, which acts in the opposite direction to **Fhr**. Note carefully that the reaction force **Frh**, the rope on the horse, does not have any effect on the stone. We also note that the force of friction, **Fes,** is not a reaction force to **Fhr**. The reaction force to **Fes** is **Fse**, the force of the stone on the earth, which according to Newton's third law is equal but opposite in direction to **Fes**. The stone will not begin to move with an acceleration until the resultant force on the stone, **Fhr - Fes**, is greater than zero. Let us say the mass of the stone is 100 kg, and we observe that as the horse draws the stone the stone starts to move with an acceleration of .5 m/s^2. Then,

   Fhr - Fes = m(stone)(**a**)(stone) = (100 kg)(.5 m/s2) = 50 kg(m/s^2) = 50 Newtons.

At this point we are going to estimate the value of **Fes**, the force of friction of the earth on the stone. Experiments have shown frictional forces are proportional to the force of gravity acting on the stone, which in this case is (100 kg) (9.8 m/s^2), or, 980 Newtons. The proportionality constant is variable because of the physical nature of the stone and the earth over which it must move. We estimate **Fes** to be

900 Newtons.   We then calculate the value of *Fhr*, the force of the horse on the rope to be,

$$Fhr - Fes = 50,$$

$$Fhr = 50 + Fes = 50 + 900 = 950 \text{ Newtons}.$$

We shall say Newton's horse has a mass of 500 kg.   From the figure the external forces on the horse are *Frh*, the force of the rope on the horse, and, *Fep*, the force of the earth pushing back on the hoofs. The sum of these two forces is the resultant force acting on the horse, and according to Newton's second law of motion the sum equals the mass of the horse times its acceleration, which equals the acceleration of the stone.

Frh + *Fep* = m(horse)a(horse) = (500 kg)(.5 m/s2) = 250 Newtons.

Since *Frh* = - *Fhr*,

$$- Fhr + Fep = 250 \text{ Newtons}.$$

$$Fhr = 950 \text{ Newtons, so, } Fep = Fhr + 250$$

$$Fep = 250 + 950 = 1,200 \text{ Newtons}.$$

The force the horse has to exert through its hoofs on the earth has the same value as *Fep*, but in the opposite direction. This force, *Fpe,* pushes the earth such that the mass of the earth times the acceleration of the earth caused by the horse's hoofs equals 1,200 Newtons.   Once again because the mass of the earth is so great, $(5.975)(10^{24})$ kilograms, the earth experiences no detectable change in its motion as the horse draws the stone.

We should note that if the horse were to pull the stone at uniform velocity, then the force of friction of the earth on the stone must equal the force of horse on the rope. This means the resultant force on the stone is 0, which means in turn the stone is either moving at uniform velocity or is a rest, according to Newton's first law of motion. We can see that the analysis of an apparently simple motion can be quite involved, and require consideration of all three of Newton's laws of motion.

It is worth emphasizing that what determines motion with acceleration is the resultant force on a body, the sum of all the forces acting at any one time. Even though Newton's horse might exert a thousand Newtons to pull the stone, no motion would occur, because of the resistance of friction between the stone and the earth. We also learned that an object at rest or in motion with uniform velocity in a straight line may have a combination of forces acting upon it with a resultant sum of zero, and thus obey the first law of motion.

Today in teaching Newton's laws of motion it is often noted that the first law of motion is a special case of the second law of motion. This occurs when the value of the acceleration is 0 in the equation for the second law,
$F = ma$. Then value of F is 0. This means a body which has 0 acceleration does not experience the action of a force, and is either at rest or in motion with uniform velocity in a straight line.

With our knowledge of Newton's three laws we can go back to Archimedes' principle and analyze the problem of buoyant force on a body floating in water. Let us study the case of an iceberg floating in artic water. Scientists know the density of ice at sea is 920 kg/m³, and the density of sea water is 1,030 kg/m³ (greater than pure water because of the salt content). We take a piece of ice and do the following experiment. We fill a container to the brim with sea water with a spout to measure the overflow when the piece of ice is forcibly immersed in the container, so as to determine the volume of the piece of ice. We find we have to counteract the buoyant force of the water to maintain the piece of ice under the surface of the water. In doing so we are creating the reaction force to the buoyant force by pushing the piece of ice under water. We find the volume of water displaced is .002 m³, which by Archimedes' principle is equal to the volume of the immersed piece of ice.

The buoyant force exerted by the sea water on the piece of ice equals the density of sea water times the volume of water displaced times a_g, the acceleration due to gravity (9.8 m/s²), since we do our experiment at sea level.

Fb (buoyant force) = (1,030 kg/m³)(.002 m³)(9.8 m/s²)

= 20.19 kg(m/s²), or, 20.19 Newtons.

The weight in air of the piece of ice equals its volume times the density of ice times a_g.

Fgi (wieght of ice) = (.002 m³)(920 kg/m³) (9.8 m/s²)

= 18.032 kg(m/s²), or, 18.032 Newtons.

Using Archimedes' principle,

weight in sea water = weight in air - buoyant force

= 18.032 Newtons - 20.19 Newtons = - 2.158 Newtons.

This shows the buoyant force is greater than the weight in air by 2.158 Newtons, so in order to keep the piece of ice immersed we have to push on the ice with a force of 2.158 Newtons to hold the piece of ice under the surface of the water

Next, we release the piece of ice to allow the buoyant force to accelerate the piece of ice to move to the surface and float.   When it floats part of its volume is above the surface, and the buoyant force of action is exactly opposed by the weight of the piece of ice thus its weight in water is 0.   Then the buoyant force of the ice when floating, **Fbfl** is found by the relation,

weight in water = weight in air – **Fbfl,**

0 = 18.032 Newtons – **Fbfl,**

Fbfl = 18.032 Newtons.

This is the equilibrium position for the piece of ice where part of its volume is above the surface of the water and part of its volume immersed.   The action force, **Fbfl**, of the water on the ice exactly equals the reaction force, **Fgi**, the force of gravity on the piece of ice, which, of course, is its weight.   The buoyant force still equals the weight of the volume of sea water displaced.   We can find the volume of water displaced by the floating piece of ice as follows.

Fbfl = 18.032 kg(m/s^2)

= (Vd ,volume displaced)(1,030 kg/m^3, density of ice)(9.8 m/s^2)

$$Vd = \frac{18.032}{(1,030)(9.8)} = .00179m^3.$$

This means .00179 m^3 of the piece of ice is below the surface of the water, or .00179 m^3/.002 m3 = 89.5% of the piece of ice is submerged when it is floating. The same reasoning applies to giant icebergs in artic waters, so when you see a huge iceberg in the artic you are really only seeing 10.5% of its total volume. This played a role in the sinking of the Titanic when it crashed into a large iceberg.

We have now had an introduction to the "Principia", and the new concepts of inertial mass, gravitational mass, quantity of motion (momentum = mv), and the various types of forces. Newton then used the concepts and the ideas of Descartes, Galileo, Wren, Huygens, and Wallis to formulate the three laws of motion. In the next chapter we will study the motion of the pendulum applying Newton's laws with the mathematics of the calculus to derive the law of the pendulum. We will then see how Newton demonstrated the equivalence of the inertial and gravitational mass of a body.

1. Newton, Is., "Philosophiae Naturalis Principia Mathematica", Londini Jussu Societatis Regiae Typis Josephi Streater, Anno MDCLXXXVII.

2. Newton, Sir Isaac, "The Mathematical Priciples of Natural Philosophy", translated into English by Andrew Motte 1729, 2 vols, Dawson of Pall Mall, 1968.

3. Dijksterhuis, E.J., "Archimedes", Princeton University Press, Princeton, New Jersey, 1987.

4. Cohen, I. Bernard, "Introduction to the 'Principia", Harvard University Press, 1972.

5. Halliday, David, and Resnick, Robert, "Physics", John Wiley & Sons, Inc., New York, 1962.

6. Cohen, I. Bernard, "The Newtonian Revolution", Cambridge University Press, 1980.

7. Einstein, Albert, "Ideas and Opinions", p.25, Crown Publishers, New York, 1964.

8. Einstein, Albert, and Infeld, Leopold, "The Evolution of Physics", Simon and Schuster, 1938.

9. Galilei, Galileo, "Dialogue Concerning the Two Chief World Systems- Ptolemaic & Copernican", translated by Stillman Drake, University of California Press, Berkeley and Los Angeles, 1962.

10.Descartes, Rene, "Principles of Philosophy", translated with explanatory notes by Valentine Rodger Miller and Reese P. Miller, D. Reidel Publishing Company, Dordrecht:Holland/Boston:USA.

11.Huygens, Christian,   "The Pendulum Clock or Geometrical Demonstrations Concerning the Motion of Pendula as Applied to Clocks   [Horologium Oscillatorium]", translated with notes by Richard J. Blackwell,   The Iowa State University Press, Ames Iowa, 1986.

12.Newton, Isaac, "The Mathematical Works of Isaac Newton", Volume VI, 1684-1691, edited by D.T. Whiteside, Cambridge University Press, London, 1974.

13.Westfall, Richard S., "Force in Newton's Physics", MacDonald, London, American Elsevier, New York, 1971.

14.Gamow, George, "The Great Physicists from Galileo to Einstein", Dover Publications, Inc. New York, 1988.

15.Asimov, Isaac, "Understanding Physics", Volume I, New American Library, New York, 1966.

16. Casper, Barry M., and Noer, Richard J., "Revolutions in Physics",   W. W. Norton & Company,   Inc., New York, 1972.

This completes Volume 2

www.ingramcontent.com/pod-product-compliance
Lightning Source LLC
Chambersburg PA
CBHW081424170526
45166CB00008B/2103